Handbook of Research on Andragogical Leadership and Technology in a Modern World

Viktor Wang
California State University, San Bernardino, USA

A volume in the Advances in Human and Social Aspects of Technology (AHSAT) Book Series

IGI Global
PUBLISHER of TIMELY KNOWLEDGE

Published in the United States of America by
IGI Global
Information Science Reference (an imprint of IGI Global)
701 E. Chocolate Avenue
Hershey PA, USA 17033
Tel: 717-533-8845
Fax: 717-533-8661
E-mail: cust@igi-global.com
Web site: http://www.igi-global.com

Library of Congress Cataloging-in-Publication Data

Names: Wang, Viktor, editor.
Title: Handbook of research on andragogical leadership and technology in a modern world / edited by Viktor Wang.
Description: Hershey, PA : Information Science Reference, [2023] | Includes bibliographical references and index. | Summary: "A strong base of reliable, verifiable knowledge is needed to foster evolution of andragogical leadership and technology across a diversity of domains. Innovative research promoting andragogical leadership and technology may add to this strong base of reliable, verifiable knowledge of this type of educational leadership needed to move our organizations in a modern world forward in the 21st century. The chapters in this volume are a cross-section of diverse perspectives, experiences, theories and philosophies related to andragogical leadership and technology across all levels of education"-- Provided by publisher.
Identifiers: LCCN 2023008185 (print) | LCCN 2023008186 (ebook) | ISBN 9781668478325 (hardcover) | ISBN 9781668478332 (ebook)
Subjects: LCSH: Adult learning--Case studies. | Educational leadership--Case studies. | Educational technology--Case studies.
Classification: LCC LC5225.L42 H367 2023 (print) | LCC LC5225.L42 (ebook) | DDC 374.0072--dc23/eng/20230313
LC record available at https://lccn.loc.gov/2023008185
LC ebook record available at https://lccn.loc.gov/2023008186

This book is published in the IGI Global book series Advances in Human and Social Aspects of Technology (AHSAT) (ISSN: 2328-1316; eISSN: 2328-1324)

British Cataloguing in Publication Data
A Cataloguing in Publication record for this book is available from the British Library.

Advances in Human and Social Aspects of Technology (AHSAT) Book Series

Mehdi Khosrow-Pour, D.B.A.
Information Resources Management Association, USA

ISSN:2328-1316
EISSN:2328-1324

Mission

In recent years, the societal impact of technology has been noted as we become increasingly more connected and are presented with more digital tools and devices. With the popularity of digital devices such as cell phones and tablets, it is crucial to consider the implications of our digital dependence and the presence of technology in our everyday lives.

The **Advances in Human and Social Aspects of Technology (AHSAT) Book Series** seeks to explore the ways in which society and human beings have been affected by technology and how the technological revolution has changed the way we conduct our lives as well as our behavior. The AHSAT book series aims to publish the most cutting-edge research on human behavior and interaction with technology and the ways in which the digital age is changing society.

Coverage

- Cyber Behavior
- Technology Dependence
- Human-Computer Interaction
- Technology Adoption
- Gender and Technology
- End-User Computing
- Technology and Social Change
- ICTs and human empowerment
- Activism and ICTs
- Computer-Mediated Communication

IGI Global is currently accepting manuscripts for publication within this series. To submit a proposal for a volume in this series, please contact our Acquisition Editors at Acquisitions@igi-global.com or visit: http://www.igi-global.com/publish/.

The Advances in Human and Social Aspects of Technology (AHSAT) Book Series (ISSN 2328-1316) is published by IGI Global, 701 E. Chocolate Avenue, Hershey, PA 17033-1240, USA, www.igi-global.com. This series is composed of titles available for purchase individually; each title is edited to be contextually exclusive from any other title within the series. For pricing and ordering information please visit http://www.igi-global.com/book-series/advances-human-social-aspects-technology/37145. Postmaster: Send all address changes to above address.

Titles in this Series

For a list of additional titles in this series, please visit: www.igi-global.com/book-series

ICT as a Driver of Women's Social and Economic Empowerment
Pankaj Dhaundiyal (CHRIST University (Deemed), India) and Sana Moid (Amity University, ndia)
Information Science Reference • © 2023 • 306pp • H/C (ISBN: 9781668461181) • US $215.00

Impact of Disruptive Technologies on the Socio-Economic Development of Emerging Countries
Fredrick Japhet Mtenzi (The Aga Khan University, Institute for Educational Development, Tanzania) George S. Oreku (The Open University of Tanzania, Tanzania) and Dennis M. Lupiana (Institute of Finance Management, Tanania)
Information Science Reference • © 2023 • 335pp • H/C (ISBN: 9781668468739) • US $225.00

Handbook of Research on Digitalization Solutions for Social and Economic Needs
Richard Pettinger (University College London, UK) Brij B. Gupta (Asia University, Taiwan) Alexandru Roja (West University of Timisoara, Romania) and Diana Cozmiuc (West University of Timisoara, Romania)
Engineering Science Reference • © 2023 • 387pp • H/C (ISBN: 9781668441022) • US $380.00

Exergaming Intervention for Children, Adolescents, and Elderly People
Shahnawaz Khan (Bahrain Polytechnic, Bahrain) Thirunavukkarasu Kannapiran (Karnavati University, India) Arunachalam Muthiah (Karnavati University, India) and Sharad Shetty (Karnavati University, India)
Information Science Reference • © 2023 • 300pp • H/C (ISBN: 9781668463208) • US $215.00

Digital Psychology's Impact on Business and Society
Muhammad Anshari (Universiti Brunei Darussalam, Brunei) Abdur Razzaq (Universitas Islam Negeri Raden Fatah Palembang, Indonesia) Mia Fithriyah (Indonesia Open University, Indonesia) and Akmal Nasri Kamal (Universiti Brunei Darussalam, Brunei)
Information Science Reference • © 2023 • 330pp • H/C (ISBN: 9781668461082) • US $270.00

Frugal Innovation and Social Transitions in the Digital Era
Muhammad Nawaz Tunio (Mohammad Ali Jinnah University, Karachi, Pakistan) and Atia Bano Memon (University of Sindh, Badin Campus, Pakistan)
Information Science Reference • © 2023 • 208pp • H/C (ISBN: 9781668454176) • US $250.00

Handbook of Research on Implementing Digital Reality and Interactive Technologies to Achieve Society 5.0
Francesca Maria Ugliotti (Politecnico di Torino, Italy) and Anna Osello (Politecnico di Torino, Italy)
Information Science Reference • © 2022 • 731pp • H/C (ISBN: 9781668448540) • US $295.00

IGI Global
PUBLISHER of TIMELY KNOWLEDGE

701 East Chocolate Avenue, Hershey, PA 17033, USA
Tel: 717-533-8845 x100 • Fax: 717-533-8661
E-Mail: cust@igi-global.com • www.igi-global.com

Quotes and Testimonials

Modern leaders must contend with the disruptive forces of technologies and navigate the social and economic unpredictability of our times, the outcomes of which can be 'for better' or 'for worse.' The mission of leaders extends well beyond their singular capacity. Leaders must understand and enact their leadership with an unshakeable positive belief in their followers. They must free the untapped potential, motivate and encourage experimentation, creativity, and innovation, and value the rich diversity in their followers - these are captured by an andragogical approach to leadership. The chapters in this book bring andragogical theory to the fore as a leadership theory and guiding philosophy for modern times. The multiple perspectives and 'use cases' developed by the authors make the book a valuable resource for the theory and practice of leadership in a challenging era.

Chinaka DomNwachukwu, Ph.D., Dean & Professor, WRWCOE College of Education, California State University, San Bernardino, USA

Leaders and teachers at all levels of education must be empowered to "do the right things, and do things right." The twenty chapters in this book will do this. For the first time since the inception of andragogy, the authors prove that the theory of andragogy, is an encompassing and powerful frame for developing and applying leadership theories and practices such as andragogical leadership and transformational leadership. This much-needed book fills a void in leadership literature. Authors from nine countries on five continents collectively and empirically address andragogical leadership and technology, capturing diverse cultural and ideological perspectives.

Dr. John A. Henschke, Past President of AAACE; Professor Emerita, University of Missouri & Lindenwood University; Inductee of the International Adult and Continuing Education Hall of Fame; Former Doctoral Student of Malcolm S. Knowles

This new handbook gathers global perspectives on a nurturing approach to leadership. With expectations for career growth and satisfaction rising, research on successfully encouraging continuous learning, improvement, and actualization is critical for leaders and organizations to thrive.

Rebecca Lubas, Dean of John M. Pfau Library, California State University, San Bernardino, USA

While applying principles of Andragogy to Leadership is not new (see McCauley et al., 2017), this book provides the most comprehensive, multi-faceted examination of Andragogy and Leadership in various contexts. This work is timely in exploring online teaching and learning, resilience and stress in higher education, and social justice, to name a few. Administrators, faculty, and learning designers alike will find this work of significant value for a positive impact on their practice.

Kathy Peno, Ph.D., Professor and Program Coordinator Adult Education Master's Program Director, Learning and Development Certificate Program, University of Rhode Island

The Handbook of Research on Andragogical Leadership and Technology in a Modern World brings together a selection of critical topics that reinforce the importance and value of andragogy. Educational philosophies and practices based on andragogy assist in facilitating the development and growth of others as well as encouraging followership and continuous improvement, both personally and professionally. These chapters illustrate that leading others takes courage and putting faith and trust in their skills so that innovation and creativity can result. This Handbook will serve as a worthwhile and valuable resource for those applying andragogy in both the leadership and technological areas.

Maria M. Witte, EdD, Associate Dean of the Graduate School, Professor of Adult Education, Auburn University, AL USA

This book is the best one in print on this topic. It covers traditional leadership topics by applying the art and science of helping adults learn, namely andragogy comprehensively, giving the reader new ideas/food for thought and options to move forward to learn to survive and thrive in the 21st Century organizations that are impacted by the volatile technological advancements, geographical power struggles, the continued swaying of ideologies, and above all, the moral leadership theories and practices. This book makes the bold conclusion that other leadership theories and practices flow from the beautifully well-reasoned theory of andragogy. It shows the researchers how to research the complexities of leadership theories and techniques from the andragogical angle. The book's authors make meaningful connections between andragogy and traditional leadership theories and practices. Upper-division undergraduate students, graduate students, primarily doctoral students, and faculty researchers will come to many realizations by applying "andragogical leadership and technology" to practice. Finally, I highly recommend adapting this book critically and flexibly to lead your organizations andragogically rather than pedagogically.

William J. Rothwell, Ph.D., the author of well over 300 refereed publications, including 158 books, a Distinguished Professor at Penn State University, and a consultant for over 50 multinational companies. He has been to China 83 times, to Singapore 32 times, and has visited every continent (except Antarctica) to do training, consulting, college teaching, and research. He has chaired 120 Ph.D. committees at Penn State.

List of Contributors

Table of Contents

Detailed Table of Contents

Nearly all educational institutions use learning management systems (LMS). Post-pandemic figures suggest that the uptake of LMS is about 98%. This means that most educators, especially those in post-schooling education, will use, and be responsible for designing, an online course site. While there is a tendency to use course sites as a repository and convenient communication medium, there is an unmet need to design these sites in a way that facilitates learning. As online teaching and learning become more prolific, especially in higher education and adult learning contexts, it is increasingly necessary to design course sites that facilitate learning as opposed to course sites that are just repositories. Just like the approaches that a teacher uses in classroom teaching, the teacher's underpinning philosophy will impact the way they will design their course sites on an LMS or website. Despite technology and online being present for over two decades, course design and teaching in the online environment remains trapped in traditional behaviorist and content-driven approaches.

Against the backdrop of emerging technologies, organizations are, now more than ever before, needing to develop agile and innovative mindsets within organizations. To achieve agility and innovation, the humans in the organizations—employees and leaders—must have the capabilities to be self-directed, and they must be driven to identify meaningful problems in the organization and engage in collaborative problem-solving to address the problems. Developing these capabilities rests mostly on the organizational environment as created by the leadership and with the opportunities presented through human resource development. The authors maintain it is useful to anchor in andragogy leadership approaches alongside opportunities for organizational learning but to not discount the value of pedagogy when used appropriately.

Over the past half-century, the word "andragogy" has migrated from the fringe vocabulary to the standard lexicon of adult education. It has identified common psychological traits in "non-traditional" students and distinguished its methods from "pedagogy," the teaching of children. This chapter will explore andragogy's apparent strengths and arguable weaknesses in terms of its socio-economic assumptions and the social class interests it may wittingly or unwittingly serve. It will conclude with a critique.

This chapter focuses on various factors requiring changes in higher education leadership, such as societal pressures, availability of financial support, labor market characteristics and expectations for job-ready college graduates, and the tension between institutional processes and the degree to which internal/external stakeholders are involved in decision-making processes. In this light, leadership development has become critically important in creating the necessary cadre of professionals committed to the mission and vision of institutions of higher education they lead. As a clearer focus on effectiveness and differential impact on the emerging leadership skills of participants is still needed, this chapter proposes a theoretical framework infusing andragogical principles into leadership development.

Andragogical leadership is a term yet to be coined. The evidence from four interviews with experts indicates that it might be a promising solution to grasp the fast and disruptive transformation processes companies are currently facing. Based on the research results, the author argues that andragogical leadership is a set of responsibilities that might at best be taken over by passionate experts in a shared leadership approach. The genuine implementation of andragogical leadership either requires or goes hand in hand with a transformation toward self-directedness, a participatory culture, and the democratization of an organization. Future research requires theoretical discussions and practical experiences contributed by experts from business studies, humanities, and psychology. The relevance of andragogical leadership and its potential to be a solution to urgent questions was proven within this chapter.

Educational leadership's implications significantly contribute to blended higher education development in the context of information technologies. However, educational leadership in blended higher educational contexts has yet to be addressed and is not explored widely. This study uses a systematic review to explore educational leadership in blended higher education contexts. It is inevitable that education leadership steadily assumes new roles and responsibilities in blended higher educational contexts. This study aims to provide research guidance for educators, school administrators, and policymakers.

This chapter reports designing, implementing, and evaluating a Leadership and Management in ELT course based on an andragogical approach for adult learners in the teacher education department of a state university in Turkey. Participants were seven adult learners aged 28-35 undertaking the course, which lasted 14 weeks with 3 hours of weekly teaching. The course was delivered online, and the course content included leadership and management theories and approaches using videos, books, and a collection of articles. Data were collected using qualitative data collection tools; participants were requested to produce reflective writings, keep a leadership portfolio, and respond to end-of-the-course evaluation. Content analysis of the data revealed that adopting an andragogical approach to teaching leadership and management in ELT has been highly useful in preparing the participants for the leadership and management working environment they were either concurrently involved in and/or had aspirations to be in their professional lives.

The Russian and Mexican Revolutions resulted in distinct socioeconomic programs during the 20th century; yet they produce comparable trajectories of political leadership and cultural discourse on adult educational reform. Using Soviet and Mexican educational materials from the revolutionary era, it is possible to compare the policies and practices produced in revolutionary contexts to the U.S. andragogy established through study of Mexico and Russia. The adult outcomes of schooling became a site of negotiating cultural hegemony that revolutionary groups used to design radically productive and disruptive teaching and learning theory, while at the same time indirectly modeling the progressive reforms employed in the U.S. Yet, since the dawn of the 21st century, both have struggled to refine their approach to adult learning in the context of the digital turn because of how the networked flow of information works to both support and undermine the institutionalized tenets of the revolutionary ethos.

The study's aim is to ascertain the role of contextual influences on leader development among natural scientists in Uganda differ from those of non-scientist leaders in the Ugandan context. The study design, methodology, and approach include a retrospective mixed-methods multi-case comparative case study to examine the leader development lifespan experiences of technical expert scientists holding leadership

Within the global community, the only constant is change. In Spring 2020, as COVID-19 became a global pandemic, countries closed their borders and closed schools, businesses, and governments. Universities that already had significant online presences and available courses/programs could be reassured that at least some of their programming would continue with minimal interruption. Effective university leaders were able to identify paths forward for their institutions that encompassed their education and were much more able to support faculty, staff, and students with the abrupt change. The authors look at professors as learning leaders, the leadership in higher education institutions, as well as leadership in the private sector. They examine how andragogical leadership can be utilized to shift current educational practices and consider how we might shape learning environments in order to foster more robust learning environments. But essentially, they explain how education leaders develop and organize institutional learning as technology continues to play an essential role.

An effective and impactful andragogical leader must have many key characteristics and competencies. They must be open to immersing into the Metaverse with the spirit of optimism, have the ability to adapt to change, and know that innovation and creativity in organizations are pivotal to survival as human activities fuse with technological domains. These leaders will use wisdom, experiences, intellectual insights, and curiosity to drive actions to achieve the future of their institutions. They need to be capable of leading their organizations through times of uncertainty, being self-confident during the Metaverse disruption, and doing so to be courageous action drivers as well as strategic thinkers. Be the action strategist, prehistorically known as the Allosaurus.

Hybrid teaching has come to the fore post pandemic disruption. Although universities are making an effort to 'return to normal', the legacy of the pandemic, alongside the ever-present need to remain competitive, is the demand for flexibility in learning and access to online learning. Many academics are now facing the challenge of hybrid teaching. It is much more than simply using technology to transmit the lecturer's voice and classroom images between physical and virtual locations. Educators need to consider how to develop effective learning strategies that provide quality and equal opportunity. This requires creativity alongside knowledge of pedagogy and technology. It must be considered that educators need to be committed to hybrid teaching and fully engage. There are existing studies in the related domain of blended learning which can help create insights into what factors will support effective hybrid teaching. The author reports on a study of factors academics to blended learning strategies and highlights how the results of the study are useful to hybrid teaching.

Technology, particularly as it supports interactive learning communities, now plays a central role in community education. While technology affords access to information more efficiently than ever before, it also requires additional skills. For technology to play out its role effectively, community educators need to design instruction, curriculum, and programs that offer customizable learning and socially rich learning environments. Every effort needs to be made to make sure that learners have opportunities to access and learn with technology. Successful community efforts to incorporate technology tend to include these factors: local needs, local resources, local broad-based planning, partnerships among organizations, accessible and appropriate technologies, opportunities for community building and socializing, and locally meaningful content and outcomes. Two community education foci are examined in detail: women's empowerment and health literacy.

Adult learning in today's society is primarily focused on the needs of adults as they function in their communities and professions. Adult learning principles and theories have been the foundation of adult education practice for over a century. However, a review of the historical context reveals that the technology-infused learning landscape of today is dramatically different from its predecessor. This makes a reexamination of these ideas important for their future application. This chapter will reexamine the adult learning principles of transformative learning, andragogy, and critical reflection in this technology-infused world and propose a new paradigm and corresponding practices for the new learning landscape in communities and professions.

Foreword

The discourse on leadership in education, particularly about educating adults, has become ever more relevant as our education classrooms have expanded into a global enterprise. *Handbook of Research on Andragogical Leadership in a Modern World* is more than a book; it is a mirror held up by practitioner-scholars for educationists to examine our praxes from a broad variety of vantage points. Framing a solid philosophy of education in our teaching and learning profession seems a logical first premise when we consider leadership. That such a leadership can be informed by time honored *andragogy* (Knowles, 1964; Henschke, 2008, 2011), of course, offers a rational approach to further insight and action on issues of education for adults in contemporary transnational and cross-cultural settings. Clearly, as professionals in our field, we ought to hold up a theoretical lens through which to evaluate the very learning models and design and delivery approaches we embrace.

In our profession, we live in a context wherein such grounding is harder to come by because we all too often remain at the abstract discourse level and do not share our actual experiences of teaching in our diverse environments. Grasping the principles of andragogy and applying these principles in our analysis, design, development, implementation, and evaluation of education programs are not often aligned parts within a whole practice. Moreover, we allow a medium, such as technology, for example, to drive decisions about our design and delivery of instruction. And may it be said that with the emerging media's bells and whistles, we have become enabled to more efficiently repeat the same ineffective instructional methods that have not served students well in the first place. Without a sound grounding in a philosophy and its correspondingly aligned teaching principles, we do not have a selection strategy to determine which medium lends itself best to which learning tasks. Particularly with the rapidly changing technology and wealth of *new and improved tools*, which are offered in the burgeoning marketplace of global eLearning, we may find ourselves at a loss to determine our functions, values, and responsibilities. Yet, we are called to make decisions for the delivery of instruction that meets the instructional intent *and* addresses the adult learners' needs. Hence, a philosophical grounding *and* a theoretical lens through which to evaluate our praxis seems fundamental. The authors in *Andragogical Leadership in a Modern World* bring us both.

Hailing from nine countries on five continents, we hear a chorus of voices of researchers and practitioners that share their thoughts on 21st Century leadership in education from culturally contextual and ideologically differentiated perspectives. These are themes that have been with many of us for some time (Strohschen & Elazier, 2007; Strohschen et al., 2008; Strohschen, 2011; Strohschen & Associates, 2016; Strohschen & Elazier, 2019). While critiques and nuanced interpretations of andragogy prevail in this book, there nonetheless is found harmony in the voices of chapter authors. Within these chapters, a *shared* vision pivots on the leading forth, the *educere* (Bass & Good, 2004) by educators of the kind of knowledge, skills, and attitudes essential to adults in the modern world. We hear in each chapter that this *educere* ought to benefit teacher and learner when mutually facilitated within a humanist approach toward emancipation and empowerment *for and with* teachers and students. Perhaps prerequisites for such forward movement are critical thinking; or strong technology skills are key; or radical system change is the answer; or perhaps supporting the development of resilience and self-efficacy need be a part of the educat*ing* we do.

The authors bring us much to consider, compare, and contrast and with that, perhaps, shake up our dichotomizing views of leadership and philosophy; of content and delivery mode; of context and process. In our profession, we have, perhaps, managed to apply the very principles about educating adults that described our discipline in the first place as different from pedagogy. As such, the practice of humanist adult education for carefully and mindfully assessing each learner's need with respect to the learning goal, style, and setting preference is a timeless approach to ensure that diverse global learners are strengthened to meet the challenges of the present and the future. Toward such ends, *Handbook of Research on Andragogical Leadership in a Modern World* amplifies the voices of our chorus of colleagues around the world and beckons us to arrive at our own examined and informed praxis, a praxis which honors adult learners' experience, goals, and ability as we *lead forth* within our respective teaching and learning journeys.

Gabriele Strohschen
DePaul University, USA

REFERENCES

Bass, R. V., & Good, J. W. (2004). Educare and educere: Is a balance possible in the educational system? *The Educational Forum*, *68*(2), 161–168. doi:10.1080/00131720408984623

Henschke, J. A. (2008). A global perspective on andragogy. In *The Commission of International Education* (pp. 43-94). American Association for Adult and Continuing Education.

Henschke, J. A. (2011). Considerations regarding the future of andragogy. *Adult Learning*, *22*(2), 34–37. doi:10.1177/104515951102200109

Knowles, M. (1970). *The modern practice of adult education: Andragogy vs pedagogy*. Association Press.

Strohschen, G. (2011). Framing a philosophy for 21st Century global recurrent education: Considerations on the role of the adult educator. In V. Wang (Ed.), *Information Communication Technologies and Adult Education Integration* (pp. 118–134). Information Science Reference. doi:10.4018/978-1-61692-906-0.ch008

Strohschen, G., & Associates. (2016). *The metagogy project: A theorem for a contemporary adult education praxis*. American Scholars Press.

Strohschen, G. (2019). The metagogy theorem: A framework for educating adults. *International Forum of Teaching and Study*, *15*(1), 3-15.

Strohschen, G., & Elazier, K. (2007). The 21st Century adult educator: Strategic and consultative partner. *Assumption University Journal, 1*(1), 42–53.

Strohschen, G., Elazier, K., Yu, T. T. C., Chuprina, L., Sun, L., & Molina Lopez, M. L. (2008). Globalization across cultures and the role of the 21st Century adult educator: Perspectives from Europe, North America, Latin America, and Asia. In J. Groen & S. Guo (Eds), *Thinking beyond borders: Global ideas, Global values* (pp. 424-429). Canadian Association for the Study of Adult Education (CASAE).

Preface

Leadership is much more than mere influence. It is about releasing the creative energy of the followers in a guided way to engage the process of change and work towards mutually beneficial goals (Henschke, 2020; Knowles et al., 2020). The leader who energizes and inspires their followers in such a way is an andragogical leader.

The nature of the andragogical leader is grounded in positive assumptions about people. The faith of andragogical leaders in their subordinates manifests in the delegation of challenging responsibilities and growth opportunities. It is starkly obvious to subordinates that they are valued by their leaders. Andragogical leaders involve their followers in every step of the planning process, assessing needs, formulating goals, designing lines of action, carrying out activities, and evaluating results. The followers of andragogical leaders are nurtured to be self-directed. Should the follower need help, the andragogical leaders are willing to do so, mentoring and helping the follower to grow. Because andragogical leaders tap into intrinsic rather than extrinsic motivation in their followers, low-achieving subordinates become high-achieving employees when they discover the excitement of self-directed learning under the influence of an andragogical leader. Andragogical leaders unshakably believe in the potential of followers. They know their followers will do superior work and they are convinced that their followers are the best employees in the organization. Andragogical leaders wield well the power of the self-fulfilling prophecy. Andragogical leaders are in tune with the unique strengths, talents, interests, and goals of followers; they value individuality and know that the best outcomes are achieved when learning strategies are tailored to fit the individual learning styles, paces, starting points, needs, and interests of all the subordinates. Andragogical leaders are comfortable with and actively seek pluralistic culture. In stark contrast to the missing controlling leader who seeks conformity, andragogical leaders seek to help each person become what Maslow (1970) calls a self-actualizing person. Led by leaders following an andragogical approach, followers are highly engaged and motivated to always improve. They are innovative and take risks to experiment, assured by the safe and trusting environment created by the andragogical leader. Failure is seen positively – an opportunity to learn and not fear punishment.

At a time when change is ever-present, the approach of andragogical leaders fosters a culture of continual improvement and drives innovative organizations rather than static organizations. Innovation and creativity in organizations are pivotal to survival as human activity fuses with technological domains.

Although technology has been around for a long while now, in schools and other social domains, it remains the 'instructional stepchild'. It is valued as an enrichment tool but is used only as time allows and after the "real" instruction is finished. Exploiting the potential of technology to use it in innovative and creative ways that change how and what humans do for the better lies within the influence of the andragogical leader. In schools, the andragogical leader plays an essential role in the digital transformation of education. Leading by example, when andragogical leaders model the use of technology, they also model creativity, innovation, and failure as a positive learning experience. Andragogical leaders should model effective technology used daily. That demonstrates to the followers that they value the efficacy of technology in performing everyday tasks and makes it evident that the andragogical leaders are personally embracing the initiative. In this way, they will drive the organization's learning and training in the organization and subsequently the integration of technology. A good starting point for initiating staff development and learning in technology integration is to identify subordinates who have already embraced technology and are excited by its potential and tap into the employee's strengths and needs.

The chapters in this volume are a cross-section of diverse perspectives, experiences, theories, and philosophies related to andragogical leadership and technology across all levels of education. This book is written for policymakers, managers, industry professionals, researchers, academicians, scholars, practitioners, instructors, and students.

ACKNOWLEDGMENT

Special thanks go to the international editorial advisory board members and reviewers who have reviewed all the chapters around the clock during the blind review process to ensure the quality of every chapter accepted for publication in this volume. Some of the world's leading scholars have contributed chapters to this volume. Some of the editorial board members have contributed chapter(s) to this book. Their insights and collegiality are sincerely appreciated. Last but not least, I thank IGI Global, whose editorial and production team has continuously supported my work in this field and other fields. I thank our users and university librarians for recommending this unique book to scholars and practitioners from all disciplines in Education and Social Sciences.

Viktor Wang
California State University, San Bernardino, USA

REFERENCES

Henschke, J. A. (2020). *Facilitating adult and organizational learning through andragogy: A history, philosophy, and major themes*. Information Science Reference.

Knowles, M. S., Holton, E. F. III, Swanson, R. A., & Robinson, P. A. (2020). *The adult learner the definitive classic in adult education and human resource development* (9th ed.). Routledge. doi:10.4324/9780429299612

Maslow, A. H. (1970). *Motivation and personality* (2nd ed.). Harper & Row.

Chapter 1
LMS Course Design for Adult Learning:
Heutagogy, Andragogy, Pedagogy

Viktor Wang
California State University, San Bernardino, USA

Geraldine Torrisi-Steele
Griffith University, Australia

ABSTRACT

Nearly all educational institutions use learning management systems (LMS). Post-pandemic figures suggest that the uptake of LMS is about 98%. This means that most educators, especially those in post-schooling education, will use, and be responsible for designing, an online course site. While there is a tendency to use course sites as a repository and convenient communication medium, there is an unmet need to design these sites in a way that facilitates learning. As online teaching and learning become more prolific, especially in higher education and adult learning contexts, it is increasingly necessary to design course sites that facilitate learning as opposed to course sites that are just repositories. Just like the approaches that a teacher uses in classroom teaching, the teacher's underpinning philosophy will impact the way they will design their course sites on an LMS or website. Despite technology and online being present for over two decades, course design and teaching in the online environment remains trapped in traditional behaviorist and content-driven approaches.

INTRODUCTION

Malcolm Knowles predicted that the teaching of adults in the 21st century would be delivered electronically (Knowles, 1975; Knowles, Holton, III, Swanson, & Robinson, 2020). His prediction was accurate and logical given that mature learners, constrained by multiple work/family responsibilities are unable to meet the demands of traditional classrooms locked into specific times and places. What Malcolm Knowles perhaps did not predict was the rapid transformation of education delivery to the online medium

DOI: 10.4018/978-1-6684-7832-5.ch001

that was catalyzed by the pandemic. According to UNESCO, around 1.5 Billion students were online during the pandemic (https://www.unesco.org/en/covid-19/education-response). If the jump to online learning showed us one thing, then it was that education systems and educators were, in the majority, not well-prepared to use technologies to the best effectiveness. The pandemic experience has shown that there is still much to learn about effective teaching online and harnessing technologies effectively to maximize learning opportunities.

The de-institutionalization of education in the form of open and independent learning systems that Knowles also predicted is underway. This creates a need for learners to develop appropriate self-directed learning skills – a capability that is more urgent than ever in the current context. Students in open and independent learning systems without having learned the skills of self-directed inquiry will experience anxiety, frustration, and, often failure, and so will their teachers (as cited in Wang, 2005). Self-directed learning skills will help mature learners learn in virtual environments where they have no face-to-face meetings with course instructors.

The uptake and diffusion of online learning are greatly assisted by the development of learning management systems (LMS). LMS can be understood as "information systems that facilitate e-learning by supporting teaching and learning, but that also can perform administrative tasks and facilitate communication between instructors and students" (Klobas & McGill, 2010 cited in Holmes & Prieto-Rodriguez, 2018, p. 21). LMS such as Moodle, Blackboard, and Canvas are commonplace in post-secondary settings such as universities and in vocational education for both on-campus and distance students. During the pandemic, LMS played a crucial role in enabling the leap to the online learning medium (Gamede, Ajani, & Afolabi, 2022). Although LMS are ubiquitous, educators often fail to take full advantage of the capabilities of the technologies to facilitate the most effective learning (Noval, 2018). One reason why educators may not take advantage of the LMS tools is that they may not properly understand the affordances of the technology (Holmes & Prieto-Rodriguez, 2018). However, at the very foundation of teaching approaches is teaching philosophy. It is well-acknowledged that teaching philosophy deeply influences approaches, and the kinds of strategies teachers implement. Designing and delivering courses through LMS is no different. "Technology choices depend on and will amplify our teaching philosophy, so examining our assumptions about teaching is essential when designing appropriate learning interventions" (Hickey, 2014). In the present chapter, we consider two camps of philosophies: Pedagogical teaching (i.e., traditional approaches) and the andragogical teaching. We posit that andragogical teaching is necessary to escape the confines of the content-driven mindset that dominates LMS use towards developing innovative approaches, for the design and delivery of courses in LMS environments that effectively engage and result in deep learning, especially for adult learners.

The aim of the chapter is two-fold: to highlight how andragogical approaches are beneficial to learning in the context of LMS and online learning, and to encourage educators to examine their own philosophies against the andragogical and pedagogical frames, thus raising their self-awareness of their philosophies. Without understanding where they stand in relation to andragogical and pedagogical frames it will be difficult for educators to implement the 'right' instructional design and to implement effective and appropriate teaching methodologies in virtual environments.

THEORETICAL FRAMEWORK

Traditional teaching methodologies (e.g., Tyler's 1949 four fundamental questions for curriculum development) were derived from Skinner's behaviorism (1968). Behaviorism's influence on instructional technology "led to the design of piecemeal instruction with immediate feedback and reinforcement, drill and practice procedures, and self-paced programmed instruction (as cited in Dewald, 2003). As Dewald (2003) further indicated, behaviorist instructors try to measure learning outcomes. They also want student learning outcomes to be observable and measurable. Educators who base their practice on behaviorist assumptions tend to believe that if they fail to measure and observe student learning outcomes, students have failed to learn. Bloom's (1956) taxonomy of educational objectives (based on behaviorism) informed educators' practice for decades. However, Bloom's taxonomy has little room for instructors to involve learners in the instructional design process. Traditional teaching may work well with children or pre-adults.

While the education of children centers on pedagogy (the art and science of teaching children), the education of adults should also refer to andragogy (the art and science of helping adults learn). The juxtaposition of pedagogy and andragogy represents the essential issue of teacher-centered methods versus learner-centered education. As Knowles (1970) argued that adults have andragogical learning characteristics whereas children have pedagogical learning characteristics, his comparison of the two models generated much controversy in the field. It is important to recognize that not every adult learner is a self-directed learner. Therefore, teacher-centered methods are still useful especially when an adult is entering a totally new content area (Dewald, 2003). However, no one can doubt the benefits of a learner-centered education when adult learners are self-directed and task-oriented in either traditional or electronic classrooms. And no one can doubt the value of helping adults move from teacher-directed to learner-directed education (Smith, 2012). When learners are capable of self-directed learning in an online learning environment, educators need to work within Knowles' andragogical model to maximize learning. Making the switch from a pedagogical to an andragogical model is essentially a paradigm shift for educators (Weimer, 2012). Figure 1 contains Knowles' (1995) comparison of the two teaching models.

Table 1. Knowles' comparison of two models

Pedagogical Model	Andragogical Model
The learner is dependent on the teacher to decide what will be taught and how it will be taught.	The learner prefers to be self-directed, so the learner and teacher should plan together what will be learned.
The learner has little experience to contribute to the learning effort, so the goal is to transmit knowledge through lectures, assigned readings, and presentations.	The learner brings a wealth of experience to the learning situation, and therefore learning should include elements such as experiments, discussion, case studies, and simulations.
Learning is organized by the logic of the subject matter.	Learning is organized around a task to be completed or problem to be solved.
Learners are considered ready to learn when they attain prescribed ages.	Learners are considered ready to learn when they have a need to know something, and learners want to be able to apply what they learn to their lives.
Motivation is inspired by external rewards such as punishment, grades, and pressure from teachers and parents.	Motivation is inspired by internal incentives such as self-confidence, better quality of life, and curiosity.
The teacher is responsible for content design, determining coverage, and efficient transmittal methods.	The teacher, as facilitator, is responsible for process design, creating the climate for learning, and making resources available.

It may be helpful to clarify the philosophical foundations underlying the preceding discussion—behaviorism and humanism. Behaviorism began with Pavlov's (1927) classic studies of animals' responses to stimuli and reinforcement. Scholars such as Tyler, Thorndike, Watson, and Skinner elaborated on the basic idea that human beings responded to stimuli with certain behavior, were rewarded or punished for this response, and this constituted learning. This philosophy emphasizes the importance of the environment or external forces in shaping desired behavior. Human behavior is believed to be tied to prior conditioning. Learners respond to a stimulus, and this is the essence of learning. Teachers who follow this philosophical perspective provide as many stimuli as possible in order to solicit responses from students. Once correct responses are obtained, students are rewarded. By conditioning, feedback, and practice, basic skills and life skills can be developed. To instructors who are behaviorists, the aim of education is to develop survival skills and change behavior (Lee, 2010). Behaviorism is the foundation for systematic instructional design models that emphasize accountability. Competency-based teacher education, programmed instruction, computer-assisted instruction, performance-level testing, and behavior modification programs were developed with this philosophy as a basis. In the 1960s, the above-mentioned modes of behaviorist education became popular in North America. Even to this day, behaviorism has been widely used in K-12 education, higher education, and adult education. Critics of this philosophy believe that learning is too complex for behavior control. Hence, many adult educators have turned to humanism.

The basis of humanism can be traced back to classical China, Greece, and Rome (Elias & Merriam, 1995, 2005). In the 1950s and 1960s, the work of Maslow and Rogers led to the popularization of humanism. Other key supporters of this philosophy are Erasmus, Rousseau, Knowles, and Tough. The notion of self-directed learning was derived from humanism. This philosophy is based on the assumption that human nature is essentially positive, and that each person possesses unlimited potential and innate goodness. Intellectualism is viewed as what distinguishes humans from animals. While humans are capable of intellectualism, animals are not. Autonomy, dignity, and freedom are seen as sacred. Humanism places emphasis on personal growth, self-direction in the learning process, individual potentiality, and self-actualization by using facilitation, self-direction, and teamwork. Maslow's hierarchy of human needs was derived from humanism in that it addressed self-actualization. Instructors who believe in humanism consider themselves learning facilitators, or consultants, linking students to learning resources. Rogers believes that we cannot teach another individual, but rather, we must facilitate his or her learning. Individualized instruction, learning projects, and sensitivity training are all practices that use humanism as a philosophical foundation. The curriculum is seen as a vehicle for meeting needs. Critics of this philosophy believe that important societal goals may not be included in learning (Lee, 2010).

Behaviorism drives pedagogical assumptions of learning, and humanism drives andragogical assumptions of learning. However, we need to be careful not to create a dualism. Many pedagogical practices embrace humanistic assumptions. Teachers need to be well versed in both pedagogical and andragogical principles so that they know which philosophy to apply in terms of helping mature learners learn in the traditional classroom or in virtual environments. Although humanism works well with adult learners, behaviorism also has a place in the education and training of mature learners. The dialectical relationship addressed earlier in this article is worth more thought as we are engaged in helping adult learners learn in the 21st century.

LMS Course Design and Delivery

Online learning and the advent of technology in education are filled with promises of transforming education. Overall, the promised transformation has failed to be realized. Part of the reason is that the focus has been on the 'technology before pedagogy' as if the technology itself possesses some intrinsic quality that will improve learning (Cowling & Birt, 2018). Often the sales pitch for LMS will emphasize that the LMS will facilitate the digital transformation of learning in the institution. While the emphasis is on the fact that the LMS possesses functionality that can be used to create transformation, the more important factor is often underplayed – that the LMS is merely a tool, and it is the way it is used that makes the most difference. A large part of the reason for the failure to realize the transformation of education through the use of technology is, in part, that educators tend 'stick' with what has worked in the past, and in part, they may be simply overwhelmed by the demands of the classroom and don't have the 'luxury' of time to explore further (Hickey, 2014), and in part because of the inertia of traditional teaching methods long-embedded in teaching. The situation is made worse by unfamiliarity with the nature of the technology.

In the traditional classroom or online approaches, the course is designed and delivered around clearly defined goals and outcomes. The structure of the course is rigidly defined and reflected in the way readings, learning resources, lectures, and activities are sequenced and presented in the LMS course. The educator provides shape and direction to the learning experience by forming groups and providing assessment assignment focuses or guidelines (Siemens, 2013). The traditional model of teaching is visualized in Figure 1.

In the traditional or pedagogical approach to teaching, instructors direct and define learning for the learners and the "value of the real-world familiarity is diminished" (Burton et al., 2013). The approach can be useful when learners are immature or lack enough mastery of the discipline to do any more than follow directions. However, as the individuals develop, they turn towards self-direction and unless the educator's approach also changes, there results in a misalignment of teaching approach and learner needs (Burton et al., 2013).

Figure 1. Concepts in traditional approaches

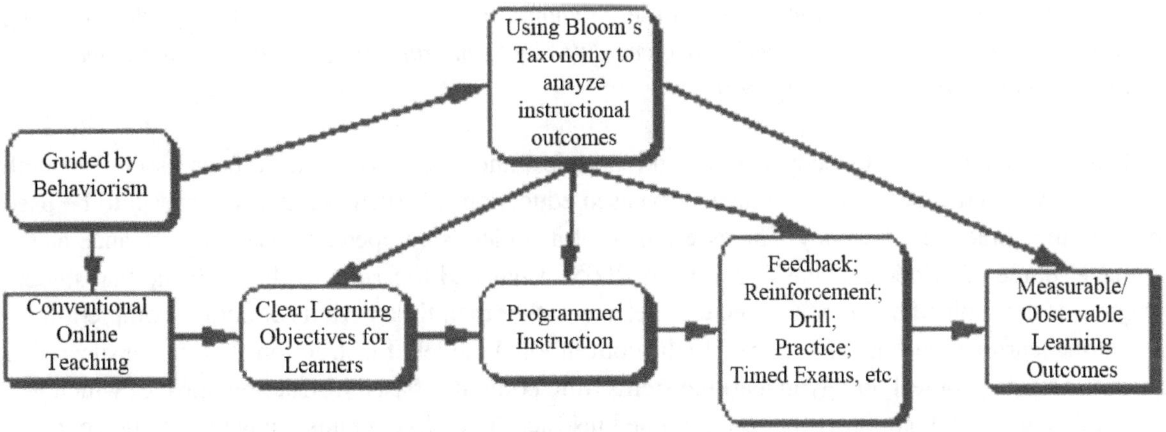

The model presented in Figure 1 illustrates the dynamic interaction of variables that contribute to measurable/observable learning outcomes. Instructors have been using this model in their traditional and electronic classrooms since Skinner advanced and popularized behaviorism. Since online teaching occurred in the early 21_{st} century, most universities and colleges including some businesses and government agencies simply have moved this model to computer screens. Since Bloom (1956) developed his Taxonomy of Educational Objectives, educators have been using Bloom's Taxonomy to develop instructional outcomes. The goal of this model has been very clear: learning objectives for learners must lead to measurable and observable learning outcomes. Instructional methods focus on teacher-centered methodologies. Teachers believe that it is their responsibility to provide stimuli and learners respond to those stimuli. If students provide correct answers, their behavior is rewarded. That is, instructors, prescribe instruction and learners assume a submissive role of following their instructors (Knowles, Holton, & Swanson, 1998, 2005, 2011).

Further, instructors treat learners as dependent learners (Grow, 1991; Wang, 2007). They believe in external rewards such as punishment, grades, and pressure from teachers and parents. Instructors believe that they are authorities. Therefore, they create instructional designs according to their own prior experience. Learner input is not taken into consideration. For those adults who follow a behaviorist philosophy of learning, this model can work when adult learners have practical deadlines to meet in a short period of time. Adults new to a subject area may also benefit from this approach to teaching and learning. This model works best, especially in adult basic education where mature learners may not be self-directed in learning as they may not even have entry-level knowledge in terms of learning a new subject. For example, since the emphasis in vocational education is on identifying the skills to perform in an occupation, teaching those skills, and requiring a certain standard of performance, this model can prove to be effective. According to Elias and Merriam (2005), education in this arena of practice is concerned with the outcomes rather than the process of learning, exit rather than entrance requirements, and criterion-referenced evaluation rather than norm-referenced evaluation. This model works well in certain cultures such as Confucius-heritage societies where all learners are required to pass standardized tests to be matriculated into middle schools, high schools, or universities. Elias and Merriam (2005) argue that,

Criterion-referenced evaluation is an important concept in behavioral psychology. In criterion-referenced evaluation, the learner's process or accomplishments are compared to a fixed standard or criterion of mastery rather than to the performance of other students. It is based on the assumption that learning objectives can be predetermined, and that given sufficient time and proper reinforcements nearly all students can meet the objectives. (p. 99)

Learning outcomes (knowledge, skills, and attitudes, and/or behaviors) have been associated with competency-based education. In competency-based education, instructors must set criteria to be used in assessing student competency. The assessment of a student's competency used performance as the primary source of evidence (Elias & Merriam, 2005). Elias and Merriam further indicate that student progress is determined by demonstrated competency rather than time periods or course completion.

Online learning lends itself very well to the conventional model of instruction. This is not surprising given the heritage of technology in learning stems from computer-based instruction, most of which was drill and practice and other forms of programmed instruction. LMS contains many tools which enabled a structured and directed approach to learning. A sequential module or topic structure is implicit in most

LMS and allows collections of resources such as lectures, readings, videos, and PowerPoint slides to be categorized and presented sequentially. Most LMS has the functionality to prevent students from moving ahead or from seeing subsequent content through adaptive release functionality. Of course, the instructor can constrain what content is available and when it is available simply by loading content when they deem it is needed. Using the LMS to load resources in a sequential and controlled manner is probably one of the simplest tasks for educators to accomplish. In addition, the LMS toolkit includes simple tools for creating quizzes for 'drill and practice' type exercises. While it's possible to create electronic quizzes and other tests that are more open or check deeper understanding, it is usually a time-consuming task, requiring creative approaches – the tools include multiple choice, drag interactions, and fill-in-the-blanks which lend themselves best to simple drill and practice. Creating more sophisticated or open-ended interactions can require greater knowledge and skill of the technical aspects of the platform and can also be time-consuming. Hence, for practical (besides pedagogical reasons), the LMS course and delivery in practice tend to follow conventional models.

For adult learners, there is a danger of over-reliance on the conventional model of instruction when designing and delivering courses. Adult learners need to be recognized as being at the center of the educational process and should experience a learning environment that reflects the democratic spirit of individualism to use Elias and Merriam's terms (2005). Andragogy is thus better aligned with the needs of adult learners.

Andragogy, connected with adult learning, is learner-centered education as opposed to teacher-driven and the actual learning progression should include the entire emotional, psychological, and intellectual being (Burton et al., 2013). The model of Andragogical Online Teaching illustrated in Figure 2 demonstrates a more powerful learning process for adult learners in the online learning environment.

Figure 2. Innovative teaching model

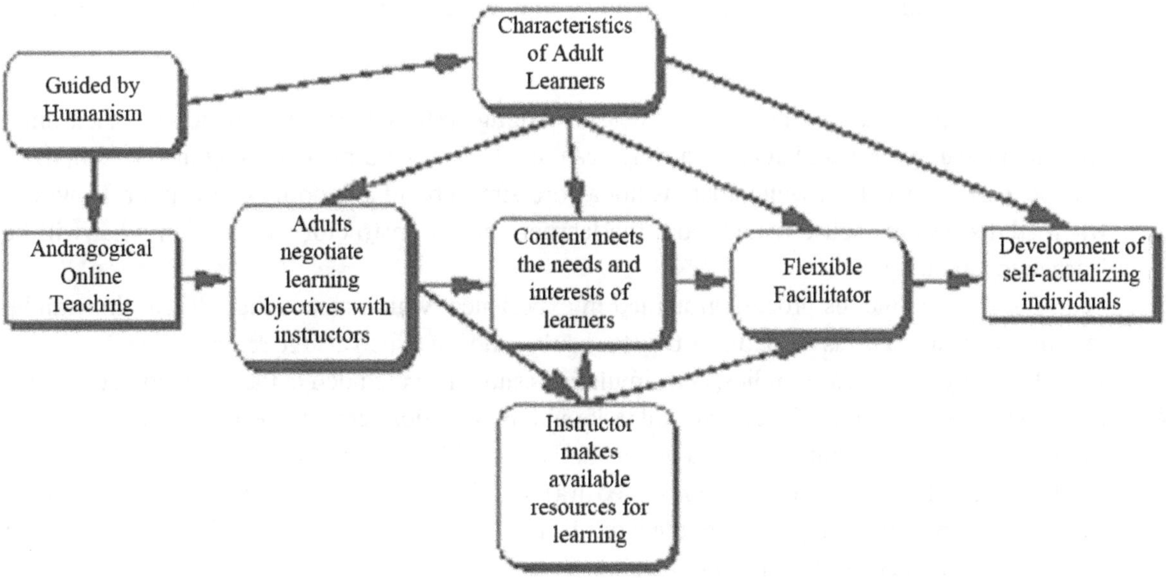

Guided by humanism, the andragogical model (Figure 2) promotes cooperation and communication among individuals as a vehicle for interpersonal growth; cooperation is a necessity for making democracy work (Elias & Merriam, 2005). The goal of this model is the self-actualization of individual learners rather than measurable or observable learning outcomes in behaviorist terms. Instructors use this model to assist individuals to grow and develop toward self-actualization and toward becoming fully functioning persons (Maslow, 1954; Elias & Merriam, 2005). Elias and Merriam (2005) further address this model by saying,

The notion of growth, development, and change is integral to much of the psychological literature on adult development. Adults are no longer viewed as finished products at the age of sixteen, eighteen, or twenty-one. Rather, adulthood is a period of change, psychologically, socially, and physiologically. Adult educators have begun to respond with activities designed to take into account adult development. One of the assumptions underlying andragogy is related to meeting the needs of adults and many of these needs are the direct outgrowth of the developmental tasks salient to the various changing stages of adulthood. (p. 144)

Instructors who have adopted this humanistic approach tend to emphasize the psychological growth of learners more than the content of programs. Many adult educators who have adopted this approach value the prior experience of adult learners, and are committed to educating the whole person—these are not readily measurable in behaviorist terms (Elias & Merriam, 2005). Since the advent of this model in the early 1970s, adult education programs have begun to center on the needs and interests of the learners and this approach is dedicated to the development of human beings. Since adults are self-directed or potentially self-directed in their learning, rigid approaches such as immediate feedback, drill, and informational lectures are not appropriate. Educators are seen as facilitators and consultants in the learning process. Upon the basis of this model and Table1, Knowles, Holton, and Swanson (1998, 2005, 2011) have specified the following educational implications for instructors in the traditional and electronic classrooms:

1. Adults need information and involvement before learning the how, the what, and the why of learning.
2. Although it can be argued that any learning has the effect of building autonomy in a person, there may be learning events in which there is not a core aim to build autonomy in a learner. However, when adults become self-directed, both the learner and the instructor should plan what will be learned together.
3. Learning is a continuous process grounded in experience, which means that all learning can be seen as relearning. This is particularly true for adults who have such a large reservoir of experience. Therefore, discussion, case studies, and simulations should be included in the instruction of adults.
4. Learning is related to one's developmental tasks and life situations create a need to know. Therefore, adult educators need to anticipate and understand adults' life situations and readiness for learning.
5. Adults learn best when information is presented in a real-life context. Experiential learning approaches have the dual benefit of appealing to the adult learner's experience base as well as increasing the likelihood of performance change after training.
6. Adults are more motivated by internal motivators than external motivators. Instructors should help arrange learning that will help adults with a problem or issue that is important in their life.

The andragogical approach to LMS course design and delivery moves beyond the structured, tightly teacher-controlled approach of conventional approaches. Rather than the teacher prescribing all the content, dictating sequence, and constraining learning towards meeting rigidly specified objectives, the teacher takes on the role of facilitator. Choice and negotiation of learning are important parts of the adult learner environment. Learners can negotiate assessment and grading and should have a voice in course content selection. Day and Baskett (1982) suggest andragogy is more than a theory, it is "an educational ideology rooted in an inquiry-based learning and teaching paradigm" (p. 150). Allen (Allen) provides a succinct summary of the 'andragogically' orientated principles of online course design:

1. Tailoring course design to students' needs, life experiences, and interests.
2. Help learners construct knowledge rather than transmit knowledge.
3. Foster peer-to-peer and peer-to-instructor interaction.
4. Create authentic learning environments and assessments.

LMS and the online tools they integrate with are useful to support more than drill and practice activity – for example, most LMS have the functionality or will integrate with software for developing portfolios and creating websites (Allen, 2016). Furthermore, chat and video-conferencing tools, many of which integrate with the LMS, provide an opportunity for creating a 'community of learning' (Allen, 2016). A community of learning approach to LMS course design and delivery enables adults to co-create the learning environment, enables them to share life experiences and needs, and helps the construction rather than the transmission of knowledge. Importantly, LMS and the integrated tools are not contained to a desktop computer. Integration with mobile devices including tablets and smartphones provide endless opportunity for authentic learning 'in the field' (Narayan, Herrington, & Cochrane, 2019).

The andragogical online teaching model revolves around the andragogical model popularized by Knowles in the early 1970s. Knowles' andragogy is established around the following deductions: (1) self-directedness, (2) need to know, (3) use of experience in learning, (4) readiness to learn, (5) orientation to learning, and (6) internal motivation (Burton et al., 2013). Burton et al. research drew a parallel between Knowles' evolving transference about the education angle in andragogy to education explorations on the evolution of learning through technology (2013).

Depending on what to teach online, the conventional teaching model guided by behaviorism can prove to be effective in many ways. But, when it comes to helping adults learn online, it is more advisable for instructors to utilize the andragogical online teaching model because it focuses on releasing the energy of adults (Knowles, Holton, & Swanson, 1998, 2005, 2011) so that they could achieve full development of their individual potentiality. Again, this is not to say the andragogical model is superior to the pedagogical model or that the pedagogical model is inferior to the andragogical model. Rather, the goal is to use a model in virtual learning environments that can help maximize learning. The more we understand the differences between these two models, the better we can make decisions about which model (or combination of models) to adopt and adapt to either the traditional classroom or the virtual learning environments. Both models have been in existence for their own particular reasons and they are both derived from both pedagogical and andragogical assumptions of learning. The dialectical relationship between the two models does help educators of children and adults make a sound decision as to which model to employ to teach children or help adult learners learn to maximize learning outcomes. To employ the innovative teaching model, the following quote discusses what andragogical teaching and learning embody.

Burton et al. (2013) voiced:

Moreover, the broadening of the mind leads to Knowles' other three postulations – readiness to learn orientation to learning, and internal motivation. For instance, constructivists hypothesize that learners construct information by their experiences and understandings. True andragogy allows learners to review and learn through educational websites while employing the six postulations. Genuine andragogy practitioners, support education as informal yet vital breaks to garner learning and education. Learners' knowledge and ideas must be broadened via practice and independence for them to grow and develop, and learners could include faculty, students, and administrators. The significance of andragogy comes in the contextualization of adult learning theory in concrete teaching and learning methodologies that includes continuous learning through social network sites.

Andragogy as a framework for adult online learning is gaining momentum. Greene and Larsen (2018) propose the concept of virtual andragogy as the new paradigm for online adult learners. Greene and Larsen (2018) point out that adult learning is one of the fastest education areas but yet there is little attention to what is needed to create learning environments that meet the needs of these adult learners – a peculiar situation given that andragogy has been around for more than 180 years (Greene & Larsen, 2018). Greene and Larsen (2018) state:

"Rather than trying to modify pedagogical truisms, a new framework, *Virtual Andragogy*, is proposed to meet this need. Building upon the body of work of seminal education, *Virtual Andragogy* offers an innovative, grounded theory for andragogic ally- appropriate, constructivist-focused educational experiences that seek to serve adult students through authentic, transformational learning (p. 1376). More recently, a third '..gogy' is emerging as a philosophy for adult education – heutagogy.

Heutagogy

Heutagogy is gaining popularity and is seen as an ongoing development in the continuum of pedagogy to andragogy (Narayan, Herrington, & Cochrane, 2019)

Heutagogy was coined in 2000 but it would be at least 15 years after that, when social media tools came into their own, before it was brought to attention in the more mainstream (Narayan, Herrington, & Cochrane, 2019). The central tenet of heutagogy is self-determined learning and the emphasis is on learner agency.

"Heutagogy, a form of self-determined learning, is a holistic, learner-cantered approach to learning and teaching, in formal and informal situations. The theory is grounded in humanistic and constructivist principles and brings together numerous threads of early learning theories into a composite picture of the learning that is suitable for and much needed in today's educational systems (Blaschke & Hase, 2016). Wheeler (2012) suggests that "the ethos of heutagogy extends to learner choice, where students can create their own programs of study. Heutagogy is self-determined learning i.e., self-managed (learners decide what and how learning takes place), whilst andragogy is self-directed (the teacher-student relationship is strongly present with the teacher facilitating and guiding the learning) (Kenyon & Hase, 2001).

The heutagogical approach to learning stands at the end of a continuum that progresses from pedagogy to andragogy, with self-directedness as a key attribute of heutagogy (Knowels, 1975; Blaschke, 2012; Canning, 2010 as cited in Beaven et al., 2014) (Figure 3) Movement along the continuum is marked by ever-increasing learner autonomy, greater flexibility in course structure, and a decrease in tutor control.

Figure 3.

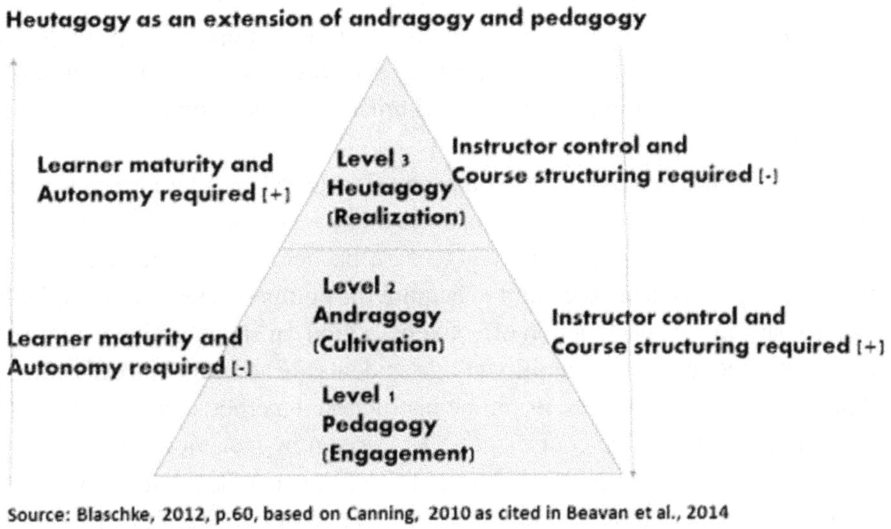

Drawing on Mezirow (1997), and Blaschke (2012) Beaven et al. argue that heutagogical framework accelerates learners' cognitive development, strengthening their capacity for critical reflection (as cited in Beaven et al., 2014). Heutagogy represents a decreased degree of educator involvement – it is about the capability of 'knowing how to learn'.

Design elements of a huetagogical learning environment include learner-defined learning contracts, the learner as the driver of a flexible curriculum, learner-directed questions, and flexible and negotiated assessment (Blaschke, 2012). Self-reflection is a critical component of huetagogical learning as it enables lifelong learning (Blaschke, 2012). The affordances of technology are well-suited to self-directed learning. The Web enables learners to create their own pathways and encourages the active creation of learning pathways. Social media affords connectivity to others, reflection, and information sharing. Mobile devices support collaboration, communication, as well as data collection, analysis, and sharing in context.

CONCLUSION

Adult learners constitute a significant portion of post-secondary education. Given the pace of societal change, knowledge creation, and technological advancement, a continuation of learning is crucial to adult learners. However, continuing to learn can be challenging for adult learners because they frequently need to juggle competing commitments of family and work while they continue to learn. The advent of digital technologies provides tools that enable flexibility of time and place. For educators, the development of LMS has simplified and streamlined the use of technologies for e-learning. The LMS helps to create more sophisticated platforms for distance education because they embody many functionalities and tools that can be used for engaging learners in many different activities and modes of learning. Despite advances in technology, the heritage of distance education and instructional technology cannot be ignored. Distance education historically relied on study guides and books of readings and instructional design for computer-based learning was based on behaviorist principles. This heritage is very detectable in today's use and design of LMS courses. Instructors tend to prescribe learning objectives, the whole teaching process is characterized by a teacher-centered approach, and student learning outcomes are measurable and observable. As leaders in adult education try to popularize principles of andragogy (the art and science of helping adults learn), especially self-directed learning, the instructional design seemed to have made the switch from behaviorism to humanism to develop the potential for adult learning (Hase & Ellis, 2001).

For example, the mission statement and vision of this Southern New Hampshire University (SNHU) reflect the principles of adult learning: SNHU uses distance learning methodologies to provide adult learners throughout the world with opportunities to earn bachelor's, masters and doctoral degrees in recognized academic programs to enable them to acquire the competencies and to apply the knowledge and skills needed to participate and function effectively in modern society. Since adult learners are the center of this university's education, no traditional-age students are admitted into this university. To keep its educational promises to adult learners, its online instructional methods must reflect andragogical assumptions of learning. Are the online courses at SNHU driven by behaviorism or conventional online teaching? The answer is "both". Although scholars and practitioners talk about andragogical approaches to teaching online, it is hard to establish a learner-focused approach to learning.

Shifting to the learner focus can be especially challenging in learning environments where the presence of technology dominates. Mishra and Koehler (2006) highlighted the complexities of teaching effectively online when they stated that effective teaching with technology requires knowledge across each of the 'pedagogical' domain, the technological domain, and the content domain (the word 'pedagogical' was used more broadly here to mean knowledge of learning and teaching). Understanding learning and teaching is a significant determinant of how educators use technologies to achieve the desired learning within a particular context.

Teaching philosophy is the lens through which educators understand learning and teaching. Hence, the importance of creating awareness of, and understanding the pedagogical, andragogical, and huetological stance cannot be understated. The authors hope that the present chapter has contributed to this understanding and has prompted readers' reflections on where their teaching beliefs sit. At this point, it is imperative to emphasize that pedagogy, andragogy, and heutagogy are not mutually exclusive. Rather they can be conceived of as a continuum (Narayan, Herrington, & Cochrane, 2019) and the educator should select, according to the learner's needs, which philosophy is most appropriate to enact.

Unfortunately, most instructors still cannot go beyond the conventional online teaching model guided by behaviorism and characterized by measurable and observable learning outcomes. Instructors depend heavily on Bloom's Taxonomy of Educational Objectives to delineate the learning process (Wang & Farmer, 2008). Researchers found that most distance education programs were highly teacher-directed and dominated by pedagogical rather than andragogical approaches (Hase & Ellis, 2001). Nowadays researchers seem to emphasize the importance of using social constructivism and interactive learning environments to increase the control of the learner over learning. If researchers admit that learning is an internal process and self-directed learning is the goal of adult learning, then the development of self-actualizing individuals cannot be measured or observed. The ultimate goal in learning should increase learners' control over their own learning. How can instructors do this when they are separated from their learners in the online learning environment? The andragogical online teaching model offers a viable option for today's instructors. For example, instructors can use this model to increase learners' internal motivation through self-directedness. As Shuttler and Burdick (2006) indicate, "learning is difficult to gauge from the results of a rote quantitative examination: this practice is seemly better suited to assessing memorization." They advocate that course and programmatic curricula should include highly participative and engaging activities and assessments that assist the faculty member in facilitating active learning practices. In the conventional online teaching model, an instructor assumes the role of an unchallengeable figure. As online education is evolving quickly in the 21st century, the teacher moves from an acting authority figure to become . . . "the guide, the pointer-out, who also participates in learning in proportion to the vitality and relevance of his [or her] facts and experiences" (Mezirow, 1991).

In the present chapter, the authors sought to draw attention to pedagogy, andragogy and heutagogy within the context of the design and delivery of courses in LMS. LMS can provide the flexibility of learning and the technologies necessary for implementing pedagogical, andragogical and huetagogical approaches. However, the LMS functionality is under-exploited and the promise of LMS for more effective education has not been realized (Zanjani, Edwards, Nykvist, & Geva, 2017), especially in adult learning contexts. Research into what engages students in LMS and how staff perceives the LMS features is sparse (Holmes & Prieto-Rodriguez, 2018). LMS courses remain, for the most part, repositories of course content. As put forward by Greene and Larsen (2018), concepts such as virtual andragogy may facilitate a paradigm shift in the design of LMS and other online courses:

Lessons learned from the experiences of crafting and teaching online courses for adult students are analyzed to offer an examination of strategies that can empower instructors, instructional designers, and administrators. The goal is to allow for the presence of the instructor, fidelity of implementation, and emotional/cognitive support for the student. (Greene & Larsen, 2018, p. 1376)

The design of LMS courses aside, it is also worth remembering that the nature of the technological tool will enable or constrain its use. While the discussion in this chapter has focused on how a teacher's approaches influence how the LMS is used, it is also important to draw the attention of those who design the functionalities of LMS to attend to the attributes of pedagogical, andragogical, and huetagogical approaches. Those designing and building LMS need to take into consideration how the LMS functionalities support various teaching approaches. There is some criticism that some of the social tools in LMS are not as effective or useful as other mainstream tools, and that there are limitations in the LMS that prevent the teacher from using them in ways they require (for example, a group interaction tool might not be able to support larger group numbers (Holmes & Prieto-Rodriguez, 2018).

LMS are commonplace in post-secondary education and adult learners make up a significant portion of learners using LMS. Unfortunately, LMS are not being used to their most effective extent and behaviorist approaches dominate LMS course design and use. The authors posit that providing the most effective and engaging education for adult learners requires a shift from a technology centric world view. Instead, technology use and design should rest on an understanding of pedagogy, andragogy and heutagogy on the part of the educators, and even on the part of those designing and building LMS. To achieve this there is a need for critical reflection on both practice and tools.

REFERENCES

Allen, S. (2016). Applying adult learning principles to online course design. *Distance Learning, 13*(3), 25–31.

Blaschke, L. (2012). Heutagogy and lifelong learning: A review of Heutagogical practice and self-determined learning. *IRRODL*, 57-71. https://files.eric.ed.gov/fulltext/EJ979639.pdf

Blaschke, L., & Hase, S. (2016). Heutagogy: A holistic framework for creating twenty-first-century self-determined learners. In B. Gros, Kinshuk, & M. Maina (Eds.), *The future of ubiquitous learning.* Springer doi:10.1007/978-3-662-47724-3_2

Bloom, B. S. (Ed.). (1956). *Taxonomy of educational objectives*. Longman.

Burton, S., Bessette, D., Brown-Jackson, K., & Grimm, F. (2013). Transforming the mental paradigm on digital literacy through understanding andragogy and maximizing continuous education. *ICERI2013 6th International Conference of Education, Research, and Innovation.*

Cowling, M., & Birt, J. (2018). Pedagogy before technology: A design-based research approach to enhancing skills development in paramedic science using mixed reality. *Information (Basel)*, *29*(9), 29. doi:10.3390/info9020029

Day, C., & Baskett, H. K. (1982). Discrepancies between intentions and practice: Reexamining some basic assumptions about adult and continuing professional education. *International Journal of Lifelong Education*, *1*(2), 143–155. doi:10.1080/0260137820010205

Dewald, N. H. (2003). Pedagogy and andragogy. In E. A. Dupuis (Ed.), *Developing web-based instruction: Planning, designing, managing, and evaluating for results* (pp. 47–68). Neal-Schuman.

Elias, J. L., & Merriam, S. B. (1995). *Philosophical foundations of adult education*. Krieger.

Elias, J. L., & Merriam, S. B. (2005). *Philosophical foundations of adult education* (3rd ed.). Krieger.

Gamede, B., Ajani, O., & Afolabi, O. (2022). Exploring the adoption and usage of learning management system as alternative for curriculum delivery in South African higher education institutions during Covid-19 lockdown. *International Journal of Higher Education*, *11*(1), 71–84. doi:10.5430/ijhe.v11n1p71

Greene, K., & Larsen, L. (2018). Virtual andragogy: A new paradigm for serving adult online learners. *International Journal of Digital Society*, *9*(2), 1376-1381.

Grow, G. O. (1991). Teaching learners to be self-directed. *Adult Education Quarterly*, *41*(3), 125–149. doi:10.1177/0001848191041003001

Hase, S., & Ellis, A. (2001). Problems with online learning are systemic, not technical. In J. Stephenson (Ed.), *Teaching & learning online: new pedagogies for new technologies* (pp. 27–30). Routledge.

Hickey, G. (2014). The importance of learning philosophies on technology selection in education. *Journal of Learning Design*, *7*(3). Advance online publication. doi:10.5204/jld.v7i3.184

Holmes, K., & Prieto-Rodriguez, E. (2018). Student and staff perceptions of a learning management system for blended learning in teacher education. *The Australian Journal of Teacher Education*, *43*(3), 21–34. doi:10.14221/ajte.2018v43n3.2

Jonassen, D. H. (1991). Objectivism versus constructivism: Do we need a new philosophy paradigm? *Educational Technology Research and Development*, *39*(3), 5–14. doi:10.1007/BF02296434

Kenyon, C., & Hase, S. (2001). Moving from andragogy to Heutagogy in vocational education. *Research to Reality: Putting VET Research To Work. Proceedings of the Australian Vocational Education and Training Research Association (AVETRA) Conference.*

King, K. P. (2009). Teachers of adults in the global community. In V. Wang (Ed.), *Strategic approaches towards curriculum development for adult learners in the global community* (pp. 91–119). Krieger.

Knowles, M. S. (1970). *The modern practice of adult education: Andragogy versus pedagogy*. Association Press.

Knowles, M. S. (1975). *Self-directed learning: A guide for learners and teachers*. Association Press.

Knowles, M. S. (1989). *The making of an adult educator: An autobiographical journey*. Jossey-Bass.

Knowles, M. S. (1995). *Designs for adult learning*. American Society for Training and Development.

Knowles, M. S., Holton, III, Swanson, & Robinson, P. A. (2020). *The adult learner the definitive classic in adult education and human resource development (9th ed.)*. Routledge.

Knowles, M. S., Holton, E., & Swanson, A. (1998). *The adult learner*. Gulf.

Knowles, M. S., Holton, E., & Swanson, A. (2005). *The adult learner* (6th ed.). Elsevier Butterworth Heinemann. doi:10.4324/9780080481913

Knowles, M. S., Holton, E., & Swanson, A. (2011). *The adult learner: The definitive classic in adult education and human resource development (7th ed.)*. Taylor & Francis.

Lee, K. (2010). Philosopher or philistine? In V. Wang (Ed.), *Assessing and evaluating adult learning in career and technical education* (pp. 27–52). doi:10.4018/978-1-61520-745-9.ch003

Mezirow, J. (1991). *Transformative dimensions of adult learning*. Jossey-Bass.

Mishra, P., & Koehler, M. J. (2006). Technological pedagogical content knowledge: A framework for integrating technology in teachers' knowledge. *Teachers College Record, 108*(6), 1017–1054. doi:10.1111/j.1467-9620.2006.00684.x

Narayan, V., Herrington, J., & Cochrane, T. (2019). Design principles for heutagogical learning: Implementing student-determined learning with mobile and social media tools. *Australasian Journal of Educational Technology, 35*(3), 86–101. doi:10.14742/ajet.3941

Noval, J. (2018). A qualitative study on the usage of a learning management system by allied health faculty. *Journal of Allied Health, 47*(3), 72–182. PMID:30194823

Pavlov, I. P. (1927). *Conditioned reflexes*. Routledge.

Rogers, C. R. (1951). *Client-centered therapy*. Houghton-Mifflin.

Shuttler, R., & Burdick, J. (2006). Creating a unified system of assessment. In M. Hricko & S. L. Howell (Eds.), *Online assessment and measurement: Foundations and challenges* (pp. 165–173). Information Science Reference. doi:10.4018/978-1-59140-720-1.ch008

Skinner, B. F. (1968). *The technology of teaching*. Appleton-Century-Crofts.

Smith, R. O. (2012). Fostering transformative learning online. In E. W. Taylor & P. Cranton (Eds.), *The handbook of transformative learning: Theory, research, and practice* (pp. 408-422). Jossey-Bass.

Tyler, R. (1949). *Basic principles of curriculum and instruction.* University of Chicago Press.

Wang, V. (2005). Perceptions of teaching preferences of online instructors. *Journal on Excellence in College Teaching, 16*(3), 33–54.

Wang, V. (2007). How contextually adapted philosophies and the situational role of adult educators affect learners' transformation and emancipation. In K. P. King & V. Wang (Eds.), *Comparative adult education around the globe: International portraits and readings of the history, practice philosophy, and theories of adult learning* (pp. 139–150). Zhejiang University Press.

Wang, V., & Farmer, L. (2008). Adult learning methods in China and Bloom's taxonomy. *International Journal for the Scholarship of Teaching and Learning, 2*(2), 1–15. doi:10.20429/ijsotl.2008.020213

Weimer, M. (2012). Learner-centered teaching and transformative learning. In E. W. Taylor & P. Cranton (Eds.), *The handbook of transformative learning: Theory, research, and practice* (pp. 439–454). Jossey-Bass.

Yuan, L., Powell, S., & Cetis, J. (2013). *MOOCs and open education: Implications for higher education.* Cetis White Paper Creative Commons Attribution 3.0 UK. http://publications.cetis.ac.uk/2013/667

Zanjani, N., Edwards, S., Nykvist, S., & Geva, S. (2017). The important elements of LMS design that affect user engagement with e-learning tools within LMSs in the higher education sector. *Australasian Journal of Educational Technology, 33*(1), 19–31.

KEY TERMS AND DEFINITIONS

Andragogy: Originally used by Alexander Kapp (a German educator) in 1833, andragogy was developed into a theory of adult education by Eugen Rosenstock-Huessy and was popularized in the US by American educator Malcolm Knowles. Knowles asserted that andragogy (Greek: "man-leading"). In the tradition of Malcolm Knowles, andragogy is a specific theoretical and practical approach, based on a humanistic conception of self-directed and autonomous learners and teachers as facilitators of learning.

Behaviorism: Also called the behaviorist approach, was the primary paradigm in psychology between 1920s to 1950 and is primarily concerned with observable behavior, as opposed to internal events like thinking and emotion.

Heutagogy: Sees the learner as the major agent in their own learning, which occurs as a result of personal experiences.

Humanism: A variety of ethical theory and practice that emphasizes reason, scientific inquiry, and human fulfillment in the natural world. Origin: 1805–15; human+ism.

LMS (Learning Management System): Information systems that facilitate e-learning by supporting teaching and learning, but that also can perform administrative tasks and facilitate communication between instructors and students" (Klobas & McGill, 2010 cited in Holmes & Prieto-Rodriguez, 2018, p. 21)

Participatory Literacy: The ability to contribute to blogs, wikis, social networking and sharing sites, virtual worlds, and gaming environments, which rely upon creativity, reasoning, focus, critical thinking, and capacity to collaborate.

Pedagogy: The word comes from the Greek παιδαγωγέω (*paidagōgeō*); in which παῖς (*país*, genitive παιδός, *paidos*) means "child" and άγω (*ágō*) means "lead"; literally translated "to lead the child". The art, science, or profession of teaching; *especially children.*

Chapter 2
Learning in Organizations:
Pedagogy, Andragogy, and Technology

Viktor Wang
California State University, San Bernardino, USA

Geraldine Torrisi-Steele
Griffith University, Australia

ABSTRACT

Against the backdrop of emerging technologies, organizations are, now more than ever before, needing to develop agile and innovative mindsets within organizations. To achieve agility and innovation, the humans in the organizations—employees and leaders—must have the capabilities to be self-directed, and they must be driven to identify meaningful problems in the organization and engage in collaborative problem-solving to address the problems. Developing these capabilities rests mostly on the organizational environment as created by the leadership and with the opportunities presented through human resource development. The authors maintain it is useful to anchor in andragogy leadership approaches alongside opportunities for organizational learning but to not discount the value of pedagogy when used appropriately.

INTRODUCTION

A few months before the writing of the present chapter, in November 2022, OpenAI, a research and deployment company focusing on artificial intelligence, launched ChatGPT. ChatGPT is a conversational AI based chatbot that can answer questions, as well as "answer follow-up questions, admit its mistakes, challenge incorrect premises, and reject inappropriate requests" (OpenAI, 2023).

At the time of writing, typing "ChatGPT" as a google search brings up top news stories that very clearly reveal the depth, and nature, of reaction to the technology (Figure 1).

DOI: 10.4018/978-1-6684-7832-5.ch002

Figure 1. Top stories on March 3, 2022, when "ChatGPT" is Google searched

In education, there are obvious implications of technologies like ChatGPT for academic integrity in student work. The author, working in an educational institution finds herself amid the frenzy of concern sweeping all levels of education: Should ChatGPT be banned? Should it be used? If so, how? How should we design assessments to reduce the risk of using ChatGPT to teach? And so on…

The release of ChatGPT has triggered an avalanche of discussion and concern, amid the uncertainty of what this 'new technology' really means for business, industry, education, and human society. Leaders are, without doubt, grappling with what directions their respective organizations should be taking with ChatGPT.

While ChatGPT is 'hot off the press' as a newer and unfamiliar technology, the reactions surrounding its launch are not unfamiliar to the 21st Century. Rapid change, often triggered by new technologies, is characteristic of our times, and without a doubt, leaders are being constantly challenged. What happens with ChatGPT and what its impact on society, industry, and education will be mostly rest in the hands of our leaders and the human capabilities of the organizations they lead. These leaders will need to develop a vision, and influence decision-making and policy related to how ChatGPT can be embraced and used, what the parameters of use are, and/or when and why it should be rejected. Realizing the vision and moving forward to the benefit of humankind with emerging, and established, technologies is dependent on the human capabilities that make up the organization. Against the background of the present discussion, it's easy to see why quality organizational learning is of consequence.

Quality organizational learning extends well beyond the traditional notion of job training to learn the skills to do a particular job. The traditional idea of training surfaced during the industrial revolution when employees were mostly factory workers who carried out procedural or sequential tasks for using equipment or undertaking repetitive tasks efficiently. The 21st-century employee obviously does much more than this. In addition, the structure of organizations is steadily (and necessarily) moving from hierarchical to flatter organizational structures with distributed leadership models. To ensure foundations for quality organizational learning, human resource development (HRD) and human resource management (HRM) practitioners must consider learning theories as a basis for the design and implementation of learning strategies and initiatives. Only then will the opportunity for employees to maximize their learning potential be realized. A better understanding of the various learning theories will result in better decisions regarding learning experiences and more desirable outcomes (Knowles, Holton III, & Swanson, 2005). Without a doubt, learning theories will contribute to long-run gains in our human capital (Torres-Coronas & Gasco-Hernandez, 2006).

In the present chapter, the authors discuss the urgent need for quality learning in organizations, especially those undergoing digital transformation. The authors adopt Bennett and McWhorter's (2021) conceptualization of quality learning in 21st-century organizations and argue that andragogically dominant approaches are necessary for quality organizational learning in the 21st century. The authors explain andragogy and pedagogy by contrasting the two perspectives and arrive at the conclusion that andragogy and pedagogy while different, can and should co-exist. The aim of the chapter is to bring to the fore andragogy as an effective overall guiding philosophy for learning and human resource development in organizations. It is hoped that the chapter will help trigger critical reflection on andragogical stances for learning in organizations and subsequently contribute to more effective designs for human resource development and leadership.

CONTEMPORARY ORGANIZATIONS, DIGITAL TRANSFORMATION, AND QUALITY LEARNING

Because digital technologies are almost fully integrated with human activity, quality learning must be understood against a digital backdrop. Most organizations, including educational institutions, are in the throes of digital transformation, with varying degrees of success. With the rapidly evolving plethora of technologies across all domains of human activity, individuals in, and out of, organizations are having to undergo a type of 'digital transformation' themselves because they need to 'upskill' or 'reskill' in the 'technology era' (Bennett & McWhorter, 2021).

Quality learning in digitally rich organizational environments enables employee development along these dimensions:

1. New literacies including data literacy, technology literacy, and digital literacy.
2. Soft/interpersonal skills in both face-to-face interaction and in technology-mediated interactions.
3. Learning agility, i.e., the willingness and ability to change learning modes; adaptability to new conditions, and importantly the ability to predict what needs to be learned in advance, i.e., employees are 'learning ready'?
4. Changing norms, exploring alternatives – the pandemic forced a change in norms such as work-from-home policies. The challenge post-pandemic is to explore alternatives and find what norms are worth keeping and which are not.

5. Cultural and social understandings and trust. The sense of belonging is a deep-seated human need regardless of physical or virtual environments. Social bonding, essential to well-functioning organizations can be challenging. It is imperative that leaders consider what fosters a sense of belonging and trust in technology-mediated environments (Bennett & McWhorter, 2021).

These characteristics of quality learning align well with the andragogical approach to learning and to leadership. Before elaborating further, it is useful to explain the andragogical and pedagogical approaches to learning.

ANDRAGOGY AND PEDAGOGY

Andragogy and Pedagogy are two different stances on learning and teaching. Pedagogy is grounded in the education of youth while andragogy in helping adults to learn. In the pedagogical approach, the trainer of the teacher is responsible for the student's learning. The teacher determines what will be learned, how it will be learned, and, when and where it will be learned. The student offers no input into how learning occurs and needs only to follow the direction of the teacher (Knowles, Holton III, & Swanson, 1998; Knowles, Holton III, Swanson, & Robinson, 2020).

In contrast to pedagogy, is adult learning or andragogy. The concept of helping adults learn evolved in Europe for quite some time before it was introduced to North America by adult education leaders. It was a German grammar schoolteacher by the name of Alexander Kapp who coined the term andragogy in 1833 to differentiate it from the theory of youth learning or pedagogy i.e., the art and science of teaching children. Andragogy is defined as the art and science of helping adults learn (Knowles, Holton & Swanson, 1998, 2005). Art here refers to "style," and science here refers to "method." Scholars in the field of adult education helped popularize the principles of andragogy in the field of adult education and training. First, Lindeman (1926) identified several key assumptions about adult learners. Table 1 shows a summary of Lindeman's key assumptions about adult learners.

Table 1. Lindeman's key assumptions about adult learners

1. Adults are motivated to learn as they experience the needs and interests that learning will satisfy.
2. Adults' orientation to learning is life-centered.
3. Experience is the richest source for adults' learning.
4. Adults have a deep need to be self-directing.
5. Individual differences among people increase with age.

In the late 1960s and early 1970s, Knowles added one more assumption related to motivation to learn. Knowles popularized this model and devoted his life to adult education and training in North America (Bash, 2003). Table 2 provides a summary of the andragogical model.

Andragogy flows directly from humanism (Nuckles, 2000) which postulates that humans have unlimited potential for learning and that their learning will eventually become self-directed. Because andragogical leaders such as Rogers (1951, 1961, 1969), and Knowles (1970, 1973, 1975, 1979, 1984,

Table 2. The andragogical model

1. The need to know. Adults need to know why they need to learn something before undertaking to learn it.
2. The learners' self-concept. Adults have a self-concept of being responsible for their own decisions, and for their own lives. Once they have arrived at that self-concept they develop a deep psychological need to be seen by others and treated by others as being capable of self-direction.
3. The role of the learners' experiences. Adults come into educational activities with both a greater volume and a different quality of experience from youths. By virtue of simply having lived longer, they have accumulated more experience than they had as youths.
4. Readiness to learn. Adults become ready to learn those things they need to know and be able to do in order to cope effectively with their real-life situations.
5. Orientation to learning. In contrast to children's and youths' subject-centered orientation to learning, adults are life-centered in their orientation to learning.
6. Motivation. While adults are responsive to some external motivators, the most potent motivators are internal pressures.

1986, 1998, 2005) firmly believe that instructors cannot teach another individual directly, the role of an instructor of adults has been defined as that of a helper. In other words, an instructor of adults in learning organizations has to be a facilitator. This is why andragogy has been defined as the art and science of helping adults learn. The word "helping" is used here instead of "teaching" to differentiate it from youth learning. According to Knowles (1998, 2005), a helper is also a facilitator. A facilitator is required to perform a different set of functions that require a different set of skills. A facilitator should not be a knowledge transmitter or information presenter. A facilitator can be a process designer and a manager who requires relationship building, needs assessment, involvement of students in planning, linking students to learning resources, and encouraging student initiative (Knowles, 1998, 2005).

Rogers (2002) agrees with Knowles and goes on to say that there needs to be a relationship between the teacher-learner that mirrors a form of the counselor-client relationship. Rogers was a clinical psychologist, and this counselor-client relationship is typical in the arena of clinical psychology. According to Rogers (2002), a teacher should facilitate learning and Rogers identifies the traits of a good facilitator as including:

Readiness and genuineness:

The facilitator is genuine and real. The facilitator is being honest and representing oneself.

Prizing, acceptance, and trust:

The facilitator cares for the learner while recognizing the learner is an individual. It is based on trust.

Empathic understanding:

The facilitator can put himself/herself in the student's own shoes and understand how the student might be feeling. It is a way of understanding how the student feels about the educational process.

According to these roles of instructors, learners are expected to negotiate curricular priorities with their instructors because if learners do this, they will know exactly what to learn, how to learn, and when to learn. This strategy of instruction falls in line with the first principle of the theory of andragogy, "the need to know." Further, learners are encouraged to involve themselves in the planning process with their instructors. Because adult learners have a rich reservoir of prior experience, prior experience is expected to be used as a learning resource. This is the second principle of the theory of andragogy. In view of other principles of the theory of andragogy such as self-directed learning, readiness to learn, orientation to learning, and motivation to learn, adult learners often enter into learning contracts negotiated with their instructors. A learning contract can specify what learners should learn. And because adult learners can teach themselves, a learning contract makes sense with adult learners. The learners can assist with the creation of what they are going to learn; they can assist with the types of learning activities in which they will participate. By doing this, they achieve their learning goals (Gailbrath & Gilley, 2007). Knowles (1980) confirms that learning contracts can provide an outline of learning plans that can be used for different learners; these contracts may accommodate the diversity of learners. Knowles (1975) created a template for a learning contract. This included the following information:

- Learning objectives
- Learning resources and strategies
- Target date for completion of objectives
- Evidence of accomplishment of objectives
- Criteria and means of validating evidence

Pedagogy and/or Andragogy?

In the 1960s and 1970s, Knowles and his associates distinguished between pedagogy and andragogy (Knowles, Holton III, & Swanson, 1998, 2005; Knowles, Holton III, Swanson, & Robinson, 2020). They suggested that children should be taught pedagogically, and adults should be taught andragogically (Jarvis, 2002).

In adult education, training, and learning organizations, it's been debated, for many years, about whether pedagogy or andragogy should be the favored approach. Interestingly, even andragogical leaders do not agree with one another regarding how adult learners in today's organizations should be taught. However, while different, pedagogy and andragogy are not mutually exclusive and can, in certain situations co-exist. Knowles made it clear that an essential feature of the theory of andragogy is flexibility (Knowles, Holton III & Swanson, 1998). To better understand the concept of flexibility and conceptualize how andragogy and pedagogy can co-exist, Wang's (2006) presents a useful matrix to help scholars and researchers discern the multilateral relationships between andragogical instructors and adult learners:

To achieve the flexibility to which Knowles' refers, both andragogical instructors and adult learners can stay in any of the quadrants in Wang's graph, (developed from Blanchard et al.'s Situational Leadership II Model, 1985) depending on learners' need for direction and need for support. In the graph, it is made evident that pedagogical model co-exists with the andragogical model in situations when adult learners have high need for direction and a high need for support. Only when adult learners have a low need for direction and a low need for support, can andragogical model be implemented. It is important to realize that adults are not always able to be independent and self-directed learners just by virtue of being adults. Learning independence and self-direction represent a maturity in learning which is greatly influenced by experiences and opportunities. Of course, other factors other than learners' need for support and learners' need for direction need to be taken into consideration. For example, learners' learning styles and preferences can also be a factor in determining the use of the theory of andragogy or pedagogy. Wang's graph helps us to understand that andragogy and pedagogy can (and should) co-exist in a complementary relationship, along a continuum. However, it is important to also recognize

Figure 2. Wang's matrix for multilateral relationships between andragogical instructors and adult learners

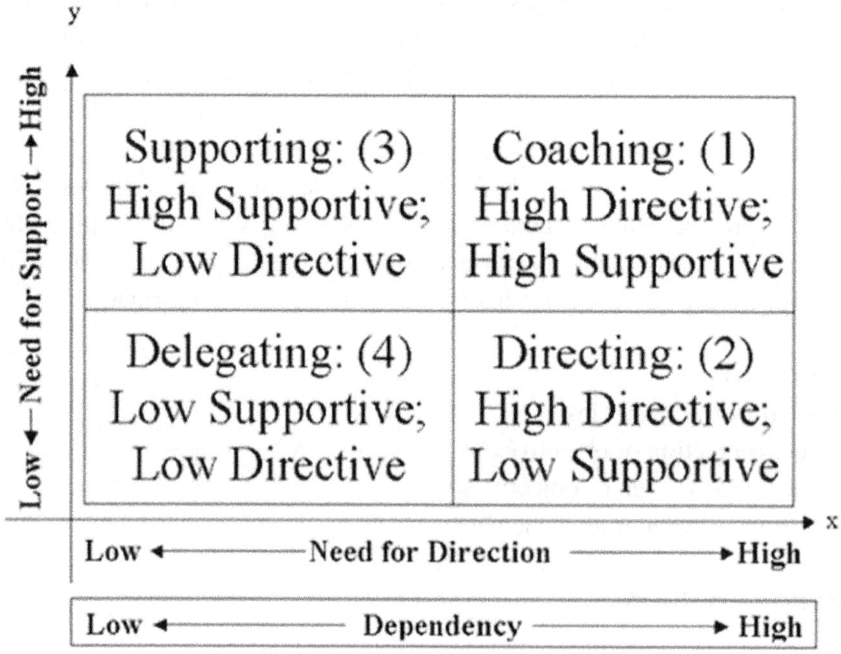

that when learners are in high dependency, high support quadrants, the teacher must set up the conditions that eventually enable the learners to become independent and self-directed learners – if today's organizations are to survive and thrive.

WHY IS ANDRAGOGY IMPORTANT FOR ORGANIZATIONS?

In today's organizations, the pedagogical model of learning and leadership development cannot be the dominant or only approach to organizational learning and to leadership development. Human resource development, including leadership development, must be aimed at achieving the andragogical learning environment. The reasons for this crystallize when one considers the trends that have become apparent in discussions of leadership in recent years. The trends in leadership theory mirror the changing socio-technological environment. These trends are effectively summarized by Jeanes (2021, p. 8):

- From heroic leaders to leadership as a process.
- Collaborative and relational understandings of leadership
- Complexity
- Emphasis on social capital in leadership development
- A shift from specific skills and competencies to a 'mindset'

These trends argue Jeanes (2021) "lie at the heart of the andragogical mindset approach" (p. 9). Hence, leadership development should aim to develop an andragogical mindset in the leader. The andragogical leaders will bring the andragogical mindset to bear on human resource development, setting the scene for andragogical learning throughout their organizations.

WORKING TOWARDS INDEPENDENT AND SELF-DRIVEN LEARNING IN ORGANIZATIONS

Andragogical learning is independent, self-driven learning. When employees are self-driven and independent, organizations can maximize the opportunity to be agile and resilient in the face of rapid and unexpected change. In an ever-changing environment, it is important for the organization to develop a dynamic (rather than static) knowledge resource, and because of the complexities of the 21st century, organizational knowledge depends on all individuals within the whole organization, not just a select few. Employee contribution to organizational knowledge can be maximized when employees are valued for their individual experiences and are motivated to learn rapidly with aim of solving important organizational challenges.

In an andragogical learning environment, employees are encouraged to, and likely want to be, self-directed. Learners may identify learning objectives and are encouraged to write them in their own way. In addition, learners have flexibility in how and when the objectives will be achieved.

The learner can manage his/her time and adapt according to other projects or commitments. Again, having the learner state the date for completion, allows for their 'buy-in' of the deadlines. The evidence of accomplishment of objectives may be different for various learners. One learner may produce a written paper; others may demonstrate either in presentations or role-play and receive feedback from others. The criteria and means of validating evidence include the ways in which the evidence of accomplishments will be assessed and by whom (Knowles, 1975). Adult learners develop a sense of commitment and ownership when they have input into the content and design of their learning activities and goals. The strong level of 'buy-in' helps to increase the level of performance and ultimately achieve the goals (Gailbrath & Gilley, 2007). Further, adult learners are encouraged to solve real-life problems because their learning tends to be task-oriented instead of subject-centered. Because of their orientation to learning, adult learning can also be considered "contextual." Hence, if the environment is conducive to buy-in from employees, employees as adult learners will be motivated and self-directed to solve the 'real-life' problems encountered within the organization.

Understanding self-directed learning is essential to understanding the requirements of employee development within organizations. Based upon the research of Houle (1961), Tough (1967, 1971) outlined a detailed explanation of self-directed learning. He described it as a type of study and the actual term he used was "self-planned learning". (Merriam and Caffarella). Later, Merriam and Caffarella (1999) went on to describe self-directed learning as a learning process and people take the initiative to "plan, carry out and evaluate their own learning experiences" p.293. Later, more studies were conducted on how adults initiate and design their own learning. Tough (1971) stated that there were many opportunities even outside of higher education institutions that provided learning situations.

Self-directed learning can play a major role in human resource development in organizations (Merriam, Caffarella, & Baumgartner, 2007). It would be useful for human resource development practitioners to understand how organizations can encourage employees to be self-directed learners so that the organizations may foster an environment that promotes self-directed learning (Clardy, 2000). Clardy conducted a study of 56 employees who represented five organizations. He wanted to look at job-related self-directed learning projects and to see what types of projects encouraged self-directed learning. Three types of projects that encouraged self-directed learning emerged from this study. First, induced self-directed learning projects include learning new skills due to the employee thinking that he or she needs to develop a skill. Second, voluntary self-directed learning project due to the employee wanting to learn something, and third, synergistic self-directed learning project due to the organization providing conditions that spark the employee's willingness to learn. Guglielmino's 1997 operational definition of self-directed learning includes "a complex of attitudes, values, and abilities that create the likelihood that an individual is capable of self-directed learning" (Merriam, Caffarella, & Baumgartner, 2007).

Organizations should seek and remove any barriers that may hinder or prevent learning from occurring. There are two main categories of barriers to adult learning: External or situational and internal or dispositional (Falasca, 2011). Some examples of external or situational include the effects of aging. Adults may have lost some of their vision or hearing which may affect their capacity for learning. Health issues such as chronic fatigue or pain may affect the learning process. Another example Falasca (2011) points out includes role characteristics; the family structure may have changed which may impact the adult learner's ability to attend classes. Internal or dispositional barriers include the feeling of anxiousness as the adult learner considers embarking into the learning community. Adults may remember their pedagogical schooling years and want to revert to a teacher-centered way of learning rather than the andragogical way of a learner-centered approach. The educator needs to let the adult learner know that it is okay to take control of learning and make choices and to have input on how the learning activities are structured (Forest III and Peterson, 2006). To overcome any barriers to learning, it is important that a safe, positive, and supportive learning environment is created. The organization can help to create this by recruiting educators who are good listeners and supportive of the needs of the learners. Educators should be knowledgeable about potential barriers to learning and ways to overcome them, (Falasca, 2011). Allowing adult learners to be active in diagnosing their own learning needs and activities may help make the learning relevant and establish a sense of readiness to learn.

For organizations to continuously improve and position themselves in the global marketplace, it is important for employees to be learning too. When organizations can encourage self-directed learning, the role of technologies in supporting flexible and self-directed learning must be considered.

However, if andragogical leaders or adult learning in today's learning organizations fail to do what they are expected to do according to the andragogical model, they sometimes inappropriately revert to purely the pedagogical model. Certain learning situations predetermine whether the andragogical model or the pedagogical model is realistic for even adult learners. For example, if adult learners have a high need for direction and structure and a high need for support, andragogical instructors must become coaches or knowledge presenters and adult learners must assume a submissive role of simply following their "pedagogical" instructors because adult learners at this time are dependent learners (Grow, 1991; Wang, 2006; Blanchard, Zigarmi, & Nelson, 1985). Blanchard (1985) adapted the leadership model called Situational Leadership II. The model was originally developed by Hersey and Blanchard in the late 1960s (Hersey & Blanchard (1969a). The model outlines the various styles a leader may adopt when attempting to influence followers. It indicates that the styles may change according to which developmental level the follower currently is. Wang's model expands it to refer to learners and instructors respectively. Blanchard et al. (1985) state that there are 4 leadership styles: S1 which is high directive and low supportive, S2 which is described as a coaching approach and is high directive and high supportive, S3 is more of a supporting approach and is a high supportive and low directive and S4 is a low supportive and low directive. Depending on where the follower is on the developmental scale influences the approach the leader will take in leading the follower to complete a task. If the follower is at a low level on the development scale, the leadership approach will be a high directive and low supportive style (S1). This can clearly be transferred to developing adults in the learning organization.

Confessore and Kops (1998) discuss five characteristics of organizations that are conducive to self-directed learning. These are:

1. Tolerance for errors, support of experimentation and risk-taking, and an emphasis on creativity and innovation;
2. The use of participative leadership style and delegation of responsibility to organizational members;
3. Support for learning initiatives that are linked to the organization's goals and values;
4. Encouragement of open communication and of information systems that provide for collaboration and teamwork and that use both internal and external learning resources;
5. Provision of opportunities and situations for individual learning. (p. 371)

TRENDS

The issue of andragogy and pedagogy will continue to spark research in today's organizations. Learning is a contextual phenomenon and subsequently as organizational environments continue to evolve it is necessary to keep researching and exploring theories such as andragogy and pedagogy in the new contexts to determine what mix and strategies will work best. The introduction of technologies into organizations and into learning environments is one event that necessitates revisiting the theories of andragogy and pedagogy to determine how technologies can be used in ways that leverage or even transform organizational learning. Most recently, interest and research in virtual worlds, even 3D immersive worlds, and organizational learning is surfacing. A newer term, Virtual Human Resource Development (VHRD) is appearing in Human Resource Management literature. "Virtual worlds are online communities in which users take the form of avatars (3D graphic representations) to interact with each other in a simulated environment" (Mancuso, Chlup, & McWhorter, 2010, p. 681). VHRD has been around for over a decade. A decade ago, some were already considering how the theory of andragogy could be used in virtual environments for effective adult learning. A study by Mancuso, Chlup, and McWhorter in 2010 lent weight to the argument that virtual worlds have promise in adult learning, but the researchers also (rightly) pointed out that adult learning and human resource development in virtual worlds is substantially different from learning and human resource development in the real world. Hence, it is necessary to reflect and explore further the theory and practice of human resource development in virtual worlds.

The advent of the pandemic has not only re-ignited interest in VHRD but has led to serious consideration of VHRD as a valuable approach to organizational learning that allows for flexibility over geographical locations. Although the world is apparently returning to pre-pandemic activity, organizational leaders face a future built on the disruption of the pandemic along with the knowledge that we are not immune from similar disruption in the future. Bennet and McWhorter (2021) provide an insightful reflection on the future of human resource development and organizational learning in virtual worlds:

"Crisis will come around again, and digital transformation due to the fourth industrial revolution is afoot. We mark 2020 as the year millions of people joined the world of VHRD, with some sobering and inspiring clues as to what lies ahead…we renew the call for HRD programs to incorporate technology beyond instructional software and online training [and urge] greater focus on data analytics with an emphasis on high touch and the human need to belong…VHRD is necessary to help organizations…thrive in a world increasingly powered by AI, robotics and a multitude of unseen intranets"

CONCLUSION

Undeniably, now more than ever before organizations pivot on human resources development through learning that facilitates agility and collaboration. Employees must be well-prepared as life-long, self-directed learners. The complexity and ever-changing social and economic context of our times means that organizational problem-solving and decision-making cannot be left to a select few 'leaders'. Employees need to have 'buy-in' to organizational problems and problem-solving. They must have the readiness to learn and deal with the 'real-life' situations they experience in the organization. To develop these capabilities, employees must be treated by others as being 'up to the task'. Leaders who are andragogical in their approach will create an environment that enables their employees to feel valued and thrive in a self-directed learning environment. Andragogical leaders will bring the andragogical approach to HRD and provide opportunities for employees, as adult learners, to contribute meaningfully to organizational goals. It is important to reiterate that the andragogical approach does not exclude entirely the pedagogical approach. When employees are not well-prepared for andragogical learning, pedagogical approaches can be employed to steadily scaffold toward self-directed learning.

Furthermore, in some cultures, the pedagogical model seems to work well with both instructors and adult learners in their organizations. For example, in Confucius-Heritage societies, the pedagogical model which features teachers as the unchallengeable authority figures and students as absolute followers is dearly welcomed (Wang & Bott, 2003-2004, p. 47). Wang and Torrisi-Steele (2019) discuss the absence of an andragogical approach to teaching in China and how apparently leaders in China have not recognized how going from a pedagogical approach to an andragogical approach could indeed be the catalyst to the development of Chinese leaders.

Organizations are moving away from the pedagogical way of teaching where the teacher provides the information and knowledge to the student. Rather organizations are moving toward ways in which teamwork, problem-solving, communication and motivation skills are taught. These skills cannot be learned in a classroom-type setting and handed down from a teacher, rather they need to be driven by competencies and they should be learner-centered rather than teacher-centered. Teachers facilitate rather than teach. Learning can be called learner-centered if it is connected to the learner's already acquired foundation of knowledge. It helps if the teacher has already determined how much knowledge the learner already has prior to facilitating the learning.

In 2000 the International Labor Office, the Centre of Labor Market Studies, and the University of Leicester, England provided information to the China Enterprise Confederation on the organization's perceptions of the quality of education and training programs. 465 organizations responded (in China). The results highlighted confirmation that organizations agreed that as well as a higher level of basic skills, they required from prospective employees a wider variety of skills: Organizational skills, critical thinking, problem-solving, communication, and teamwork Dahlman, Zhihua Zeng, & Wang, (2007). Furthermore, the Definition and Selection of Competencies (DeSeCo) Project of the OECD conducted a study that outlined the main competencies needed to have a well-functioning society (Rychen & Salganik, 2003). Twelve countries were studied. It was concluded that the Asian countries were focused on the importance of creativity and critical thinking skills believing that employees needed these skills to help the organizations continuously innovate and compete in the global marketplace (Dahlman, Zhihua Zeng, & Wang, 2007).

That capabilities such as critical thinking, problem-solving, and self-directed learning are essential to century 21st century organizations and society is uncontested. But how to best develop these capabilities remains problematic. Traditionally, human resource development has been approached pedagogically and leaders in the past were directing more than facilitating. It is difficult to overcome the inertia of this heritage. Viewing the challenges of learning in organizations through the lens of andragogy but not discounting pedagogy is conducive to evolving the nature of organizational learning towards 21st-century requirements.

REFERENCES

Bash, L. (2003). *Adult learners in the academy*. Anker Publishing Company, Inc.

Bennett, E., & McWhorter, R. (2021). Virtual HRD's role in crisis and the post Covid-19 professional lifeworld: Accelerating skills for digital transformation. *Advances in Developing Human Resources*, *23*(1), 5–25. doi:10.1177/1523422320973288

Blanchard, K., Zigarmi, P., & Zigarmi, D. (1985). *Leadership and the one-minute manager: Increasing effectiveness through situational leadership*. William Morrow.

Brookfield, S. D. (1986). *Understanding and facilitating adult learning*. Jossey-Bass.

Brookfield, S. D. (1995). *Becoming a critically reflective teacher*. Jossey-Bass.

Clardy, A. (2000). Learning on their own: Vocationally oriented self-directed learning projects. *Human Resource Development Quarterly*, *11*(2), 105–125. doi:10.1002/1532-1096(200022)11:2<105::AID-HRDQ2>3.0.CO;2-5

Confessore, S. J., & Kops, W. J. (1998). Self-directed learning and the learning organization: Examining the connection between the individual and the learning environment. *Human Resource Development Quarterly*, *9*(4), 365–375. doi:10.1002/hrdq.3920090407

Dahlman, C., Zeng, Z. D., & Wang, S. (2007). *Enhancing China's competitiveness through lifelong learning*. The World Bank.

Ellinger, A. D. (2004). The concept of self-directed learning and its implications for human resource development. *Advances in Developing Human Resources*, *6*(2), 158–177. doi:10.1177/1523422304263327

Falasca, M. (2011). Barriers to adult learning: Bridging the gap. *Australian Journal of Adult Learning*, *51*(3), 583–588. http://find.galegroup.com/gtx/start.do?userGroupName=gale15691&prodId=AONE

Forrest, S. III, & Peterson, T. (2006). It's Called andragogy. *Academy of Management Learning & Education*, *5*(1), 113–122. doi:10.5465/amle.2006.20388390

Galbraith, M. W., & Gilley, J. W. (1984). Using self-directed learning contracts to improve performance and instruction. *Performance & Instruction Journal*, *23*(8), 9–10. doi:10.1002/pfi.4150230809

Grow, G. O. (1991). Teaching learners to be self-directed. *Adult Education Quarterly, 41*(3), 125–149. doi:10.1177/0001848191041003001

Guglielmino, L. M. (1997). *Contributions of the self-directed learning readiness scale (SDLRS) and the learning preference assessment (LPA) to the definition and measurement of self-direction in learning.* Paper presented at the First World Conference on Self-Directed Learning, Montreal, Canada.

Hersey, P., & Blanchard, K. H. (1969a). Life-cycle theory of leadership. *Training and Development Journal, 23*, 26–34.

Houle, C. O. (1961). *The inquiring minds.* University of Wisconsin Press.

Jarvis, P. (2002). *The theory & practice of teaching.* Kogan Page.

Jeanes, E. (2021, March). A meeting of mind(sets). Integrating the pedagogy and andragogy of mindsets for leadership development. *Thinking Skills and Creativity, 39*, 100758. https://www.sciencedirect.com/science/article/abs/pii/S1871187120302327. doi:10.1016/j.tsc.2020.100758

Knowles, M. S. (1970). *The modern practice of adult education: Andragogy versus pedagogy.* Association Press.

Knowles, M. S. (1975). *Self-directed learning: A guide for learners and teachers.* Association Press.

Knowles, M. S. (1979). Andragogy revisited part II. *Adult Education Quarterly, 30*(1), 52–53. doi:10.1177/074171367903000104

Knowles, M. S. (1980). *The modem practice of adult education: From pedagogy to andragogy.* Follett.

Knowles, M. S. (1984). *Andragogy in action.* Jossey-Bass.

Knowles, M. S. (1986). *Using learning contracts.* Jossey-Bass.

Knowles, M. S., Holton, III, Swanson, & Robinson, P. A. (2020). *The adult learner the definitive classic in adult education and human resource development* (9th ed.). Routledge.

Knowles, M. S., Holton, E., & Swanson, A. (1998). *The adult learner.* Gulf Publishing Company.

Knowles, M. S., Holton, E., & Swanson, A. (2005). *The adult learner the definitive classic in adult education and human resource development* (6th ed.). Elsevier Butterworth Heinemann.

Knowles, M. S., & Hulda, F. (1973). *Introduction to group dynamics.* Follett.

Lindeman, E. C. (1926). *The meaning of adult education.* New Republic.

Mancuso, D., Chlup, D., & McWhorter, R. (2010). A study of adult learning in a virtual world. *Advances in Developing Human Resources, 12*(6), 681–699. doi:10.1177/1523422310395368

Merriam, S. B., & Caffarella, R. S. (1999). *Learning in adulthood: A comprehensive guide.* Jossey-Bass.

Merriam, S. B., Caffarella, R. S., & Baumgartner, L. M. (2007). *Learning in adulthood: A comprehensive guide* (3rd ed.). Jossey-Bass.

Nuckles, C. R. (2000). Student-centered teaching: Making it work. *Adult Learning*, *11*(4), 5–6. doi:10.1177/104515959901100403

Open, A. I. (2023, March 3). *Introducing ChatGPT.* https://openai.com/blog/chatgpt

Rogers, C. R. (1951). *Client-centered therapy.* Houghton-Mifflin.

Rogers, C. R. (1961). *On become a person.* Houghton-Mifflin.

Rogers, C. R. (1969). *Freedom to learn.* Merrill.

Rogers, C. R. (2002). The interpersonal relationship in the facilitation of learning. In R. Harrison, R. Reeve, A. Hanson, & J. Clarke (Eds), *Supporting lifelong learning* (pp. 25-29). Perspectives on learning, 1. Routledge-Farmer and Open University.

Rychen, D. S., & Salganik, L. H. (2003). *Key Competencies for a Successful Life and a Well-Functioning Society*. Hogrefe Publishing.

Torres-Coronas, T., & Arias-Oliva, M. (2006). *Encyclopedia of human resources information systems: Challenges in e-HRM.* http://www.etse.urv.es/~ttorres/

Tough, A. (1967). *Learning without a teacher: Educational Research Series 3.* Ontario Institute for Studies in Education.

Tough, A. (1971). *The adult's learning projects: A fresh approach to theory and practice in adult learning*. Ontario Institute for Studies in Education.

Wang, V. (2006). *Essential elements for andragogical styles and methods: How to create andragogical modes in adult education.* PEARSON Education.

Wang, V., & Bott, P. A. (2003-2004). Modes of teaching of Chinese adult educators. *The New York Journal of Adult Learning*, *2*(2), 32–51.

Wang, V., & Torrisi-Steele, G. (2019). Transformation of higher education in China: A teaching methods perspective. In M. Khosrow-Pour (Ed.), *Handbook of research on challenges and opportunities in launching a technology-driven international university* (pp. 131–160). Information Science Reference. doi:10.4018/978-1-5225-6255-9.ch008

KEY TERMS AND DEFINITIONS

Andragogy: Andragogy refers to the art and science of helping adults learn. The word "helping" is heavily emphasized to differentiate the theory of andragogy from the theory of youth learning. Some scholars refer andragogy a set of assumptions; others refer it to as a set of guidelines. Still, others refer it to as a philosophy. However, Knowles refers it to as a theory, which has been widely accepted in the field of adult education and training. According to andragogical leaders in North America, the theory of andragogy sparked a revolution in adult education and training simply because previously every learner was taught pedagogically.

Confucius-Heritage Society: Confucius saw growing disorder in his lifetime. In response, he developed a philosophy to maintain the status quo in societies. His philosophy has been translated into teaching, "let a teacher be a teacher and let a student be a student." What this means is that teachers are leaders and students are followers. Their roles are well-defined. In other words, in Confucius-Heritage societies such as Japan, China, South Korea, and Singapore, pedagogy is deeply rooted in their cultures. Probably andragogy has only a small place in these societies.

Emerging Technologies: A new technology or an innovative application of existing technology. The presence of emerging technologies in any context creates disruption i.e. it may change radically the way something is being done, or make possible a new thing or way of doing something.

Flexibility: Flexibility is used figuratively, meaning to change easily in response to situations. In this chapter, it refers to being able to move freely from the pedagogical model to andragogical model and vice versa.

Humanism: Humanism originated in China. It literally means teacher of the humanities. It is further believed that humanism would develop autonomous and responsible individuals. Humanistic instructors see themselves as facilitators, helpers, and partners in the learning process. They establish a context for learning and serve as a flexible resource for adult learners. Humanistic teachers must trust students to assume responsibility for their learning (as cited in Wang & Sarbo, 2004, pp. 209-210).

Learning Contract: A document spelling out learner expectations within an organizational context. Learning that has as its purpose improving one's competence to perform in a job or in a profession must take into consideration the needs and expectations of organizations, professions, and society. Learning contracts provide a means for negotiating reconciliation between these external needs and expectations and the learner's internal needs and interests.

Pedagogy: Pedagogy refers to the art and science of teaching children. The word "teaching" is emphasized because the role of an instructor is defined as that of a knowledge dictator. A knowledge dictator is supposed to lecture heavily, and children are supposed to assume the role of following their instructors submissively. This kind of education is top-down. It is teacher-directed education and learning is subject-centered. Learners' prior experience is of little importance. What counts in this kind of teaching/learning situation is the experience of the instructors. Although popular with the education and training of children, some mature children may prefer andragogy to pedagogy. For example, some high schoolers may be capable of teaching themselves once they are more experienced with a subject matter.

Principles: A principle refers to a standard, such as a guide to behavior, or rule.

Chapter 3
Andragogy and Higher Education in Late Capitalism: A Critique

Howard A. Doughty
Seneca College, Canada

ABSTRACT

Over the past half-century, the word "andragogy" has migrated from the fringe vocabulary to the standard lexicon of adult education. It has identified common psychological traits in "non-traditional" students and distinguished its methods from "pedagogy," the teaching of children. This chapter will explore andragogy's apparent strengths and arguable weaknesses in terms of its socio-economic assumptions and the social class interests it may wittingly or unwittingly serve. It will conclude with a critique.

INTRODUCTION

Confronted as we are by profound changes in everything from our understanding of neurology (mental health), ecology (environmental sustainability), and the implications of particle physics and cosmology for our understanding of the universe and, perhaps, the hope of finding a meaningful place in it (spiritual sublimity?), it is no great revelation to find ourselves in basic agreement with the far-famed British philosopher, Alfred North Whitehead (1861-1947) who, in 1931, expressed awareness of the fleeting nature of human cultures, the ideas they are said to embody, and the words people use to describe them. So, Whitehead realized that confidence could be had in the validity of declarative statements "only when the timespan of cultural change was greater than the span of individuals" (Whitehead, 1931).

Contemporarily, from learned debates about cultural relativism to street-level concerns about "fake news," "conspiracy theories," and the ethics of artificially intelligent plagiarism and Chat-GPT essay-writing, our sense of disconnection from an absolute, eternal, non-negotiable *truth*, or at least in the merit of the pursuit thereof, has never (at least from our admittedly provincial, presentist perspective) seemed more elusive. Deprived of (or liberated from) the comfort/confinement to be found in unquestioned, received wisdom about (and faith in) eternal verities, our age of endless uncertainly and socially

DOI: 10.4018/978-1-6684-7832-5.ch003

constructed reality has led us to believe that the recognized and revered certainties of our or any other community, whether they relate to practical actions, empirical facts, ethical preferences, or supernatural (metaphysical, theological) speculations, are at best tentative, conditional, contingent, very likely transitory, certainly incomplete, and quite possibly wrong.

So it is that andragogy, the core concept in this chapter and in this book, may be taken as a passing fancy—once a neologism, serviceable in the present moment, often elevated to "a badge of identity" (Brookfield, 1986), and methinks, destined eventually to be an archaic cultural remnant.

We may hope that its fancy will not pass too soon, of course, lest the contributors to this volume be complicit in an elegant practical joke played upon them/ourselves. We may have, I think, some confidence that it will be around for a while; but, we may also hope that its demise is so distant that our comments on it will not be obsolete before it is dropped from our ever-shifting and sifting lexicological (dis)order. After all, no less an expert than Patricia Cross (1981) long since told us that "although the word andragogy makes a neat contrast with the more familiar and traditional pedagogy, the contrast appears difficult to maintain." And yet, here we are still talking about it forty-plus years later.

The fragility of terms should, however, be kept in mind. Some create splashes in various pans and speedily disappear; others endure for centuries. So, Ray (2013) reminded us a decade ago that not only do neologisms have discernible patterns of creation, evolution, and presumably ultimate extinction, but they (or their proponents) also compete with others for recognition and dominance. Accordingly, while "pedagogy" (childhood learning) seems to have held sway in that particular field and seems also to have successfully colonized others up to and including post-secondary (though chiefly undergraduate) education, and while "eldergogy" (Yeo, 1982) has apparently tussled unsuccessfully with "geragogy" (Battersby, 1987; Johnson, 2016, Wright, 2016) for recognition as the teaching of older adults, newer terms continue to burst forth from the crowded, confused, and confusing jabberwocky of educational theory.

One such is "metagogy." Whether advanced by identifiable human beings (Propenko, 2009) or artfully/artificially intelligent and potentially posthumans (mplsbohemiam, n.d.; Rollermation, 2011), its supporters boast that it not only "is inclusive of an andragogic approach, but also includes conventional pedagogical approaches as they are appropriate" (Peterson & Ray, 2013, p. 83)—more a "mix-and-match" than a classic Hegelian dialectic. Wagers on their relative longevity will be made only by the most adventuresome.

In this chapter, I plan to treat the concept of andragogy lightly insofar as debates about its target audience's demographics, motivations, self-concepts, intellectual capacities, and the like are concerned. That is because I am less concerned with the educational theory *per se* than I am in the political economy of education. Abstract concepts may float in the clouds, but the attitudes, actions, and pertinent effects of their use in the world as we find it have material consequences. Andragogy, like most words and the activities they describe are designed and deployed by detectable political and economic sectors and interests. While psycho-social factors cannot be neglected, the premise of this chapter is that, although they may explain the beliefs and behavior of individuals and groups, those individuals and groups are always to be found within a firm and fixed ideological and institutional environment. That environment, furthermore, is constituted by the overall mode, means, and relations of knowledge production and dissemination. Our ability to understand that environment is, I respectfully submit, the key to comprehending the patterns of culture that are ultimately responsible for everything from the allocation of research funding for projects in technology and the sciences (highly valued) and those in the humanities (very much less so) to the current "culture wars" that pits critical race theory against the apparent epidemic of student anxieties and mental health issues on campus in the struggle for academic legitimacy defined in budgetary dollars.

Education is, moreover, not only a domain wherein the hegemonic culture exerts control over who gets taught what, when, where, why, and how, but it is also a contested and conflicted theatrical setting wherein struggles for control are played out. Psychology and sociology may suss out the struts and frets of individual actors in their contractual classroom hours upon the stage. All the particularities and singularities of costuming, make-up, voice, and gesture take place within, and can be understood best when we are aware of the overall structure of, the theater, the dimensions of the stage, the set designs, the dramatic conventions of the Acts and Scenes, the rules of performance, the uses of lighting and the acoustics that are shared by playwrights, actors and audiences alike.

After all, only when all are familiar with the basic orders, tropes, and techniques can the cast of characters, their actions (and inactions), noises (and silences) be comprehensible both to those who watch and listen and to those who move and speak. What applies in the literary imagination also applies in practical action. The foundational explanations of education (and all other identifiable human patterns and arrangements of living) are ultimately to be found in political economy—the study of political authority, economic wealth, and their combined ability to shape people and events and to limit and expand the range of possibilities open to all.

ORIGINS

Standard treatments of andragogy typically mention that the word was most likely "coined" in Europe by Kapp, widely introduced in America by Lindeman, and most recently and powerfully popularized in the West by Knowles" (Ekoto & Gaikwad, 2015). I will vary only a little from this familiar account. This chapter begins in earnest with a mention of two disremembered or misremembered German educators—one of whom emigrated to the United States—and a third, a first-generation German American who, while paid homage as the man who introduced American educators to andragogy, remains securely tucked into the past, having been surpassed by more contemporary, less controversial minds. They are representative of andragogy's untidy early days.

Each played a notable part in andragogy's history. All have been more or less eclipsed. They are Alexander Kapp (ca. 1799-1869), Eugen Rosenstock-Huessy (1888-1973), and Eric C. Lindeman (1885-1953), the orphaned son of German immigrants whose background as a manual laborer and whose relative lack of formal education made him an example and an exemplar of the value and virtue of adult education.

My purpose is not to preface a discussion of andragogy as it has come to be known in North America over the past sixty years with a portrayal of exotic characters who may paint colorful images of its past, but who are no longer relevant to the topic as it has become embedded in education's lexicological landscape in the late twentieth and early twenty-first centuries. Quite the contrary, by rapidly recounting their approaches to the subject and taking my cue, in part, from Gabriele Strohschen (2020), I want to encourage inquiries into the *purpose* of the higher education, for which we develop our techniques, generate our rationales, and add to our vocabularies. The questions of the "who," "when," "where." "what" and "how" of education is held here to be secondary to, and derivative from, the "why." Whose interests does education serve? What is education *for*?

I hope to highlight the conspicuous absence of Kapp and Rosenstock-Huessy and the subtle glossing over of the radical impulses of Lindeman from the contemporary narrative in order to illustrate how fully early andragogic theories and practices have been subsumed within the political economy of late capitalism. By redefining and refining the different and sometimes the opposite explicitly political aspects of its European ancestors to suit its current purposes, andragogy has been deployed in two directions—each at variance from what its ancestors intended. In doing so, I hope to clear the way for a more comprehensive discussions than those normally encountered within the domain of andragogy as it is currently thought and practiced. I hope, as well, to contrast current practice with possible alternatives within the domain of adult education being advanced by critical educators at odds with the corporate agenda.

Scaffolding

Strohschen (2020) argues that our approaches: pedagogic, andragogic, geragogic, metagogic, heutagogy (Kenyon & Hase, 2001), or what you will, "scaffold" our discourses about ideologies, values, and missions of adult education and allows our arrival at meaningfully examined and contextually relevant educational program design and delivery. Upon what foundation does this scaffolding rest? How recently has it become firm? When will it crack?

I argue that a concrete pad has been quite recently laid, but the original blueprints or, more accurately, the several artists' sketches are far removed from the current project. Moreover, in rejecting and abandoning the inchoate notions of those who came before and in realigning the morals and methods of andragogy, much has been lost, whether from a viewpoint that can usefully be labelled "idealistic" in both the ontologically conservative or the practically progressive senses of the term.

The departure from the aspirations of the first explorers is, of course, the result of demographic and technological trends—the one toward mass and, more recently, universal education and the other toward the technological mediation of teaching and learning in the interest of what is called neoliberalism as ideology and as late capitalist practice.

As for Kapp, Rosenstock-Huessy, and Lindeman, I wish to highlight their conspicuous absence or the downplaying of their legacies in their contemporary narrative in order to illustrate how fully andragogic theory and practice have been located within the political economy of late capitalism and defined and refined to suit its purposes. In doing so, I hope to open the floor for a more comprehensive discussion than those normally encountered in current thought and practice. I want, as well, to contrast current practice with possible alternatives within the domain of adult education. A genuinely helpful research agenda would be one in which the agitating potential of adult education would be unleashed should the time come when it is no longer so tightly tethered to the political and administrative priorities of Human Resources Departments (HRD) in enormous business, government, private foundation, and public educational institutions, along with their apologists, legitimizers, and advocates in the mass and social media.

Alexander Kapp

The concept (if not the theory and practice) of andragogy, has a lengthy (if not a continuous and consistent) history. By all accounts, the term was first used by Alexander Kapp, a German gymnasium teacher who admired Plato immensely, retrieved the attitudes and abstractions evident in *The Republic,* and applied them to the circumstances of the early modernizing economy in the disintegrating Central European imperial orders and the still early industrializing decades of the nineteenth-century.

Plato, of course, had elaborated through his allegedly contemporaneous notes from the discussions of his mentor, Socrates, a rather novel metallurgical metaphor which justified the existence of innate social differences of social class and caste.

In his/their telling, men are born with gold, silver, and brass (sometimes iron) in their souls. Men of gold are guided by reason and are fit for leadership; men of silver are guided by spirit and are able administrators/guardians of the state; and men of bronze are guided mostly by their animal appetites and suited to be farmers, artisans, petty merchants and traders, and manual laborers only.

The good society, therefore, is one in which the hierarchy of gods-given qualities and character traits are properly arranged. The golden philosopher-kings properly dictate statutes and policies; silver auxiliary guardians are trusted with the execution and enforcement of the rules; and a necessarily larger portion of the population dutifully carry out their appointed tasks as the humble economic base of society. Women and slaves, of course, are studiously ignored and absent from explicit inclusion in the pyramid of power. The ideal republic of Plato's Greece, therefore, more closely approximates Sparta—a militaristic, proto-fascist state, than the Athens in which Socrates and Plato were born, (mostly) lived, and certainly died—in the case of the former at the hands of the *demos*, the "mob"—a fate (though self-chosen) which his acolytes understood to be proof of the accuracy of his opinion that "the people" were incapable of governing and that an inappropriately egalitarian distribution of power foretold the doom of the just society.

Plato was clearly no democrat; however, to his credit, he did not endorse a hereditary class system. The gods, after all, worked in mysterious ways and performed wonders by sometimes inserting a golden soul into the body of a hod-carrier, a farrier, or (in Socrates' case) a soldier and a stonemason, while also occasionally putting bronze into the soul of a son of an aristocrat. In the just society, there was apparently room for upward and downward mobility, but only on the basis of innate virtues. It was also arguably tonic to cleanse the incompetent and self-indulgent from the ruling classes in order to refresh it with worthy souls from the lower socio-economic ranks. Periodic minor shake-ups in the stratification system were acceptable; however, allowing those with the *wrong* stuff to rise to power was not. And allowing a divinely determined meritocracy of quality to be overthrown by insurrectionist ambition, mob rule, violence, or the popular vote was worse.

Plato positioned character above all in the assessment of human beings, and education was deemed proper if it developed and applied the inherent qualities of each person to the task for which their nature fitted them. As Loeng (2017, p. 1) explains, Kapp was convinced that "the forming of character is the foremost value for human beings." Attention must be paid to bringing out and refining those qualities that conform to a person's proper occupation. So, Kapp's interest was focused on "inner qualities, rather than outer, objective competencies." He did not so much develop a "theory of adult learning" as a program of socialization/indoctrination concentrating on the manners and morals that conformed to the congenitally assigned social roles. If the proper qualities were encouraged to flourish, both the person and the polity would flourish.

Kapp captured his views on education in a book with the then-typically brutal title, *Platon's Erziehungslehre, als Pädagogik für die Einzelnen und als Staatspädagogik. Oder dessen praktische Philosophie.* In it, he did not take pains to explain how adults must be taught differently. His purpose was to justify his belief that, upon maturity, they could and should continue to be taught at all.

By the turn of the previous century, Kapp's Platonism—its explicit commitment to what even some honored sons of the European Enlightenment often called a "natural aristocracy" may have at least temporarily fallen on hard times (it had largely been replaced by a so-called "social" Darwinism), but the term he invented to distinguish the teaching of children from the teaching of adults in what he presumed to be the important process of lifelong learning and the reinforcement of inclinations toward responsible leadership and due deference remained.

Eugen Rosenstock-Huessy

Andragogy was resuscitated about a century ago by Eugen Rosenstock-Huessy. According to Loeng (2013) once again, Rosenstock-Huessy's "work related to andragogy in adult educational literature. [He was] a leading force in the first theoretical academic reflections on adult pedagogical issues in the 1920s [as well as] an inspiration for adult educators in the inter-war and post-war period." He is, like many others, at least familiar to expert historians and specialists in American educational theory, but not to the majority of contemporary working andragogic educators.

Almost as far from Plato as possible, but still within the acceptable space of an elastic notion of academic thinking, Rosenstock-Huessy, began as a law professor before migrating into sociology, history, theology, and labor education. A self-described "impure thinker" (Rosenstock-Huessy, 1970), he lived an unconventional life in which he championed social reform and spiritual regeneration.

Rosenstock-Huessy embraced the view that adult education should provide nothing less than "intentional spiritual guidance," a simultaneous consideration of "wisdom" and "practice," and a dialectic in which the "humanistic" and "vocational" sides of education would be collapsed in andragogy. It would not be merely a new approach to teaching and learning in which the life experience and maturity of the student would be taken into account in the techniques for the transmission of knowledge. Instead, it would be a moral and political process in which students and teachers would be self-consciously involved in the dynamic project of personal growth and social change. Its curriculum would emerge from the community and require a common commitment to solving "problems where the answers are not known in advance" (Leong, 2013, p. 244). His particular commitment was to giving industrial workers a voice in determining the future of their society and an increased awareness of their place within it. In effect, he turned Kapp upon his head.

Having experienced and been deeply affected by his time as a German officer on the western front is World War I (where he organized courses for troops on "broader topics" than instruction in patriotism) Rosenstock-Huessy left his post as a law professor at Leipzig and founded *Die Akademie der Arbeit* (the Academy of Labor) in Frankfurt at the same time as the Frankfurt School of Social Research was beginning as the fount of three generations of critical theorists from T. W. Adorno (1903-1969) and Max Horkheimer (1895-1973) to Jürgen Habermas (b. 1929) to Albrecht Wellmer (1933-2018) and their many, diverse beneficiaries. With the ascension of Adolf Hitler, he found refuge in the United States, where he spent most of the rest of his life.

Unable to reconcile with the German academy after World War II, Rosenstock-Huessy declined invitations to return to Europe except as a guest lecturer. Unlike his contemporaries Erich Fromm (1900-1980) and Herbert Marcuse (1898-1979), he never won the influence as a public intellectual that he believed he deserved. He spent the bulk of his academic life (1935-1957) at Dartmouth College in New Hampshire where, according to Cristaudo, Fiering & Leutzch (2015, p. 9), "he found in the college classroom a new, private mission as educator." His sweeping accounts of human civilization are currently being prepared for publication and his many hours of lectures remain available on DVD. Whatever his audience, however, Rosenstock-Huessy's mission and his definition of andragogy were, as Leong (2018) succinctly puts it, "aimed at developing the adult's sensibility for the spirit of the age and at motivating him/her to take action with the purpose of improving society."

Eric C. Lindeman

The progressive element in adult education could hardly find a more representative case study than the life and work of Eric C. Lindeman. Admitted to the University of Michigan as a mature student at the age of 22, he distinguished himself as an excellent writer of eclectic taste and skills—poet, playwright, polemicist, essayist as well as a proficient cross-disciplinary scholar whose early professional career saw him in roles from a youth worker with the YMCA and the Red Cross to the Chair of the American Civil Liberties Union Commission on Academic Freedom. He joined the New York School of Social Work in 1926 and published his famous book, *The Meaning of Adult Education* the same year.

In it, he identifies some of the main themes that would preoccupy theorists of andragogy in the 1960s and beyond. Emphasizing the importance of life experience as a motivator and resource for adult learners, he discounted the importance of classrooms and curricula, and preferred to think and act according to the life needs of the student. His notion of self-direction, therefore, did not imply the search for vocationally relevant programs of skills acquisition as dictated by labor market needs, but a holistic approach in which every day-life situations, not preconceived subject matter with measurable learning outcomes, were the guides to discovering what should be studied and how it could contribute to the whole personality and life-growth.

Tellingly, in contrast to later experts who insisted on separating young people and adults into two categories in which teacher/subject-centered education was appropriate for the former, but self-directed learning (now too often understood as achieving work-related job training) was best for the latter, Lindeman saw no necessity for division. If contemporary metagogues urge a synthesis of pedagogy and andragogy, Lindeman may have the answer: don't divide them in the first place.

Lindeman, of course, is nominally revered as a godfather of andragogy. His principal book is often venerated as "the Bible" of andragogy. It appears, however, that its message has become more generally praised than specifically practiced in the real world of education. It may occasionally be read as a ritual of passage or saluted as a symbol of an honorable legacy, but it is less often embodied in the educational workplace or discussed in its more prestigious journals, at conferences, and especially among legislators, employers, and donors. His eventful life had its flaws and personal failings, but his daughter's early characterization as a "radical and scholar" (Leonard, 1991) describes him better than the often-used honorific, the "elder statesman" of adult education. His is not the kind of presence that is commonly welcomed in Centers of Teaching and Learning, in Professional Development Programs, and on Presidential Committees (with or without token faculty and student representation) exploring ways for colleges and universities to "build back better" as a "new normal" is negotiated in the alleged aftermath of the ongoing SARS-CoV-2 pandemic.

Relying on American figures from Ralph Waldo Emerson (1803-1882) to social workers Jane Addams (1860-1935) and Mary Parker Follet (1868-1933) as well as his friend, John Dewey (1859-1952) and European educators such as the Danish pastor, pamphleteer, philosopher, and school reformer N. F. S. Grundtvig (1783-1872) for inspiration and insights, his "philosophy" can be succinctly summarized in two quotations:

Adult education is a co-operative venture in non-authoritarian, informal learning the chief purpose of which is to discover the meaning of experience, a quest of the mind which digs down to the roots of the preconceptions which formulate our conduct; a technique of learning for adults which makes education coterminous with life and hence elevates learning to the level of an experiment. (Lindeman, 1925, p. 3).

and

Orthodox education, may be a preparation for life but adult education is an agitated instrumentality for changing life ... [It] will become an agency of progress if its short-term goal of self-improvement can be made compatible with a long-term experimental but resolute policy of changing the social order. (Lindeman 1926).

ANDRAGOGY 2.0

Just as quick sketches of the pioneers can cast light on the origins of andragogy, so also can consideration of current trends in andragogic education be illuminated by glimpsing its chief proponents. In fact, in the interests of economy, it is possible to focus on just one: the iconic modernizer and popularizer, Malcolm Knowles (1913-1997).

Malcolm Knowles

Over a half-century ago, the current principal interpretation of andragogy was successfully launched by Malcolm S. Knowles (1968; 1970). So closely was Knowles identified with it, that he often felt obliged to acknowledge that he didn't invent it. Andragogy had long pre-dated Knowles' independent development of a parallel system but, it has been said, Knowles was only informed of the European origins and practice or of the early American encounter with andragogy in 1984—almost two decades after his own activities in the field had begun. The surprise apparently came when Dusan Savicevic of the University of Belgrade told Knowles that andragogy had been practiced selectively in Europe for decades (Savicevic, 1991, Loeng, 2018, p. 4). From Savicevic, Knowles learned that andragogy "was not about adult schools but about helping adults with the learning process" (Hanselmann, 1951) and that considerable effort had been invested in attempts to establish a scientific basis for the approach (Pöggeler, 1957). Indeed, the activities associated with andragogy had been pursued elsewhere *sans la lettre* (so to speak) for generations. So, according to Leong (2018, p. 8), "the term andragogy did not gain a foothold in the Nordic countries" because their rich tradition of popular education already "attended to the same ideals." As innovations go, it had a deep history, if somewhat neglected on this side of the Atlantic.

The "ideals" of European and American andragogy were not, of course, identical. Though they shared some of the same principles, their foundational purposes were quite different. Whereas the European variant—whether conservative or radical—was deeply concerned with the larger society and kept the ideas of community and social class top of mind, Knowles' (re)invention of andragogy in post-World War II America naturally expressed the dominant precepts of American political culture. Writes Leong: "Knowles' andragogy is worked out in the US, and reflects individualistic values such as freedom, independence, self-realization and self-direction" whereas Rosenstock-Huessy and his andragogy is also related to society, but in a different way.

The "liberal-individualist" Americans wanted to enable individual workers and others to "pursue happiness" (American code for "property" since *The Declaration of Independence*); the radical Europeans wanted to enable working people "to regenerate themselves in their country" (Leong, 2018, p. 7); and the "conservative" Europeans wanted to harmonize the individual with the state—not a tremendously popular ambition in post-Nazi/fascist Europe.

Neither European version was radically leftist, much less revolutionary, but Rosenstock-Huessy at least acknowledged the concept of "class" and sought to enable the working classes to affirm their collective interests in nation-building within a country torn by painful and seemingly irreparable social divisions. Knowles' andragogy, in the alternative, was premised on the "liberal tradition" in America (Hartz, 1991 [1955]) and the bedrock of "possessive individualism" (Macpherson, 1962). This liberal-individualist commitment was oblivious to (or unwilling to acknowledge) structural inequities—whether based on social class, race, gender, or any of the other burgeoning number of intersectional oppressions now proliferating or finally being grudgingly acknowledged in the United States.

The American Cultural Context

Wholly within the ideological framework that then and now dominates both North America and most modern democratic capitalist societies worldwide, Knowles set andragogy against pedagogy as a preferable framework for adult education. His professed interpretation and application were avowedly non-ideological. He neither expressly sought to use education to promote an elitist social order, nor to inspire transitional social change. His stated aims and methods were consistent with the pragmatic, pluralistic, meritocratic, affluent, new post-ideological, highly technological, (post)industrial state announced by the likes of Daniel Bell (1919-2011), (1960, 1973) and John Kenneth Galbraith (1908-2006), (1958, 1967). It assumed that the future would be "challenging" as technological and social change pressed us inexorably forward and required that we be prepared to witness, take full advantage of, and ultimately rejoice in "the end of history" (Fukuyama, 1992).

This reassuring pluralist model also presumed that American society was a brilliantly designed dynamic equilibrium of countervailing powers and interests glowingly described by Max Lerner (1902-1992) as a precarious "gigantism" held together by "some sort of balance ...between the contending form of bigness ... Big Government ... Big Business, Big Labor, Big Distribution, the Big Press, the Big Church, and the Big Army. No one of these is monolithic," he went on, "each of them in turn is a tangle of conflicting forces, and each in turn has had to achieve an equilibrium within itself. Thus, the American system of power has become like a system of nebulae held together by reciprocal tensions in inter-galactic space" (Lerner, 1957, p. 398); for a contemporaneous critique of the complacency of American life at the outset of postindustrialism, see the progressively more disenchanted works of Henry S. Kariel (1924-2004), (1961, 1966, 1977, and 1985).

Despite occasional cautions (Rumberger, 1984; Tsang & Levin, 1985) that the bulk of the new jobs in the high-tech economy would not be the much-promised, well-paid, intellectually satisfying, creative, clean, opportunities connected to computerization and what Warren Bennis (1925-2014) and Phillip Slater (1927-2014) cheerfully labelled the "temporary society" (Bennis & Slater, 1968), but in the low-level service sector of fast-food servers and sales clerks, and despite the claim that automation would eliminate mainly dirty, dangerous, onerous, and mind-numbing industrial work, we have reluctantly come to recognize the dangers that automation poses for office and even professional workers. Where once the main worries of sociologists in the optimistic post-war decades concerned the question of what people would do to amuse themselves when the work week was reduced to thirty, twenty, and even ten or twelve hours per week, we now are witnessing the rise of precarious work in the "gig" economy when even college and university teaching is given over as much as 75% to "contract" academic labor and suffering the shock of the fifty- and sixty-hour work weeks needed to cover basic domestic costs. Still, the prevailing myth was that, in the words of sociologist Seymour Martin Lipset (1922-2006), the United States was already "the good society in operation" (1960, p. 403).

The comfortable complacency and incipient triumphalism of such liberal pluralists might have done well to have been tempered by an occasionally barbed Galbriathian remark or even a nod to Niebuhr's skeptical quip that liberal democracy can do little more than find "proximate solutions to insoluble problems" (Niebuhr, 1944). A little less *hubris* might have helped soften the consequences or even avoided the worst impacts of various international misadventures in South-east Asia and the Middle East, the long-delayed reckoning with structural racism, structural economic inequity, and other seeds of division that are apparent in the current crisis of confidence in fundamental political institutions from the Capitol Building to local school boards where debates over andragogy have too often been displaced by ill-informed altercations about Critical Race Theory, gender-neutral washrooms, and non-binary gender pronouns. Nonetheless, astonishing as it may seem in retrospect, such matters rarely entered discussions about andragogy. Instead, the focus was primarily, if not exclusively, on the psychological composition of adult learners and their alleged needs.

Knowles distinguished between traditional learners in the public education system—children, adolescents, youth, and young adults—progressing sequentially and smoothly through elementary, secondary, and college education—and mature adult learners who were returning to school, often after significant interludes during which they may have assumed the responsibilities of full participants in society, experienced employment and perhaps unemployment, sometimes started families, and even decided to upgrade their credentials in middle age to facilitate a voluntary or involuntary career change or to pursue some interrupted ambition.

Transforming Higher Education

With increased markets for credentials and the stimulus of such measures as the *GI Bill*, the demand for more spaces for both younger and older students in the 1950s and 1960s heated up. Not only were technological innovations transforming kitchen appliances and adding television sets to the inventory of household needs, but even the Interstate Highway System was bringing both the capacity to deploy troops to every corner of the continental and the promise of road-trip holidays to an increasing number of Americans. A more affluent lifestyle was going together with a growing need for "knowledge workers" just as the "Cold War" and a huge military budget, spurred on by the advent of the nuclear arms race and the subsequent novelty of the space race, combined to transform the entire project of postsecondary learning. Although the prestige of the legacy colleges and universities remained to cater to the best, the brightest, and the wealthiest young folk, many of them were expanded into powerful research sites. They were joined by the advent of mass education in some of the state facilities which turned into "multiversities" in the 1950s and 1960s, and to these were added a host of lesser, but still legitimate state universities and community colleges. The result was an unprecedented increase in financial investment, bricks-and-mortar/glass-and-steel construction, and a level of knowledge creation and knowledge dissemination that redefined the place and the purpose of education in America. Science demanded it.

Commerce and industry demanded it. Democratic and egalitarian impulses demanded it. And national defense demanded it. A laboratory for experiments in andragogy was born.

As Trow (2010) and others have shown, each instance of the modern reorganization of the world economy has resulted in a parallel transformation of the organization, governance, social function, demographics, and both the content and the methods of education. So, whrn the industrial revolution required that brawn be supplemented by brain at least to the extent that workers became literate and numerate, the response was the creation of a tax-funded, compulsory public elementary school system. The industrial working class was born. Then, with mass urbanization, the rise of the commercial and service economies, and the proliferation of office jobs, the completion of high school required the extension of mass education and the growth of public secondary schools.

Now, in the postmodern, postindustrial, digital era, we have tacitly agreed that tertiary education is becoming a prerequisite for employment; hence, not only have state colleges and universities grown, but all manner of technical, vocational, and online job-training institutions have proliferated with the result that adult education has become at least a temporary growth industry. We therefore beg answers to questions about how to manage the adult market.

Enter andragogy as it infiltrated "continuing education" and "lifelong learning." Unlike its European predecessors, however, the North America style and substance of ingredients for the "learning society" came into full accord with the interests of corporate HRD. It now includes such innovations as competency-based and the prioritizing of marketable employability skills that are taking over not only diploma and first-degree courses, but also post-graduate degrees, professional certifications, and a dazzling array of micro-credentials and online "badges" designed to meet specific industry training needs.

What we have been witnessing is not the current stage in the evolution of adult education as it might have been envisaged as an aspirational goal of self-help organizations such as the nineteenth-century Mechanics Institutes. It is not the fulfillment of the dream of the nineteenth or early-twentieth-century school reformers who believed deeply in the necessity of at least a basic education in the arts as well as vocational training in order to advance both social and economic progress. It is certainly not the realization of what John Stuart Mill (1806-1873) regarded as a necessity if a future representative democracy were to survive the inevitable addition of *proletarian* and otherwise unschooled tradesmen and scullery maids to the voters' rolls; nor is it akin to what Antonio Gramsci (1893-1937) had in mind when he scribbled out his notes on education from a jail cell in Mussolini's Italy almost a century ago. Gramsci (1961, pp 24-43) imagined a revolutionary egalitarian society of "worker-intellectuals" self-governing a joyful socialist society and Mill foresaw a peaceable path to democratic socialism provided that the "masses" could be more or less civilized through universal, comprehensive, and essentially *bourgeois* education. But both, in their different ways, saw adult education as a means to liberate the working classes from capitalist servility and to free all citizens to realize their best potential version of themselves by their own creative standards. For Mill, the path went through liberal democracy to evolutionary socialism; for Gramsci, it went through revolutionary socialism to authentic democracy. In neither case was the benign or paternalistic capitalist corporation conceived to be a proper or even a possible vehicle to carry us to personal or collective emancipation.

Wrote Mill (quoted in Chomsky, 2014):

The form of association, however, which if mankind continues to improve, must be expected in the end to predominate, is not that which can exist between a capitalist as chief, and work-people without a voice in the management, but the association of the laborers themselves on terms of equality, collectively owning the capital with which they carry on their operations, and working under managers elected and removable by themselves.

The Newer "Older Student"

As the American political economy grew enormously in productivity and profits in the 1950s and forward, the exclusivity of elite, liberal and professional education was replaced by the diversity of mass education—often with a vocational, technological focus. Instrumental reasoning, results-oriented outcomes, practical consequences, transactional social relations, and the concept of education as an investment all gained ground in the minds of planners and administrators. Corporate interests—both public and private—insinuated themselves into college boardrooms and campus classrooms. All pretenses of higher education as a refuge from the allegedly "real" world and any presumption that the academy provided a privileged sanctuary from which academics could legitimately (and safely) criticize the vices and violence of business, government, industry, and what was becoming known as the "mass society" were removed.

Then, by the 1990s, even this huge change began to be superseded by the turn to universal access to accredited documentation of employability. There was much talk about grade inflation and shoddy academic standards in the quest to attract students to the "experience" of higher education and the pressure to ensure that the customers were kept happy and attrition rates were kept low. There was the constant pressure to attract grants money and to raise tuition fees as public funds for education dried up. Then, there was the pressure to publish in a world with four million working professors, tens if not hundreds of thousands of variously accredited institutions, and a proliferation of predatory journals all engaged in a Hobbesian war of all against peer-reviewed all in the battle for a diminishing number of tenure-track jobs.

Andragogy was the main attempt to find a way of teaching and learning that would cope with the new competitive environment. It would go beyond meeting the nurturing and socialization needs of people from K-12 to college completion. It would provide the new set of principles and practices that would be required to deal with students whose life experience, motivations, and curricular needs had been shaped outside the culture of educational institutions. Unlike traditional students, the adult learner was understood to be reward-based, self-directed, broadly experienced, intrinsically motivated, problem-oriented, and eager to learn.

The new adult learner would be ideologically neutered and put psychologically at ease with competitive individualism, the fetishism of self-improvement, and the ethics of personal responsibility. Adult learners would be no-nonsense consumers of educational utilities. They would have no complaint about being described as "clients," "customers," and "consumers" of commercialized, commodified, and largely fungible curricula.

Likewise, the evolving multiversities and polytechnics would thrive on market-driven innovation, technological enhancement, measurable faculty performance indicators, and organizational efficiencies from machine-graded assignments to online, flexible curricular delivery systems. In this increasingly transactional approach to education, the idea of the entrepreneurial supplier and the rational consumer may have lost the intimacy of the tutorial or the seminar and may have lacked the performative ingenuity of the brilliant lecturer, but they gained the equality and mutual respect of transactional partners. The adult learner, as customer, was "always right." Education stepped out of the groves of academe to become a promoter and purveyor of knowledge, whether in up-scale research universities or the stripped-down discount department stores and strip malls of knowledge. Wherever educational "services" could be provided, education was becoming a business enterprise in which the learner invested money in the form of tuition payments and expected the institutions to supply academic credits which would be tested in the marketplace by enhanced income and prestige of future employment.

The Critics

Under pressure from the realities of competitive corporate business models, the once lofty ambitions of andragogy have come under considerable criticism. A lack of conceptual clarity and an absence of empirical evidence as to its efficacy have been apparent almost from the outset.

Efforts have been made, for example, to operationalize and quantify the claims of andragogy in terms of its understanding of the adult learner's motivations, expectations, and satisfaction with the kind of student-centered learning, but vocationally marketable knowledge that the adult learner allegedly craves. As Ekoto & Gaikwad (2016) note, however, "the study of adult learning—andragogy—has emerged as a learning framework due to its increasing popularity in the past four decades. Research on andragogy has been mostly exploratory." Consequently, production, performance, and quality control have been lacking.

Indeed, most of the research that is done has taken the form of unpublished doctoral dissertations using small-sample survey research to investigate relationships between cognitive variables (learning) and affective variables (satisfaction) to evaluate the success of inadequately defined versions of andragogy in disparate parts of the world. A disheartening end to Ekoto and Gaikwad's own highly sophisticated statistical study of "learning satisfaction" among 91 graduate students in master's and Doctoral programs in Business, Education, Public Health, and Religion in the Philippines reached only a "conclusion [that] does not directly confirm or revoke andragogy in general and its association with learning satisfaction in particular." They therefore recommended "research in a different context, with a different sampling method, a larger sample, and a using different related factors." We await the results.

Shapiro (2016), on the other hand, examined the literature, drew upon the earlier critiques of Lee (2003) and Sandlin (2005), and added his own causes to reject Knowles' totemic representation of andragogy. He lists several ways in which the case for andragogy comes up short. He summarizes some of the main concerns thus: Andragogy, he says:

- treats education as a neutral, nonpolitical activity;
- is based on a "generic" learner who tends to be white, male, and middle class;
- appears to value only one way of knowing and ignores what does not fit its theory;
- is almost entirely individualistic, ignoring the importance of context;
- does not challenge the prevailing social structure, even if that social structure promotes inequality.

These are not sturdy bases upon which to build a convincingly elaborate, universal theory. He then postulates that Knowles' unproven claims about the nature of the adult learner are problematic at best. For example, he says, chronological age may mean neither intellectual nor emotional maturity: "adults may have been around longer but that does not mean they are capable of learning from their experiences."

Indeed, over a quarter-century ago, Hanson (1996) concluded her decidedly unfriendly assessment by arguing that "andragogy is nothing else but a 'normative utopia' based on the facts that Knowles' assumptions either have not substantiated backing or are ill-informed." Her view gains credibility when close inspection is made of the efforts to substantiate Knowles' assumptions. After a diligent search, Merriam & Brockett (2007) insisted that andragogy is nowhere clearly defined as theory or practice. More recently, Caruth (2016) suggested that it doesn't matter anyway because, "though the population of adult students [has been] significantly increasing in colleges and universities, andragogy is not being used, colleges are not prepared to use andragogy, and higher education is missing on andragogical praxis." Of course, Caruth is not totally negative. As with most contributions, she took pains to recommend "further empirical research to solidify andragogy."

Part of the confusion and consternation, of course, comes from peculiar contradictions in and among andragogy's most fervent supporters. It is said, for example, that the American Management Association's master's degree program in management was structured around competency-based education and andragogy. Meanwhile, Bouchrika (2022) endorses the prescription that "adult learners become more like autodidacts or self-taught individuals. They stay," he says, "at the helm of their learning." The problem there is that it is difficult to reconcile the notion of "competency-based" (which implies an observable, well-defined measure of the "competency" that the learner has attained) and the learner's control over the content of the purported competency and the means to achieve it—especially when it is apparently located within the managerial community in the corporate sector. Autodidacts, after all, are notoriously ambiguous when it comes to setting limits and defining borders; corporate managers, on the other hand, tend to be "no-nonsense" players for whom a master's degree would seem to imply that there is something specific—perhaps even a curriculum—to be managed and mastered.

Andragogy's Absence of Self-Consciousness

As for the state of andragogy itself, there is abundant evidence that andragogy's champions and practitioners have a high regard for themselves and their success in effectively making andragogy a synonym for adult education. Meanwhile, they seem to have lost their sense of purpose and any self-awareness about what it is that they do.

Beyond chanting the assumptions of Knowles' andragogical model which Ferreira and MacLean (2017) call a "psycho-motivational cocktail" that explains very little, but fails to explain a great deal, the future of andragogy may not be as comfortable as hoped and anticipated. There was much expectation that the large new online capabilities would provide a huge market for andragogical learners. That was where self-motivated, self-directed adult learners were said to have the best chance of displaying and benefitting from the absence of a teacher-monitor and making/making up their own way. Unfortunately, there is regularly "an attrition rate of up to 80%" in that domain. Also given the fact that pre-pandemic records reveal that "most distance education students are adult learners" (Caruth, 2014), something is plainly amiss.

What that "something" may be is not clear. What is clear is that research on andragogy is unaccountably lacking. Caruth (2014) reported that, "despite its 40-year history of being considered the principal model of adult education … the majority of research completed is actually in the practice of andragogy" with little about its utility and effectiveness. Moreover, in a recent, painstakingly detailed analysis of global research on andragogy, Ahmad, Meher, Siddique & Khan (2021) revealed that, while they had "seen decades of work done by researchers in the field of pedagogy, andragogy is still in need of much attention from educators and researchers." Accordingly, in addition to finding little of use from their work, they did what all good researchers do: they recommended more research, but with a twist: they urged research into the lack of research: specifically, they insisted on the need that "a further qualitative study be conducted which should be area specific and unearth the reasons behind low research productivity in the field of andragogy."

A BETTER PURPOSE, A BETTER PATH

Perhaps agreement can be found on the proposition that some of the sentiments supporting Knowles' notion of andragogy are commendable and that the distinction between pedagogy and andragogy is sustainable (with or without the addition of the vocally unpalatable heutagogy). And who can deny the implicit and explicit core concept in andragogy—namely, that adults seeking to be educated are worthy of respect and ought not to be treated like six-year-olds or sixteen-year-olds—especially if they do not think or behave like them?

But, when we investigate more deeply, we quickly learn that andragogy shares qualities that make it like so many popular, banal, but superficially benign theories of human nature, human knowledge, and human thought. It is conceptually confused, theoretically incoherent, and empirically unsustainable. It can also be persuasively accused of being Eurocentric, misogynistic, and racist.

In addition, upon even a little critical scrutiny, there is ample evidence to suggest that the pop psychology that underwrites andragogy in any of its engaging forms is remarkably less than is advertised on the pertinent book jackets. It is similar to many other attractive, easy to comprehend, clean and tidy methods of reducing complexities to simplicities, giving them the appearance of accuracy and utility, and tempting people to form perspectives and policies upon them. I am thinking of anything from the "multiple intelligences" conjured by Howard Gardner (b. 1943) to the to the "stages of moral development" posited by Lawrence Kohlberg (1927-1987) or, worse, the stages of economic development set forth by W. W. Rostow (1916-2003) as a fraudulent intellectual justification for American imperial misadventures from Latin America to South-east Asia.

A better purpose can therefore not be served by efforts to create a more rigorous theoretical framework, a more precise vocabulary, or more extensive research designs capable of providing evidence of its empirical validity. Instead, we would be well-served by getting to the root of the matter and asking what was asked near the beginning: Whose interests does education serve? What is education *for*?

Habermas and Human Purpose

One of the most enduring and serious of the many contributions by Jürgen Habermas (b. 1929) to our understanding of ourselves, the nature of knowledge, and our educational aspirations was put forward in his early book, *Knowledge, and Human Interests* (1971). In it, he described different sorts of inquiry, related different kinds of knowledge to particular kinds of human action, and recognized that one of the chief functions of education was to enable us to exercise control over some aspect of reality: the sciences, for example, gave us the means to exercise technical control over nature while the humanities and social sciences enable us to exercise control over ourselves and other people. Power is never absolute since our knowledge is never complete; however, being able to understand—even partially and imperfectly—the external world of things and the internal and interactive worlds of sentient individuals does promote the furtherance of our interests as persons and as members of our species.

So, if Roger Bacon's early modern claim that "knowledge is power" and Lord Acton's nineteenth-century warning that "power corrupts" are useful guides in our efforts to teach and to learn, then it behooves us to pay careful attention to what we do and why we do it when engaging in the project of education. Teaching is, as Henry A. Giroux (b. 1943) frequently reminds us, is inherently both moral and political. It is moral insofar as it purports help students to distinguish between good and evil (even if that amounts to the mere preference for "right" answers over "wrong" answers in a grade three class in arithmetic) and it is political in that it purports to assist students to decide upon the most efficacious ways to enhance what is good and to inhibit what is evil.

Put otherwise, every book about the fundamentals of "management" or "leadership" takes for granted that there are people (children, employees, baseball players, military troops, parishioners) who must be "managed" and "led." The possible legitimacy of some anarchistic options needs seldom to be articulated or even acknowledged. For example, *Economics: An Introduction* by Paul Samuelson (1915-2009) is arguably the most successful college textbook ever written. It is now in its 19th edition with multiple printings of each one. Since its initial publication in 1946, it has sold over four million copies in over forty languages. It helped its author win a Nobel Prize as well. Its publication, however, would have been unlikely and its success impossible had his neo-Keynesian/neo-classical interpretation of economics not been consistent with the dominant ideology of its day. Honest, self-aware scholars, and critics alike must understand that there is no such thing as "raw" data in studies; the data are deliberately and selectively chosen according to methodological criteria that are never pristine. There are also no such things as wholly untainted conclusions; their paradigmatic acceptability is inherent on the research questions we pose.

Knowles' andragogy, too, comports well with the dominant neoliberal theories of this day. Its hidden biases and unsupported assumptions fit nicely with long-standing American traditions and aspirations. It serves the interests not only of the formidable enterprise of formal schooling, but of on-site job training, technological upgrading, and corporate-style brainstorming sessions and joint management-worker committees to co-opt and deflect concerns about issues as diverse as ever-controversial standards of diversity and as potentially catastrophic as climate change and environmental degradation—to say nothing of workers' rights on or off the "shop floor."

Emancipatory Options

Enter Edward Thompson (1924-1993). If any single historian could be credited with changing the style and substance of social history in the twentieth century, it would be hard to pass over E. P. Thompson. His *The Making of the English Working Class* (1963) and his many other books on historical theory (1978) and on topics from eighteenth-century law (1975) to nineteenth-century romantics (1955), and even a science-fiction novel (1978) made him one of the most important writers of our time.

Peace activist and poet, scholar and polemicist, Thompson was always on the periphery of the academy. He spent the best part of his life in adult education, notably at the University of Leeds and the University of Warwick, until he resigned in protest over the commercialization of the institution (Thompson, 1970). Thereafter, although he had no difficulty securing speaking engagements, he was marginalized in terms of the "profession."

Just as he found his mission as a historian in "rescuing the poor stockinger, the Luddite cropper, the 'obsolete' hand-loom weaver, the 'utopian' artisan, and even the deluded followers of Joanna Southcott, from the enormous condescension of posterity" (1968, p. 13), so also did he find in adult education the opportunity to bring together his knowledge and his political views. The British Workers' Educational Association had joined with the "Extra-Mural Department" at the University of Leeds to offer a program that would, in Thompson's terms, allow the university an opportunity to fulfill its "highest task"; namely, to address "the educational need of a class in society denied by economic circumstances or social environment a fair share in the use of other institutions of higher education" and to do so with a "specific emphasis—'education for social purpose'—on making this section or class more effective in social activities" (as cited in Searby, 1970, p. 5).

Thompson promoted and personally fulfilled the promise to encourage a "class-conscious attitude … in the pursuit of truth and in social activity [and] of the interests of their own class in its struggle for emancipation" (as cited in Searby, 1970, p. 6).

Thompson's commitment to the British working class differs from the experience of like-minded people in different situations and with different traditions; names such as Antonio Gramsci, Paulo Freire (1921-1997), and Henry A. Giroux spring prominently to mind. Unfortunately, self-help organizations such as the nineteenth-century Mechanics Institutes and the Workers' Educational Association may not be available everywhere and always, but they do present a genuine alternative to andragogy as it has devolved and deformed in North America.

True, many trade unions take deserved pride in programs to educate their leadership and membership in the techniques of monitoring the employers' contract compliance, making full use of the grievance procedure under various collective agreements, and (increasingly) fostering awareness of issues from ecology to the current enthusiasm for Diversity, Equity, and Inclusion. These are important practical activities and assist mightily in instrumental efforts to win what justice is available under the collective agreements that govern wages and working conditions is a shrinking segment of society. Even here, however, the same problems apply as they do in management's appropriation of andragogy to achieve its aims of bureaucratizing the imagination of people whose need to dissent demands more than an interactive suggestion box and collaborative committee to provide a pressure valve (and an opportunity to co-opt opportunistic working-class heroes to the lower rungs of the corporate mobility ladder).

The point is that, as long as andragogy (or whatever word, phrase, or slogan next gains popularity) lacks critical consciousness and the active engagement of the student in the creation, realization, and internalization of that conscious awareness, so-called student-centered teaching and learning remain no less the socializing and ideological agents of the late capitalist labor process than the company newsletter or the major print and broadcast media. The structural subordination of the adult worker in the workplace (including teachers, regardless of any delusions of professionalism) equally suppresses intellectual and political autonomy. They remain as much members of the "totally administered society" as Adorno and Horkheimer (1947) described them and just as "one-dimensional" as Marcuse (1964) diagnosed them to be. Or as Sheldon S. Wolin [1922-2015] (2017) more recently added, they are manipulated citizens of a "managed democracy" that is little other than a politics of "inverted totalitarianism," a more detailed update on "friendly fascism" (Gross, 1980).

Learning that consciously and conscientiously inculcates complicity in dominant social norms and the acceptance of mastery of whatever marketable communication capabilities and arithmetical skills may remain necessary in a culture replete with ChatGTP, mini-calculators, and governance by algorithm is surely little improvement over the alleged faults of traditional pedagogy. What andragogy appears to provide is the rhetoric, but not the reality of meeting the needs of autonomous adults. It repeats the educational indoctrination that results in the infantilization of the adult learner and the proletarianization of the teacher—whether in junior kindergarten or in a supervisory role for post-doc fellows.

The degree to which a truncated version of andragogy has been enthusiastically adopted by precisely those corporate interests its progressive forbears meant to democratize has been shown by critics such as Michael Collins (1995, p. 87). By conflating ideals of student-centered learning with competency-based education, adding in some HRD notions such as "team management," "partnerships," and "quality circles," the employers and managers can achieve their "primary purpose [of] consensus formation [using] strategies [that] confirm predetermined institutional interests."

The Courage to Teach

Over a quarter-century ago, Michael Welton (1995, p. 5) observed that the "andragogical consensus (anchoring the study of adult education in methods of teaching and understanding the individual adult learner) [that had been] solidified by Malcolm Knowles and others in the 1960s and 1970s, [had] fallen apart at the seams." Welton made four charges:

1. adult education has abandoned its once vital role in fostering democratic social action;
2. the discipline of adult learning was based on shaky foundations;
3. the contemporary modern practice of adult education is governed by an instrumental rationality that works to the advantage of business, industry, and large-scale organizations;
4. consequently, the guiding principle of adult education, self-directed learning, is conceptually inadequate to serve the interests of the poor, oppressed, and disenfranchised in North American society.

If any remaining good can be found in Knowles, it is probably a remnant of something he absorbed from Lindeman, but it is equally improbable that it can be salvaged and extracted from the layers of corporatizing that has sullied the entire enterprise. At the same time, although the term itself should likely be retired, the initial impulses of the earlier progressives can surely be restored and is being restored and extended by people such as Henry A. Giroux, possibly the most prolific writer today in the tradition of Paulo Freire, whose latest book, *The Pedagogy of Resistance* (2022) is merely the most recent of more than seventy volumes over the past (see https://www.henryagiroux.com/).

At this time, it would be appropriate to offer a rallying cry answering the question "what is to be done?" The current global polycrisis precludes any simple response. It also precludes despair. Pessimism is simply an unaffordable luxury today.

So, the best that I can do is to repeat the call for a renewed agency on the part of adult (and other) educators and activists. It was made thirty years ago by Parker Palmer (1991). It is this: "Courage."

ACKNOWLEDGMENT

I wish to express my gratitude to Brian Flack, Tony Goulem, and Frann Harris, whose advice, when followed, made this a better chapter and, when neglected or ignored, left it a worse one. I must also express my thanks to Tony Iacovino, Joe Grogan, and Maria Vasilodimitrakis for their ongoing encouragement, as well as the anonymous reviewers who did their best to restrain my impulses and idiosyncrasies.

REFERENCES

Ahmad, S., Meher, S., Siddique, N., & Khan, M. A. (2021). *Global research on andragogy: A bibliometric analysis*. Library Philosophy and Practice.

Battersby, D. (1987). From androgogy to gerogogy. *Journal of Educational Gerontology*, *2*(1), 4–10.

Bell, D. (1960). *The end of ideology: On the exhaustion of political ieas in the fifties*. Free Press.

Bell, D. (1973). *The coming of post-industrial society*. Basic Books.

Bouchrika, I. (2022). *The andragogy approach: Knowles' adult learning theory principles*. https://research.com/education/the-andragogy-approach

Brookfield, S. D. (1984). *Understanding and facilitating advanced learning*. Jossey-Bass.

Caruth, G. D. (2014). *Meeting the needs of older students in higher education*. Participatory Educational Research (PER). http://www.partedres.com/archieve/issue_1_2/3-per_14-09_volume_1_issue_2_page_21_35.pdf

Chomsky, N. (2014). What is the common good? *Toward Freedom*, 9. https://towardfreedom.org/story/archives/special-reports-archives/noam-chomsky-what-is-the-common-good/

Collins, M. (1995). Critical commentaries on the role of the adult educator: From self-directed learning to postmodernist sensibilities. In M. R. Welton (Ed.), *In defense of the lifeworld* (pp. 71–98). SUNY Press.

Cristaudo, W., Fiering, N., & Leutzsch, A. (2015). *Introduction: Eugen Rosenstock-Huessy (1888*-1973*). Culture, Theory & Critique*, *56*(1), 1–12. doi:10.1080/14735784.2014.979086

Cross, P. (1981). *Adults as learners*. Jossey-Bass.

Ekoto, E., & Galkwad, P. (2015). The impact of andragogy on learning satisfaction of graduate students. *American Journal of Educational Research*, *3*(11), 1378–1386. doi:10.12691/education-3-11-6

Ferreira, D., & MacLean, G. (2017). Andragogy in the 21st century. *Language Research Bulletin ICU Tokyo*, *32*, 10–18.

Fukuyama, F. (1992). *The end of history and the last man*. Free Press.

Galbraith, J. K. (1958). *The affluent society*. Houghton Mifflin.

Galbraith, J. K. (1967). *The new industrial state*. Houghton Mifflin.

Gramsci, A. (1961). *Selection from the prison notebooks*. International Publishers.

Gross, B. (1980). *Friendly fascism: The new face of power in America*. M. Evans.

Habermas, J. (1971). *Knowledge and human interests*. Beacon Press.

Hanselmann, H. (1951). Andragogik: Wesen, Möglichkeiten, Grenzen der Erwchsenenbildung [Andragogy: Essence, Potentiality and Limits of Adult Education]. Rotapfel Verlag.

Hanson, A. (1996). The search for a separate theory of adult learning: Does anyone really need andragogy? In R. Edwards, A. Hanson, & P. Raggatt (Eds.), *Boundaries of adult learning* (pp. 99-108). Routledge.

Hartz, L. (1991). *The liberal tradition in America*. Harper. (Original work published 1955)

Henschke, J. A. (2011). Considerations regarding the future of andragogy. *Participatory Educational Research*, *1*(2), 21–35.

Horkheimer, M., & Adorno, T. W. (1972). *Dialectic of enlightenment*. Herder & Herder.

Johnson, L. (2016). Geragogy. In S. L. Danver (Ed.), *The SAGE encyclopedia of online education* (pp. 504–508). SAGE Publications.

Kariel, H. S. (1961). *The decline of American pluralism*. Stanford University Press.

Kariel, H. S. (1966). *The promise of politics*. Prentice-Hall.

Kariel, H. S. (1977). *Beyond liberalism: Where relations grow.* Chandler & Sharp.

Kariel, H. S. (1989). *The desperate politics of postmodernism*. University of Massachusetts Press.

Kenyon, C., & Hase, S. (2001). Moving from andragogy to heutagogy. *ERIC 18*(1). https://eric.ed.gov/?idED456279

Knowles, M. S. (1968). Andragogy, not pedagogy. *Academic Leadership, 16*(10), 350-352. 386.

Knowles, M. S. (1970). *The modern practice of adult education: Andragogy versus pedagogy*. Association Press.

Leonard, E. (1991). *Friendly rebel: A personal and social history of Eric C. Lindeman*. Adamant Press.

Leong, S. (2018). Various ways of understanding the concept of andragogy. *Cogent Education*, *5*(1), 1496643. Advance online publication. doi:10.1080/2331186X.2018.1496643

Lerner, M. (1992). *America as a civilization: Life and thought in the United States today*. Simon and Schuster.

Lindeman, E. C. (1925). *What is adult education?* [Unpublished manuscript]. Columbia University, Butler Library Lindeman Archive.

Lindeman, E. C. (1926). *The meaning of adult education*. New Republic.

Lipset, S. M. (1960). *Political man: The social bases of politics*. Doubleday.

Loeng, S. (2013). Eugen Rosenstock-Huessy: An andragogical pioneer. *Studies in Continuing Education*, *35*(2), 241–253. doi:10.1080/0158037X.2012.749850

Loeng, S. (2017). Alexander Kapp: The first known user of the andragogy concept. *International Journal of Lifelong Education*, *36*(20), 1–15. doi:10.1080/02601370.2017.1363826

Macpherson, C. B. (1962). *The political theory of possessive individualism: Hobbes and locke*. Oxford University Press.

Marcuse, H. (1964). *One-dimensional man*. Beacon Press.

Merriam, Sh. B., & Brockett, R. G. (2007). *The profession and practice of adult education*. Jossey Bass.

mpksbohemian [screen name]. (n.d.) *Metagogy.* http://www.verbotomy.com/verbottle.php?jargonism_id=54

Niebuhr, R. (1944). *The children of light and the children of darkness.* https://www.semanticscholar.org/paper/The-Children-of-Light-and-the-Children-of-Darkness%3A-Niebuhr/516e53cd752cae99195320cfeeb038b1e95ca06c

Palmer, P. (1991). The courage to teach. *The National Teaching and Learning Forum, 1*(2), 2-7.

Peterson, C. M., & Ray, C. M. (2013). Andragogy and metagogy: The evolution of neologisms. *Journal of Adult Education, 42*(2), 80–85.

Pöggeler, F. (1987). *Einfürung in die Andragogik: Grundfragen der Erwachsenenbildung* [Introduction to Andragogy: Fundamental Issues in Adult Education]. Düsseldorf, Germany: A. Henn Verlag.

Prokopenko, E. (2009). *Metagogical analysis of pedagogical apparatuses.* https://estherspace.wordpress.com/2009/01/17/metagogical-analysis-of-pedagogical-apparatuses

Rollermation [screen name]. (2011). *Metagogy I.* http://elearningportal,drupalgardens.com.comment/6

Rosenstock-Huessy, E. (1970). *I am an impure thinker.* Argo Books.

Rumberger, R. W. (1984). The job market for college graduates, 1960-1990. *The Journal of Higher Education, 55*(4), 433–454. doi:10.2307/1981442rr

Savicevic, D. M. (1991). Modern conceptions of andragogy: A European framework. *Studies in the Education of Adults, 23*(3), 179–201. doi:10.1080/02660830.1991.11730556

Searby, P. (1993). Edward Thompson as a teacher: Yorkshire and Warwick. In J. Rule & R. Malcolmson (Eds.), *Protest and aurvival: Essays for E. P. Thompson.* Merlin.

Strohschen, G. (2020). By means of critical theory: Informed emancipatory education—An essay on realities and possibilities. In V. Wang (Ed.), *Handbook of research on ethical challenges in higher education leadership and administration* (pp. 296–308). Information Science Reference. doi:10.4018/978-1-7998-4141-8.ch016

Thompson, E. P. (1955). *William Morris: Romantic to revolutionary.* Lawrence & Wishart.

Thompson, E. P. (1968). *The making of the English working class.* Penguin. https://www.academia.edu/30385746/E_P_Thompson_Making_of_the_English_Working_Class_Vintage_1966_1_

Thompson, E. P. (1970). *Warwick university ltd: Industry, management and the universities.* Penguin. https://files.libcom.org/files/2022-09/warwick_uni_ltd_ep_thompson.pd

Thompson, E. P. (1975). *Whigs and hunters: The origins of the Black act.* Allen Lane.

Thompson, E. P. (1975). *The poverty of theory and other essays.* Merlin.

Thompson, E. P. (1978). *The sykaos papers.* Bloomsbury.

Trow, M. (2010). *Twentieth-century higher education: Elite to mass to universal.* Johns Hopkins. doi:10.56021/9780801894411

Tsang, M. C., & Levin, H. (1985). The economics of overeducation. *Economics of Education Review, 4*(2), 93–104. doi:10.1016/0272-7757(85)90051-2

Welton, M. R. (Ed.). (1995). Introduction. In M. R. Welton (Ed.), In defense of the lifeworld (pp. 1-10). SUNY Press.

Whitehead, A. N. (1931). On foresight. In W. B. Graham (Ed.), *Business adrift.* McGraw-Hill.

Wolin, S. S. (2017). *Democracy incorporated: Managed democracy and the specter of inverted totalitarianism - new edition*. Princeton University.

Wright, P. (2016). Helping older adults conquer digital tablets. *Ger ontechnology, 14*(2), 78-88. doi:10.4017/gt.2016.14.3.005.00

Yeo, G. (1982). "Eldergogy": A special approach to education for elders. *Lifelong Learning, 5*(5), 4–7.

KEY TERMS AND DEFINITIONS

Andragogy: Coined in the early nineteenth century, it was meant to identify the need for adult education and, later, to give voice to the needs of adult workers to improve their social standing and participate in the building of a society that would reflect their class interests; more recently, it has come to embrace a narrower vision with the goal of catering to adult students seeking employability skills that conform to the needs of late capitalist corporations.

Corporatism: A term used to refer to a pattern of social organization in which institutions such as colleges and universities come to adopt a labor process that mimics that of large-scale business, with reliance contract workers, commodification of curricula, and a market mentality that stresses consumer demand, innovation, and entrepreneurship over traditional academic values and any form of critical theory.

Late Capitalism: A fluid term that has been in use for over 100 years, but now commonly refers to a political economy featuring a digital, high-tech, and postindustrial capitalism in which financial capital is dominant, income inequality grows, large corporations dominate market mechanisms, and a postmodern culture prevails.

Pedagogy: Originally the theory and practice of educating children, it became associated with the understanding of education in general; as such it provided a foil for those eager to carve out a professional role for exclusively "adult" education. While the social purpose of childhood education has been commonly understood (if highly contested), the aims of adult education remain fundamentally controversial.

Social Class: Contrasted with sociological stratification, which measures inequality in income, influence, and status among individuals and groups, social class identifies structural relations of power in relation to the overall mode of production of knowledge as well as goods; it permits a better understanding of the interests and ideologies of dominant classes (owners and managers) and the ways in which education serves their political and economic interests by making middling and working class people politically submissive and economically dependent.

Chapter 4
Andragogy and Leadership in Higher Education:
A Framework for the Implementation of Leadership Development Programs

Marius Boboc
Cleveland State University, USA

ABSTRACT

This chapter focuses on various factors requiring changes in higher education leadership, such as societal pressures, availability of financial support, labor market characteristics and expectations for job-ready college graduates, and the tension between institutional processes and the degree to which internal/external stakeholders are involved in decision-making processes. In this light, leadership development has become critically important in creating the necessary cadre of professionals committed to the mission and vision of institutions of higher education they lead. As a clearer focus on effectiveness and differential impact on the emerging leadership skills of participants is still needed, this chapter proposes a theoretical framework infusing andragogical principles into leadership development.

INTRODUCTION

Given the range of challenges facing higher education, ranging from curricula to students, faculty, staff, shared governance, the overall value added to society, etc. (Zusman, 2005), leadership in colleges and universities has been the center of attention for quite some time now. Living in a global post-industrial society characterized by ever-increasing information consumption (Taylor & Machado-Taylor, 2010) has had a direct impact on higher education as a sector designed to serve the public good. Access, attrition, affordability, and accountability are the four main areas of focus for college and university leaders in the U.S. today (Berdahl, Altbach, & Gumport, 2011). To an increasing degree, the same applies to more countries around the world, as resources and support for higher education are under increasing scrutiny. In this light, theoretical frameworks focused on leadership development underscore authentic (Cooper,

DOI: 10.4018/978-1-6684-7832-5.ch004

Scandura, & Schriesheim, 2005), distributive (Jones, Lefoe, Harvey, & Ryland, 2012), or visionary characteristics (Taylor & Machado-Taylor, 2010). Situated in the larger context of globalization in the technological era of the 21st century, higher education institutions around the world need to balance a myriad of local needs with increasing requirements stemming from a focus on global citizenship and its reliance on knowledge, skills, and dispositions that bring all of us together (Archer & Zhang, 2021). Moreover, in light of the COVID-19 pandemic and its impact on instruction and overall work in virtual environments, rethinking social mobility and its correlation to cultural capital and diversity implies recalibrating policy and leadership in higher education (Challen, 2020). In this chapter, the author propose a conceptual framework intended to outline variables and relations among them (Miles & Huberman, 1994) as they support the effective implementation of leadership development programs built on andragogic principles, thus addressing a gap in current practice.

CHALLENGES IN HIGHER EDUCATION AND THE NEED FOR EFFECTIVE LEADERS

In synchronicity with the evolution of the higher education sector, the concept of leadership acquired additional significance over time. Leadership can be thought of as a complex process aimed at establishing a relationship toward a common goal within a particular context (Perruci, 2011, as cited in Sowcik, Andenoro, McNutt, & Murphy, 2015). Charisma-led leadership was centered on the idea that individuals are born to lead. Scientific management followed, elevating leadership to include processes designed to help individuals become efficient leaders. The next iteration was focused on the need for a more engaged/ engaging leader, who would become transformational through leadership development (Kelly, 2019, as cited in Roux, 2020). Increasingly convoluted layers of multi-sector attributes that come into play to activate leadership skills across a plethora of environments require flexible planning and execution of strategic moves that factor in scenarios dealing with an unpredictable future highlighting diminishing resources and a stronger competition for them. At the same time, too much information, coupled with data abuse could become a particular contextual parameter that may end up delaying or blocking change by potentially leading to a lack of trust in leadership and/or planning processes (Mack, 2015). An analysis of the evolution of organizations points to greater complexity and maturity over time, leading up to the current stage where a full range of organizations have become more aware of their societal impact, thus emphasizing stronger ethical considerations supporting enhanced community-focused collaboration (Roux, 2020). In this light, organizations have become more agile and invested in providing value-added products/services/solutions to customers (individuals, groups, or other organizations), while deploying a less hierarchical structure to secure a competitive advantage by more pro-active decision making (Heerwagen, 2016, as cited in Roux, 2020).

As colleges and universities are becoming more complex institutions that have grown significantly in terms of the spectrum of their specialized services offered to co-located communities, they have to address the needs for flexible, pro-active leadership to cope with societal expectations of entrepreneurship, globalization, and marketization (Dopson et al., 2019). Emerging structural changes perceived as disruptive/innovative require concerted efforts focused on preserving the integrity of higher education's mission and commitment to the public good, while sustaining the spaces created to establish communities of learners, conduct research, and engage in pedagogical practices that address authentic, real-life issues (Keohane, 2013). As mentioned earlier, the various pressures that the higher education sector faces include societal factors, financial support, labor market characteristics, inclusive of recent changes in modes of instruction delivery and overall expectations related to employability held by graduates.

Current trends that impact the way we could envision a just world tomorrow have to factor in the ever-widening economic disparities over the past few decades, with direct correlation to environmental concerns, employment, access to schooling/knowledge co-construction, healthcare, as well as opportunities for civic and political involvement, thus bringing about social justice (Noble, 2015). In this light, leadership is the set of principles, policies, and structures that constantly attempt to balance current capacity to manage change with current and future needs that trigger change. Given the complex and dynamic nature of leadership as a construct bridging agency, place, and purpose, integrating leadership, management, and administration contrast transactional, transrelational, and transformative approaches to leadership based on several themes ranging from workplace culture to cooperation style to power source and to organizational structure and its viability/sustainability, etc. (Branson, Marra, Franken, & Penney, 2018). Defining effective leadership is constantly evolving in order to deal with sustainability challenges around the world (Satterwhite, McIntyre Miller, & Sheridan, 2015). An eclectic emerging leadership style would connect with any and all of these three approaches to leadership on several of the aforementioned themes at any given time, thus indicating that leadership maturity validates knowledge, skills, and dispositions as they get activated by various contextual factors, both challenges and opportunities. Maturing a leadership style relies on constant reflection and analysis of data points as well as decision making structures and processes to identify the interplay among contextual factors, participating agents, and organizational mission, culture, and expectations.

Leadership development has become critically important in creating the necessary cadre of professionals committed to the mission and vision of institutions of higher education they lead. Programs aimed at preparing current and future leaders in colleges and universities have been researched for quite a while, yet a clearer focus on effectiveness and differential impact on the emerging leadership skills of participants is still needed (Reyes et al., 2019). Effective leaders want to stay abreast of innovation, be engaged in pursuing ways to work closely with peers and employees, while staying ahead of competition (London, 2002). Principles of Andragogy provide a basis for leadership development programs that can precipitate effective leaders.

Principles of Andragogy

Under these circumstances, using the principles of andragogy to create user-focused, flexible leadership development programs for higher education administrators could lead to greater return on investment and a more positive campus/workplace environment. Perceived as an intellectual project aimed at providing a unifying perspective on the emerging academic discipline of adult education, thus meeting "the need of the field to define itself and create clear boundaries" (St. Clair & Kapplinger, 2021). The deliberate attempt to support adult learners by way of specific parameters that define instructional acts in which they engage led to the development of Knowles' (1970) theoretical framework that casts andragogy as the unifying construct. More recent conceptualizations outline the following assumptions about adult learners (Knowledge, Holton, & Swanson, 2011):

1. **Self-Concept:** Based on an understanding of one's autonomous learning and supported by self-direction and self-awareness.

2. **Experience:** A combination of life and work experiences represent a vital component of the platform supporting co-constructing knowledge.
3. **Readiness to Learn:** Relying on prior knowledge, skills, and experience, adult learners increasingly employ applications of their knowledge/skill development to advance their instructional needs.
4. **Orientation to Learning:** Clearer and stronger connections to real-life situations engage adult learners by grounding their knowledge/skill/disposition development in problem-solving.
5. **Motivation:** The more adult learners are able to monitor and refine their approach toward learning, the more internal motivators support that process.
6. **The Need to Know:** Validity and relevance of information used to develop knowledge/skills are increasingly perceived as facilitating factors in the instructional process.

College and university leaders are experienced professionals with a wide range of expertise areas, attributes, talents, needs, expectations, and aspirations. As such, tapping into who they are individually could inform the design of a leadership program intended to benefit the collective of professionals they lead. Andragogy emphasizes adult learners' characteristics derived from their ability to self-direct and use their prior experiences, awareness of readiness to learn as well as what needs to be learned, in addition to their orientation/disposition toward learning, and motivation to learn (Tipton & Wideman, 2021).

Based on the andragogic foundation outlined earlier, leadership development programming should apply a growth mindset (Dweck, 2016) to negotiate how executive, strategic decision making capitalizes on the importance of assuming leadership roles, how the latter are different from managerial circumstances, and how to empower future leaders by cultivating their innate qualities (Recigno & Kramer, 2022), which has not been done deliberately to date. By blending the informal and formal traits of leadership development curricula, learners' balance among knowledge, skills, and dispositions, shaped by prior life and work experiences, enhances the potential for real dialogue and change (Bloomquist & Georges, 2022), as outlined in the following section.

Designing a Cyclical Leadership Development Program

An integrated leadership development program proposed by London (2002) includes the following components: a) needs analysis; b) assessment of talent at various organizational level; c) skill gap analysis; d) early identification of talent; e) support for development; f) program/curriculum planning; g) support for development; and h) ongoing assessment and development. Adapting the model to insert andragogical principles in order to frame leadership development in the following particular descriptive ways is intended to capture the essence of a cyclical approach to sustaining leadership capacity (see Figure 1). Proposing a theoretical framework of this caliber is derived from an attempt to shift a mode of theorizing from static to dynamic, while allowing for variables to bring to life the different ways in which the process of leadership development can take shape (Makadok, Burton, & Barney, 2018).

As we move from left to right along the visual aid above, the procedural steps in the first block to the left relate to the following:

1. Conduct a needs assessment to determine capacity, requirements, and support for leadership development. If the leadership development program is a first of its kind, the initial needs assessment should lead to the identification of leadership-focused issues in the past and how they were dealt with, thus leading to a comparison between desired and observable outcomes. Concurrently, an

analysis of current leadership initiatives, with a particular emphasis on emerging issues within those initiatives could dovetail with the apparent needs for leadership development based on previous occurrences. Finally, in order to complete the comprehensive scope of such analyses, anticipating future requirements for more effective/impactful leadership should generate a fuller representation of how to provide professional development programs tailored to particular needs, interests, expectations, and aspirations participants may demonstrate.

Figure 1. Cyclical leadership development program design matrix

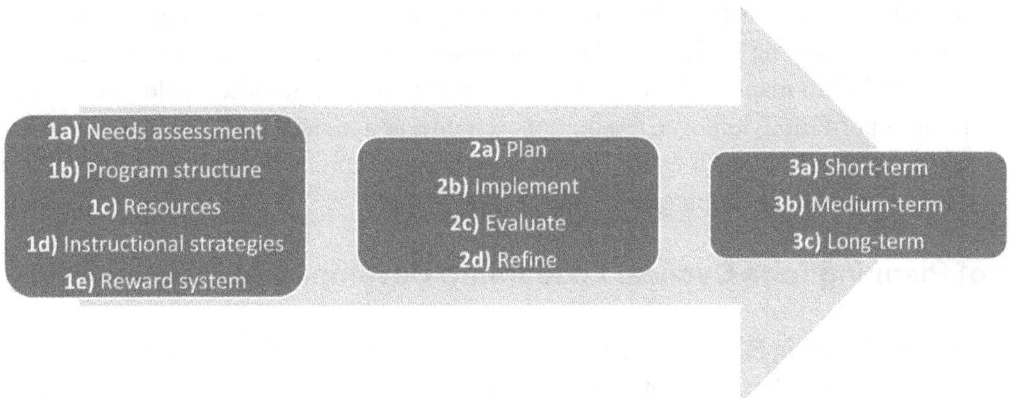

2. Taking into account the findings from the previous needs assessment stage, identify program tracks that ground particular overlapping areas of leadership development potential. Task-specific set of leadership skills could form the foundation for these program tracks, ideally leading to a developmental sequencing of content. At the same time, a programmatic structure should be put in place in a way that addresses actual needs and is flexible enough to allow for mid-way course corrections. Part of this planning process, participants could form thematic cohorts aligned with a "train the trainer" approach to leadership development. Concurrently, depending on the leadership capacity, specific topics could feature guest speakers from among the program participants, which would allow their previous life and work experience to ground constructive conversations on becoming an effective, inclusive, and transformative leader based on particular traits as adult learners.
3. Connected to an earlier point made under the needs assessment stage in the program design process, identify the capacity needed to launch the program successfully by identifying a current wide range of resources (both required and optional, internal, and external). At the same time, any resource that is not available but could have a demonstrable positive impact on the overall integrity of the program should be ranked and categorized in terms of urgency of need. Moreover, ensure a cross-sectional representation of the diversity, inclusion, equity, and belonging principles when selecting instructional resources that would prompt participants to delve into and analyze different leadership challenges and opportunities in higher education, correlated directly with their own prior life and work experiences that qualify their leadership potential.
4. Based on expected program outcomes, structure instructional strategies, inclusive of a full range of assessment protocols, to maximize participant engagement. These strategies should be tailored to

preliminary data collected from program participants related to learning preferences, perceived areas of strength as well as areas of improvement, communication styles, and intra- and inter-personal skills. Concurrently, use formative assessment techniques to gauge program effectiveness at critical points in its implementation, which should allow for curricular and/or pedagogical changes, as needed. Further details about how to break down andragogical principles to craft a leadership development program are referenced in the section below on applications of andragogic principles to such programs.

a. Create a rewards system that could validate participants' lessons learned in the program. Such rewards could also be used to mark symbolically the gradual progression in the effectiveness of leadership demonstrated by program participants. In other words, special notices could be given to "emerging leaders" or the "most community-focused impactful initiatives pursued by a leader" to map out the content tracks in the program to observable improvements in a leader's portfolio. Moreover, use the cohort model as a foundation for mentoring and networking opportunities, which could lead to greater applicability of the leadership skills refined/developed through the program.

Levels of Planning for a Cyclical Leadership Development Program Design

While the block to the left in Figure 1 represents the most comprehensive and nuanced level of planning for the cyclical leadership development program design, the middle block captures the main planning categories that apply to the process whether it is used only once or part of a series. In this light, as the visual aid below shows (see Figure 2), these planning categories could be used for each of the items in the block on the left in a way that allows for correlations to be established between any of them. In this light, the first planning phase, at the micro level, encompassing the needs assessment, creating a program structure, identifying resources, outlining instructional strategies, and assembling a rewards system should inform the implementation phase in terms of potential facilitators and/or obstacles, thus hinting to pressure points that could be addressed early on in order to ensure a seamless implementation. Similarly, the evaluation phase should rely on the entire set of data points activated during the previous implementation stage, which could lead to mid-course corrections, whenever needed. Finally, a summative evaluation at the conclusion of the leadership development program would identify things that went as planned as well as areas of improvement for any future programmatic iterations. Differently put, these four planning phases, representing a meso level, work in tandem with the components in the micro-level planning block to the left in Figure 1.

The macro-level planning in Figure 2 completes the sequence of activities intended to produce a leadership development program that is user-focused, flexible, and effective. The full range of events at this level relates to short-, medium-, and long-term considerations given to the balance between the actualized program outcomes, the proof of impact of leadership skills demonstrated by program participants, and the horizon of expectations these participants came with into the program. Here is how, in the end, the three levels in Figure 2 could interact with each other, thus highlighting how each component in the micro-level planning phase could be tied to any and all of the other components in the other two planning levels:

Figure 2. Interplay among the three levels of planning for the cyclical leadership development program design matrix

1. Create a baseline of data in about leadership development needs at the onset of the program; contrast these preliminary needs with participant feedback upon completion of the program in order to zoom in on potential discrepancies between program intent and execution/implementation; identify short-/medium-, and long-term professional development needs that follow the completion of the program so that participants can use their enhanced knowledge, skills, and dispositions to strengthen the institutional leadership profile.
2. Keep track of the initial program structure, use formative and summative evaluative data to determine fidelity of implementation, refine its structure to move from short-term to medium-term planning. Once a leadership development program has been offered a sufficient number of times, analyze facilitating factors as well as obstacles that should be addressed as the program becomes institutionalized, thus moving into long-term planning.
3. Similar to the earlier point, identify sustainability parameters for the set of resources needed (and proven) to launch a leadership development program successfully. Starting off with a range of internal resources that could be augmented by relying on external sources of support, use proof of program impact, coupled with an outline of continuing as well as emerging leadership challenges to carve out resource requirements for medium- and long-term implementation of such programs.
4. Part of the larger effort to ensure enhanced participant engagement, the range of instructional strategies used during the initial implementation of the leadership development program should be reviewed regularly to determine effectiveness. Moreover, the andragogic principles intended to inform the delivery of instruction, irrespective of modality, should be tailored to program participants so that the record of their leadership development can be used to create learning opportunities aligned with actual needs in a dynamic manner. As mentioned later in this chapter, generating different program

tracks for professionals with various leadership experience, interests, and aspirations would ensure maximum program effectiveness.

5. As far as the rewards system is concerned, cultivating the intrinsic motivation program participants may demonstrate initially would rely on a range of markers of success, both actual and symbolic, that would validate their willingness to grow professionally, while assuming greater risks as informed, well-rounded leaders upon completion of as many iterations of the program as necessary. Paying attention to contextual indicators that could be tied to both recognition and professional impact these leaders can show through their participation in the program would motivate and inspire at the same time.

Applications of Andragogic Principles to Leadership Development

Adapting from the work of Davis McCauley, Hammer, and Hinojosa (2017) on providing some structured examples of andragogy-centered instructional tools that could be used in a leadership development program, Table 1 outlines similar recommendations that consider both considerations from a developmental approach to instruction as well as a longitudinal perspective on such a program. Details related to each suggested instructional strategy, inclusive of assessment tools, follow the listing below, with a particular emphasis on employing a professional portfolio as a medium of both progress and best work included by program participants. It should be noted that the premise of the roll-out plan below relies on the assumption that a cohort-based leadership development program could be offered a number of times, each one attempting to improve it based on lessons learned, participant feedback, greater scope of work, etc. Any new iteration of the program could include a track for professionals new to leadership as well as advanced tracks for individuals with leadership experience and training. For the former group, it would be useful to include in the program design opportunities for these emerging leaders to interact, both formally/structured and informally/unstructured, with more experienced leaders in the program (as well as with notable leaders in their respective fields, involved in the program as guest speakers and/or mentors/advocates/critical partners). These emerging leaders would engage with the program curriculum via a range of activities, referenced in Table 1, and some of these instructional strategies may be new to them.

For the more experienced leaders in the program, it is recommended that they are offered scaffolding for their learning by way of tapping into some of the activities in which they were involved the first time they were part of the leadership development program. That would allow them to keep track of their performance over time, while noting personal growth, coupled with opportunities for analysis of what helped them learn versus what may have impacted negatively their original leadership development plan, all of which could be put to use in new program team settings, referenced in the descriptions in Table 1.

The description of how the aforementioned applications of andragogic principles to leadership development starts with self-concept, which is a complex concept centered on one's leadership skills' influence on followers' professional behavior situated in a common mission (Lord & Brown, 2004). Occupational self-efficacy is tied to a person's awareness of their ability to complete work-related tasks (Schyns & Sczesny, 2010). Self-efficacy appears to be supportive of learning orientation (Bell and Kozlowski, 2002, as cited in Schyns and Sczesny, 2010). Moreover, self-efficacy is perceived as a critical component of leadership development (Popper and Mayseless, 2007, as cited in Schyns and Sczesny, 2010). Research findings indicate that there is a positive relationship between self-concept, seen as self-perceived leadership attributes, and occupational self-efficacy (Schyns & Sczesny, 2010), based on which to sequence instructional opportunities in a leadership development program aimed at grounding participants in their own prior life and work experiences to align their emerging or refined leadership skills with particular job requirements. For an initial leadership development program implementation, a pre-assessment could be used to determine participants' awareness of their knowledge bases, skill sets, and associated dispositions that activate a leadership stance in a given workplace situation. Used to generate a baseline for leadership skills, potential or actual, this pre-assessment could support the development of some of the reflective essays that could expand on topics of relevance to program participants.

Table 1. Applications of andragogic principles to leadership development

Andragogic Principles	Instructional Strategies for Short-Term Implementation	Instructional Strategies for Medium-Term Implementation	Instructional Strategies for Long-Term Implementation
1. Self-concept	- Reflective essays (Davis McCauley et al., 2017) - Pre-assessment (to establish a baseline)	- Journaling - Follow-up reflective essays tied to critical leadership issues - Self-assessment of awareness of contextualized leadership skills	- Targeted journaling - Follow-up reflective essays tied to critical leadership issues - Summative self-evaluation of contextualized leadership skills (with a subset focused on any relevant context changes over time) - Case studies on select critical leadership issues
2. Experience	- Leadership portfolio and sharing of relevant experiences in class (Davis McCauley et al., 2017) - Journaling focused on a preliminary investigation of the interplay among knowledge, skills, and dispositions tied to leadership development - Critical incidents requiring the application of contextualized leadership skills	- Expanded leadership portfolio - Targeted journaling - Focused critical incidents requiring refining of contextualized leadership skills - Advocate – mentor - critical partner pairing	- Expanded leadership portfolio - Targeted journaling - Advocate – mentor – critical partner pairing focused on emerging themes in previous critical incidents requiring the application of contextualized leadership skills
3. Readiness to learn	- Survival scenarios, role-playing, and teamwork (Davis McCauley et al., 2017) - Journaling - Teamwork contracts	- Focused journaling on findings from analyses of teamwork contracts: achievements and future improvements	- Focused journaling on findings from analyses of teamwork contracts: achievements and future improvements - Prompts for leadership case studies grounded in select journal entries related to analyses of teamwork contracts
4. Orientation to learning	- Grounding class discussions in real-life situations (Davis McCauley et al., 2017) - Leadership-related concept maps - Journaling	- Refined leadership-related concept maps - Focused journaling on real-life situations that require actual or hypothetical leadership skills	- Leadership circles dealing with current leadership challenges and more general, hypothetical circumstances
5. Motivation	- Case studies and goal-meeting assignment (Davis McCauley, 2017)	- Grit journaling	- Refined grit journaling
6. Need to know	- Leader interview (Davis McCauley et al, 2017) - Leadership development plan	- Refined leadership development plan	- Further refined leadership development plan - Leadership circles dealing with an analysis of facilitators and hindrances from participants' leadership development plans

In light of the proposed developmental approach to the creation of such a leadership training program, should it be offered more than once, the instructional strategies for medium-term implementation build on the pre-assessment and reflective essays in the initial program implementation. To that end, journal entries prompted by assigned readings could be coupled with reflective essays designed to expand on critical leadership issues referenced in previous essays. Depending on the duration of the leadership development program and the scope of its involvement of established leadership peers, the self-concept principle could be addressed directly by way of a self-assessment of awareness of participants' emerging or refining leadership skills in their own particular workplace environments. For long-term implementation considerations, it is advisable to escalate the focus of the instructional strategies used so that program participants can monitor their own leadership development by means of individual and team activities and assignments. Consequently, journaling could be targeted on specific topics of interest to program participants. In a similar vein, reflective essays could blend commentary on current leadership challenges pertinent to the context of participants and recommended solutions to leadership issues in scenario-based strategic planning. Any previous self-evaluation of leadership skills should support the selection of workplace environment parameters that may have changed over time (or the entire work setting could be new), in which case the applications of leadership skills should account for a pertinent transfer of leadership skills. As a way to disseminate findings or lessons learned from the implementation of this andragogic principle in leadership development programs over time, participants could be asked to generate case study prompts that relate to the most relevant aspect of their own leadership journey mediated by the program series.

Experience, the second andragogic principle in the leadership development program, in-class sharing of relevant experiences could set the foundation for a leadership portfolio that elevates various circumstances in the lives of program participants as timely and adequate responses to leadership challenges. Other instructional strategies during the initial implementation of the program include journaling emphasizing the interplay among knowledge bases, skills sets, and associated dispositions determined by program participants to have played a role in their leadership development. This interplay, in all of its possible configurations, could support the identification of critical incidents requiring the application of emerging leadership skills on the part of program participants. A curated selection of such critical incidents could be included in the leadership portfolio at this stage of the program implementation. Upon subsequent program iterations, once participants have a clearer sense of their leadership attributes and how they rely on their knowledge, skills, and dispositions, journaling could delve deeper than during the initial program implementation into particular skills and their supporting knowledge bases and associated dispositions sustained participants' role as impactful, transformative leaders. Similarly, the critical incidents used before could be refined to focus on specific types of circumstances that elicited particular leadership responses in a demonstrably effective manner. As mentioned earlier, relevant critical incidents and journal entries could be included in the leadership portfolio, based on which advocate – mentor – critical partner pairings could be created. In this light, well-seasoned leaders could provide guidance to program participants by negotiating what it may mean to become an effective leader, how to get at that stage in one's professional life, and how to navigate the nuances of organizational structure and change management. Once the program becomes institutionalized, the journal entries selected for inclusion in the leadership portfolio could be synchronized with the topics discussed in the advocate – mentor – critical partner pairs, thus reflecting the potential need for program participants in their roles as well-established leaders to access a support network and content repository.

Readiness to learn implies that adult learners rely on prior life and work experience to adapt to various stimuli in new learning contexts, in which they find relevance to their own growth and development both as individuals and professionals. During the initial implementation of the leadership development program, setting up teams could help ground various instructional activities into the collective experience/expertise of team members, thus elevating the opportunities to learn through role-playing and survival scenario analyses and discussions. In this light, involving team members into negotiations leading up to team contracts could document the areas of interest adult learners have in balancing leading and following as a team member, along with levels of expected performance on the team and criteria to evaluate it. Journaling, as an instructional strategy, could also reference team members' self-monitoring within the parameters of team activities and how they rise up the collective expectations for team performance. An example of a team contract is included in Appendix A, adapted from work by Johnson, Johnson, and Holubec (1998). For subsequent iterations of the leadership development program, such team contracts could be used as a foundation for focused journaling related to findings from analyses of team contracts to determine achievements, their possible facilitators and obstacles, as well as future improvements. Based on the latter point, new team contracts could be negotiated by team members as a way to connect their awareness of how they perform on a team, both as a leader and as a follower, in order to maximize team performance under the new learning circumstances. In case the leadership development program becomes well established, following a number of implementations, the practice of refining the focus of journal entries as a way to support self-monitoring and self-awareness related to leadership development could be continued. Based on the most relevant findings from these journal entries as well as analyses of team contracts, program participants could be involved in the development of case study prompts that could be used with subsequent program cohorts.

Relevance, recency, and applicability to real-life situations are part of adult learners' orientation to learning. Therefore, grounding class discussions in authentic learning contexts could be coupled with the design of outlines or maps for leadership concepts of emerging importance to program participants. Journaling could be used to connect outstanding components of class discussions to individual leadership concept maps in an attempt to support the gradual internalization of information intended to lead to developing knowledge bases, skill sets, and associated dispositions. As far as medium-term program implementation instructional strategies are concerned, leadership concept maps could be refined by more experienced program participants. Similarly, journal entries could be elevated to address particular real-life situations that required actual leadership skill as a way to anticipate how to handle future, more complex leadership challenges based on lessons learned. For institutionalized programs of this kind, participating adult learners could be involved in the formation of leadership circles that include program graduates from previous cohorts as well as well-established leaders who could serve as advocates/mentors/critical partners. These leadership circles could augment the program curriculum by engaging participants to reflect on their own professional learning and development by connecting with analyses/commentary/feedback from other leadership circle members.

The more adult learners gain knowledge and skills, the more positive their dispositions become toward their professional field and associated responsibilities, inclusive of leadership roles. In this light, learning relies on a gradual shift from extrinsic to intrinsic motivation. A combination of relevant case studies and goal-setting assignments could prompt program participants to keep track of their responses to various stimuli in a range of learning opportunities designed to support leadership development. All of the previous andragogic principles are interrelated and they collectively impact how adult learners co-construct knowledge and refine their skill sets in order to be(come) effective leaders. Under these circumstances, of particular interest is how the program curriculum and instructional strategies help participants work toward achieving their professional and leadership goals in a flexible yet sustained manner. To that end, the more challenging learning and professional opportunities area, the greater the personal gain that reflects the overall professional growth. "Grit is about holding the same top-level goal for a very long time…the more unified, aligned, and coordinated our goal hierarchies are, the better" (Duckworth, 2016, 38-39). Therefore, providing program participants with prompts to generate grit journal entries could document their motivation development journey, which could be included in medium- and long-term program implementation cycles.

Adult learners' need to know can translate into a series of interviews with well-known leaders in their respective fields, some of which could already be involved in the leadership development program by serving as advocate – mentor – critical partner. A preliminary individual leadership plan could be drawn prior to any interview, which could be used to generate probing questions for interviewee. Their answers/feedback could then support the refinement of the individual leadership plan. Should the program be reiterated, this plan could be elaborated upon as a signature assignment. For long-term program implementation purposes, analyses of major themes in individual leadership plans could be used in leadership circles, described earlier, as a way to focus on facilitators and hindrances noted by circle members.

Using Instructional Technology in the Design of Leadership Development Programs

The previous details on how andragogic principles could be used to inform the design of leadership development programs could also be tied to the implementation of instructional technology as a way to enhance participant engagement. In this light, there are various technological tools that could be used to support organically the development of future leaders in higher education. Leaders who fail to anticipate and adapt to the multitude of challenges facing the organizations they manage will run the risk of placing organizational viability and sustainability in jeopardy (Sabga, 2017). Consequently, embedding instructional technology into the fabric of such leadership development programs could serve as models for future refinement and/or implementation across a multitude of work settings.

It is well documented that online instruction, in general, has become a lot more prevalent than ever before, more so in the aftermath of the COVID-19 pandemic. While before the pandemic, there was research indicating that online instruction could be as effective as traditional, face-to-face instruction based on concerted efforts to enhance active learner engagement, while cultivating a student-focused teaching presence (Dixson, 2010; Jones, 2011). Reasons associated with increased learner engagement in virtual settings relate to the instructor's more facilitative role, students being able to control the pace of their learning more than in face-to-face environments, getting more frequent and pertinent feedback from instructors, a user-friendly interface provided by the Learning Management System (LMS), and greater emphasis on multi-modality to demonstrate mastery of content (Swan, 2002). The implication here is that the design of online or hybrid learning environments should take into account the various ways in which learners have opportunities to engage effectively with peers, instructors, curricula (Davidson-Shivers & Rasmussen, 2006; McNeil, Robin, & Miller, 2000; Nandi, Hamilton, & Harland, 2012) as well as with the LMS itself (Davidson-Shivers, 2009; Wanstreet, 2006). The impact of the pandemic on virtual learning has been documented to focus on the availability of resources, adequate preparation of instructors, access to instructional technology and specialized support services, time and task management (Chawdhry, Paullet, Baugh, & Nakama, 2022), existing strategies to pivot to online instruction, and pedagogical guidance for instructors who may be new to teaching in virtual settings (Kaqinari, Makarova, Audran, Doring, & Gobel, 2021).

In this light, while not emphasizing any particular instructional technology solution, the following considerations are designed to structure technology-enhanced leadership development programs with their participant engagement and overall sustainability in mind:

1. **LMS:** Ensure that program participants are familiar with the platform, have the necessary support services in place, and are able to interact effectively with peers, instructors, and the program content. In case it is determined that a hybrid model is better suited to the needs of participants, then course design and pedagogical characteristics should be factored into the right mix of face-to-face and virtual sessions, with clear connections between the two types, and a strong rationale for the instructional activities included in each of them.
2. **Instructor:** Determine the technology-driven abilities of instructors to engage program participants in a wide range of activities both in entirely online and hybrid modalities of delivery of instruction, resulting in a tailored series of professional development opportunities prior to the launch of the leadership development program. Equally important, provide ongoing support for instructors once the program is initiated, allowing for a free flow of information/feedback aimed at maximizing the instructional impact on participants.
3. **Learners:** Similar to the earlier point about instructors' prior preparation to teach online or in hybrid fashion, program participants' readiness should be met with orientation sessions at the onset of the program, coupled with tutorials and technical support throughout the program.
4. **Instructional Strategies:** Going back to the strategies referenced in Table 1, designed to infuse andragogic principles into the learning structure of a leadership development program, identify which of them lend themselves to delivery in virtual settings so that the design of those instructional episodes ensures participant engagement. For example, the leadership circles referenced under "orientation for learning" in Table 1 could be done in person, entirely online, or in hybrid fashion. Weighing the pros and cons of each option, with an eye toward maintaining participant interest and interaction, could lead to various choices. Thinking holistically of the entire program, such a planning process could focus on the sequence of instructional strategies so that the learning environment adds value to participants, as opposed to a convenient, expedient solution that may not take into account participants' levels of readiness and expectations for learning.
5. **Assessment:** in order to enhance the educational experiences of program participants, special emphasis should be placed on the use of a wide range of assessment strategies, inclusive of formative assessment so that social presence, cognitive presence, and teaching presence seen as critical components in the creation of a community of inquiry in the virtual environment (Garrison, Anderson, & Archer, 1999). In this light, assessment protocols could enhance all three types of presence, while allowing for the instructor to engage in ongoing or dynamic course design based on participant engagement and feedback (Boboc, 2015). As far as the leadership portfolio mentioned under "Experience" in Table 1, the following suggestions are intended to structure this complex assessment strategy in a way that captures the essence of program participants' learning by increasing portability/transfer potential, relevance, and the comprehensive representation of leadership skill gain mediated by the program curriculum and supporting instructional activities:
    a. Since the portfolio should be intended as an opportunity to showcase the best artifacts selected by program participants to connect their life and work experiences to their emerging/refin-

ing leadership skills, allow for both formal and informal assessments to inform the selection process learners will make. That will enhance self-assessment, metacognition, and a stronger collection of artifacts representative for the leaders these participants want to become. This point also underscores the need for a developmental approach to the assemblage of the portfolio, thus prompting program participants to determine which assessments from the various iterations of the program would best demonstrate their professional growth and development. In turn, this approach would reinforce the intrinsic motivation and the orientation to learning that animates participants as they negotiate meaning and relevance out of the program curriculum.

b. Generate criteria to evaluate each artifact as well as the entire leadership portfolio by involving program participants in the design process. Concurrently, negotiate relevant examples of performance levels that these evaluative criteria would be connected to. The criteria should also reflect the developmental nature of the portfolio, allowing for different ways for participants to document leadership skill gains based on clear evidence and reflective practice.

c. In terms of implementing the criteria for the leadership portfolio, consider using peer evaluators from within a given cohort to discuss artifact selection, relevance, and framing of how it demonstrates professional growth and development. At the same time, set up roundtables by inviting well-established leaders, some of which could already serve as advocate – mentor – critical partner in the program, so that they could act as expert panels to evaluate the portfolio., while providing networking opportunities.

CONCLUSION AND FUTURE RESEARCH

Given the fact that we live in a world increasingly marked by volatility, uncertainty, complexity, and ambiguity (VUCA), systemic and systematic responses to any emerging, sometimes compounding challenges requires concerted efforts aimed at protecting the public good by way of a spectrum of community-focused solutions (Bennett & Lemoine, 2014). Socio-economic challenges, climate change, technological disruption, diminishing resources, and environmental challenges compound the issues faced by societies at large. Within this less-than-encouraging picture, leadership plays a critical role in mapping out a course that would tackle emerging problems by analyzing dynamic trends, outlining scenario-based planning options, and being increasingly aware of the thin line between potential threats and opportunities. Under these circumstances, leadership development programs should be designed so that strategic planning could be focused, inclusive, authentic, grounded in the reality on the ground, and flexible (Mack, 2015). Furthermore, strategic foresight implies defining issues, identifying key themes based on which to develop scenarios intended to lead to solution implementation, out of which lessons learned could inform future iterations of the process. In this light, the future of leadership balances the individual and the collective by humanizing the workplace (Suderman & Foster, 2015). Moreover, leadership development programs should focus conversations on the definition and applications of leadership on interconnectedness and securing strategic capacity to deal with change (and manage it effectively), while engaging communities/societies as dynamic systems (Satterwhite, McIntyre Miller, & Sheridan, 2015).

The design of cyclical leadership development programs proposed in this chapter is intended to go back and forth between various stages of planning – short-, medium-, and long-term – while integrating andragogic principles. Therefore, by being immersed in significant learning opportunities, program participants could use prior life and work experiences as a foundation for the integration of knowledge, skills, and dispositions. Concurrently, these leaders will become increasingly aware of how they learn and select solutions to real-life problems that demonstrate concern and caring for communities, thus allowing them to focus on their own human dimension as adult learners as they pursue future leadership development options (Fink, 2013). By blending the main characteristics of the workplace of the future with flexible approaches to professional development, the new stage in the evolution of leadership would be expected to demonstrate integration of ideas and viewpoints, a commitment to lifelong learning, adaptive solutions to problems, ability to identify appropriate technology-based strategies to analyze data and engage in scenario-based planning, as well as awareness of the leadership journey and the various twists and turns that may feature (Roux, 2020). Furthermore, leaders who are concerned with social justice create opportunities for community members to have access to civic engagement to deal with systemic issues that prevent them from realizing their potential (Noble, 2015).

In summation, designing leadership development programs along andragogic principles could lead to a capacity-building approach to ensuring that colleges and universities are well positioned to cope with the wide range of economic, social, economic, and ecological issues that humanity faces (Jeanes, 2021). Based on recommendations from Breen and Gleason (2022), a commitment to intentionality in the design of such programs should lead to clarity of focus and scope of work for participants, who, in turn, would be(come) deliberate in their learning, supported by the integration of andragogic principles. Concurrently, context is critical in grounding participants' prior life and work experiences they bring to the table, while equipping them with flexibility in transferring leadership skills to a wide range of work environments. Identity building as a process, both for leadership educators/trainers and for leaders themselves should become part of the reflective practice in which participants in a leadership development program engage as a way to formalize gradually their own professional growth by activating various skill sets required to lead effectively.

Given the theoretical nature of the proposed framework in this chapter, future research should focus on implementation options that could lead to refinement and scaling up opportunities, thus moving the framework from concept to contextualized execution. In its current form, the main limitation stems from a perceived one-size-fits-all approach to designing leadership development programs. However, the intent behind the generic formulation of program design derives from an attempt to present a structure that is flexible enough to allow for the development of clear connections among the various facets of leadership – as an individual, a team member, an organizational executive, a professional leader, and a well-established professional in a complex and convoluted work (Hieker & Pringle, 2021). Furthermore, future lines of inquiry could focus on the impact of emerging leadership skills mediated by such programs, while acknowledging a wide range of implementation differences. A particular strand of interest in evaluating the viability and feasibility of this approach to leadership development programs could focus on participants' ability to manage change effectively, which would require a network built on various stages in the change process, coupled with clear change principles in order to reach the optimal balance between operation and innovation (Kotter, 2012). Finally, a more pragmatic question for future work is related to how a program featuring the design elements in this chapter could be institutionalized, while allowing for the dissemination of findings to inform the evolution of the blend between andragogy and leadership development. In the end, that would emphasize leaders and their teams using their awareness of internal and external parameters (facilitators and hindrances) to design strategies supporting a common vision to achieve sustainable success (Berard, 2013).

REFERENCES

Archer, D. E., & Zhang, Y. (2021). The critical role of global and culturally responsive leadership for higher education in the 21st century. *Journal of Interdisciplinary Studies in Education, 10*(1), i–v.

Bennet, N., & Lemoine, G. J. (2014). What VUCA really means for you. *Harvard Business Review*. https://hbr.org/2014/01/what-vuca-really-means-for-you

Berard, J. (2013). *Accelerating leadership development: Practical solutions for building your organization's potential*. Jossey-Bass.

Berdahl, R. O., Altbach, P. G., & Gumport, P. J. (2011). Introduction: The contexts of American higher education. In P. G. Altbach, P. J. Gumport, & R. O. Berdahl (Eds.), *American higher education in the twenty-first century: Social, political, and economic challenges*. Johns Hopkins University Press.

Bloomquist, C. D., & Georges, L. (2022). Interdisciplinary leadership: A leadership development model for scholar-practitioners. *Journal of Leadership Education, 21*(4), 58–75. doi:10.12806/V21/I4/A4

Boboc, M. (2015). Online course dynamic design informed by student response and formative assessment. In S. Koc, X. Liu, & P. Wachira (Eds.), *Assessment in online and blended learning environments* (pp. 103–124). Information Age Publishing.

Branson, C. M., Marra, M., Franken, M., & Penney, D. (2018). *Leadership in higher education from a transrelational perspective*. Bloomsbury Academic. doi:10.5040/9781350042414

Breen, J. M., & Gleason, M. C. (2022). A call for the leadership education field. *New Directions for Student Leadership, 2022*(176), 111–122. doi:10.1002/yd.20535 PMID:36565148

Challen, C. (2020). Rethinking higher education policy and leadership in the 21st century: Enhancing strategies for global citizenship and justice. *Journal of Higher Education Policy and Leadership Studies, 1*(1), 77–81. doi:10.29252/johepal.1.1.77

Chawdhry, A. A., Paullet, K., Baugh, J., & Nakama, D. (2022). How the COVID-19 shutdown impacted grades at the collegiate level? *Information Systems Education Journal, 20*(5), 34–41.

Cooper, C. D., Scandura, T. A., & Schriesheim, C. A. (2005). Looking forward but learning from our past: Potential challenges to developing authentic leadership theory and authentic leaders. *The Leadership Quarterly, 16*(3), 475–493. doi:10.1016/j.leaqua.2005.03.008

Davidson-Shivers, G. V. (2009). Frequency and types of instructor interactions in online instruction. *Journal of Interactive Online Learning, 8*(1), 23–40.

Davidson-Shivers, G. V., & Rasmussen, K. L. (2006). *Web-based learning: Design, implementation, & evaluation*. Merrill Prentice Hall.

Davis McCauley, K., Hammer, E., & Hinojosa, A. S. (2017). An andragogical approach to teaching leadership. *Management Teaching Review, 2*(4), 312–324. doi:10.1177/2379298117736885

Dixson, M. D. (2010). Creating effective student engagement in online courses: What do students find engaging? *The Journal of Scholarship of Teaching and Learning, 10*(2), 1–13.

Dopson, S., Ferlie, E., McGivern, G., Fischer, M. D., Mitra, M., Ledger, J., & Behrens, S. (2019). Leadership development in higher education: A literature review and implications for programme redesign. *Higher Education Quarterly*, *73*(2), 218–234. doi:10.1111/hequ.12194

Duckworth, A. (2016). *Grit: The power of passion and perseverance*. Scribner.

Dweck, C. (2016). *Mindset: The new psychology of success*. Penguin Random House LLC.

Fink, L. D. (2013). *Creating significant learning experiences: An integrated approach to designing college courses*. Jossey-Bass.

Garrison, D. R., Anderson, T., & Archer, W. (1999). Critical inquiry in a text-based environment: Computer conferencing in higher education. *The Internet and Higher Education*, *2*(2-3), 87–105. doi:10.1016/S1096-7516(00)00016-6

Hieker, C., & Pringle, J. (2021). *The future of leadership development: Disruption and the impact of megatrends*. Palgrave MacMillan. doi:10.1007/978-3-030-53544-5

Jeanes, E. (2021). A meeting of mind(sets): Integrating the pedagogy and andragogy of mindsets for leadership development. *Thinking Skills and Creativity*, *39*, 100758. Advance online publication. doi:10.1016/j.tsc.2020.100758

Johnson, D. W., Johnson, R. T., & Holubec, E. (1998). *Cooperation in the classroom*. Interaction Book Company.

Jones, I. M. (2011). Can you see me now? Defining teaching presence in the online classroom through building a learning community. *Journal of Legal Studies Education*, *28*(1), 67–116. doi:10.1111/j.1744-1722.2010.01085.x

Jones, S., Lefoe, G., Harvey, M., & Ryland, K. (2012). Distributed leadership: A collaborative framework for academics, executives and professionals in higher education. *Journal of Higher Education Policy and Management*, *34*(1), 67–78. doi:10.1080/1360080X.2012.642334

Kaqinari, T., Makarova, E., Audran, J., Doring, A. K., Gobel, K., & Kern, D. (2021). The switch to online teaching during the first COVID-19 lockdown: A comparative study at four European universities. *Journal of University Teaching & Learning Practice*, *18*(5), 10–21. doi:10.53761/1.18.5.10

Keohane, N. O. (2013). Higher education in the 21st century: Innovation, adaptation, preservation. *PS, Political Science & Politics*, *46*(1), 102–105. doi:10.1017/S1049096512001734

Knowles, M. S. (1970). *The modern practice of adult education: Andragogy versus pedagogy*. Association Press.

Knowles, M. S., Holton, E. F. III, & Swanson, R. A. (2011). *The adult learner: The definitive classic in adult education and human resource development* (7th ed.). Elsevier.

Kotter, J. P. (2012, November). Accelerate! *Harvard Business Review*. https://hbr.org/2012/11/accelerate

London, M. (2002). *Leadership development: Paths to self-insight and professional growth*. Lawrence Erlbaum Associates, Inc.

Lord, R. G., & Brown, D. J. (2004). *Leadership processes and follower self-identity*. Lawrence Erlbaum Associates, Inc.

Mack, T. C. (2015). Leadership in the future. In M. Sowcik, A. C. Andenoro, M. McNutt, & S. E. Murphy (Eds.), *Leadership 2050: Critical challenges, key contexts, and emerging trends* (pp. 9–22). Emerald Group Publishing Limited.

Makadok, R., Burton, R., & Barney, J. (2018). A practical guide for making theory contributions in strategic management. *Strategic Management Journal, 39*(6), 1530–1545. doi:10.1002mj.2789

McNeil, S. G., Robin, B. R., & Miller, R. M. (2000). Facilitating interaction, communication and collaboration in online courses. *Computers & Geosciences, 26*(6), 699–708. doi:10.1016/S0098-3004(99)00106-5

Miles, M. B., & Huberman, A. M. (1994). *Qualitative data analysis: An expanded sourcebook* (2nd ed.). Sage Publications, Inc.

Monaghan, C. H., & Boboc, M. (2017). (Re) Defining Leadership in Higher Education in the US. In V. Wang (Ed.), *Encyclopedia of Strategic Leadership and Management* (pp. 567–579). IGI Global. doi:10.4018/978-1-5225-1049-9.ch040

Nandi, D., Hamilton, M., & Harland, J. (2012). Evaluating the quality of interaction in asynchronous discussion forums in fully online courses. *Distance Education, 33*(1), 5–30. doi:10.1080/01587919.2012.667957

Noble, D. J. (2015). Leading for tomorrow in a world yearning for social justice. In M. Sowcik, A. C. Andenoro, M. McNutt, & S. E. Murphy (Eds.), *Leadership 2050: Critical challenges, key contexts, and emerging trends* (pp. 43–58). Emerald Group Publishing Limited.

Recigno, T. E., & Kramer, P. (2022). The development of leaders: Suggested modules for education. *The Open Journal of Occupational Therapy, 10*(4), 1–8. doi:10.15453/2168-6408.1941

Reyes, D. L., Dinh, J., Lacerenza, C. N., Marlow, S. L., Joseph, D. L., & Salas, E. (2019). The state of higher education leadership development program evaluation: A meta-analysis, critical review, and recommendations. *The Leadership Quarterly, 30*(5), 101311. Advance online publication. doi:10.1016/j.leaqua.2019.101311

Roux, M. (2020). Leadership 4.0. In J. Reams (Ed.), *Maturing leadership: How adult development impacts leadership* (pp. 7–35). Emerald Group Publishing Limited. doi:10.1108/978-1-78973-401-020201003

Sabga, N. I. (2017). *Leaders who learn: The intersection of behavioral science, adult learning and leadership* (Publication No. 10636182) [Doctoral dissertation, University of South Florida]. ProQuest Dissertations Publishing.

Satterwhite, R., McIntyre Miller, W., & Sheridan, K. (2015). Leadership for sustainability and peace: Responding to the wicked challenges of the future. In M. Sowcik, A. C. Andenoro, M. McNutt, & S. E. Murphy (Eds.), *Leadership 2050: Critical challenges, key contexts, and emerging trends* (pp. 59–74). Emerald Group Publishing Limited.

Schyns, B., & Sczesny, S. (2010). Leadership attributes valence in self-concept and occupational self-efficacy. *Career Development International, 15*(1), 78–92. doi:10.1108/13620431011020907

Sowcik, M., Andenoro, A. C., McNutt, M., & Murphy, S. E. (2015). *Critical challenges, key contexts, and emerging trends*. Emerald Group Publishing Limited.

St. Clair, R., & Kapplinger, B. (2021). Alley or autobahn? Assessing 50 years of the Andragogical Project. *Adult Education Quarterly*, *71*(3), 272–289. doi:10.1177/07417136211027879

Suderman, J. L., & Foster, P. A. (2015). Envisioning leadership in 2050: Four future scenarios. In M. Sowcik, A. C. Andenoro, M. McNutt, & S. E. Murphy (Eds.), *Leadership 2050: Critical challenges, key contexts, and emerging trends* (pp. 23–38). Emerald Group Publishing Limited.

Swan, K. (2002). Building learning communities in online courses: The importance of interaction. *Education, Communication & Information, 2*(1), 23-49.

Taylor, J. S., & Machado-Taylor, M. de L. (2010). Leading strategic change in higher education: The need for a paradigm shift toward visionary leadership. In T. Claes & D. S. Preston (Eds.), *Frontiers in Higher Education* (pp. 167–194). Brill. doi:10.1163/9789042030978_011

Tipton, W., & Wideman, S. (2021). Toward an invitational andragogy: Articulating a teaching philosophy for the andragogic classroom. *Journal of Communication Pedagogy*, *5*, 156–163. doi:10.31446/JCP.2021.2.16

Wanstreet, C. E. (2006). Interaction in online environments: A review of the literature. *Quarterly Review of Distance Education*, *7*(4), 399–411.

Zusman, A. (2005). Challenges facing higher education in the twenty-first century. *American higher education in the twenty-first century: Social, political, and economic challenges, 2*, 115-160.

KEY TERMS AND DEFINITIONS

Andragogy: Set of guiding principles and associated practices focused on adult learning.

Instructional Strategies: Techniques used by teachers/instructors to engage learners in activities designed to develop knowledge bases, skills sets, and associated dispositions.

Instructional Technology: Computer-based tools used to facilitate/enhance learning.

Leadership: Set of behaviors demonstrated by individuals at the top ranks of an organization/institution aimed at providing a sense of direction, informed by organizational/institutional vision and/or mission, managing daily operations, and ensuring ethical principles and integrity.

Leadership Development: Process by which individuals are engaged in learning opportunities in order to develop their ability to provide a sense of direction, manage daily operations, and demonstrate ethical principles and integrity.

APPENDIX

Team Contract Sample

Team Contract

Individual Goals **(general and leadership-related)**	*Ways in which I can contribute to my team's tasks*	*Team Goals* **(general and leadership-related)**
1.		
2.		
3.		
4.		
Plan for achieving my individual goals and contributing to my team's work:		
The timeline for achieving my individual goals is: Outline of team tasks: Team tasks priorities: Beginning date for team tasks: First achievement: Second achievement: Third achievement (if necessary): Final date for team tasks:		

Signatures of team members:

____________________ ____________________

____________________ ____________________

Chapter 5 Enabling Andragogical Leadership: Skillset, Toolset, Mindset, and Frame Conditions

Matthias Teine
FIEGE Logistik Stiftung & Co. KG, Germany

ABSTRACT

Andragogical leadership is a term yet to be coined. The evidence from four interviews with experts indicates that it might be a promising solution to grasp the fast and disruptive transformation processes companies are currently facing. Based on the research results, the author argues that andragogical leadership is a set of responsibilities that might at best be taken over by passionate experts in a shared leadership approach. The genuine implementation of andragogical leadership either requires or goes hand in hand with a transformation toward self-directedness, a participatory culture, and the democratization of an organization. Future research requires theoretical discussions and practical experiences contributed by experts from business studies, humanities, and psychology. The relevance of andragogical leadership and its potential to be a solution to urgent questions was proven within this chapter.

INTRODUCTION

Nowadays, companies across all industries are challenged by increasingly fast-changing business environments with digital transformation, shifts in demographics, and sustainability (resource scarcity/ climate change) being among its main drivers (Brown et al., 2018, pp. 3-6). For example, the *World Economic Forum* (WEF) (2020) predicts that by 2025 85 million jobs will be replaced by machines and that an estimated 40% of the workforce requires reskilling of up to six months. Regarding this, “94% of business leaders report that they expect employees to pick up new skills on the job” (WEF, 2020, p. 5). Kropp and McRae (2022) expect that also up to 65% of managerial tasks might be automated by 2025, too (para. 23). Considering both, the tasks of leaders will likely become more associate centered, e.g.,

DOI: 10.4018/978-1-6684-7832-5.ch005

"managing their perception of their career trajectories" (Kropp & McRae, 2022, para. 25) or acting as "training coaches" (Kovács-Ondrejkovic et al., 2019, para. 38).

The tremendous impact of such *long-term trends* is certain, "but the speed at which they unfold can be hard to predict." (Brown et al., 2018, p. 6) However, whilst these trends allow companies to plan and prepare accordingly (Brown et al., 2018, p. 38), the COVID-19 pandemic showed us the hard way that we are also living "in an environment that can shift from moment to moment" (Volini et al., 2021, p.4). A *short-term shock* happened. The WEF (2020) referred to this situation as a "double-disruption" scenario as the pandemic-induced *disruption* and economic contraction interweave e.g., with technology-driven transformation processes (p. 5). Hence, companies must be prepared for both *continuous* and *episodic* change, they should learn to *thrive on it*, and embrace it as a chance to develop sustainably (Volini et al., 2021, p. 4).

The required organizational development processes highly depend on the competence development of a company's workforce, they are mutually constitutive (Meyer & Haunschild, 2017, pp. 1, 7). Organizations must build the human capital that allows them to not only react to change but to shape it in their favor. For this purpose, companies should offer structured re-/ upskilling and give associates the freedom to take initiative e.g., to run passion projects that allow them to grow in a self-governed way (Volini et al., 2021, p. 20). Thereby, crucial information about an associate's "interests, passions, and capabilities that may otherwise remain hidden" (Volini et al., 2021, p. 20) can be revealed, too. This information could now be used to "redeploy workers against critical business priorities" (Volini et al., 2021, p. 20) as they occur. Moreover, workers will leave familiar paths as they follow their individual learning interests. In doing so, they nudge the organization to move, to get out of its comfort zone, to develop, and thereby to become more resilient to substantial change (Meyer & Haunschild, 2017, p. 7; Teine, 2022, p. 153). Having the freedom to be more self-governed in their development also requires the workers to continuously adapt in a way that they can even be thought of as entrepreneurs marketing their knowledge and competencies which *they* must reproduce goal-oriented, actively, and efficiently in a self-responsible way (Gonon, 2002, p. 327; Teine, 2022, p. 152). It is *them* ensuring that their capabilities are needed, bought, and used efficiently (Gonon, 2002, p. 327; Kovács-Ondrejkovic et al., 2019, para. 34). In this respect Kovács-Ondrejkovic et al. outline that "individuals who want to remain in demand in a changing job market must take responsibility for keeping their skills up to date and accept that job-related training is a career-long commitment." (2019, para. 5)[1] This shift in responsibility is even further emphasized by the WEF's Future of Jobs Report 2020: "The window of opportunity to reskill and upskill workers has become shorter in the newly constrained labor market. This applies to workers who are likely to stay in their roles as well as those who risk losing their roles due to rising recession-related unemployment and can no longer expect to retrain at work." (WEF, 2020, p. 6)

Anyways, companies need qualified labor and thus, it lies within their responsibility to make their workforce future-proof (WEF, 2020, p. 6; Kovács-Ondrejkovic et al., 2019, para. 5). They "must promote upskilling and reskilling to help people prepare and to avoid a two-tiered workforce in which some workers are ready for change and others are not." (Kovács-Ondrejkovic et al., 2019, para. 34) Therefore, companies should establish a leadership that promotes the formerly described processes, that nurtures a learning culture, and that does not shy away from harnessing digital technologies to e.g., carry out targeted upskilling programs or promote self-directed learning (Kovács-Ondrejkovic et al., 2019, para. 21, 37, 38). "Best-in-class companies encourage employees to adopt a continuous-growth mindset and to learn on the job, provide opportunities for people to take time off for learning, and have managers act as training coaches." (Kovács-Ondrejkovic et al., 2019, para. 38) Hence, the importance of leadership for successful change that is driven by individual and organizational learning processes becomes obvious. It is the leaders who must "strengthen others by increasing self-determination and developing competence" (Kouzes & Posner, 2012, p. 1), as employees cannot perform when they feel weak, dependent, or alienated (Kouzes & Posner, 2012, p. 1). For short: By enabling associates to learn and grow, leaders enable organizations to change and thus, to stay sustainably successful.

A leadership approach that like no other understands that learning associates are the actual enablers of organizational change and long-term success might be *Andragogical Leadership* – a term that is yet to be coined. For this reason, this chapter addresses the following research foci:

1. Providing a definition of *andragogical leadership* that integrates
    a. the concepts of *Leadership* and *Andragogy*, complemented by remarks on *self-direct learning* and *lifelong learning,*
    b. the *leader* development and *leadership* development perspective, and
    c. the dyad of *organizational development* and *individual competence development* (of leaders and associates).
2. A concise discussion on how andragogical leadership is different from established approaches like e.g., *transformational leadership* or *coaching leadership* that as well focus strongly on leaders acting as coaches or development partners of their associates.
3. An examination of the question of how companies can best support their leaders in becoming effective andragogical leaders. Here, attention will be paid to the potentials of modern technology with a twofold focus: How can e.g., digital learning technologies be harnessed for developing andragogical leaders (individual level), and what tools would help them become effective (organizational level)?

Therefore, this chapter is organized as follows: At first, critical terms and theories are outlined, including a concise literature review, to finally propose a definition of andragogical leadership. Then, the research design, which relies on expert interviews and qualitative content analysis, is described in detail. The insights from these interviews are presented consequently, followed by an overview of potential future research directions. This chapter closes with a conclusion that outlines the circumstances under which andragogical leadership might be an effective leadership approach.

TOWARD A DEFINITION OF ANDRAGOGICAL LEADERSHIP

Leadership and Leadership Development

According to Kouzes and Posner (2012), it is leaders who "take people and organizations to places they have never been before" (p. 1) as they "step up and take charge" (p. 1) when times, problems, technologies, and people change (p. 1). Effective leadership ensures that an organization reaches its objectives, that employees have everything needed to perform, and that they participate in organizational development processes (Danişman et al., 2015, p. 144). In contrast to management roles, which are bound to positions and formal authority, leadership can come without such and focuses on *enabling groups of people* to work together in *meaningful* ways (Day, 2000, p. 582). Leaders "inspire people to do things differently, to struggle against uncertain odds, and to persevere toward a misty image of a better future." (Kouzes & Posner, 2012, p. 1) It is no wonder then that leadership is discussed as a source of competitive advantage for several decades, particularly in today's innovation-driven and dynamic markets (Day, 2000, p. 581; Danişman et al., 2015, p. 144), and that organizations are investing into its development accordingly (Day, 2000, p. 581).

Leadership development expands the "collective capacity of organizational members to engage effectively in leadership roles and processes" (Day, 2000, p. 582). It fosters the *individual and collective adaptability* to complex and/ or unforeseen challenges and situations, to learn the way out of problems that arise e.g., "from the disintegration of traditional organizational structures and the associated loss of sensemaking" (Day, 2000, p. 582). This definition implies a notion of leadership covering two equally important perspectives: For the one, *leader development*, which covers the individual training of leaders to possess e.g., strong intrapersonal skills assuming that leadership will result. Here, the focus is on investing in human capital (Day, 2000, pp. 583f.). For the other, actual *leadership development* understands leadership as a social process that engages everyone and in which "each person is considered a leader" (Day, 2000, p. 583). Here, leadership is seen as an "emergent property of effective systems design" (Day, 2000, p. 583). The focus is thus on investing in social capital and using social systems (Day, 2000, pp. 583, 585f.).

Over time and in response to the respective prevailing challenges, different leadership styles arose. Amongst the most prominent styles might be *empowering*, *servant*, *shared*, and *transformational leadership*. Of these, according to a literature review by Danişman et al. (2015), only shared leadership has a strong effect on organizational performance ($r = 0.51$). Transformational leadership ($r = 0.31$) was found to have a medium effect, whilst servant ($r = 0.27$) and empowering leadership ($r = 0.04$; not statistically significant to $p = 0.05$) only have small to neglectable effects (pp. 147-149). These leadership styles rely on ideas that might also characterize andragogical leadership, e.g.,

- From *shared leadership*, mutual influence, leading one another to shared goals, and shared responsibility (Martin et al., 2018),
- From *transformational leadership*, being honestly concerned with the *followers'* self-actualization and fostering their innovativeness and creativeness (Bass, 1999 p. 11),
- From *servant leadership*, empowering and developing people, encouraging self-directed decision making and fostering autonomy, and the belief in each person's abilities and their capability to learn (van Dierendonck, 2011, pp. 1230, 1233), and
- From *empowering leadership*, also enabling the employees' self-determination and autonomy to foster creativity, performance, and innovation (Schröder et al., 2021).

In the following, the theoretical basis of andragogical leadership is outlined to get a better understanding of what might distinguish andragogical leadership from e.g., shared, transformational, servant, or empowering leadership.

Andragogy and Self-Directed Learning

Andragogical leadership builds upon the *educational theory* of Andragogy and links it to Deci & Ryan's Self-Determination Theory, a *motivation theory* (Wang & Bain, 2016; Deci & Ryan, 1993). Following, both theories and their core assumptions will be introduced briefly:

Andragogy provides a theoretical framework for adult teaching and learning (Wang & Bain, 2016, p. 252) that finds its roots in five core assumptions (Knowles et al., 2005, pp. 39f.):

- Adults are intrinsically motivated to learn, as learning satisfies some of their substantial needs and interests. Addressing these is thereby the starting point of all related activities.
- Adults learn life-centered, not subject-centered.
- Adults learn primarily by the analysis of experience, their richest source of learning.
- Adults want to learn in a *self-directed* way, and they want to step in a process of mutual inquiry with their 'teacher'.
- Adults are utterly individual in their learning preferences, which is why highly individual approaches are needed.

In andragogy, learners take on ownership of their own learning in a self-directed way (Wang & Bain, 2016, p. 269). Self-directed learning, as a key trait of andragogy, can be defined as "a process in which the individual takes the initiative to learn with or without other people's help, recognizes their learning needs, sets their goals, identifies the resources, chooses and implements suitable learning strategies and assesses the achieves results" (Cadorin et al., 2012, p. 154, in reference to Knowles, 1975). Self-directed learners are motivated by e.g., "the need of [an] appraisal, self-esteem, fulfillment, the desire to grow, the satisfaction of meeting set goals, the need to learn specific knowledge and intellectual curiosity" (Cadorin et al., 2012, p. 154).

This hypothesis, however, must be seen critically, as "There are several indications that many adults do not want to learn continuously or their life long, nor do they prefer self-directed learning." (Kraft, 1999, p. 838) Also, Knowles et al. (2005) estimate this assumption to be the weakest as "the learners have a history of not being successful in similar learning situations" (Knowles et al., 2005, p. 158). Thus, they emphasize that though their "motivation is high because they are trapped in low-wage jobs and are anxious to improve their lives…their prior experiences with this type of learning could be a significant barrier to learning if self-directed learning is thrust upon them too quickly" (Knowles et al., 2005, pp. 158-161).

This might indicate that andragogy's assumption about self-directed learning presumes that a learner's basic needs according to the *Self-Determination Theory* are fulfilled (Deci & Ryan, 1993, pp. 229f.; Wang & Bain, 2017, p. 261):

- *Autonomy* (experienced integration and freedom to choices),
- *Competence* (perceived ability to the completion of a task), and
- *Relatedness* (connection to e.g., peers, colleagues).

If these are met, intrinsic motivation and self-regulation are enhanced, which are enablers of self-directed learning (Deci & Ryan, 1993, p. 229; Wang & Bain, 2017, p. 261).

These explanations lead to two possible conclusions: On the one hand, andragogical leaders must be enabled to shape their associates' working and learning environment in a way that their basic needs are met, and that self-directed learning is supported (Wang & Bain, 2017, p. 261). On the other hand, the learners must develop and possess *learning how to learn competence*[2] to successfully take on their role and responsibility as entrepreneurs of their own labor, skills, and knowledge (European Parliament & Council of the European Union, 2006, L 394/16).

Proposed Definition of Andragogical Leadership

Merging the previous paragraphs, andragogical leadership shall be understood as a leadership approach, in which the *organizational capacity to solve complex, crucial, and unforeseen problems* is determined by leaders whose main responsibility lies in *genuinely supporting the associates with acquiring the respectively needed competencies* whilst taking their *individual development aspirations* into account. What is important with this definition?

- Andragogical leadership is suited for companies that want to transform sustainably. It helps companies enhance their performance in the long run but does not focus on improving short-term profitability, productivity, or efficiency. Consequently, leadership is seen, metaphorically, as a people-focused change agent and not an institution that aims to increase gains from 'business as usual. This would rather be a managerial task.
- Leaders are not in a distinguished position, but they are partnering with their associates on eye level. Their contribution is to *enable* meaningful learning by shaping the needed frame conditions (they are architects of the organization) and to *support* their associates grow and self-actualize – at best to the company's favor. Consequently, associates are seen as capable and trustworthy experts in charge to solve the company's problems.
- Associates learn life-centered, based on meaningful problems, and competence-based. Ideally, leaders match the associates' learning to those of the company's emergent problems they identify with. Consequently, this learning contributes to both the organization, as it generates demanded solutions and the individual, as it favors their self-actualization.

To achieve this, companies must invest in shaping the *cultural, technical, and organizational* frame conditions, e.g., by nurturing a company-wide growth mindset, installing modern learning platforms, and implementing processes & policies that support leaders to become andragogically effective. Self-evidently, the leaders themselves must be enabled, too. Therefore, companies should offer e.g., awareness campaigns, training, coaching, and experience exchange platforms. If properly done, leadership will then ensure to the favor of both that the individuals' continuous, self-directed, and meaningful learning is inseparably tied to an organization's sustainable development, so that they mutually grow.

RESEARCH DESIGN

The applied qualitative research design has an explorative and theory-generating character and focuses on a holistic understanding of the data and its interpretation in the respective context (Lamnek, 2010, pp. 4, 20). The applied data generation and data evaluation methodologies are explained in the following.

Data Generation With Problem-Centered Expert Interviews

Problem-centered interviews were used to address the theory generation for this chapter. According to Witzel (2000), problem-centered interviews "aim to gather objective evidence on human behavior as well as on subjective perceptions and ways of processing social reality" (para. 1). Therefore, they utilize a *problem-centered orientation*, *object orientation*, and *process orientation* as described following:

1. **Problem-Centered Orientation:** The researcher utilizes his/her own comprehensive professional knowledge (Teine, 2018, p.45) and previous knowledge/observations about the interviewees' orientations and actions toward the research subject, to understand their explanations and to "continue the problem-centered questioning and re-questioning" (Witzel, 2000, para. 4). In parallel to this broad and differentiated generation of data material the interviewer is "working on understanding the subjective view of the respondent while gradually making communication more precisely address the research problem" (Witzel, 2000, para. 4). Thus, the questioning becomes continuously focused on the actual research questions as the researcher progressively understands the interviewees' point of view (Teine, 2018, p. 45). What does this mean in practice?
    a. The questions at the beginning of an interview are very open and aim to understand the general viewpoint of the experts, as the following example illustrates: "Where would you draw the line between what lays in the freedom of the employees and where this freedom has its limits because it doesn't lay within the interests of the employer anymore?" (P1, para. 2)
    b. As an interview progresses, the questions become more focused on details and understanding nuances. An example would be: "How would you differentiate between the role of a coach and a mentor, and do you think that a leader could realistically fulfill either of these roles?" (P3, para. 36)
2. **Object Orientation:** Emphasizes "methodical flexibility in face of the different necessities of the objects being observed" (Witzel, 2000, para. 4), meaning that the method can and must be applied in a way that suits the research subject best. Hence, there are no process or frame conditions that must be adhered to *strictly* (Teine, 2018, para. 45). This does not free the researcher from being prepared or adhering to methodological standards (Witzel, 2000, para. 3f.).
3. **Process Orientation:** The researcher must maintain a communication process characterized by trust, in which the interviewees feel accepted and taken seriously. This is a precondition for them to open up, to speak unprotected and in cooperation with the researcher, and to self-reflect (Witzel, 2000, para. 4). This meant for this chapter that every expert was contacted personally and the publication process and how their anonymity will be safeguarded was made transparent. Also, the reason why the author asked *them* to participate as outlined to emphasize their being recognized as experts. Moreover, the interview style was *never confronting* but questioning by asking for their position on alternative hypotheses like e.g., "Would this, what you just said, be your best idea, or don't you think that we should" (P4, para. 19).

During the interview redundancies and contradictions are welcomed as they facilitate interpretation and express ambivalence. The focus is thus not on a based question-answer-based inquiry or isolated answers to isolated questions (Witzel, 2000, para. 4). Comprehending and reassuring questions or formulations like "Aren't we back at our starting point now?" (by the author; P3, para. 38) or "I really see this differently!" (by the expert; P2, para. 9) make great examples here.

Short questionnaires on socio-demographic information, interview guidelines including lead questions, tape recordings, and their full transcripts, and postscripts that "entail an outline of the topics discussed" (Witzel, 2000, para. 9) and comments on the "situative and nonverbal aspects" (Witzel, 2000, para. 9) are the central instruments of problem-centered interviews (Witzel, 2000, para. 5-9). For the research conducted for this chapter, the *interview guidelines* and the *full interview transcripts* were of particular importance. The guidelines provided orientation and structure, and they ensured comparability of the interviews whilst the transcripts are authentic and precise records of the communication process (Witzel,

2000, para. 8, 9). They ensured that no crucial information got lost and allowed for an interpretation in form of a hermeneutic circle (Lamnek, 2010, pp. 56-60; cf. also Figure 1). However, a specialty here is that the interviews were carried out as so-called *semi-structured interviews*. This means that the guidelines were seen as "questionnaires that are administered verbally" (Fylan, 2005, p. 66), so that the "conversation can take many directions before all the areas you want to address are covered" (Fylan, 2005, p. 66). Given the comparatively complex research questions and the coding-based data analysis approach, the authors also followed Fylan's recommendations that the "semi-structured interviews should contain more structure" (2005, p. 66). Here, the questions were asked in a story-telling approach meaning that they were not asked in a fixed order, but they were arranged to drive the interviewee's story forward best (Witzel, 2000, para. 18; Teine, 2018, p. 46). Moreover, this allowed learning about the expert's individual knowledge and experiences best. Though the interviewees are experts in the same field they have different specializations, which were exploited efficiently this way. Taken together, this allowed comprehending different points of view on the common research questions.

Anyways, the interviews were conducted with experts that brought both theoretical knowledge and practical experiences regarding leadership (development), andragogy, self-directed/ lifelong learning, or related fields. The convenient sample consisted of German mid-level leaders from operations and corporate functions plus experienced experts in the aforementioned fields. This allowed the inclusion of different, probably contradictory perspectives, e.g., on how different job profiles (routine work and physical labor vs. strategic, project-related work) might reason a lower or higher acceptance of the andragogical leadership practices. All participants were recruited from the author's professional network and are personally well-known to him. The interview participants (P) were:

- **P1:** Experienced middle management leader in an HR function at an international logistics provider, located in a business unit overhead, professional coaching background.
- **P2:** Middle management leader with a focus on Lean Management, located in a big distribution center of an international logistics provider, close to daily shop floor operations.
- **P3:** Experienced leadership and organizational development expert in a car electronics-producing business unit of a renowned internationally active corporation.
- **P4:** Expert in organizational and leadership development particularly in shop floor areas, working for an internationally renowned company producing power tools.

Three of the candidates were female, and one male. For the sake of anonymity, they will be referred to as males. All interviews were limited to a duration of 45 minutes to keep focus, on average they took 38 minutes.

Data Evaluation Using Qualitative Content Analysis

The interviews were evaluated by applying a qualitative content analysis approach (Mayring, 2000). Hence, the evaluation had an explorative rather than hypothesis-testing character. The interviews were not evaluated as individual cases. Instead, the aim was to work out knowledge and viewpoints, relevance structures, reality constructions, and patterns of interpretation shared between the experts (Meuser & Nagel, 1991, p. 452). Statements by individual experts that do not seem representative or shared by others are used to challenge the common understanding and to open a room for discussion (Meuser & Nagel, 1991, p. 452). However, the applied analysis approach brings together elements of two approaches:

- The *Interpretative Analysis of Guideline-led Expert Interviews* (Meuser & Nagel, 1991), a flexible approach that aims to understand the said in its respective context and thus must be adapted to the data material and research objective (pp. 452f.). The focus is on both 1) theories and concepts with an explanatory character for a system but also 2) how experts shape a certain system with their actions (Meuser & Nagel, 1991, p. 454).
- The *Qualitative Content Analysis* (Mayring, 2000) in its *structuring* and *summarizing* forms. This well-established approach can be applied to "all sort of recorded communication" (para. 4) e.g., transcripts of interviews, protocols of observations, and similar. According to Mayring, qualitative content analysis defines itself "as an approach of empirical, methodological controlled analysis of tests within their context of communication, following content analytical rules and step-by-step models, without rash quantification." (para. 4)

Both are rule-based approaches with so-called categories at the center of analysis (Mayring, 2000, para. 8; Meuser & Nagel, 1991, pp. 459-462). Whilst for Meuser & Nagel (1991) categories are the results of reducing the data material step-by-step to headline-like paraphrases that represent shared experiences, observations, interpretations, values, etc. (p. 460), for Mayring (2000) "The aspects of text interpretation, following the research questions, are putted [sic] into categories, which were carefully founded and revised within the process of analysis (Feedback loops)." (para. 8) Thus, categories are inductively developed and deductively applied (Mayring, 2000, para. 8). Here, a combined approach consisting of six consecutive phases was applied as outlined by Teine (2022) (pp. 121-124):

1. **Transcription:** All interviews were fully transcribed and rehearsed to get familiar with the material.
2. **Deductive Category Development:** An initial set of categories was defined based on the referred theories and the interview guideline.
3. **Coding, and Definition of Lead Examples and Coding Rules:** Those text elements that fall under the definition of a category were coded. Those elements that represent a category best were set as *lead examples*. Based on the commonalities of all texts coded with a category, *coding rules* were defined. Only text elements that adhere these coding rules were put into the respective category. This was a continuous, iterative process taking several loops.
4. **Inductive Category Development:** Text elements that do not fall under one of the existing categories became lead examples of a new category and defined its initial coding rules.
5. **Coding the Complete Data Material:** Given that the transcripts were coded one after another, a pre-final set of categories, lead examples, and coding rules was given once the last transcript was coded. Now, the steps 1 to 4 were carried out again until no further modifications could be made.
6. **Presentation of Results:** The research results are presented in this chapter, and thus the research questions were answered by summarizing all codes per category as one coherent theory.

The first five steps could be represented as follows (Figure 1; in accordance with Teine, 2022, p. 126, Abbildung 14).

Here, *four categories* (namely: Rationale & Purpose, Leadership & Learning, Enabling, and Outlook & Questions for the future) with *10* additional *subcategories* contain a total of *119 codes*. The subcategories *Leadership tasks & responsibilities* (24 codes), *Cost/benefit ratio* (16), *Assumptions* (15), *Status quo* (12) and *Frame conditions* (12) envelope most of the interview material.

Figure 1. Relation of the category system's maturity to the progress of the coding process

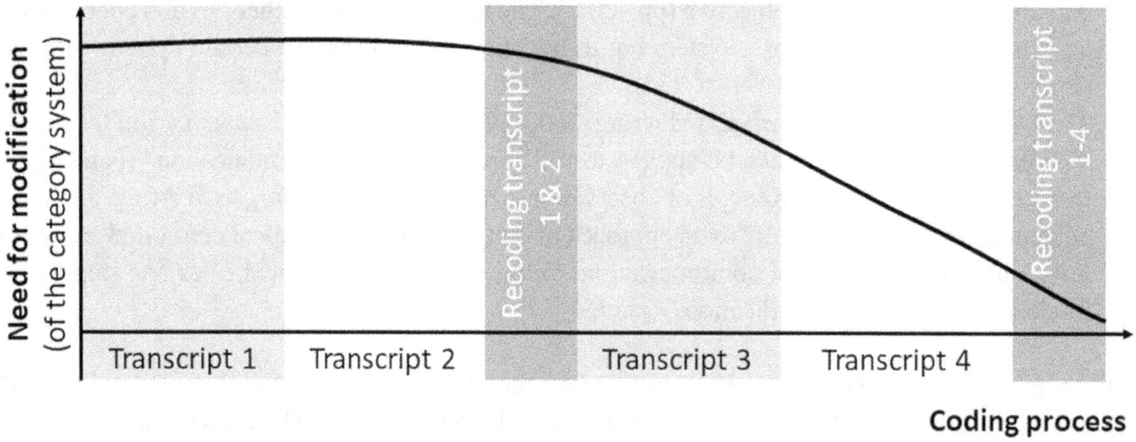

INSIGHTS FROM THE INTERVIEWS

In the following, the results of the expert interviews will be presented. Given that all interviews were conducted in German, the following insights and quotes are translations by the author. The experts will be referenced as listed under the headline *Data Generation*, i.e., (P1, para. 3) indicates that the referenced statement could be found in the third paragraph of the interview transcript of participant 1.

The results are displayed following the structure of the category system i.e., by their (sub)categories – not in order of the introduced research questions.

As a starting point of the interview, the author asked all experts to visualize the same situation: "Imagine an organization, in which andragogical leadership is lived practice. Associates define their own learning objectives and their approach to achieving those based on their individual preferences. Leaders are partners on eye level that create opportunities to build competencies based on experiences. They give freedom and shape the therefore needed frame conditions. At the same time, organizations learned to harness these individual competence development processes to continuously adapt to their changing environments. A win-win situation. Too idealistic or not that far away from practice, as you experience it?" This initial question elicited the first responses that then led to the following results:

Asked for the rationale and purpose, all experts believe in the *need for andragogical leadership*, initially defined as a leadership style that supports self-directed learning, and that brings competitive advantages to an organization. Though the experts can give neither a measure for the cost/benefit ratio of andragogical leadership and the complex cultural change that comes with it (P1, para. 27) nor an incentive for leaders to implement this approach (P2, para. 13), they expect benefits like e.g.,

- Overcoming skilled labor shortage (P1, para. 1),
- Higher work force productivity (P1, para. 27),
- Addressing the younger generations striving for self-actualization (P3, para. 43),
- A better fit between job and skill profiles (P4, para. 6),
- Harnessing the employees' knowledge for process improvements (P4, para. 28), and

- Maintaining the employees' motivation and commitment (P4, para. 28).

In practice, the experts experience different realities (*status quo*). Whilst P1 describes andragogical leadership as an "idealistic concept that must and will become a practice in future" (P1, para. 1), P4 already made some first-hand experiences (being led this way; P4, para. 2, 4) and is far from calling it a Utopia (P4, para. 2). However, as a leader himself, P1 speaks of positive experiences only when applying andragogical leadership practices (P1, para. 9). P2 and P3 possess similar points of view. Asked for an estimation of the status quo, P3 rated a 3.2 on a scale from *1 – fully disagree* to *5 – fully agree* that andragogical leadership is lived practice in his company (P3, para. 3). P2 emphasized the significant efforts needed to reach this target state and that fundamentals are missing, still (P2, para. 1).

Regarding the current practice, P1 stresses the missing awareness for this topic, and that learning (i.e., attending seminars) is mostly seen as an incentive to keep the associates happy – across all levels (P1, para. 25). This might be due to a lack of role models (P1, para. 17) and leaders missing a big picture they could align the development of their associates with (P1, para. 13). P3's organization already offers a plethora of digital learning offers that would allow for self-directed learning and they also have learning consultants in place that support learners harnessing these opportunities (P3, para. 5). However, learners are not enabled to e.g., define and pursue learning objectives themselves, and their leaders lack the respectively needed supportive mindset, which is why they e.g., do not support learning activities that are not matching the employee's original job scope (P3, para. 5, 7). Such a denial of learning seems short-sighted, as P4 illustrates based on his own experiences. He experienced a supportive leader that allowed him to dive deeper into topics that did not match his job but kept motivating him intrinsically, and thus he felt well-appreciated (P4, para. 4). This led him to shape projects that allowed others to learn and gather experiences in a similar way, with astonishing results and only positive feedback by the leaders as the participating associates unbound their potentials and motivation (P4, para. 36).

The experts entirely shared the *assumption* that people strive for appreciation, acknowledgment, and self-actualization, and that leadership must address these needs to achieve goals and to keep the associates motivated (P1, para. 10f.; P2, para. 17; P3, para. 11; P4, para. 6, 8). Everyone is believed to be capable and motivated by contributing to the bigger picture with whatever task they perform and however little their contribution might be (P1, para. 19, 23; P4, para. 36). If people are not given the opportunity to contribute, to feel effective/ competent, and to fulfill tasks that help them prosper, chances are high that they lose motivation, that they become genuinely unhappy, and that as a result, they leave the company (P4, para. 8, 28, 36). Hence, in the long run, only a leadership that supports associates in their self-actualization by providing development opportunities and a job/ task in which they can feel competent and grow is economically sustainable (P4, para. 6). Here, P3 also outlines that people do often not learn for the sake of learning itself but that they want to bring in their newly acquired knowledge and competences, and that they want their learning efforts to be acknowledged/ rewarded (P3, para. 11). However, people differ and so does the intensity and approach with which they strive for self-actualization and continuous development (P1, para. 23; P2, para. 13), which is why leaders must account for the individuality of every associate.

Given the high expected efforts of such a leadership, the *cost/benefit ratio* was discussed with the experts as well. Here, however, their opinions are somewhat contradictory. P1 pleads for openness to any learning and development aspirations an associate might bring, but he also emphasizes that it is a leader's role then to weigh the costs (e.g., time investment) with the benefits (e.g., improved productivity) (P1, para. 1). Though this calculation is quite easy when it comes to e.g., a German project manager learning English because he must carry out international projects, it gets increasingly difficult when a company is e.g., about to decide whether it financially supports an associate studying for his Bachelor's degree (P1, para. 3). Is it likely that this associate will generate new business cases or revenue channels? Will his or her new knowledge help be disruptively innovative? (P1, para. 3) Also, particularly if the investment is comparatively high, a leader might estimate the associate's passion for the defined learning objective. Will he or she successfully complete e.g., studies? Is he or she really about to perceive his or her dreams and will bring competitive advantages for the company? (P1, para. 3) Or is he or she just looking for an escape from the actual work? (P3, para. 11) P1 raises these questions as he, like P3 does as well (P3, para. 11), and is convinced that given the plethora of high-quality free online learning resources, anyone can learn anything if he/ she is actually motivated (P1, para. 5). Hence, the remaining question is, if the respective associate is not willing to invest his leisure time, why should the company invest resources, if there is no direct benefit for its business? (P1, para. 5). A possible answer might be that any (also workplace-unrelated) learning might lead to new impulses for the company and might foster its innovativeness. This, however, is an uncalculatable lottery as P1 outlines (P1, para. 7, 17).[3] P3 on the opposite estimates that in 20% of the cases when learning is solely rooted in the associates' interests it leads to something directly beneficial for the company (P3, para. 9). An example could be an HR colleague who is interested in software developing, which might be a demanded competence in the company (P3, para. 9). Given that a company must adapt to its environment and markets, it would not be wise to fully support all of the associates' aspirations (P3, para. 18f.), e.g., if the mechanic wants to become a vet (P4, para. 4). Such cost/benefit calculation look at the whole company, but often leaders make decisions only with their team, department, or site in mind. Here, P4 pleads for leaders always look at the whole company. Though it might be difficult for a department to let go of a competent colleague, it is beneficial for him/ her and the company when there is a better matching position available elsewhere. Otherwise, the colleague might even leave the company because there are no development or self-actualization opportunities available, which would be the worst scenario for everyone (P4, para. 8).

In summary, the experts plead for a culture and structures that give plenty of freedom for the associates to develop and prosper but their self-actualization aspirations should, at least roughly, be in-line with the company's goals and values to be mutually beneficial (P2, para. 5; P3, para. 19; P4, para. 14, 16). Else, it might even make sense to support the associate in finding a better matching position in another company.

Besides answering such somewhat generic questions, the interviews focused mainly on the role and responsibilities of andragogical leaders, covered by the leadership and learning category. Here, it was remarkable that P1 and P2 argued based on an understanding of andragogical leadership as a new leadership style or a role to be taken over formally, whilst P3 and P4 explicitly wanted to think outside organization charts and outlined that the tasks and responsibilities associated with andragogical leadership could also be taken over by e.g., a passionate team member following a shared leadership approach (P3, para. 25, 29, 33; P4, para. 18, 20). P3 also suggested that such a role might be taken over by a so-called learning consultant, an expert within the organization who meets the learners on eye level and guides the associates' learning processes (P3, para. 33). P4 suggests a similar, peer-to-peer-based concept. Given that formal leaders might not carry out this responsibility passionately, he proposes that e.g., a colleague who brings the needed expertise and would fulfill this role by heart should have it – either as a stand-alone role or as a supporter of the formal leader (P4, para. 18, 20).

If andragogical leadership is not bound to a formal role, it might be thought of as a set of *tasks and responsibilities* within a shared leadership approach. Those would find their focal point in the associates, their skills, and their individual development needs and aspirations (P1, para. 1). A core task of an andragogical leader might be e.g., to analyze an associate's strengths and matching them with existing development opportunities (P1, para. 1).

A major topic of the interviews was whether leaders could take over the role of a coach for the associates. In such a role, leaders meet their associates on eye level to support them in self-assessing their e.g., career status, identifying development goals, and outlining a plan to match/ achieve these (P1, para. 1, 3; P2, para. 1, 5). They would thereby help the associates find a way to self-actualize (P4, para. 4). Andragogical leaders ensure that the individual development goals are aligned with the company's goals and offer the respective guidance (P1, para. 3, 7). Such an orientation might be particularly helpful for young associates or those who are not clear about where to go yet (P1, para. 9). It is important though to leave enough room for the associates to make their own decisions as too narrowly defined e.g., methods could have a demotivating effect (P2, para. 31). Moreover, the leaders must also be open for career and development paths that were not foreseen yet but might nevertheless be beneficial to the company (P1, para. 3). P1 came up with the following example: An associate might be interested in reworking salary structures though this was not on the organization's agenda yet. However, given that the company is striving to strengthen its employer branding and that salary is an important asset here, the leader could match both (P1, para. 3).

However, the term coaching must be seen critically in this context as coaches should only work in the interest of their clients. Leaders, if they were truly coaches, would only care about their associates pursuing their goals – even if they do not match the company's interests. This is a conflict for leaders, as they are typically installed to act in the company's interests, they must weigh costs and benefits (P2, para. 5). P2 becomes even more concrete and outlines that he has neither the right nor the competence to be a "Life-Coach", as he called it (P2, para. 9). His job is to be the best support possible during the working hours, and to have a positive impact on the associate's life in his very area of responsibility (P2, para. 9). Thus, *the coaching notion is not suitable to describe the andragogical work of leaders* (P1, para. 7). What is key to this leadership approach though is to strongly involve the associates in aligning their development and the company's goals (P1, para. 13).

In addition to this discussion, the experts pinpointed the following tasks and responsibilities for andragogical leaders:

- Making the company's purpose and strategy (its 'big picture') tangible for the associates to then derive meaningful learning objectives together so that the associates feel to contribute (P1, para. 13; P3, para. 17).
- Offering the associates space and tools to help them identify their strengths and potentials, to find out about themselves and what is important to them when it comes to their career. Thus, they can grow in a guided, safe, but nevertheless impactful way (P1, para. 23) – in all their uniqueness, which might be a particularly demanding responsibility.
- Feedback on the learning and development aspirations of the associates to the organization (P1, para. 25).
- Making the purpose of a certain task or role visible to an associate and showing how it contributes to the whole (P2, para. 9). "My job as a coach is to show my associate how awesome his job is. To genuinely tell him that clocking in at 6 o'clock and working is the best 8-hour investment he could have made. When I fail at this, I failed as a leader." (P2, para. 9)
- Help keep the associates' motivation high during the learning process, particularly during hard times (P2, para. 35).

- Finding opportunities together with the associates to ensure that their newly acquired knowledge is utilized and that they feel competent and self-efficient (P3, para. 17).

In summary, the most important asset of an andragogical leader, as emphasized by P4, is his or her passion for people (P4, para. 4), to foster their autonomy, their self-determination, their empowerment (P4, para. 36). "In my opinion, a leader must bring the passion for people and to support their development. Thus, when you want to live such a leadership role, you need the respective skills, and you should find fulfillment in it." (P4, para. 4) Being an enabler of andragogy must "fit the individual values" (P3, para. 31) and it must be seen critical that "leaders must constantly bend themselves" (P3, para. 31) to new approaches and demands. A person that can passionately carry out the respective activities should do so (P3, para. 31f.). "Whoever has the dream to support people in their self-actualization should be enabled to get the needed position." (P4, para. 4) Alternatively, associates could search for one or more colleagues that might act as their development/ career partners as the conflicts of interest might be, in some cases, too big between the formal leader and the associate (P3, para. 37).

Given this background, the focus must not only be on the leaders but also on the associates and their *degrees of freedom and responsibility* when it comes to their learning. Speaking from his own experiences, P1 pleads for not being too predetermining (P1, para. 9). Whilst the leaders should give in the business goals, it is the learner's responsibility to derive and define learning objectives (P1, para. 37). This presupposes that both the leaders and the associates are enabled to carry out this process respectively (P1, para. 37). P2 agrees that the overall responsibility for the learning process and its success lies within the associates. Given that learning is always demanding and time-consuming the learners must commit to the endeavor (P2, para. 5, 33, 35). In his view, it is the commitment that makes a difference, as e.g., the internet provides enough free resources. "When you're really burning for something, you can do it, particular in Germany," he says (P2, para. 5). P3 possesses a similar point of view and stresses: "If I would be motivated, I'd do it in my leisure time. This trade-off shows how important it really is to me as I could also spend my evenings using online tutorials. I'm not dependent on my employer, knowledge isn't the problem anymore." (P3, para. 11) Particularly when the own aspirations do not fit the company's goals at all, P4 does not see the company in charge to provide the respective freedom for learning in full intensity (P4, para. 16).

Enabling and empowering andragogical leadership requires certain *frame conditions*. A core aspect might be that to be effective andragogical leaders, the leaders themselves must experience this leadership approach firsthand (P1, para. 17). The top management must act as a role model in providing freedom and delegating responsibility e.g., by only predefining a strategy but making it the responsibility of the subsequent leadership levels to define goals for operations (P1, para. 21). P2 adds that the company must of course allocate sufficient *time and budget*, and he emphasized that these are the most crucially needed resources (P2, para. 27). For the individual, having such resources is a key enabler to achieve defined learning goals, to feel supported, and to stay motivated (P2, para. 21, 29). This creates the needed commitment and ensures that the daily business/ operations, which is typically the highest priority, does not obstruct the learning (P2, para. 33; P4, para. 4). Therefore, leaders that can be considered 'People people', that possess a strong intrinsic motivation to help others develop, are needed (P4, para. 4) and are a way more important enabler for andragogical leadership than e.g., having a Learning Management System in place (P2, para. 27). Also, the company needs a sophisticated understanding of the associates, their needs and values, and how changing frame conditions affects their opportunities to grow and their desire to either stay with the company or to leave (P4, para. 8). One of such frame conditions might be to allow the associates to invest one day per week (or: 20% of their time) into their own development and in pursuing their self-actualization (P4, para. 16). The acceptance for such cultural artifacts and for experimenting with them rises as their concrete benefit becomes tangible for those in charge to shape the frame conditions (P4, para. 36). Here, P4 outlines that it might be the 'easy' and 'obvious' things (e.g., that associates can easily suggest projects that improve practice at site whilst offering learning opportunities) that in the end will lead to having a lived learning culture where people feel heard, acknowledged, and fostered (P4, para. 36).

But what learning and development activities might andragogical leaders foster? The experts mentioned *examples* showing that learning and development are more than attending seminars. For example, the modernization of workplaces or the introduction of new methods naturally comes with the need to further qualify the associates (P1, para. 19; P2, para. 31). In such a case, only a few associates are affected e.g., when processes, hard- and software are modernized for specific functions (P1, para. 19). The opposite would be e.g., the introduction of Lean principles to a whole company, which might disrupt well-established routines. Such a scenario would require widespread enabling (e.g., that everyone knows basic vocabulary to talk on a level with customers) but also local lighthouse projects that proves the concept and qualify the affected colleagues similarly (P2, para. 31, 41; P4, para. 36). Both examples find their roots in the company changing, but the motivation to develop further might also lay within the individual to go new paths. Here, the well-established concepts of job shadowing and job rotation might help an associate to find out about interesting career opportunities and to match expectations with reality (P4, para. 24).

A learning consultant might support the associates during this process, to help them sort and find matching opportunities, or he might also connect learners with ongoing change projects that match their passions and learning needs. Also, such learning consultants might help to pick the right content from a library of online resources (Learning Management System) (P4, para. 24).

All these aspects require a proper enabling of the learners, the leaders, and the organization in terms of knowledge and mindset (P3, para. 5). "On the one side there are associates not knowing how to define learning objectives themselves, and on the other side there are leaders defining the objectives for their associates based on the company's business objectives, but they do not know how to foster self-determination." (P3, para. 5) However, enabling a whole company is time-consuming and thus it is not possible to simply switch to another leadership approach from one day to another (P1, para. 35). The opposite is the case, it needs a participatory process in which all associates (shop floor workers to top management) shape new structures and nurture a culture that accounts for the broad variety of needs and specialties from the individuals, departments, and business units (P4, para. 18).

A starting point might be to assess the *learners'* maturity level against the target state (P3, para. 9). Are they aware of their responsibility? Do they know that it is them who oversee their own learning? (P1, para. 25) Do they possess the skill set and mindset needed to drive their own development self-dependently? (P3, para. 9) If not, are there structures in place that help them acquire such skills and attitudes e.g., learning consultants? (P3, para. 5) And what about the *leaders*? Are they capable of sensitizing and enabling their associates respectively? (P1, para. 25) Do they have the needed degrees of freedom, competence, and (formally) the responsibility to act as truly andragogical leaders? (P1, para. 21) Does the company provide training, mentoring, or coaching to its leaders that e.g., supports them in living up to its own standards and aspirations, to be genuine mentors and coaches themselves, and harness a broad suite of diagnostic tools and development measures? (P1, para. 23; P4, para. 22). Also, organizations need to support their leaders in enduring their insecurities, which arise when the employees they promote outgrow themselves (P2, para. 1). Additionally, organizations must understand that if they want authentic, genuine, passionate leaders they should allow them to focus on those topics where they can actually contribute (P3, para. 31).

Though the experts articulated a plethora of knowledge and experience, some topics remain to be unanswered. Thus, they provide an outlook and questions to be answered in the future, including:

- Defining meaningful KPIs for successful learning & development and leadership (P1, para. 39).
- Are modern leadership approaches like shared leadership or andragogical leadership suitable for big corporations? If not, what will a competitive organization look like in the future? Will they break into smaller units that live a different leadership approach compared to now? (P3, para. 45)
- Given P4's experiences that white-collar workers often have more freedom and opportunities to learn self-directedly, he asks: "What is the significant difference here, that companies do not provide both with the same opportunities?" (P4, para. 26) A reason might be that productivity on the shop floor level is simply easier to measure, and thus, decreased productivity (as time is spent on learning) becomes immediately tangible (P4, para. 26).

FUTURE RESEARCH DIRECTIONS

The research conducted for this chapter had a strong explorative character and followed a transparent, rule-based approach in both data generation and evaluation. This led to insightful, multi-faceted research results that complement the currently still little body of literature. Nevertheless, given the limitations in the research design (e.g., small sample size, sample design) and the fact that andragogical leadership is a term and concept yet to be coined, further research must be conducted. Against this background, taking the findings presented in this chapter into account, the following research is proposed addressing leader and leadership development alike:

1. Theoretical discussions, literature reviews, and qualitative research with experts to further specify the responsibilities of leaders taking this role, including a definition of required traits, competencies, and frame conditions needed to become effective. This includes recommendations for e.g., targeted leadership development approaches.
2. Qualitative and quantitative studies that critically evaluate and discuss the potential of andragogical leadership as a new theory, particularly in differentiation to established approaches and based on empirical evidence.
3. Quasi-experimental designs following participatory approaches and case studies for modeling of an organization's design (structure, processes, culture) that allows for, enables, and benefits most from effective andragogical leadership. Here, the focus might be on leadership development.

Additionally, research on the maturity of both leaders and associates and whether they can be assumed to already possess the competencies needed to cope with their changing roles is needed. This might be complemented by studies on how they can be enabled best.

As research on leadership is characterized by a high degree of complexity and lies within different faculties like e.g., business administration, higher education, and psychology, discussions, and studies accounting for different perspectives and following diverse methodological approaches are needed.

CONCLUSION

It is unlikely that the pace of change individuals and organizations are currently facing will slow down. Thus, both must continuously adapt to stay competitive, and in doing so individual competence development and organizational development are inseparably tied together (Teine, 2022, pp. 271f.). Therefore, also both sides carry a share of the responsibility to be successful. Companies must provide sufficient freedom and support for their associates' development aspirations and their desire to make a meaningful contribution (Volini et al., 2021, p. 6). And the associates are required to be self-responsible learners that step into the driver's seat of their own lifelong learning journey, following their passions but also constantly reequipping themselves with competencies in need (Gonon, 2002, p. 327). This indeed requires them to possess *learning how to learn competencies* (European Parliament & Council of the European Union, 2006, L 394/16). It is the role of leadership to match both sides.

These observations are well in line with an increasing number of companies nurturing learning cultures that foster self-directed learning, particularly with the idea to utilize the results of non- and informal learning (Meyer & Haunschild, 2017; Teine, 2022, p. 271), and participatory approaches gaining attention in corporate learning (Teine, 2018, 2022). Given the organizations' dependency on their associates' development, organizational development and thereby work might become continuously democratized in the end (see also Teine, 2022, pp. 270-273). Based on the experts' answers, andragogical leadership might be the answer to the questions and challenges that arise as companies enter such democratization processes.

The andragogical leadership role is not exclusive to formal leaders. Following the idea of shared leadership, the tasks and responsibilities of an andragogical leader should at best be taken over by a person that is competent and passionate about them. Also, andragogical leadership does not necessarily possess new claims and consequently, it is not a new leadership theory. However, andragogical leadership prioritizes, emphasizes, and facets a leadership that supports learning and development in a unique way. The research results indicate that effective andragogical leadership might be the urgently needed solution for companies finding themselves in demanding transformation processes. Hence, organizations should invest in the development of their leaders (individual) and leadership (system, structure) alike. In this regard, the research results did not indicate any unknown benefits of harnessing digital tools to leverage e.g., the leader development (see also Teine, 2022, pp. 36-41 for an overview of potential benefits and use cases of digital learning). Self-evidently, the targeted utilization of digital tools might support leaders to become more efficient and effective and to grow into their new roles more easily. Thus, well-designed digitization likely supports companies in transitioning to an andragogical leadership.

Moreover, andragogical leadership requires a certain understanding of organizational performance and respectively defined goals (e.g., short-term revenue orientation vs. adaptability; P1, para. 33; P2, para. 5), a suited structure (e.g., shared leadership, or a certain company size; P3, para. 29-33, 45), and a participatory culture with humanistic tendencies (e.g., P4, para. 4, 6) to become effective. Accounting for the research results, andragogical leadership might be most effective for a company that can be characterized as what Laloux (2014) described as *evolutionary* and *post-modern* organizations.

For evolutionary organizations achieving a clear and noble purpose is more important than profit, growth, or market shares – they strive for holism and community, and thereby they most likely also attract people that seek to discover their true vocation (Laloux, 2014, p. 51). They strive for meaningful contributions and achievements, and they are driven by trust and confidence in their employees (Laloux, 2014, pp. 44f., 51). Hence, hierarchy, bureaucracy, and control are seen as symptomatic for anxious,

Ego seeking managers (Laloux, 2015, p. 51). Post-modern organizations want to be pluralistic, and it is assumed that people strive for belonging and close, harmonious relationships – and value those over results (Laloux, 2014, pp. 30f.). Such organizations design processes bottom-up and try to consider everyone's point of view. Power and hierarchy are, at best, to be avoided or distributed equally whilst leaders are seen as generous, empathic, and careful servants to their employees (Laloux, 2014, pp. 31f.).[4] A high degree of mutual trust, decentralization, and empowerment characterize the post-modern organization, which is only possible when the executive and middle management give up control and share their power (Laloux, 2014, p. 33). This in turn requires a set of shared values and a strong (leadership) culture, as they give orientation and guidance, particularly when hard rules are abundant (Laloux, 2014, p. 32f.). The purpose of leadership in these organizations is to find the best possible compromise for all stakeholders with a long-term perspective as sustainable results outweigh short term profitability (Laloux, 2014, p. 34).

Besides its "ideologic" (P1, para. 5) and maybe even "romantic" (P3, para. 40) assumptions, andragogical leadership is ultimate – as any other leadership approach as well – a mean that suits the purpose of achieving a company's long-term goals in the most efficient, effective, and sustainable way (e.g., P1, para. 19; P2, para. 5; P3, para. 19, 21; P4, para. 6). Anyways, andragogical leadership might currently be the best solution to bringing sustainable business development in changeful times and human-centricity (i.e., a genuine focus on empowerment/ democratization, shaping an atmosphere that allows associates to prosper, and fostering self-actualization) together, and finally it might even show that there is no contradiction in serving both equally.

ACKNOWLEDGMENT

This research received no specific grant from any funding agency in the public, commercial, or not-for-profit sectors.

REFERENCES

Aepli, M., Angst, V., Iten, R., Kaiser, H., Lüthi, I., & Schweri, J. (2017). *Die entwicklung der kompetenzanforderungen auf dem arbeitsmarkt im zuge der digitalisierung. Schlussbericht. Arbeitsmarktpolitik No. 47*. SECO Publikation.

Bass, B. M. (1998). Two decades of research and development in transformational leadership. *European Journal of Work and Organizational Psychology*, *8*(1), 9–32. doi:10.1080/135943299398410

Borrego, M., Douglas, E. P., & Amelink, C. T. (2009). Quantitative, qualitative, and mixed research methods in engineering education. *Journal of Engineering Education*, *98*(1), 53–66. doi:10.1002/j.2168-9830.2009.tb01005.x

Brown, J., Gosling, T., Sethi, B., Sheppard, B., Stubbings, C., Sviolka, J., Williams, J., Zarubina, D., & Fisher, L. (2018). *Workforce of the future. The competing forces shaping 2030*. PricewaterhouseCoopers.

Cadorin, L., Suter, N., Dante, A., Williamson, S. N., Devetti, A., & Palese, A. (2012). Self-directed learning competence assessment within different healthcare professionals and amongst students in Italy. *Nurse Education in Practice*, *12*(3), 153–158. doi:10.1016/j.nepr.2011.10.013 PMID:22112861

Danişman, S., Tosuntas, S. B., & Karadag, E. (2015). The effect of leadership on organizational performance. In E. Karadag (Ed.), *Leadership and organizational outcomes* (pp. 143–168). Springer. doi:10.1007/978-3-319-14908-0_9

Day, D. V. (2000). Leadership development: A review in context. *The Leadership Quarterly*, *11*(4), 581–613. doi:10.1016/S1048-9843(00)00061-8

Deci, E. L., & Ryan, R. M. (1993). Die selbstbestimmungstheorie der motivation und ihre bedeutung für die pädagogik. *Zeitschrift für Pädagogik*, *39*(2), 223–238.

Erpenbeck, J., & Sauter, W. (2013). *So werden wir lernen! Kompetenzentwicklung in einer welt fühlender Computer, kluger wolken und sinnsuchender Netze*. Springer Gabler. doi:10.1007/978-3-642-37181-3

European Parliament & Council of the European Union. (2006). *Recommendation 2006/962 – Key competences for lifelong learning*. Author.

Fylan, F. (2005). Semi-structured interviewing. In J. Miles & P. Gilbert (Eds.), *A Handbook of research methods for clinical and health psychology* (pp. 65–77). Oxford University Press.

Gonon, P. (2002). Der betrieb als erzieher – Knappheit als pädagogische herausforderung. *Zeitschrift für Pädagogik*, *48*(3), 317–335.

Knowles, M. S., Holton, E. F., & Swanson, R. A. (2005). *The adult learner*. Elsevier. doi:10.4324/9780080481913

Kouzes, J., & Posner, B. (2012). *The leadership challenge. How to make extraordinary things happen in organizations*. The Leadership Challenge, A Wiley Brand.

Kovács-Ondrejkovic, O., Strack, R., Antebi, P., López Gobernado, A., & Lyle, E. (2019, November 5). *Decoding Global Trends in Upskilling and Reskilling*. https://www.bcg.com/publications/2019/decoding-global-trends-upskilling-reskilling

Kraft, S. (1999). Selbstgesteuertes lernen. Problembereiche in theorie und praxis. *Zeitschrift für Pädagogik*, *45*, 833–845.

Kropp, B., & McRae, E. R. (2022, January 13). *11 Trends That Will Shape Work in 2022 and Beyond*. https://hbr.org/2022/01/11-trends-that-will-shape-work-in-2022-and-beyond

Laloux, F. (2014). *Reinventing Organizations. Ein Leitfaden zur Gestaltung sinnstiftender Formen der Zusammenarbeit*. Vahlen.

Lamnek, S. (2010). *Qualitative sozialforschung*. Beltz Verlag.

Martin, J., Cormican, K., Samaio, S. C. B., & Wu, Q. (2018). Shared leadership and team performance: An analysis of moderating factors. *Procedia Computer Science*, *138*, 671–679. doi:10.1016/j.procs.2018.10.089

Mayring, P. (2000). Qualitative content analysis. *Forum Qualitative Social Research*, *1*(2), 20.

Meuser, M., & Nagel, U. (1991). ExpertInneninterviews – vielfach erprobt, wenig bedacht: Ein beitrag zur qualitativen methodendiskussion. In D. Garz & K. Kraimer (Eds.), *Qualitativ-empirische Sozialforschung: Konzepte, Methoden, Analysen* (pp. 441–471). Westdt. Verlag. doi:10.1007/978-3-322-97024-4_14

Meyer, R., & Haunschild, A. (2017). Individuelle kompetenzentwicklung und betriebliche organisationsentwicklung im kontext moderner beruflichkeit – berufspädagogische und arbeitswissenschaftliche Befunde und Herausforderungen. *bwpat, 32*, 1-20.

Schröder, S. H., Baldegger, U., & Klösel, K. (2021). Empowering leadership als erfolgsfaktor in der innovation. *Leadership, Education, Personality. An Interdisciplinary Journal, 3*, 9–20.

Sloane, P. F. E., & Dilger, B. (2005). The competence clash – Dilemmata bei der übertragung des 'Konzepts der nationalen Bildungsstandards' auf die berufliche Bildung. *bwpat, 8*.

Teine, M. (2018). A prototypical participatory design-process: Bringing digital learning and user experience together. In I. Bouchrika & P. Vu (Eds.), *Learner experience and usability in online education* (pp. 36–60). IGI Global. doi:10.4018/978-1-5225-4206-3.ch002

Teine, M. (2022). *Zur partizipativen gestaltung von (digitalen) lernangeboten in der betrieblichen (weiter-)bildung. Eine empirische modellentwicklung*. Universität Paderborn.

van Dierendonck, D. (2011). Servant leadership: A review and synthesis. *Journal of Management, 37*(4), 1228–1261. doi:10.1177/0149206310380462

Volini, E., Schwartz, J., Eaton, K., Mallon, D., Van Durme, Y., Hauptmann, M., Scott, R., & Poynoton, S. (2021). *The social enterprise in a world disrupted. Leading the shift from survival to thrive. 2021 Deloitte Global Capital Trends*. Deloitte.

Wang, V., & Bain, B. (2016). Andragogical leadership and management in the workplace. In V. Wang B. Bain, J. Hope, & C. Hansman (Eds.), Educational leadership and organizational management (pp. 249-273). Information Age Publishing.

Witzel, A. (2000). The problem-centered interview. *Forum Qualitative Social Research, 1*(1), 22.

World Economic Forum. (2020). *The Future of Jobs Report 2020*. Author.

ADDITIONAL READING

Gonon, P. (2002). Der Betrieb als Erzieher – Knappheit als pädagogische Herausforderung. *Zeitschrift für Pädagogik, 48*(3), 317–335.

Knowles, M. S., Holton, E. F., & Swanson, R. A. (2005). *The adult learner*. Elsevier. doi:10.4324/9780080481913

Laloux, F. (2014). *Reinventing organizations. Ein Leitfaden zur Gestaltung sinnstiftender Formen der Zusammenarbeit*. Vahlen.

Meyer, R., & Haunschild, A. (2017). Individuelle kompetenzentwicklung und betriebliche organisationsentwicklung im kontext moderner beruflichkeit – berufspädagogische und arbeitswissenschaftliche Befunde und Herausforderungen. *bwpat, 32*, 1-20.

Teine, M. (2022). *Zur partizipativen Gestaltung von (digitalen) Lernangeboten in der betrieblichen (Weiter-)Bildung. Eine empirische Modellentwicklung*. Universität Paderborn.

Wang, V., & Bain, B. (2016). Andragogical leadership and management in the workplace. In V. Wang, B. Bain, J. Hope, & C. Hansman (Eds.), *Educational leadership and organizational management* (pp. 249–273). Information Age Publishing.

KEY TERMS AND DEFINITIONS

Andragogical Leadership: Andragogical Leadership shall be understood as a leadership concept, in which the *organizational capacity to solve complex, crucial, and unforeseen problems* is determined by leaders whose main responsibility lies in *genuinely supporting and enabling the associates acquiring the respectively needed competencies* by *solving emerging problems that match their development aspirations.*

Andragogy: A theory primarily coined by M. S. Knowles defining key assumptions about adult learners, with self-directedness probably being the most prominent. Wang and Bain outline that this theory "has the potential to encourage adults to become self-directed learners, to inspire them to realize actualization, especially when psychological needs are met that foster and support internal well-being and motivation" (2017, p. 251).

Competence: In contrast to the term *competency*, which refers to a specific and quite narrowly defined set of capabilities needed e.g., to perform a certain task, the term *competence* refers to a more holistic concept, "A fundamental disposition that enables an individual to act appropriately and ethically correctly (= adequately) in the situation." (Sloane & Dilger, 2005) They are the precondition to acting creatively and self-organized in open, complex, dynamic, and sometimes even chaotic situations (Erpenbeck & Sauter, 2013, p. 32).

Leader Development: Emphasizes the "purposeful investment in human capital" (Day, 2000, p. 584) resp. the "individual-based knowledge, skills, and abilities associated with formal leadership roles" (Day, 2000, p. 584), typically including e.g., self-awareness, self-regulation, and self-motivation (Day, 2000, p. 584).

Leadership (Concept): Leadership in contrast to Management is not bound to formal authority and aims to "generally enable groups of people to work together in meaningful ways" (Day, 2000, p. 582). It is neither limited to a position, nor to an organization (Day, 2000, p. 582). Speaking of *followers* rather than *employees* of leaders further stresses the difference between Leaders and Managers (cf. Day, 2000, p. 583).

Leadership (Skill): Leadership might be defined as a set of skills, and anyone might possess them. Thereby, leadership is a result, an effect, "an emergent property of effective systems design" (Day, 2000, p. 583), and thus anyone might be considered a leader.

Leadership Development: Focuses on building social capital, i.e., "networked relationships among individuals that enhance cooperation and resource exchange in creating organizational value" (Day, 2000, p. 585). Commitment, trust, and respect coin the respective relationships, which can be viewed from a structural (e.g., transaction of resources), relational (e.g., mutually developed trust and trustworthiness), and cognitive (e.g., shared visions, common values) point of view (Day, 2000, p. 585).

Self-Directed Learning: Supports the learner's self-determination regarding his/ her goals, the allocated time for and timing of the activities, identifying suitable resources, choosing an appropriate method, structuring the learning, assessing the outcomes, and partnering with others (Aepli et al., 2005, pp. 2f.). Thus, self-directed learners are in charge of their own learning, there are autonomous in defining what for, how and where they are learning (Aepli et al., 2005, p. 10). This indeed includes that a self-directed learner can autonomously decide to attend highly instructor-led courses as this might suit his needs best (Knowles et al., 2005, p. 186). Possessing this competence must therefore be seen as a precondition for successful and sustainable lifelong learning.

ENDNOTES

1 In this chapter, the term *career* is meant to describe an *individual employment biography* that matches the persons interests, competences, and individual frame conditions (e.g., part-time working parents, reduced management responsibility after work-induced exhaustions, …) in the best way possible at any given point of time and that has been driven in a self-directed manner. It is *not* meant to describe a *higher, faster, further mentality.*

2 'Learning to learn' is the ability to pursue and persist in learning, to organise one's own learning, including through effective management of time and information, both individually and in groups. This competence includes awareness of one's learning process and needs, identifying available opportunities, and the ability to overcome obstacles in order to learn successfully. This competence means gaining, processing and assimilating new knowledge and skills as well as seeking and making use of guidance. Learning to learn engages learners to build on prior learning and life experiences in order to use and apply knowledge and skills in a variety of contexts: at home, at work, in education and training. Motivation and confidence are crucial to an individual's competence. (…) This requires effective management of one's learning, career and work patterns, and, in particular, the ability to persevere with learning, to concentrate for extended periods and to reflect critically on the purposes and aims of learning. Individuals should be able to dedicate time to learning autonomously and with self-discipline, but also to work collaboratively as part of the learning process, draw the benefits from a heterogeneous group, and to share what they have learnt. Individuals should be able to organise their own learning, evaluate their own work, and to seek advice, information and support when appropriate." (European Parliament & Council of the European Union, 2006)

3 For a company with e.g., 10.000 associates allocating an annual budget of 1.000 Euros to each of its associates (a comparatively small amount) the costs would sum up to 10 Mio. Euros. Assuming that the company is hoping for product innovations, process improvements, or similar, it is recommended to allocate the available budget to as many people as possible (so to say to diversify the risk of a total loss of this investment). P1 estimates the chances that one of these persons comes up with an idea worth more than 10 Mio. Euros (or, many people with ideas that sum up an equal amount) to be comparatively low (P1, para. 7, 17).

4 Post-modern organizations live in a contradiction, e.g., seeking for consensus in extremely large groups typically leads to stagnation and, in consequence, fights for power in the background. Hence, to function, they must set and enforce rules, which is authoritarian and might neglect the

individual. Also, the post-modern perspective often tears down old structures but is bad in coming up with working alternatives (Laloux, 2014, p. 31f.).

Chapter 6
Educational Leadership in Blended Higher Educational Contexts

Li Ming
Beijing Language and Culture University, China

Zhonggen Yu
Beijing Language and Culture University, China

ABSTRACT

Educational leadership's implications significantly contribute to blended higher education development in the context of information technologies. However, educational leadership in blended higher educational contexts has yet to be addressed and is not explored widely. This study uses a systematic review to explore educational leadership in blended higher education contexts. It is inevitable that education leadership steadily assumes new roles and responsibilities in blended higher educational contexts. This study aims to provide research guidance for educators, school administrators, and policymakers.

INTRODUCTION

As blended higher education evolves, educational leadership has attracted growing attention in academic and practical areas in the past two decades (Siddique et al., 2011). It has been recognized that educational leadership is a necessary and sufficient condition for school performance and student outcomes (De Nobile, 2018). The implications of educational leadership significantly contribute to the development of blended higher education in the context of information technologies (Garrison & Kanuka, 2004). It has highlighted that educational leadership can enhance the teaching-learning process (Strom & Porfilio, 2019). How educational leadership responds to changing education environments in the post-COVID-19 pandemic era is worthy of exploration (Comstock et al., 2021). Researchers and educators have made remarkably persisting efforts to carry out reforms for all students (Price & Mansfield, 2021).

DOI: 10.4018/978-1-6684-7832-5.ch006

However, educational leadership in blended higher educational contexts has yet to be addressed and is not explored widely (Day et al., 2016). Emphasis on its assessment and evaluation has been given less importance, especially in blended education (Habiballah et al., 2021). As a critical factor in the success of blended higher education, educational leadership determines school management and the role of teachers (Kovačević et al., 2020; Li & Yu, 2022). A lack of educational theory, practice, and policy make it difficult to ensure an education center's success and a smooth transfer from face-to-face classroom learning to blended learning (Wan et al., 2018). Exploratory studies are needed to adapt to the current educational environments in flux (Wenner & Campbell, 2017).

The authors thus seek to explore educational leadership in blended higher education contexts. The study pivots on the research questions: What is the new role of educational leadership in blended higher educational contexts? Can educational leadership directly affect students' academic achievement in blended higher educational contexts? What challenge does educational leadership encounter in blended higher educational contexts? To examine the roles of educational leadership in blended higher educational contexts, the requirements and significant challenges of educational leadership which benefits school management and students' academic success, are explored.

LITERATURE REVIEW

Educational Leadership

Educational leadership refers to an essential factor that significantly contributes to providing the teacher with an effective teaching process to offer students an optimal educational climate, which leads to successful learning achievements (Billingsley, 2007; Bush & Glover, 2014). Leadership in higher education can inspire faculty members to efficiently perform their work and give directions to realize specific learning goals or outcomes (Szeto, 2015). The realization of educational leadership can reduce dissatisfiers, such as working conditions and educational policy, so that teachers can provide their best work (Drew, 2010; Fields et al., 2019). Students can be motivated to achieve the required learning task and the desired goals (Hohepa, 2013; Johnson & Fournillier, 2022). Educational leadership plays a crucial role in enhancing student academic performance, especially those underachieved students (Bertrand & Rodela, 2018). Interestingly, although educational leadership has yet to be universally approved (Nworic, 2012), most definitions of leadership in blended higher educational contexts reflect the significant three aspects: influence, intention, and exercised influence (Spillane, 2004).

Educational leadership in blended higher educational contexts occurred worldwide and adapted to the continuously changing demands of blended learning environments (Martin & Marion, 2005). Learning environments undergo swift dynamics of change (e.g. traditional face-to-face, online, and blended learning) and pose a need for new competencies of leadership in the context of blended learning and managing (Habiballah et al., 2021; Taylor & Machado, 2006). Educational leadership in blended higher educational contexts positively impacts teachers' and students' perceptions of integrating web-based online learning in teaching (Eadens & Ceballos, 2022). On the other hand, it strengthens school culture, forms a collaborative structure, develops intellectual stimulation, establishes consensus, and models behaviour (Ng, 2008). In addition, the characteristics and practice of educational leadership in blended managerial and learning environments still need further research (Fields et al., 2019). Educational leadership, which conveys enthusiasm and inspiration to the working staff in the blended spaces, remains further research (McGinity et al., 2022).

Developments and Challenges

There have been diverse research perspectives on educational leadership since 1970 (Karadag, 2020). Educational leadership is a popular research subject associated with administrators and schools (Boyce & Bowers, 2018). Studies on educational leadership extend its concept into a broader framework, exploring the instructional and managerial aspects (Yang, 2021). These studies focus on instructional-centred leadership and differentiate between principals (or the sole instructional leaders) and team leadership (school leadership) (Price & Mansfield, 2021). Along with the newer conceptualization of educational leadership, several different leadership within the field of educational leadership are incorporated into research for learning frameworks (Berestova et al., 2020). Recently, educational leadership research dedicated to measuring leadership practices correlated with the educational leadership for learning conceptual framework, as leadership roles are not a single school leader but everyone engaging in leadership functions within higher education (Leithwood, 2021). Current studies have devoted themselves to what is necessary to ensure students' academic success and how teachers perceive educational leadership for learning ((Berestova et al., 2020).

However, the research findings of educational leadership demonstrate that there may be little direct effect on students' success within the field of blended higher educational contexts (Heck & Hallinger, 2014). There may be a need for more significant evidence that educational leadership behaviours significantly affect students' learning outcomes in measuring the effectiveness of educational leadership (Hendriks & Stein, 2012). The vast majority of educational leadership literature demonstrates that there can be indirect effects of educational leadership on students' academic performance due to teacher participation in leadership practices and shared academic climate (van der Gaag et al., 2006). Thus, there is a need for a better exploration of educational leadership pathways, including teachers.

Furthermore, leadership often needs clarification given higher educational contexts (Drew, 2010). It is necessary to examine meaningful ways to make educational leadership strategies more effective within the academic community (Martin & Marion, 2005). In addition, educational leadership does not always promote institutional functioning in influencing decisions and guiding people, as management involves implementing and conveying institutional policies (Leithwood, 2021). A meaningful understanding of new characteristics of educational leadership is helpful for faculty members to perform better in their work (McGinity et al., 2022). However, only some studies analyze current new leadership characteristics within a specific context (the blended higher educational contexts) (Chance, 2022).

Thus, we proposed the following alternative questions:

RQ1: What is the new role of educational leadership in blended higher educational contexts?
RQ2: Can educational leadership directly affect students' academic achievement in blended higher educational contexts?
RQ3: What challenge does educational leadership encounter in blended higher educational contexts?

METHODS

For the present study, a quantitative approach was adopted to examine the past literature on educational leadership in blended higher education contexts. We conducted a two-step method to ensure fair outcome reporting through a rapid assessment. Because manual censorship tended to be subjective when the researchers analyzed and interpreted related research findings (He et al., 2017). Thus, the researchers conducted a mixed-method systematic review combined with a traditional bibliometric analysis to identify the articles. The two steps were as follows:

This study searched online databases and obtained plentiful literature by a rapid evidence assessment review method. This study removed unrelated literature by extensively searching online databases. The researchers formed research themes based on clustering results after clustering the literature. To minimize selection or publication bias, the researchers included different online databases (Web of Science and Scopus databases). The researchers searched the three databases and obtained 449 results by keying in "educat* leadership" (topic) and "educat* leadership" OR challenge* (topic) and "educat* leadership" OR role* (topic) and "educat* leadership" OR character* (topic) and "educat* leadership" OR "academic achievement*" (topic) and "educat* leadership" OR "learning outcome*" (topic), ranging from the inception to 12 November 2022. The researchers obtained the results (N= 449) from plain texts.

This study excluded and included the obtained literature according to the Preferred Reporting Items for Systematic Review and Meta-analysis Protocol (PRISMA) (Figure 1) (Moher et al., 2009). Several international researchers collaborated to determine guidelines for a systematic review to enhance the representativeness of the included studies. The systematic review included each section of a review study: title, abstract, introduction, literature review, methods, results, discussion, and conclusion. The inter-rater reliability (k=0.93) is based on the systematically reviewed results. Two researchers prudently reviewed the literature. If both of them could not agree, a third reviewer would be invited to determine the selection. The studies could be included if they (1) came to convincing conclusions and were rigidly designed, (2) focused on educational leadership, or (3) could provide sufficient information or a definitive source for a review. The studies would be excluded if they (1) were irrelevant to the research based on titles or abstracts, (2) were irrelevant to educational leadership, (3) were not belong to the field of education, (4) were non-academic reports, letters, or editorial collections, (5) could not provide full-texts; or (6) had no abstracts. Finally, the researchers included 22 peer-reviewed articles (please see the note above the references) for the study.

RESULTS

RQ1: What Is the New Role of Educational Leadership in Blended Higher Educational Contexts?

Blended learning environments significantly depart from traditional administration and leadership policy in higher education (Garrison & Kanuka, 2004; Woods & Roberts, 2019). Blended learning effectively integrates face-to-face and web-based online learning experiences, which requires a fundamental recognition of leadership dynamics (Woods & Roberts, 2019). Educational leadership is inherently about improving the teaching and learning relationship. Unlike in the past, educational leadership may necessarily rethink and redesign the learning and teaching process in the context of blended education (Connolly et al., 2019). Blended higher educational contexts endue leadership with new responsibilities to create transformative environments that can move students to high levels of learning (Woods & Roberts, 2019). Meanwhile, enlisting and guiding faculty members' talents and energies will be a prime concern of educational leadership in blended higher educational contexts (Smith et al., 2017).

Figure 1. A flow chart of the literature inclusion and exclusion based on PRISMA

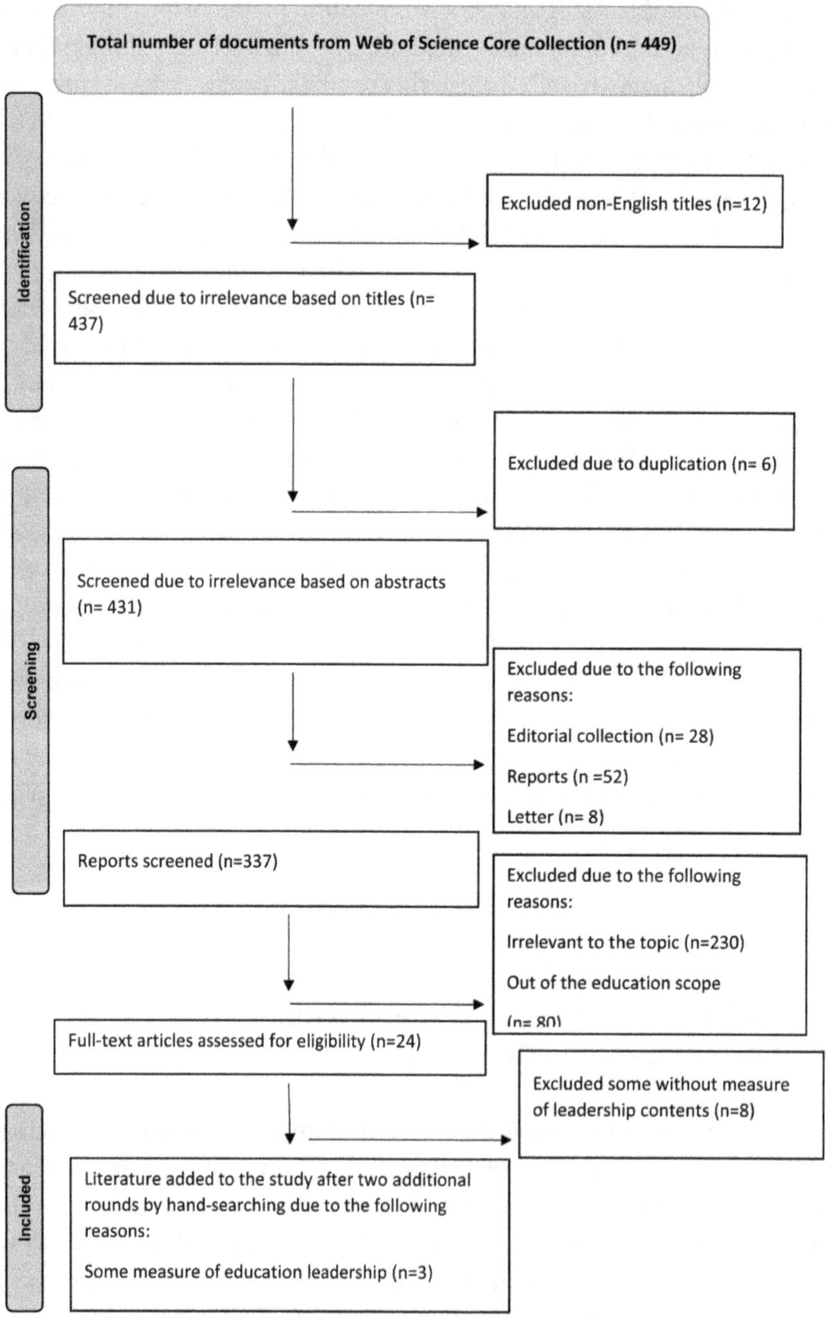

The new role of educational leadership also involves strategic direction and administration considering the higher educational goals (Douglass, 2018). Unlike a traditional higher education model, blended education includes scheduling. It supports the new structures and processes of a blended learning-teaching model, which requires educational leadership in blended higher educational contexts. The paradigm of strategic planning changes sufficiently to support the blended learning environments (Hallinger et al., 2022). For example, education leadership needs to redesign the watchword of higher education's promise in viewing the potential of information technology to support meaningful educational experiences (Garrison & Anderson, 2003). In blended higher education contexts, education leadership plays a significant role in developing operations to prop up more formal blended learning-teaching approaches based on higher education interests (Grant, 2009).

Education leadership requires a more effective and efficient approach to facilitating teaching activities (Cruz-González et al., 2021). Issues about the front end of leadership implementation in blended higher educational contexts fall into administrative support (Grant, 2017). In essence, leadership directs faculty members toward realizing shared objectives or outcomes (Murphy, 2019). Leadership is required in blended higher education to empower its followers to accomplish the required task and motivate them to perform their job effectively (Hallinger, 2018). An essential educational leadership quality requires introducing new ideas of teaching and enhancing excitement in teaching, as technology-mediated education may cause technology anxiety and teacher resistance (Johnson et al., 2012; Osika et al., 2009)

RQ2: Does Educational Leadership Directly Affect Students' Academic Achievement in Blended Higher Educational Contexts?

Results reveal that educational leadership significantly affects student achievement in blended higher educational contexts (De Nobile, 2018). Since the onset of the novel Coronavirus, educational leadership is slowly gaining prominence in improving student learning outcomes (Leithwood, 2021). The potential of educational leadership is closely correlated with student performance and guides professional development in student language learning (Garg & Gera, 2019). The overall relationship-related results indicate that educational leadership may not directly impact student academic achievement in blended higher educational contexts (Berestova et al., 2020). The relationship between educational leadership and student achievement can also depend on the leadership level (Price & Mansfield, 2021). When educational leadership studies are examined in the context of blended education, it is found that educational leadership has a medium-lever positive effect on student academic success (Kythreotis et al., 2010; Torrance & Humes, 2015).

However, the effect on student academic achievement of educational leadership in the blended higher education contexts appears destined to continue discussing (Boyce & Bowers, 2018). Because the explicit and implicit influence and the level of educational leadership are still ongoing discussions, education leadership theories would develop further (Roach et al., 2011). Whether this effect on student achievement is direct or indirect, many studies have proved that educational leadership has a positive effect (Özdemir et al., 2022). Although some studies demonstrate that educational leadership directly affects student academic performance, few studies support that there is an indirect effect on student academic performance. It can be generally identified that the direct effect of educational leadership on student achievement is unreliable (Choi et al., 2018; Fields et al., 2019).

RQ3: What Challenge Does Educational Leadership Encounter in Blended Higher Educational Contexts?

Blended Higher Education Demands

Educational leadership may encounter the challenges of education demands in blended higher educational contexts nowadays (Henderson, 2017). Blended learning is pervading higher education and practical education leadership to confront existing situations in higher education (Hohepa, 2013). Some higher education leaders are notorious resisters to adapt and change (Bertrand & Rodela, 2018). Higher education institutions, especially universities, may need to fully realize and accept the idea of dynamic education leadership, which can hinder the development of education leadership (Young, 2002). There is a need to remind that education leadership, especially in universities, is challenged to position to meet the demands of blended higher education (Acquaro, 2019). Blended learning environments in higher education require higher quality to meet the correlated demands and expectations for student learning experiences and outcomes (Anderson et al., 2018). Indeed, learning environment transformation requires education leaders to realize faculty members' benefits fully (Anderson et al., 2012). In essence, there are demands for education leaders to take responsibility for developing technology-based practices and sustainable learning in blended higher education (Archbald, 2016).

Education Leadership Practice

Blended higher educational contexts may bring about expanded leadership practice which leads school leaders to focus more on the core of education (Barth, 2013). Educational leadership may touch with the technical core of educational settings, which are not typically part of traditional leadership domains (Bertrand & Rodela, 2018). The additional practices also include: how to set directions which can identify specific and short-term goals, how to make high-performance expectations, and form a shared vision, how to stimulate the professional capacities of faculty members, establish trusting relationships with staff, and provide support for staff members, secure accountability such as internal accountability among staff and external accountability through fulfilling the requirements of staff members (Campanotta et al., 2018; Carver, 2016; Ceballos et al., 2020).

Technology Issue

Appropriately implementing information and communication technology in conventional teaching and learning settings in higher education becomes an ordeal for education leadership (Clayton et al., 2013). The challenge faced by policymakers and administrators is to find ways of shaping the experience of students and teachers through different forms that can be mediated by information and communication technology (Jacobson et al., 2015). This experience is essential to develop effective teaching models in blended higher education contexts (Cosner et al., 2015). Education leadership focuses on practical tips and practices that can further explore an instructional approach to shape blended synchronous learning based on students' and teachers' experiences (Cheng & Szeto, 2016). Blended higher education would break down without an instructional approach to examine blended synchronous learning, examine information technology gaps, and validate solutions. Education leadership needs to play a leadership role in applying technology and removing barriers to blended higher education development (Cunningham

& Lochmiller, 2020). Information and communication technology represents a potential challenge to academic faculty and leaders of education leadership (Diamond, 2014). Because learning technologies play a significant role in blended higher education, it is a question for higher education institutions to reveal the advantages of the transformational changes of blended learning and teaching format (Dolph, 2017). Leaders and academic faculty need to consider how to establish a viable and dynamic operation of higher education and reasonably change the traditional structure (Duke, 2019).

DISCUSSION

Adding the New Roles of Educational Leadership

Blended higher education contexts foster an environment that addresses the effective change-capability of education leadership (Fairman & Mackenzie, 2015). Facing individual student and faculty needs complexities, education leadership may increase relative skills shortage (Fraenkel et al., 2012). The requirements for an ability to adapt flexibly and operate are critical in the current complex leadership environment (Galey-Horn, 2020). In addition, the ability to extend outwards is considered a requirement relating to preparedness to take risks and deal with changes (Gerstenschlager & Barlow, 2019). Knowing how to problem-solve is a need for leadership development in a blended higher education environment (Grissom et al., 2019). Indeed, a recurring theme of education leadership development is the ability to deal with change and be clear-eyed in the context of rapid changes (Harris & Spillane, 2008). Academic and administrative leaders should make sense of the ambiguity of concomitant education at the current stage (Drew, 2010).

Blended higher education contexts give the emerging role of educational leadership as knowledge manager (Martin & Marion, 2005). Leadership in blended higher education maintains knowledge-based management and process. Knowledge management implies that today's higher education is in a new knowledge era in which leadership plays a significant role in maintaining an inviting environment to maximize worker ability (Drew, 2010). Knowledge-managing education leadership is necessary to the competitive edge in blended higher education contexts. Education leadership should be able to identify knowledge gaps and analyze their components; therefore, leaders can determine potential threats to the organization and seek appropriate resources (Martin & Marion, 2005). Moreover, they should be trained to be sensitive to knowledge management.

Blended higher education contexts appeal to accountability, social equity and collaboration of education leadership (Martin & Marion, 2005). Compared with before, education leadership takes much responsibility for managing the school or university. In addition, education leadership is most responsible for encouraging information sharing within blended higher education contexts (Drew, 2010). Educational leaders should pay more attention to social equality. They focus on improving practices in blended teaching and learning environments (Martin & Marion, 2005). Collaboration is emphasized with school people's different levels and duties (Galey-Horn, 2020). Thus, education leadership attaches more importance to relations and connections between subordinates, teachers, and government officials. Education leadership directs working with different people to improve the schools in blended higher education contexts.

Blended higher education contexts call for the new role of education leadership to improve individual ownership of educational leaders (Galey-Horn, 2020). Because some educational leaders seldom play a leading role in the university or the school, they cannot get their share of the responsibilities in the school missions (Galey-Horn, 2020). The directive and monitoring role of education leaders tends to impede education development. This kind of education leadership can significantly improve teachers' professional development and students' academic success (Zhang et al., 2017)). Successful education leadership should allow leaders to develop individual ownership (Gerstenschlager & Barlow, 2019). It is beneficial to develop education leadership healthier. With the increasing use of information and communication technology, all educational communities' facets should be considered to make higher education development more robust. Education leadership should keep a high level of performance to promote educational development.

Blended higher education contexts require leadership to prepare for a highly competitive global environment (Taylor & Machado, 2006). Education leadership must inevitably find ways to keep organizational capacities with ever-increasing environmental demands and opportunities, which is a big responsibility for its new role (Harris & Spillane, 2008). Educational leaders must carry duties for the development of innovative strategies in blended higher education contexts. It is recognized that successful strategic planning and management are the heart of the ongoing performance of education leadership (Galey-Horn, 2020). Therefore, higher education institutions and organizations should improve the current situation of education leadership, which should be promoted a more proactive approach to strategic management. In the meantime, education leadership should employ leaders with the best and brightest minds due to the complexities of modern leadership demands (Galey-Horn, 2020).

Blended higher education contexts need leadership to transition education to multiple delivery modes (Nworie, 2012). Higher education leadership tends to vary significantly in areas of online learning during the epidemic period (Galey-Horn, 2020). Higher education institutions may rely on information and communication technologies to conduct distance education alongside traditional learning in the post-pandemic era (Galey-Horn, 2020). Effective education leadership can ensure the success and sustainability of online education along with conventional learning environments (Nworie, 2012). The roles and responsibilities of educational leadership are to guide institutions to act as an intermediaries and provide suitable types of programs to meet the needs of students and teachers (Nworie, 2012).

Education Leadership Role in Student Academic Success

Education leadership is a significant support for student academic success at the policy level (Karadag, 2020). School leaders, policymakers, education systems, and practitioners are accountable for student academic success in education leadership (Hashim, 2020). Education leadership affects learning through strategic actions that facilitate the structural and academic processes of student learning (Leithwood et al., 2020). In addition, the indirect effect of education leadership on student learning outcomes can be measurable (McAdamis, 2010). Although there are measurement limitations, the leadership of blended higher education is the key performance indicator in student academic achievements (Ni et al., 2019). It cannot be denied that universities still account for a small portion of influencing student learning outcomes compared with social-economic status or family background (Galey-Horn, 2020).

Principals or school leaders contribute to improving student learning through their effects on education leadership at school levels, such as school culture, academic structures and processes, and teachers (Leithwood et al., 2019). Although principals or school leaders do not teach students, their effects on student academic success are achieved via their impact on faculty staff (Hollingworth et al., 2018). For example, education leadership can influence student academic achievements using principal involvement, which can improve school culture and climate (Karadag, 2020). In addition, positive education leadership can clear academic mission and capacity and enhance collective teacher efficacy and staff motivation, leading to staff professional learning (Honig et al., 2017). Furthermore, education leadership can deploy organizational structure and resource allocation to impact student learning.

Education leadership can enhance student learning outcomes through personal antecedent (e.g., principal knowledge and beliefs) effects (Karadag, 2020). The personal values of principals or school leaders can be considered when researchers examine the potential enhancing factor of student learning outcomes (Hayes & Irby, 2020). It is believed that personal antecedents of school leaders are relevant in enacting education leadership in practice, which results in improved student learning. Education leadership embodies the personal antecedent of leaders (thinking, expectations, beliefs, and experience), shaping the normative statement in which all students can learn.

Education leadership can affect student academic achievements through the impacts of environmental conditions (Hayes & Irby, 2020). Environmental conditions include internal school conditions (e.g., teacher quality, university size and level, student composition, university settings, and context conditions (e.g., socio-economic situation) which moderate environmental conditions on student academic achievements (Leithwood, 2021). Thus, striking university characteristics and decisive education leadership may improve student learning outcomes. Practitioners or researchers can explore the most suitable leadership style to enhance student learning, as there may be no single or the most suitable style for student learning in blended higher education contexts (Hashim, 2020). From the perspective of comprehensive effects, each university is in its way of school improvement and enhancement of student learning. Moreover, education leadership can be more direct in challenging circumstances.

There are several ways for education leadership to improve student learning outcomes (Galey-Horn, 2020). Improving student learning outcomes is the core of education leadership practice in blended higher education contexts (Karadag, 2020). The high-performing education leadership should primarily focus on the quality of teaching. For example, education leadership for student learning should pay great attention to getting the right people to become teachers (Nworie, 2012). It is necessary to construct a system to ensure that education leadership can direct qualified teachers to achieve their full potential in teaching. Successful leadership can prepare new teachers for entry into their professions and provide them with multiple opportunities which ensure teachers are effective in blended teaching (Karadag, 2020).

Education leadership can improve student learning outcomes through school leaders' monitoring and dialogue (Ni et al., 2019). Monitoring enables education leadership to visit online or face-to-face teaching and learning activities and provide timely feedback (Galey-Horn, 2020). Student learning progress and outcome data can be obtained via monitoring, which has become an indispensable part of education in blended higher education contexts (Ni et al., 2019). In addition, education leadership can facilitate student learning outcomes through dialogue. Dialogue can allow teachers and students to discuss their unique learning needs. It is beneficial for school leaders to support pedagogic development and the performance trends across the university in blended higher educational contexts (Galey-Horn, 2020).

Education Leadership Responding to Significant Challenges

There are several ways by which education leadership can be full of potential and dynamism (Garrison & Kanuka, 2004). Education leadership must be prepared to answer the challenges in blended higher educational contexts in the following ways: policy support and planning (Hashim, 2020). Universities and higher education institutions may draft policies to guide technological innovation and preserve the conventional values of higher education to enhance leadership (Leithwood, 2021). When thoughtfully designed, education leadership can help work for a favourable learning environment in blended higher educational contexts (McAdamis, 2010). The innovative use of information and communication technology can contribute to the influence of education leadership, which eventually helps build a normative environment in blended higher education (Hollingworth et al., 2018). Moreover, universities and higher education institutions may draft policies to produce different types of pedagogy to adapt leadership to blended higher educational contexts (Honig et al., 2017). School leaders or principals should recognize that it is essential to connect face-to-face and online components with online instructors (Hayes & Irby, 2020). Additionally, education leadership may focus on rethinking the organizational and overarching pedagogical structures to achieve a continuity of experience (Orphanos & Orr, 2014).

Education leadership has its position in blended higher educational contexts to consider social justice for students and requires more planning (Orr et al., 2019). In essence, teaching changes due to digital activities produce differences; therefore, education leaders should consider a more complex conceptualization of higher education (VanGronigen et al., 2018). Blended higher educational contexts require profound changes in ideas, contents, and pedagogical methods of education (Tahir et al., 2020). Education leadership may shoulder extra responsibility for students and teachers and notice that their voices are heard. To address this issue, school leaders may provide a supportive environment for students (Perrone & Tucker, 2019). As a result of the blended learning experience, some students may face difficulty with tools or a lack of digital literacy, which is required to be successful in the learning process (Woulfin, 2018). It suggests setting up a supervising and directing system of education leadership to offer specific policy treatment, such as increasing the fundraising channel. Thus, education leadership needs to be further perfected (Tingle et al., 2019). Education leadership may commit to addressing social justice issues, breaking down barriers, building trust, strengthening relationships, and opening more educational opportunities (Sweeney & Mausbach, 2019).

The change in the new roles of educational leadership might bring adjustment, conflict, or role ambiguity (McQuay, 2002). School administrators and teachers may face overload and redesign of work. In power-centred roles, principals tend to lack the necessary knowledge and skills to make the right decisions, leading to dispersed leadership (Bishop & Mane, 2004). Therefore, educational leadership in blended higher education may seek teacher participation in decision-making to accumulate formal authority (Boateng, 2012). Effective educational leadership can promote participation in school decisions. Although decisive educational leadership may offer comprehensive support to demonstrate organizational values and school culture, it also requires avoiding being morally compulsory (Bishop & Mane, 2004). Greater capacities of educational leadership may lead to greater productivity (McQuay, 2002). Therefore, educational leadership may identify best practices to strengthen job security for teachers.

Furthermore, in blended higher educational contexts, educational leadership may intensely focus on academic freedom (Orozco & Allison, 2008). Educational leadership, which is going through a transition—has certain advantages—provide intellectual stimulation and creates an environment of academic freedom for teachers and scholars (Ruan, 2021). Transitional educational leadership can influence insistence on academic freedom and establish an excellent academic atmosphere for individual scholars (Ruan, 2021). Educational leadership is essential in encouraging teachers to be more productive in their academic areas (Sifuna, 2012).

As a process of educational leadership development, the purpose of education may justify itself as a necessary provision for andragogy (Henschke, 2011). Andragogy refers to adult education emphasizing self-directed learning (Hartree, 1984; Knowles, 1984). Andragogy is different for child learners based on traditional pedagogy (Knowles, 1996). In blended higher educational contexts, educational leadership should advocate pedagogical concepts that advocate autonomous learning in light of improving education capability. Meanwhile, teachers can understand individual learners as a facilitator of learning (Hartree, 1984). Under a blended teaching mode, knowledge transfer appeals education should prepare adult learners for self-aided study (Chan, 2010). Educational leadership should face the changes in teaching space accordingly and consider which it evolves have much to offer given the mixture of face-to-face education with information technology-based online education (Hartree, 1984). Furthermore, there are still some constraints to implementing blended education, such as suitable infrastructure. Educational leadership should consider how it will adapt to the significant needs to manage the change from traditional education to blended education and respond to the needs of learners, teachers, scholars, and principals themselves (Ward, 2003).

CONCLUSION

Major Findings

It is inevitable that education leadership steadily assumes new roles and responsibilities in blended higher educational contexts. Strong education leadership and clear policy can make sure faculty members and students adapt to blended teaching and learning environments. In addition, educational leadership dramatically affects students' academic achievements in blended higher educational contexts. However, educational leadership will encounter challenges in blended higher educational contexts in facilitating a blended higher learning experience, providing an inviting school learning environment, and guiding direction to meet teachers' and students' perceptions.

Limitations

There are several limitations to this study. First, studies cited in this research may be insufficient due to the databases and keywords used in this systematic review. Second, the study on education leadership is limited by the researchers' resource availability. Third, the research only focuses on English articles about education leadership in blended higher educational contexts. Some significant findings might be neglected, and this might result in the bias of the research findings in this study.

Future Research Directions

The future study will use qualitative and quantitative research methods to examine education leadership development programs related to student learning outcomes. The researchers will explore how students participate in academic development programs and confirm that education leadership can be strengthened. This future study aims to examine how education leadership competencies can better adapt and be applied in blended academic contexts. Meanwhile, the researchers will analyze if education leadership can better support and ensure a meeting of the technical and teaching methods, which may provide teachers and students with an inviting learning and teaching environment.

Another direction of future research on education leadership may reveal various levels of education leadership among universities due to increasingly more complicated institutions and sectors. At the same time, the researchers will explore productive ways to examine the effectiveness of education leadership in blended higher educational contexts. In addition, the researchers will examine the differences among the professional roles of education leadership, which may provide some potential implications for practice and policy. Considering the different standards of education leadership in traditional and blended higher educational contexts, the researchers will study a framework for addressing the needs of various education leadership roles. Thus, future studies might guide school leaders or policymakers to develop or revise standards.

The future study will discuss how to improve the quality of education leadership through empirical support. The quality of education leadership is an important variable in securing positive university outcomes and student academic performance. In addition, the researcher will explore how to widen leadership participation, as there needs to be more evidence about the most effective way to enhance the quality of education leadership within blended higher educational contexts. The future study will involve some strong managerial leadership and examine the sustainable progress to achieve as the role of education leadership is second only to classroom teaching in school development.

REFERENCES

Acquaro, D. (2019). Preparing the next generation of educational leaders: Initiating a leadership discourse in initial teacher education. *International Studies in Educational Administration*, *47*, 107–124.

Anderson, E., Winn, K. M., Young, M. D., Groth, C., Korach, S., Pounder, D., & Rorrer, A. K. (2018). Examining university leadership preparation: An analysis of program attributes and practices. *Journal of Research on Leadership Education*, *13*(4), 375–397. doi:10.1177/1942775117735873

Anderson, S. E., Mascall, B., Stiegelbauer, S., & Park, J. (2012). No one way: Differentiating school district leadership and support for school improvement. *Journal of Educational Change*, *13*(4), 403–430. doi:10.100710833-012-9189-y

Archbald, D. (2016). System-Level Instructional Leadership—A District-Level Leadership Case: Implementing PLCs in Schools. *The International Journal of Educational Leadership Preparation*, *11*(2), 116–151.

Barth, R. S. (2013). The time is ripe (again). *Educational Leadership*, *71*(2), 10–16.

Batagiannis, S. C. (2011). Promise and possibility for aspiring principals: An emerging leadership identity through learning to do action research. *Qualitative Report*, *16*(5), 1304–1329.

Berestova, A., Gayfullina, N., & Tikhomirov, S. (2020). Leadership and Functional Competence Development in Teachers: World Experience. *International Journal of Instruction*, *13*(1), 607–622. doi:10.29333/iji.2020.13139a

Bertrand, M., & Rodela, K. C. (2018). A framework for rethinking educational leadership in the margins: Implications for social justice leadership preparation. *Journal of Research on Leadership Education*, *13*(1), 10–37. doi:10.1177/1942775117739414

Billingsley, B. S. (2007). Recognizing and supporting the critical roles of teachers in special education leadership. *Exceptionality*, *15*(3), 163–176. doi:10.1080/09362830701503503

Bishop, J. H., & Mane, F. (2004). The impacts of career-technical education on high school labor market success. *Economics of Education Review*, *23*(4), 381–402. doi:10.1016/j.econedurev.2004.04.001

Boateng, C. (2012). *Restructuring vocational and technical education in Ghana: The role of leadership development*. Academic Press.

Boyce, J., & Bowers, A. J. (2018). Different levels of leadership for learning: Investigating differences between teachers individually and collectively using multilevel factor analysis of the 2011-2012 Schools and Staffing Survey. *International Journal of Leadership in Education*, *21*(2), 197–225. doi:10.1080/13603124.2016.1139187

Bush, T. (2007). Educational leadership and management: Theory, policy and practice. *South African Journal of Education*, *27*(3), 391–406.

Bush, T., & Glover, D. (2014). School leadership models: What do we know? *School Leadership & Management*, *34*(5), 553–571. doi:10.1080/13632434.2014.928680

Campanotta, L., Simpson, P., & Newton, J. (2018). Program quality in leadership preparation programs: An assessment tool. *Education*, *138*(3), 219–228.

Carver, C. L. (2016). Transforming identities: The transition from teacher to leader during teacher leader preparation. *Journal of Research on Leadership Education*, *11*(2), 158–180. doi:10.1177/1942775116658635

Ceballos, M., Buckridge, H., & Taylor, R. T. (2020). Educational Leadership Students and Mixed Reality Experiences: Building Student Confidence to Communicate with Parents and Teachers. *The International Journal of Educational Leadership Preparation*, *15*(1), 58–71.

Chan, S. (2010). Applications of andragogy in multi-disciplined teaching and learning. *Journal of Adult Education*, *39*(2), 25–35.

Chance, N. L. (2022). Resilient leadership: A phenomenological exploration into how black women in higher education leadership navigate cultural adversity. *Journal of Humanistic Psychology*, *62*(1), 44–78. doi:10.1177/00221678211003000

Cheng, A. Y., & Szeto, E. (2016). Teacher leadership development and principal facilitation: Novice teachers' perspectives. *Teaching and Teacher Education*, *58*, 140–148. doi:10.1016/j.tate.2016.05.003

Choi, E. H., Kim, E. K., & Kim, P. B. (2018). Effects of the educational leadership of nursing unit managers on team effectiveness: Mediating effects of organizational communication. *Asian Nursing Research*, *12*(2), 99–105. doi:10.1016/j.anr.2018.03.001 PMID:29614322

Clayton, J. K., Sanzo, K. L., & Myran, S. (2013). Understanding mentoring in leadership development: Perspectives of district administrators and aspiring leaders. *Journal of Research on Leadership Education*, *8*(1), 77–96. doi:10.1177/1942775112464959

Comstock, M., Supovitz, J., & Kaul, M. (2021). Exchange quality in teacher leadership ties: Examining relational quality using social network and leader-member exchange theories. *Journal of Professional Capital and Community*, *6*(4), 395–409. doi:10.1108/JPCC-01-2021-0002

Connolly, M., James, C., & Fertig, M. (2019). The difference between educational management and educational leadership and the importance of educational responsibility. *Educational Management Administration & Leadership, 47*(4), 504–519. doi:10.1177/1741143217745880

Cosner, S., Tozer, S., Zavitkovsky, P., & Whalen, S. P. (2015). Cultivating exemplary school leadership preparation at a research intensive university. *Journal of Research on Leadership Education, 10*(1), 11–38. doi:10.1177/1942775115569575

Cruz-González, C., Rodríguez, C. L., & Segovia, J. D. (2021). A systematic review of principals' leadership identity from 1993 to 2019. *Educational Management Administration & Leadership, 49*(1), 31–53. doi:10.1177/1741143219896053

Cunningham, K. M., & Lochmiller, C. R. (2020). Content-specific leadership: Identifying literature-based implications for principal preparation. *Journal of Research on Leadership Education, 15*(4), 261–282. doi:10.1177/1942775119845004

Day, C., Gu, Q., & Sammons, P. (2016). The impact of leadership on student outcomes: How successful school leaders use transformational and instructional strategies to make a difference. *Educational Administration Quarterly, 52*(2), 221–258. doi:10.1177/0013161X15616863

De Nobile, J. (2018). Towards a theoretical model of middle leadership in schools. *School Leadership & Management, 38*(4), 395–416. doi:10.1080/13632434.2017.1411902

Diamond, A. (2014). Pre-service early childhood educators' leadership development through reflective engagement with experiential service learning and leadership literature. *Australasian Journal of Early Childhood, 39*(4), 12–20. doi:10.1177/183693911403900403

Dolph, D. (2017). Challenges and opportunities for school improvement: Recommendations for urban school principals. *Education and Urban Society, 49*(4), 363–387. doi:10.1177/0013124516659110

Douglass, A. (2018). Redefining leadership: Lessons from an early education leadership development initiative. *Early Childhood Education Journal, 46*(4), 387–396. doi:10.100710643-017-0871-9

Drew, G. (2010). Issues and challenges in higher education leadership: Engaging for change. *Australian Educational Researcher, 37*(3), 57–76. doi:10.1007/BF03216930

Duke, D. L. (2019). Judgment and the preparation of educational leaders. *Journal of Research on Leadership Education, 14*(3), 191–211. doi:10.1177/1942775117752455

Eadens, D. W., & Ceballos, M. (2022). Educational Leadership Preparation and Professional Roles: Are We Serving the Needs of Leadership Roles Along the Leadership Continuum? *Journal of Research on Leadership Education, 19427751221076416*. Advance online publication. doi:10.1177/19427751221076416

Fairman, J. C., & Mackenzie, S. V. (2015). How teacher leaders influence others and understand their leadership. *International Journal of Leadership in Education, 18*(1), 61–87. doi:10.1080/13603124.2014.904002

Fields, J., Kenny, N. A., & Mueller, R. A. (2019). Conceptualizing educational leadership in an academic development program. *The International Journal for Academic Development, 24*(3), 218–231. doi:10.1080/1360144X.2019.1570211

Fraenkel, J. R., Wallen, N. E., & Hyun, H. H. (2012). *How to design and evaluate research in education* (Vol. 7). McGraw-hill.

Galey-Horn, S. (2020). Capacity-building for district reform: The role of instructional-coach teams. *Teachers College Record, 122*(10), 1–40. doi:10.1177/016146812012201003

Garg, N., & Gera, S. (2019). Gratitude and leadership in higher education institutions: Exploring the mediating role of social intelligence among teachers. *Journal of Applied Research in Higher Education, 12*(5), 915–926. Advance online publication. doi:10.1108/JARHE-09-2019-0241

Garrison, D. R., & Anderson, T. (2003). *E-learning in the 21st century: A framework for research and practice.* London: Routledge/Falmer. doi:10.4324/9780203166093

Garrison, D. R., & Kanuka, H. (2004). Blended learning: Uncovering its transformative potential in higher education. *The Internet and Higher Education, 7*(2), 95–105. doi:10.1016/j.iheduc.2004.02.001

Gerstenschlager, N. E., & Barlow, A. T. (2019). Transitioning from practicing teacher to teacher leader: A case study. *Teacher Development, 23*(1), 18–35. doi:10.1080/13664530.2018.1515106

Grant, C. (2009). Towards a conceptual understanding of education leadership: Place, space and practices. *Education as Change, 13*(1), 45–57. doi:10.1080/16823200902933529

Grant, C. (2017). Distributed leadership in South Africa: Yet another passing fad or a robust theoretical tool for investigating school leadership practice? *School Leadership & Management, 37*(5), 457–475. doi:10.1080/13632434.2017.1360856

Grissom, J. A., Mitani, H., & Woo, D. S. (2019). Principal preparation programs and principal outcomes. *Educational Administration Quarterly, 55*(1), 73–115. doi:10.1177/0013161X18785865

Habiballah, S., Bibu, N., & Danaiata, D. (2021). Educational Leadership and ICT Implementation in Israeli Arab Sector-towards a Model of Hybrid Leadership. *Revista de Management Comparat Interna-tional, 22*(1), 74–86. doi:10.24818/RMCI.2021.1.74

Hallinger, P. (2013). A conceptual framework for systematic reviews of research in educational leadership and management. *Journal of Educational Administration, 51*(2), 126–149. doi:10.1108/09578231311304670

Hallinger, P. (2018). Surfacing a hidden literature: A systematic review of research on educational leadership and management in Africa. *Educational Management Administration & Leadership, 46*(3), 362–384. doi:10.1177/1741143217694895

Hallinger, P., & Kovačević, J. (2022). Mapping the intellectual lineage of educational management, administration and leadership, 1972–2020. *Educational Management Administration & Leadership, 50*(2), 192–216. doi:10.1177/17411432211006093

Hanafi, Y., Taufiq, A., Saefi, M., Ikhsan, M. A., Diyana, T. N., Thoriquttyas, T., & Anam, F. K. (2021). The new identity of Indonesian Islamic boarding schools in the "new normal": The education leadership response to COVID-19. *Heliyon, 7*(3), e06549. doi:10.1016/j.heliyon.2021.e06549 PMID:33851049

Harris, A., & Spillane, J. (2008). Distributed leadership through the looking glass. *Management in Education, 22*(1), 31–34. doi:10.1177/0892020607085623

Hart, A. W. (1995). Reconceiving school leadership: Emergent views. *The Elementary School Journal, 96*(1), 9–28. doi:10.1086/461812

Hartree, A. (1984). Malcolm Knowles' theory of andragogy: A critique. *International Journal of Lifelong Education, 3*(3), 203–210. doi:10.1080/0260137840030304

Hashim, A. K. (2020). Coaching and districtwide improvement: Exploring the systemic leadership practices of instructional coaches. *Teachers College Record, 122*(10), 1–44. doi:10.1177/016146812012201005

Hayes, S. D., & Irby, B. J. (2020). Challenges in preparing aspiring principals for instructional leadership: Voices from the field. *International Journal of Leadership in Education, 23*(2), 131–151. doi:10.1080/13603124.2018.1562102

He, Q., Wang, G., Luo, L., Shi, Q., Xie, J., & Meng, X. (2017). Mapping the managerial areas of building information modeling (BIM) using scientometric analysis. *International Journal of Project Management, 35*(4), 670–685. doi:10.1016/j.ijproman.2016.08.001

Heck, R. H., & Hallinger, P. (2014). Modeling the longitudinal effects of school leadership on teaching and learning. *Journal of Educational Administration, 52*(5), 653–681. doi:10.1108/JEA-08-2013-0097

Henderson, L. (2017). 'Someone had to have faith in them as professionals': An evaluation of an action research project to develop educational leadership across the early years. *Educational Action Research, 25*(3), 387–401. doi:10.1080/09650792.2016.1173566

Hendriks, M., & Steen, R. (2012). Results from school leadership effectiveness studies (2005–2010). In J. Scheerens (Ed.), *School Leadership Effects Revisited* (pp. 65–129). Springer. doi:10.1007/978-94-007-2768-7_4

Henschke, J. A. (2011). Considerations regarding the future of andragogy. *Adult Learning, 22*(1), 34–37. doi:10.1177/104515951102200109

Hohepa, M. K. (2013). Educational leadership and indigeneity: Doing things the same, differently. *American Journal of Education, 119*(4), 617–631. doi:10.1086/670964

Hollingworth, L., Olsen, D., Asikin-Garmager, A., & Winn, K. M. (2018). Initiating conversations and opening doors: How principals establish a positive building culture to sustain school improvement efforts. *Educational Management Administration & Leadership, 46*(6), 1014–1034. doi:10.1177/1741143217720461

Honig, M. I., Venkateswaran, N., & McNeil, P. (2017). Research use as learning: The case of fundamental change in school district central offices. *American Educational Research Journal, 54*(5), 938–971. doi:10.3102/0002831217712466

Jacobson, S., McCarthy, M., & Pounder, D. (2015). What makes a leadership preparation program exemplary? *Journal of Research on Leadership Education, 10*(1), 63–76. doi:10.1177/1942775115569420

Johnson, N. N., & Fournillier, J. B. (2022). Increasing diversity in leadership: Perspectives of four Black women educational leaders in the context of the United States. *Journal of Educational Administration and History, 54*(2), 174–192. doi:10.1080/00220620.2021.1985976

Johnson, T., Wisniewski, M. A., Kuhlemeyer, G., Isaacs, G., & Krzykowski, J. (2012). Technology adoption in higher education: Overcoming anxiety through faculty bootcamp. *Journal of Asynchronous Learning Networks, 16*(2), 63–72. doi:10.24059/olj.v16i2.240

Karadag, E. (2020). The effect of educational leadership on students' achievement: A cross-cultural meta-analysis research on studies between 2008 and 2018. *Asia Pacific Education Review, 21*(1), 49–64. doi:10.100712564-019-09612-1

Knowles, M. (1996). *Andragogy: An emerging technology for adult learning*. Academic Press.

Knowles, M. S. (1984). *Theory of andragogy*. Jossey-Bass.

Kovačević, J., Mujkić, A., & Kapo, A. (2020). Examining school leadership in a transitional context: A mixed-methods study of leadership practices and school cultures as mechanisms of educational change. *Educational Management Administration & Leadership*. Advance online publication. doi:10.1177/1741143220971286

Kythreotis, A., Pashiardis, P., & Kyriakides, L. (2010). The influence of school leadership styles and culture on students' achievement in Cyprus primary schools. *Journal of Educational Administration, 48*(2), 218–240. doi:10.1108/09578231011027860

Leithwood, K. (2021). A review of evidence about equitable school leadership. *Education Sciences, 11*(8), 377. doi:10.3390/educsci11080377

Leithwood, K., Harris, A., & Hopkins, D. (2020). Seven strong claims about successful school leadership revisited. *School Leadership & Management, 40*(1), 5–22. doi:10.1080/13632434.2019.1596077

Leithwood, K., Mascall, B., & Strauss, T. (2009). What we have learned and where we go from here. In K. Leithwood, B. Mascall, & T. Strauss (Eds.), *Distributed leadership according to the evidence* (pp. 269–282). Routledge. doi:10.4324/9780203868539

Leithwood, K., Sun, J., & McCullough, C. (2019). How school districts influence student achievement. *Journal of Educational Administration, 57*(5), 519–539. Advance online publication. doi:10.1108/JEA-09-2018-0175

Li, M., & Yu, Z. (2022). Teachers' satisfaction, role, and digital literacy during the COVID-19 pandemic. *Sustainability, 14*(3), 1121. doi:10.3390u14031121

Martin, J. S., & Marion, R. (2005). Higher education leadership roles in knowledge processing. *The Learning Organization, 12*(2), 140–151. Advance online publication. doi:10.1108/09696470510583520

McAdamis, S. (2010). Central office and school leaders create a conversation. *The Learning Professional, 31*(3), 24.

McGinity, R., Heffernan, A., & Courtney, S. J. (2022). Mapping trends in educational-leadership research: A longitudinal examination of knowledge production, approaches and locations. *Educational Management Administration & Leadership, 50*(2), 217–232. doi:10.1177/17411432211030758

McQuay, P. (2002). *A Discussion Paper on Vocational Technical Education in the United States of America*. Academic Press.

Moher, D., Liberati, A., Tetzlaff, J., & Altman, D. G. (2009). Preferred reporting items for systematic reviews and meta-analyses: The PRISMA statement. *Annals of Internal Medicine, 151*(4), 264–269. doi:10.7326/0003-4819-151-4-200908180-00135 PMID:19622511

Murphy, G. (2019). A systematic review and thematic synthesis of research on school leadership in the Republic of Ireland: 2008–2018. *Journal of Educational Administration, 57*(6), 675–689. Advance online publication. doi:10.1108/JEA-11-2018-0211

Ng, W. L. (2008). Transformational leadership and the integration of information and communications technology into teaching. *The Asia-Pacific Education Researcher, 17*(1), 1–14. doi:10.3860/taper.v17i1.346

Ni, Y., Rorrer, A. K., Pounder, D., Young, M., & Korach, S. (2019). Leadership matters: Preparation program quality and learning outcomes. *Journal of Educational Administration, 57*(2), 185–206. Advance online publication. doi:10.1108/JEA-05-2018-0093

Nworie, J. (2012). Applying leadership theories to distance education leadership. *Online Journal of Distance Learning Administration, 15*(5), 1–16.

Orozco, L., & Allison, B. (2008). Looking Within: Examining Positive Relationships and Healthy Organizational Cultures in Departments of Educational Leadership. *Educational Leadership and Administration: Teaching and Program Development, 20*, 64–76.

Orphanos, S., & Orr, M. T. (2014). Learning leadership matters: The influence of innovative school leadership preparation on teachers' experiences and outcomes. *Educational Management Administration & Leadership, 42*(5), 680–700. doi:10.1177/1741143213502187

Orr, M. T., Hollingworth, L., & Beaudin, B. (2019). Performance assessment for school leaders: Comparing field trial and implementation results. *Journal of Educational Administration, 58*(1), 38–59. doi:10.1108/JEA-05-2018-0087

Osika, E. R., Johnson, R. Y., & Buteau, R. (2009). Factors influencing faculty use of technology in online instruction: A case study. *Online Journal of Distance Learning Administration, 12*(1), 1–14.

Özdemir, N., Gümüş, S., Kılınç, A. Ç., & Bellibaş, M. Ş. (2022). A systematic review of research on the relationship between school leadership and student achievement: An updated framework and future direction. *Educational Management Administration & Leadership*. Advance online publication. doi:10.1177/17411432221118662

Perrone, F., & Tucker, P. D. (2019). Shifting profile of leadership preparation programs in the 21st century. *Educational Administration Quarterly, 55*(2), 253–295. doi:10.1177/0013161X18799473

Price, D. C., & Mansfield, K. C. (2021). Leadership in the time of COVID: Connecting community resources to meet the needs of North Carolina students. In *Frontiers in Education* (p. 313). Frontiers. doi:10.3389/feduc.2020.615101

Roach, V., Smith, L. W., & Boutin, J. (2011). School leadership policy trends and developments: Policy expediency or policy excellence? *Educational Administration Quarterly, 47*(1), 71–113. doi:10.1177/0011000010378611

Ruan, N. (2021). Accumulating academic freedom for intellectual leadership: Women professors' experiences in Hong Kong. *Educational Philosophy and Theory*, *53*(11), 1097–1107. doi:10.1080/00131857.2020.1773797

Siddique, A., Aslam, H. D., Khan, M., & Fatima, U. (2011). Impact of academic leadership on faculty's motivation, and organizational effectiveness in higher education system. *International Journal of Academic Research*, *3*(3).

Sifuna, D. N. (2012). Leadership in Kenyan public universities and the challenges of autonomy and academic freedom: An overview of trends since independence. *Journal of Higher Education in Africa/Revue de l'enseignement supérieur en Afrique*, *10*(1), 121-137.

Smith, D., Kelly, D., & Allard, C. (2017). Dialogic spaces: A critical policy development perspective of educational leadership qualifications. *International Journal of Leadership in Education*, *20*(4), 393–415. doi:10.1080/13603124.2015.1116608

Spillane, J. P. (2004). Educational leadership. *Educational Evaluation and Policy Analysis*, *26*(2), 169–172. doi:10.3102/01623737026002169

Strom, K., & Porfilio, B. (2019). Critical hybrid pedagogies: A self-study inquiry into faculty practices in a blended educational leadership EdD program. *E-Learning and Digital Media*, *16*(1), 1–14. doi:10.1177/2042753018811871

Sweeney, D., & Mausbach, A. (2019). Creating Powerful Principal and Coach Partnerships. *Educational Leadership*, *76*(6), 30–35.

Szeto, E. (2015). Community of Inquiry as an instructional approach: What effects of teaching, social and cognitive presences are there in blended synchronous learning and teaching? *Computers & Education*, *81*, 191–201. doi:10.1016/j.compedu.2014.10.015

Tahir, L. M., Musah, M. B., Al-Hudawi, S. H. V., & Daud, K. (2020). Becoming a Teacher Leader: Exploring Malaysian In-Service Teachers' Perceptions, Readiness and Challenges. *Education & Science/Egitim ve Bilim*, *45*(202).

Taylor, J., & Machado, M. D. L. (2006). Higher education leadership and management: From conflict to interdependence through strategic planning. *Tertiary Education and Management*, *12*(2), 137–160. doi:10.1080/13583883.2006.9967164

Tingle, E., Corrales, A., & Peters, M. L. (2019). Leadership development programs: Investing in school principals. *Educational Studies*, *45*(1), 1–16. doi:10.1080/03055698.2017.1382332

Torrance, D., & Humes, W. (2015). The shifting discourses of educational leadership: International trends and Scotland's response. *Educational Management Administration & Leadership*, *43*(5), 792–810. doi:10.1177/1741143214535748

van der Gaag, M., Hoffman, T., Remijsen, M., Hijman, R., de Haan, L., van Meijel, B., van Harten, P., Valmaggia, L., de Hert, M., Cuijpers, A., & Wiersma, D. (2006). The fivefactor model of the positive and negative syndrome scale II: A ten-fold cross-validation of a revised model. *Schizophrenia Research*, *85*(1), 280–287. doi:10.1016/j.schres.2006.03.021 PMID:16730429

VanGronigen, B. A., Cunningham, K. M., & Young, M. D. (2018). How exemplary educational leadership preparation programs hone the interpersonal-intrapersonal (i2) skills of future leaders. *Journal of Transformative Leadership & Policy Studies*, *7*(2), 1–11. doi:10.36851/jtlps.v7i2.503

Wan, S. W. Y., Law, E. H. F., & Chan, K. K. (2018). Teachers' perception of distributed leadership in Hong Kong primary schools. *School Leadership & Management*, *38*(1), 102–141. doi:10.1080/13632434.2017.1371689

Ward, K. (2003). Faculty service roles and the scholarship of engagement. *ASHE-ERIC Higher Education Report*. Jossey-Bass Higher and Adult Education Series. Jossey-Bass.

Wenner, J. A., & Campbell, T. (2017). The theoretical and empirical basis of teacher leadership: A review of the literature. *Review of Educational Research*, *87*(1), 134–171. doi:10.3102/0034654316653478

Woods, P. A., & Roberts, A. (2019). Collaborative school leadership in a global society: A critical perspective. *Educational Management Administration & Leadership*, *47*(5), 663–677. doi:10.1177/1741143218759088

Woulfin, S. L. (2018). Mediating instructional reform: An examination of the relationship between district policy and instructional coaching. *AERA Open*, *4*(3). doi:10.1177/2332858418792278

Yang, J. (2021). Comparing Chinese and US Educational Leadership Styles: A Review of the Literature. *Performance Improvement Quarterly*, *34*(3), 247–260. doi:10.1002/piq.21371

Young, G. (2002). 'Hybrid' Teaching Seeks to End the Drive Between Traditional and Online Instruction. *The Chronicle of Higher Education*, A33.

KEY TERMS AND DEFINITIONS

Blended Learning (BL): A growing popular and effective approach incorporating web-based online methods and traditional learning.

Challenges: There are some things that could be improved in improving educational leadership in blended higher educational contexts.

Educational Leadership (EL): One of the essential factors to the success of current efforts to improve student learning. EL mainly comes from school administrators or school-board members.

Higher Educational Contexts: The provision of service functions to students by providing the theoretical knowledge and practical experiences for readiness to begin work. In higher educational contexts, the potential of academic and social interaction impacts end outcomes and the way learning takes place.

Student Learning Outcome: One of the central performance indicators which tend to become measures of learning activities and the quality of the educational service offered to students.

Chapter 7
Exploring the Potential of an Andragogical Approach to Teaching Leadership

Yasemin Kırkgöz
Cukurova University, Turkey

ABSTRACT

This chapter reports designing, implementing, and evaluating a Leadership and Management in ELT course based on an andragogical approach for adult learners in the teacher education department of a state university in Turkey. Participants were seven adult learners aged 28-35 undertaking the course, which lasted 14 weeks with 3 hours of weekly teaching. The course was delivered online, and the course content included leadership and management theories and approaches using videos, books, and a collection of articles. Data were collected using qualitative data collection tools; participants were requested to produce reflective writings, keep a leadership portfolio, and respond to end-of-the-course evaluation. Content analysis of the data revealed that adopting an andragogical approach to teaching leadership and management in ELT has been highly useful in preparing the participants for the leadership and management working environment they were either concurrently involved in and/or had aspirations to be in their professional lives.

INTRODUCTION

The importance of effective leadership and management for the desired functioning of schools and educational institutions is widely acknowledged in the twenty-first century (Dimmock, 2003; Kırkgöz, 2007; Miller, 2001). Bush (2008), one of the pioneering researchers in educational leadership, maintains that "there is growing recognition that the quality of leaders and leadership, is critical if schools are to produce the best possible outcomes for their learners and their stakeholders" (p. ix). In addition, the demands of today's business world require leaders with knowledge and skills to effectively lead

DOI: 10.4018/978-1-6684-7832-5.ch007

and manage people and institutions. This has led many countries to introduce various opportunities for leadership development for practicing and aspiring leaders as a key dimension of school improvement.

Leadership refers to an ability to influence and motivate people in the same organization for the effectiveness and success of the organization (House, Javidan, Hanges, & Dorfman, 2002). Management comprises a set of activities directed towards efficient and effective utilization of organizational resources in order to achieve organizational goals (Sapre, 2002 as cited in Bush, 2008, p.1). Effective leaders perform various leadership responsibilities including delegating tasks, establishing strategic direction, communicating goals, developing visions based on shared values, motivating the followers, and leading change (Coleman & Glover, 2010; Day, Harris, & Hadfield, 2001; Dipaola, 2003). In the absence of effective leaders, the requirements of the current century simply cannot be realized. To be effective, it is important that leaders are well prepared and educated. The issue of how educational leaders can be well educated for their leadership roles can only be addressed by designing and implementing effective leadership preparation programs (Rathert & Kırkgöz, 2017).

I brought leadership and management to the field of ELT in 2014, and since then I have been delivering a Leadership and Management in ELT Course at a postgraduate level. This chapter reports my experience of designing, implementing, and evaluating a Leadership and Management in ELT course based on Andragogical Approach for adult learners in the Teacher Education department of a state university in Turkey. The theoretical framework presents the six assumptions of andragogy that are used as the guiding principles for the design and development of the course. It also explores how andragogical mindsets play an important role in adult learning and considers how the integration of andragogical mindsets perspectives contributes to the development of leadership education. The methodology section of the study gives an outline of the andragogical mindset-based approach that is adopted in the leadership development of adults. As such, the course focuses on process and cognitive development, and capacity-building skills, which are increasingly regarded as appropriate for developing leaders (Kennedy, Carroll, & Francoeur, 2013). The chapter concludes by presenting an argument for the value of andragogical style leadership and management development.

BACKGROUND

What constitutes the background to the present research is my interest in bringing the topic of leadership and management into the field of English Language Teaching (ELT). Since 2014, I have been delivering a course entitled Leadership and Management in ELT at the ELT Department of a state university in Turkey at the postgraduate level to doctorate students. Over the years, I have realized the need to update the course content. This section presents an overview of the evaluation process of the course covering the years 2014-2020 by collecting information from the graduates in order to increase the quality of the course. It briefly mentions needs analysis with the participants of the current course, and finally, it introduces innovations that were incorporated into the revised course content because of this process.

Course Evaluation

The main goal of course evaluation was to obtain feedback from the past course participants regarding their learning experience in terms of how well the Leadership and Management in ELT course prepared them for the leadership and management work environment. A short course evaluation survey was sent out to past course members via an email. The evaluation questions focused on: Did the course adequately help facilitate them to carry out their leadership and managerial tasks? Was the selection of topics relevant for their work environment? What suggestion would they offer for the course improvement? The participants' responses to these questions were important in assisting me to improve the course for the benefit of future course participants (Denson, Loveday, & Dalton, 2010).

An evaluation of the previous courses created a need for a reexamination of the course content. The 12 past participants who responded to the course evaluation survey agreed that overall, they were satisfied with the quality of the course in terms of the selection of topics. However, they pointed out that less attention had been given to practical tasks that would prepare them to handle daily problems. They stated that they encountered numerous educational problems, and suggested that a course designed to address the gap between leadership and management theories and practices would be more relevant for the prospective leaders. This mismatch between the course participants' real needs and what was offered in the program clearly necessitated important adjustments in my Leadership and Management in ELT course.

Needs Analysis

In addition to obtaining feedback from the past students who were registered into the course between 2014-2020, I carried out a needs analysis with my new cohort of leadership students at the beginning of the academic year 2021-2022 to find out their specific leadership related needs and expectations from the program. Needs analysis is considered to be the first step in the process of course design, and it refers to the "systematic analysis of what learners need in order to operate in the target communication situation." (Woodrow, 2018, p. 21). Needs analysis involved conducting individual interviews with 12 of the registered participants into the course. The findings of needs analysis led to four major innovations to be introduced in the revised Leadership and Management in ELT course, described in the following section.

Innovations Into the Revised Course

The four components of the innovation introduced into the course executed during the 2012-2022 academic year included: "Andragogy, online education, creating more active learning environment and introducing problem scenarios.

Andragogy

Torrisi-Steele (2020, as cited in Wang & Torrisi-Steele, 2021) observes that in many countries, higher education leadership positions are occupied by academics, who often have very little, if any, experience or training in leadership practices, decision making processes, even of ethical decision-making in their capacity as leaders processes. Their "ascension" to positions of "power" typically emerges by way of academic achievements or another circumstance such as an interpersonal connection with the administrators. (Wand & Torrisi-Steele, 2021). Wand and Torrisi-Steele (2021) illustrate how the lack of practice in the theory of leadership can have negative implications.

This general picture of leadership appointment is also true for Turkey, where no specific leadership training is provided to academics. In Turkey, most leaders in higher education are experts in their disciplines, such as a language teacher educator, yet they do not have a degree in leadership. Nor are they given leadership training to be appointed for their leadership posts (Rathert & Kırkgöz, 2017). Due to a lack of an existing leadership program in the Turkish higher education context, as in many other countries, I wanted to build my course on sound theoretical principles and achieve a more responsive course by incorporating new insights from the relevant literature on adult learning theories. My review of the literature clearly demonstrates that andragogy, described in the theoretical framework part of the article, would be highly relevant for this purpose.

Online Education

The second innovation was a shift to online education from the face-to-face education and use online learning tools. Course evaluation between 2014-2020 delivered face-to-face clearly demonstrated the need to upgrade innovation into the course. In fact, the shift to online education was an inevitable consequence of the outbreak of Covid-19 in the delivery of educational programs. Information and communications technology (ICT) has been rapidly advancing over the last several decades. This growth in technology has led to a change in the responsibility of leaders and managers at all levels of education. Today's leaders need to keep abreast of technical progress and adapt to use technology in order to maintain effective communication with the staff, and carry out technology related tasks effectively.

Clearly, since 2014, when I first started to deliver the Leadership and Management in ELT course, there have been considerable advancements in the use of technology, particularly after the outbreak of the COVID-19 pandemic in March 2020, when education experienced a paradigm shift from face-to-face teaching to distance/online education format, globally; with this in mind, new online platforms and social media tools started to be used as an important element in distance education (Kırkgöz, Ersoy, & Savran, 2020).

It has been argued that online education offers opportunities for implementing andragogy in higher education. As put by Papo (2011) online education has brought out a new look at andragogically oriented approaches to learning by applying new technologies. Cercone (2008) examines how online learning can facilitate learning preferences of adult learners, such as self-direction, one of the key principles of andragogy. Adult learners can use various technological tools and accomplish learning tasks through exploration at their own pace and anywhere (Ferri, Grifoni, & Guzzo, 2020). This facility can provide the learner with an exploratory learning environment helping the development of self-driven learning (Kooi, 2008).

Another benefit of the use of technology in online education is that it creates an opportunity to develop community among learners. This need for technologically supported learning environment is also of particular relevance for adult learners. Accordingly, in the present study, an educational platform, Teams, was employed to provide course materials as well as YouTube videos. The Teams' facility was used for interaction among the participants offering them an opportunity to study materials and examples of good leadership practices, as will be described in the Methodology section of the article.

Creating More Active Learning Environment

Another finding emerging from the course evaluation and needs analysis was the need to create a more active learning environment, in which learners could connect theory with practice. Andragogy involves using experience that adults have built over the years in processing new knowledge (Halx, 2010). Taking this point further, Hussain (2013) reports that "adults are assumed to be mature intellectually and socially having their own concept and experience of life and learning formally or informally" (p. 124). Active learning has been defined as creating learning environments, in which learners deeply participate in the learning process. In their study, Wang, Hitch, and Torrisi-Steele (2022) established the possibility of active learning in online environments. Marschall and Davis (2012) suggest that adult learners be given opportunities to use their life experience by actively involving themselves in the learning process. In this way, adults can enhance their critical thinking and problem solving abilities, which are considered highly valued twenty-first century skills. This characteristic of adult learners has been brought into the course by promoting active learning as a teaching approach to prepare prospective leaders for the present and future workforce more effectively.

Introducing Problem Scenarios

Along with creating active learning opportunities for the prospective leaders, efforts have been made to incorporate problem scenarios into the course activities to enhance prospective leaders' problem-solving abilities, using the philosophy of Problem-based learning (PBL). In fact, PBL is a teaching approach in which real-world problems are utilized to promote student learning of concepts. The use of PBL yields several benefits such as the development of critical thinking skills and problem solving abilities. PBL offers opportunities for working in groups; as such, it develops students' communication skills. (Duch, Groh, & Allen, 2001).

THEORETICAL FRAMEWORK

The theoretical framework of the study has been informed by andragogical approach and mindsets to be elaborated in the ensuing section:

ANDRAGOGICAL APPROACH

Andragogy can be defined as an "art and science of helping adults learn" (Knowles, 1970), and it comprises "activities intentionally designed for the purpose of bringing about learning among those whose age, social roles or self-perception, define them as adults" (Merriam & Bierema, 2013). The six assumptions of andragogy principles include the learners' self-concept, the role of experience, readiness to learn, orientation to learning, motivation, and the need to know (McCauley, Hammer, & Hinojos, 2017).

According to the "self-concept" assumption of andragogy, adult learners have a well-established self-concept, and the learner is self-directing and autonomous in their approach to learning. The "role of experience" is the second assumption of andragogy, and it holds that adults have an accumulation of unique life experiences that they can draw from and apply information to different situations for effective learning to take place (Knowles, Holton, & Swanson, 2015). "Readiness to learn" is the third assumption of andragogy, which holds that adult learners have a desire and readiness to learn, especially when the learning is highly relevant to real-life situations they face. Based on the readiness to learn assumption, adults desire information that is immediately applicable to their everyday experiences at work and in their daily lives (Knowles et al., 2015).

"Orientation to learning" is the next assumption of andragogy, which holds the view that adults have a task-centered, problem-focused orientation to learning. To address this assumption, lessons can be carried out to allow participants to put their learning into immediate applications in real-life situations. "Motivation" is another assumption of andragogy, which holds the view that adults have a tendency to learn through intrinsic motivation such as goal attainment, the personal payoff for a higher quality of life, job satisfaction, or high self-esteem. The sixth assumption of andragogy is "the need to know", which holds the view that adult learners need to know what they will be learning, why this learning is important, and how they will learn. As mentioned by Tough (1979), adults must understand why learning is important to them personally and how such learning applies to their own life.

The application of the aforementioned andragogical principles particularly comes to the fore when considering the role of leadership development for adults. In teaching leadership, the study employs the six-assumption model of andragogy (Chan, 2010; Dachner & Polin, 2016) with the aim of providing greater opportunities to tailor the unique and varying needs of adult learners. Along with it, the present study applies the andragogical mindsets approach as it is considered closely relevant for handling leadership and managerial issues, and hence it is recommended to be integrated into any level of leadership development programs from less experienced to highly experienced learners (Jeanes, 2021).

Andragogical Mindsets

In the context of andragogy, mindsets can be considered to be "ways of thinking and meaning making developed over time through contextual interactions and personal relationships" (Stewart & Woldoko, 2016), and it is often related to leadership development (Fornaciari, & Lund Dean, 2014; Francoeur, 2013; Gosling & Mintzberg, 2003). In the present study, each of the six principles highlighted earlier will be approached as andragogical mindsets similar to the case of Gosling and Mintzberg (2003) who utilized the 'five minds of a manager' as a five mindsets framework to help reconcile competing challenges faced by practicing managers. These include the reflective mindset, the analytic mindset, the worldly mindset, the collaborative mindset, and the action mindset that address the management of the self, organizations, context, and relationships and change respectively. The reflective mind considers experiences; the analytic mindset calls for deep analysis of the systems and their underlying assumptions; the worldly mindset concerned with the experience of others.

Accordingly, the question investigated in this study is: "What is the impact of applying andragogical mindsets perspectives on the leadership development of adult learners?" The six assumptions of andragogy principles to be elaborated in the ensuing part of the chapter are used as a guiding framework for the design and development of the Leadership and Management in ELT course to create individualized learning opportunities for adult participants.

METHODOLOGY

The Leadership and Management in ELT course was delivered via an online platform, *Teams*, to support postgraduate students' learning and leadership development at a state university in Turkey. All the virtually held lectures, questions, and answer sessions were audio recorded. In case the participants missed a lesson, they could view its recording in their personal accounts. The use of the online platform enabled the building of an online learning community among the participants, and it allowed group work and pair work discussions to be done via the Teams' breakout room facilities.

Participants

The participants were seven adult learners; three males and four females, aged between 28 and 35. They were undertaking the Leadership and Management in ELT postgraduate course, which lasted 14 weeks with 3 hours of weekly teaching. The participants were either in the leadership role or in the position of aspiring leaders with a desire to transition into a leadership role.

The Framework of The Leadership and Management in ELT Course

The andragogical mindsets framework guided the design and the delivery of the present Leadership and Management in ELT course, and the creation of instructional materials, including problem scenarios based on real-life situations, and different reflective tasks. The course content included leadership theories and approaches on various topics. The list of the topics covered in the course is provided in the Appendix. The selection of the materials included books, videos, and a collection of relevant articles. A keynote speaker with a specialty in leadership was invited to deliver a virtual talk. Such an approach to teaching leadership is intended to help me, as the course designer, to manage the challenge of meeting the needs of the course participants with varying levels of leadership and professional experience, thereby fostering the participants' knowledge of leadership along with their personal leadership development.

In the implementations of the course, each of the six characteristics of the andragogical mindsets has been taken into consideration. The following section presents the implementations of each characteristic of andragogy in the course delivery.

Self-Concept

To address the learner's self-concept and the need to take control of their own learning experiences, the course participants in the present study were required to complete a series of reflective essays after covering each topic. To illustrate, they were asked to write about what their personality is like, and how it influences their leadership, the values and ethical stance adopted by the leaders in their educational institutions, and their stories about them. They were given an opportunity to identify great leaders from the national and international context, and explore their achievements, leadership behaviors, and personal competencies.

In addition, to encourage the development of their self-concept and autonomy, the course participants were assigned to create a leadership portfolio. Both of these assignments; the reflective essays and the leadership portfolio, helped participants in refining their self-concept in relation to who they are as leaders currently, and the types of leaders they wish to be in the future. In this way, the participants of the course were encouraged to engage in deep self-reflection.

The Role of Experience

It has been argued that adults have more experience, which can be drawn upon in the learning process (Jeanes, 2021). In the present Leadership and Management in ELT course, most participants had some leadership experiences within a school or university setting to draw upon. Hence, as the course lecturer, I drew out the experience that they brought to class discussions by encouraging them to reflect upon their own leadership/followership experiences in relation to course material. The following are examples of typical tasks, which ask the participants to draw on personal experiences.

Think of a time your school managers negotiated a serious problem in your organization. Provide us with the strategies used by your managers and tell us how you would negotiate them with your colleagues if you were the manager.

How would you describe decision-making by giving an example from your own teaching context? What made your decision-making process successful or unsuccessful?

In addition, the participants were given an opportunity to exchange their experiences with each other, thereby deepening the mutual learning process. To achieve this objective, the use of Teams' breakout room facilities allowed discussions to be carried out in various forms, in pairs or in groups.

Readiness to Learn

To capitalize on the assumption of andragogy, which holds that adult learners have the readiness to learn, I demonstrated how the subject matter in the course is personally beneficial to course participants. To illustrate, we discussed the advantages of teamwork and performed scenarios related to teamwork in order to give the participants the experience of how leaders can utilize teams to enhance organizational effectiveness. By performing such activities, the participants developed a readiness for learning about the topics. In the following task, the participants were asked to address a question that is immediately applicable to their everyday experiences at work:

What measures do you take to minimize your stress? Are there additional measures that you might adopt?

Orientation to Learning

To meet the needs of the course for the participants with different levels of life experience, I used a variety of approaches to make the concept of leadership "real." For example, I used examples of leadership in schools or higher education institutions that the participants are likely to be familiar with. I invited a keynote speaker with a specialty in leadership. They were given some leadership problems for deep analysis of the complexity of management systems, or they were requested to formulate their own problems to solve. In addition, I encouraged them to relate lessons in the course to their own past or present work experiences, which corresponded with their real-life learning orientation, as illustrated in the following example:

If you were the one responsible for decision-making in a difficult and sensitive situation, what steps would you follow?

Motivation

It has been argued that adults are more intrinsically motivated, and thus they are more likely to be in the learning process (Jeanes, 2021). To capitalize upon this principle of andragogy, I made sure that the selection of the course content would be motivating for them, and be most relevant to the participants' needs and interests (Merriam & Bierema, 2013). Keeping a leadership portfolio was one of the major sources of motivation to be completed by the end of the 14-week course. The leadership portfolio accommodated the participants' weekly reflective essays expressing what they learned from the weekly topics and how they relate the topic to their own experience, their solution to problem scenarios, and other related tasks. Additionally, I assigned a major goal accomplishment assignment, which required the course participants to collaboratively *prepare a professional development program for novice leaders.*

The Need to Know

As mentioned earlier, this principle of andragogy holds the view that adults need to know what they will be learning, why this learning is important, and how they will learn. To address this principle, I explained the rationale for each learning topic, task, problem-solving activity, and others, and the reason why each is important to them personally. As recommended by Domysheva and Kopylova (2021) during the initial class hour, the course content was given, and the course objectives were clearly explained. I made sure that practical activities and problem-solving tasks complementing each topic reflected real-life situations and had relevance to adults' job-related contexts. Briefly, the first-class session was used for the participants' learning orientation.

Data Collection Tools

Data were collected mainly using qualitative data collection tools; participants were asked to write reflections following the discussion of weekly topics, produce end-of-the-course reflections, and keep a leadership portfolio, in which to keep all their work. Data from these sources were collected systematically during the course.

Data Analysis

Qualitative data were analyzed using inductive content analysis following a three-stage procedure (Strauss & Corbin, 1998). First, data from the reflective writings were reviewed to determine the themes in relation to the influence of the andragogical approach on the participants' development of leadership and management capacity. Then, the identified themes were compared with data from the end-of-the-course reflections. They were confirmed and some necessary modifications were made. Finally, member checking was used with each participant to ensure that data interpretation had been accurately performed. In addition, each participant is labeled as Participant 1, Participant 2, and Participant 7. Excerpts from qualitative data are used in the findings section to support research findings.

FINDINGS

It has been found that the application of the andragogical mindsets principles has been highly useful in preparing the participants for the leadership and management working environment they were either concurrently involved in and/or had aspirations to be in their professional lives. Specifically, three themes were identified as a result of data analysis. The use of the andragogical mindsets principles helped the participants to *bridge between theory and practice*; it promoted *reflective thinking* and contributed to the participants' *growth mindset*. Each of these findings is presented below, and supported with excerpts from the participants.

Bridge between Theory and Practice

The participants were able to make connections between the theory covered in the course and everyday practices, which enabled them to make abstract leadership and management concepts meaningful and real. They stated that leadership concepts did not remain in their mindset as abstract thinking; rather, they were able to relate them to practice thanks to the hands-on activities and problem scenarios they were involved in analyzing. The following excerpt from reflective writing of a participant represents the general opinion:

Honestly, this course has cleaned my glasses. Now I can see more clearly and more logically leadership and management issues. In my daily life and in my job, I had experienced some of the topics we studied in this course just I had not known what they were. Now I have a clear idea about them and I can deal with them easily. (Participant 4)

In the following excerpt, the participant refers to Bush's (2003) typology of management models and tries to find a connection between the managerial model and her everyday experiences:

As an ELT teacher in Turkey, I have come across many such principles in my career. They are so busy with formalities that they have no time to improve their institution. They are scared to leave their comfort zone. Once, I thought it was a good idea to have a class in the schoolyard where I could teach colors by letting my students see the colors around. However, my principal was almost shocked to hear that fearing that something bad might turn out and we could not account for it. (Participant 3)

The *participative leadership* is another leadership model proposed by Bush (2003, 2008). In the following excerpt, the participant gives a brief description of the model, and relates it to the manager in his institution:

In this model, participation in decision-making is the core that makes it democratic. Any member of an institution might be the leader, which bonds the staff together. There is less burden on the leader as the roles are shared among members, which creates ownership. However, one negative side of this type of leadership is the slowness of the decision-making process. We have this type of leadership in our institution. Our leader is an instructor like us. She knows the process, our problems and our needs. Before deciding on any issue, we all come together and discuss it in a democratic environment. She sits among us in a circle that shows that she supports us. (Participant 2)

A similar example is given below in relation to another participant's reflections on *transactional leadership*:

In this leadership type, teachers and management interact just for transactions. It is an exchange process, in which the authoritarian behavior and power of a principal might be observed. This type of manage-

ment does not create long-term commitment within an institution. I had such a leader once and I never wanted to see him or walk past his office door. He was too stuck with paperwork and I felt that we were not humans but robots performing tasks at set hours in set places. (Participant 7)

In the following excerpt from reflective writing, the participant makes a clear reference to moral leadership, then, she expresses her ideas of disapproval of such a leadership type:

Leaders within this model focus more on the consistency of values, beliefs, or thoughts allegedly held by a leader. The main point is whether there is consistency between principles and practice. Does the leader justify their decisions in moral terms? Leaders who are inconsistent and unpredictable disturb me. I had managers who were like the changeable weather of İstanbul. You could never expect what would come next. This would create stress in me, which would result in my explosion, and I used to argue with them fiercely. I could never get on well with them. I was always persona-non-grata as a result. (Participant 6)

Reflective Thinking

As has previously been established, the capacity for reflection and learning from experience is by no means certain for those who are adults with more experience (Jeanes, 2021). One way of attaining effective reflective abilities is through reflective writings (Freese, 2006), which are considered to be useful tools for people to evaluate their theories and beliefs on a particular issue (Tsang, 2003) such as leadership and management theories. In relation to this, writing by reflecting on experiences, can help potential leaders and managers to think more critically on their own and others' leadership and management practices.

Amongst the sophisticated questions that the course participants were encouraged to reflect upon and keep in their portfolios were "What have l learned from this topic", and "Why is this topic significant for me?' and 'How can I relate them to my own experiences?' Analysis of the participants' reflective writings evidenced that the participants found the topics covered in the course highly rewarding to their emerging leadership as illustrated in the following excerpt:

Learning the difference between leader-manager and different models of leadership, we have gained insights into how we were managed in the past and what kind of leadership model we are now exposed to. We have learned that effective leadership has a profound effect not only on the motivation of the staff but also on the success of the students. (Participant 4)

In the reflective writing below, the participant refers to the managerial model and its positive aspect:

The managerial model reminds one of a lot of paperwork, formality, and hierarchy imposed by superiors within an institution. In this type of model, the so-called leaders tend to stick to bureaucracy without much, if any, vision. The positive side of this type of leadership is that it is easier to implement, just order the members of the institute and they have to obey. (Participant 2)

The participants' reflections were related not only to theoretical understandings but also to the relevance of tasks and assignments to their evolving leadership capacity. They expressed that they were able to gain ample practical knowledge through well-designed tasks, as illustrated in the reflections of the following participants:

In addition to gaining theoretical knowledge, we had a practical assignment, which required us to prepare a model program for leaders. To do this assignment, we used our knowledge to produce a training program for novice leaders. (Participant 7)

In one of our individual presentations, it is stressed that decision-making and managing conflicts are two main responsibilities of a leader. If a leader wants to tackle these two important duties, s/he must take micro-politic issues and inter-group relations into consideration by keeping his/her values in mind. In addition, in order to overcome some problems regarding demotivation or lack of interest in their organization, leaders have to provide some "hygiene" and "satisfiers" to their staff. In an organization, these motivating elements are indispensable. (Participant 5)

In the following excerpt, the participant gives a detailed reflection on the overall content of the course in her end-of-the-year reflection. As noted, she refers to many aspects of topics covered in the course such as *distributed leadership, stress, time management,* and *teamwork,* which she regards to be highly relevant to her current and future work context:

Distributed leadership is another topic that we discussed in detail. It can be inferred that distributed leadership by giving responsibility or sharing duties with the staff makes them motivated and creates a sense of ownership. Through the end of the semester, we examined how to work through teams in organizations. It is highlighted that developing team effectiveness requires getting the highest performance from the staff by using their strengths and improving their weaknesses. The importance of timing meetings, chairing the meeting, and how to keep records of the meetings were analyzed and discussed. (Participant 7)

Bain, Packer, and Mills (1999) proposed a scale of reflection, which represents different levels of reflections. The scale comprises levels of "reporting, responding, relating, reasoning and reconstructing", in which reporting and responding correspond to the least critical levels whereas "reconstructing" corresponds to the highest level of criticality. In the present study, an analysis of the participants' reflection writings demonstrates that most writings seem to correspond to "relating", "reasoning", and "reconstructing" levels of reflection, consecutively. That is, the participants had a tendency to reflect at the levels of "relating" leadership and management concepts to their experiences; their reflective writings also corresponded to the "reasoning" level, whereby they sought a deeper understanding of leadership issues, and "reconstructing", whereby they took a position on an issue, and/or expressed the personal significance of their learning leadership and management topics.

The Growth Mindset

Fundamental to andragogic learning is to develop and evolve meaning. As argued by Jeanes (2021), "the andragogic approaches require us to be 'in the right frame of mind;' in other words, a 'growth mind' to be effectively engaged with and benefit from developmental opportunities" (p. 103). A closer examination of the qualitative data clearly demonstrates the growth mindset, namely, developing and evolving the meaning of the leadership and management issues for the participants, as evidenced by the reflective writings and end-of-the-course evaluation. Below is an example from the participant's end-of-the-year reflection illustrating this point clearly:

Before taking this course, I was thinking that leadership is in people's characteristics and behavior. Now that I have completed the course, I have learned that we can improve it by training people through special strategies. We should start from the individual's private life. I have learned that leadership is a skill that could be improved. In addition, I have learned that in order to motivate others to participate in a group or team we should give them opportunities to express themselves and give some responsibilities to them, consult with them, and value their ideas. (Participant 5)

The growth mindset, as is implied in the andragogical mindsets, recognizes that in a complex and rapidly changing world one has to tackle challenges that might not be in one's comfort zone. In a similar vein, throughout the course, it has been found that the participants were committed to successfully completing the given tasks, projects, and reflective writings, all of which contributed to their growth mindset, leading to enhanced leadership competence. During the process of evolving meaning in the field of leadership and management, the participants critically evaluated their developing mindset, as illustrated below:

The experiences related to being a leader of an institution and managing it in the field of education were unachievable for me before taking this course. However, after I started to take this course, I understood that I already had many experiences regarding leadership and management topics. Leadership course strengthened the idea that each teacher is the leader of his/her class, and s/he needs to make decisions even each minute of the lessons. In addition to being the leader of our own classes, there is also the possibility of being appointed as a manager in our institutions; thus, the course we have taken and the issues we have covered will have great importance in our future careers. (Participant 7)

For me, leadership mainly focused on the person leading the team or organization, and her or his relationship with followers, in which subordinates are controlled, influenced and managed by a single individual leader. This course has suggested that leadership is an activity that can be shared or distributed among members of a group or organization. Actually, for me, it has opened up new lines of thinking about leadership and I have learned methods and strategies that can improve my leadership skills. (Participant 1)

The design of the course through andragogical mindsets approach also has had much to offer to adults in developing their leadership knowledge and capacity. The growth mindset has proven to be effective in the present course. The frequent use of the problem scenarios and related tasks also contributed to their andragogical mindsets in leadership development. Undoubtedly, the opportunities for reading, presentation on a chosen topic, and discussions increased over time, and the capacity for reflection and learning from engaging tasks has been well appreciated, as in the excerpt below:

The course materials and the assignments were all to the point and improved all of us in this learning journey. Especially, the books chosen covered all the topics that we need to know in order to be successful leaders and criticize the leaders we currently have. During the first several weeks of the semester, by using Tony Bush's book, we built a foundation regarding leadership. After covering the topics in this book, we started to go through the book of Coleman and Glover (2010), which mostly focused on the value aspect of leadership. In this book, we had a chance to present on a chosen topic. In this way, not only our learning has become permanent but also, we learned the topics in detail. Moreover, we were recommended to read some articles. By reading these articles and writing our reflections on them, we all gained a deeper understanding of the issues such as model proposals for the training of leaders. (Participant 6)

To sum up, it is true that leadership can occur throughout most of our lives, but leadership and managerial education is ordinarily associated with adults. As highlighted by Jeanes (2021), the applicability of mindset particularly comes to the fore when developing leadership and managerial education with adults. As seen from the preceding discussion, the adoption of the andragogical leadership mindset approach that formed the conceptual foundation of the leadership development course helped the participants to develop their competence and capacity in terms of analysis, collaboration, and reflection on their own experience and exploring the experiences of others.

Findings suggest that prospective leaders can learn optimally through andragogical approaches. In this respect, findings support earlier studies such as by Jeanes (2021) who found the application of andragogical perspective to be closely relevant for handling leadership and managerial issues, and hence it is recommended to be integrated into any level of leadership development programs.

When the above discussed findings from the present study (2021-2022 academic year) is compared with those from the previous programs (2014-2020 academic years), it can clearly be seen that the course participants were able to see the relevance of course activities to their leadership needs in their current work environment. To be more specific, one of the major concerns raised by the participants of the previous Leadership and Management in ELT courses was that they did not see much connection between the course and everyday practices in terms of dealing with emerging educational issues. In the current course, it is evident that this particular issue was largely addressed by using problem scenarios, which was found to have formed a bridge between theory and everyday issues for the participants. Unlike the old curriculum, in which the course books and articles formed the core teaching materials, in the revised course, content was enriched by the use of technological tools such as YouTube videos and podcasts from the leaders.

Based on the findings from the present study, it can confidently be stated that the revised Leadership and Management in ELT course is more effective than the previous one because course participants perceive it directly relevant to their current and future leadership context, and different from the previous ones.

FUTURE RESEARCH DIRECTIONS

This study was carried out with a small number of participants, therefore generalizing the findings may not be possible. Future research can be conducted towards increasing the number of participants in order to obtain results that are more generalizable.

In the present study, qualitative data collection tools were used, which provided rich source of information. Future research can collect quantitative data from different research instruments such as a questionnaire to complement the present research findings.

In addition, while the present research supports andragogically-based learning, wide scale implementation and evaluation of andragogically-based model, particularly in leadership and management programs, is still needed to determine the benefits and challenges of such learning.

Finally, a follow-up study can be conducted with the course participants to observe the effects of leadership skills they have gained as part of the current course on their leadership and management practices in their real-world context. Such a study would also enable us to observe the extent to which leaders and managers can transfer those skills into their work environment.

SOLUTIONS AND RECOMMENDATIONS

The importance of leadership preparation and development is uncontested. Problems reported in the implementation of educational policy strongly suggest that leadership principles are often neglected or applied insufficiently (Stefan & Kırkgöz, 2017). This study has demonstrated that the andragogical framework can be employed to educate current and prospective leaders and managers to enhance their leadership capacity and help them cope with emerging challenges that might require analysis, collaboration, and reflection among many other capacities.

In light of available research findings, a number of recommendations can be offered:

Given that, adults are problem-centered (Kırkgöz, 2017; 2018; Merriam & Bierema, 2014), and that we live in a world of growing complexity, it is important to incorporate leadership and management-related problem scenarios from real-life situations for the emerging leaders to tackle and solve them individually and collaboratively in order to help them build their problem-solving capacity.

Potential leaders need to be encouraged to relate theoretical principles to their experiences and working context by means of reflective writing. They can also be assigned projects to explore the experiences of other leaders and learn from such experiences. An example of this would be to assign a leader interview project, which might involve interviewing leaders from different sectors; education, or industry; then, analyzing the interviews for commonalities and differences between the leaders' experiences (See Stefan & Kırkgöz, 2017, for details).

Finally, as demonstrated in the present study, Andragogical mindsets perspectives can appropriately be used for the leadership development of adult learners. Tasks and projects can be developed to help leaders put each of the six Andragogical mindsets principles into practice.

CONCLUSION

The world we live in is becoming increasingly complex, diverse, and changeable, which makes it indispensable for the twenty-first century leaders to be equipped with the capacity to bring out innovations and handle complex problems, amongst many other responsibilities, if they are to thrive in their professions and stand up as effective leaders in schools and organizations. In this Leadership and Management in ELT course, the andragogical mindset-based approach has been used to train current and prospective leaders and managers to cope with potential challenges that might require analysis, collaboration, reflection, and problem-solving capacities, among many other issues. Leadership can occur throughout most of our lives and training young leaders and managers to adopt a mindset-based approach is a necessary condition for leadership development (Heslin & Keating, 2017; Kennedy et al., 2013). In a time when we are faced with complex global challenges, developing the leaders of tomorrow and the mindsets they need is a critical issue that needs to be addressed. Training our leaders of tomorrow with known practices of today has inherent limitations (Jeanes, 2021). Adopting a mindset approach, typically found in andragogical pedagogies, and in leadership development in particular, plays a crucial role in the leadership development of adults. This study suggests that the integration of andragogical mindsets perspectives has the potential to contribute to the development of leadership education and learning in the field of ELT. There is no doubt that the framework applied in the present study can have relevance beyond the field of ELT.

REFERENCES

Bain, R., Packer, J., & Mills, C. (1999). Using journal writing to enhance student teachers' reflectivity during field experience placement. *Teachers and Teaching*, *5*(1), 51–73. doi:10.1080/1354060990050104

Bush, T. (2003). *Theories of educational leadership and management* (3rd ed.). SAGE Publications, Inc.

Bush, T. (2008). *Leadership and management development in education*. SAGE Publications, Inc. doi:10.4135/9781446213605

Bush, T., & Glover, D. (2003). School leadership: Concepts and evidence. *National College for School Leadership*, 1-42.

Cercone, K. (2008). Characteristics of adult learners with implications for online learning design. *AACE Journal*, *16*(2), 137–159.

Chan, S. (2010). Applications of andragogy in multi-disciplined teaching and learning. *Journal of Adult Education*, *39*(2), 25–35.

Coleman, M., & Glover, D. (2010). *Educational leadership and management: Developing insights and skills*. Open University Press.

Dachner, A. M., & Polin, B. (2016). A systematic approach to educating the emerging adult learner in undergraduate management courses. *Journal of Management Education*, *40*(2), 121–151. doi:10.1177/1052562915613589

Davis, H. S. (2012). Discussion as a bridge: Strategies that engage adolescent and adult learning styles in the postsecondary classroom. *The Journal of Scholarship of Teaching and Learning*, *13*(1), 68–76.

Day, C., Harris, A., & Hadfield, M. (2001). Challenging the orthodoxy of effective school leadership. *International Journal of Leadership in Education*, *4*(1), 39–56. doi:10.1080/13603120117505

Denson, N., Loveday, T., & Dalton, H. M. (2010). Student evaluation of courses: What predicts satisfaction? *Higher Education Research & Development*, *29*(4), 339–356. doi:10.1080/07294360903394466

Dimmock, C. (2003). Leadership in learning-centred schools: cultural context, functions and qualities. In M. Brundrett, N. Burton, & R. Smith (Eds.), *Leadership in education* (pp. 3–22). SAGE Publications, Inc. doi:10.4135/9781446215036.n1

DiPaola, M. F. (2003). Conflict and change: Daily challenges for school leaders. In N. Bennett, M. Crawford, & M. Cartwright (Eds.), *Effective educational leadership* (pp. 143–158). Paul Chapman.

Domysheva, S. A., & Kopylova, N. V. (2021). Overcoming non-linguistic barriers to effective speaking in Russian adult ESP classroom. *History, Pedagogics, Philology*, *27*(1), 87–94. doi:10.18287/2542-0445-2021-27-1-87-94

Duch, B. J., Groh, S. E., & Allen, D. E. (2001). Why problem-based learning? A case study of institutional change in undergraduate education. In B. Duch, S. Groh, & D. Allen (Eds.), *The power of problem based learning* (pp. 3–11). Stylus.

Ferri, F., Grifoni, P., & Guzzo, T. (2020). Online learning and emergency remote teaching: Opportunities and Challenges in Emergency Situations. *Societies (Basel, Switzerland)*, *10*(4), 1–18. doi:10.3390oc10040086

Fornaciari, C. J., & Lund Dean, K. (2014). The 21st-century syllabus: From pedagogy to andragogy. *Journal of Management Education*, *38*(5), 701–723. doi:10.1177/1052562913504763

Francoeur, J. (2013). Mindset not skill set: Evaluating in new paradigms of leadership. *Advances in Developing Human Resources*, *15*(1), 10–26. doi:10.1177/1523422312466835

Freese, A. R. (2006). Reframing one's teaching: Discovering our teacher selves through reflection and inquiry. *Teaching and Teacher Education*, *22*(1), 100–119. doi:10.1016/j.tate.2005.07.003

Gosling, J., & Mintzberg, H. (2003). The five minds of a manager. *Harvard Business Review*, *81*(11), 54–63. PMID:14619151

Heslin, P. A., & Keating, L. A. (2017). In learning mode? The role of mindsets in derailing and enabling experiential leadership development. *The Leadership Quarterly*, *28*(3), 367–384. doi:10.1016/j.leaqua.2016.10.010

House, R., Javidan, M., Hanges, P., & Dorfman, P. (2002). Understanding cultures and Implicit Leadership Theories across the Globe: An Introduction to Project GLOBE. *Journal of World Business*, *37*(1), 3–10. doi:10.1016/S1090-9516(01)00069-4

Jeanes, E. (2021). A meeting of mind(sets). Integrating the pedagogy and andragogy of mindsets for leadership development. *Thinking Skills and Creativity*, *39*(10058), 1–9. doi:10.1016/j.tsc.2020.100758

Kennedy, F., Carroll, B., & Francoeur, J. (2013). Mindset not skill set: Evaluating in new paradigms of leadership. *Advances in Developing Human Resources*, *15*(1), 10–26. doi:10.1177/1523422312466835

Kırkgöz, Y. (2007). Innovation as a curriculum renewal process in a Turkish University. In A. Rice (Ed.), *Revitalizing an established program for adult learners* (pp. 135–158). TESOL Publications.

Kırkgöz, Y. (2017). Working the problem: Finding solutions to student dissatisfaction in EAP for engineering. In T. Stewart (Ed.), *Voices from the TESOL classroom: Participant inquiries in higher education classes* (pp. 17–26). TESOL Publications.

Kırkgöz, Y. (2018). Problem-based learning in a teacher education program: A study of learning outcomes. In Z. S. Genç (Ed.), *Updating perspectives on English language teaching and teacher education* (pp. 79–92). Peter Lang.

Kırkgöz, Y., Ersoy, M., & Şavran, E. (2020, November 21-22). Exploring Pre-Service Teachers' Experiences during Onlıne Education. In *Proceedings of Global Conference on Advances in Education and Research (GCAER-2020), Tsuruoka, Japan (Online)* (pp. 37-42). Academic Press.

Knowles, M. S. (1970). *The modern practice of adult education: Andragogy versus pedagogy*. Association Press.

Knowles, M. S., Holton, E. F., & Swanson, R. A. (2015). *The adult learner: The definitive classic in adult education and human resource development* (8th ed.). Elsevier.

McCauley, K. D., Hammer, E., & Hinojos, A. S. (2017). An andragogical approach to teaching leadership. *Management Teaching Review*, *2*(4), 312–324. doi:10.1177/2379298117736885

Merriam, S. B., & Bierema, L. L. (2013). *Adult learning: Linking theory to practice*. Jossey-Bass.

Miller, D. (2001). Successful change leaders: What makes them? What do they do that is different? *Journal of Change Management*, *2*(4), 359–368. doi:10.1080/714042515

Nahavandi, A. (2008). *The art and science of leadership*. Pearson Education.

Rathert, S., & Kırkgöz, Y. (2017). Examining leadership perceptions and practices to deliberate on educational policy implementation - an interdisciplinary approach. *International Journal on Lifelong Education and Leadership*, *3*(1), 1–10. doi:10.25233/ijlel/2017-v3i1p1

Riley, K., & MacBeath. (2003). Effective leaders and effective schools. In N. Bennett, M. Crawford, & M. Cartwright (Eds.), *Effective educational leadership* (pp. 173-185). Paul Chapman.

Strauss, A., & Corbin, J. (1998). *Basics of qualitative research: Techniques and procedures for developing Grounded Theory*. SAGE Publications, Inc.

Tough, A. (1979). *The adult's learning projects: A fresh approach to theory and practice in adult learning* (2nd ed.). Ontario Institute for Studies in Education.

Tsang, W. K. (2003). Journaling from Internship to Practice Teaching. *Reflective Practice*, *4*(2), 221–240. doi:10.1080/14623940308269

Wang, V., Hitch, L., & Torrisi-Steele, G. (2022, January). Active learning online: Necessity, faculty role, and a concept model for course design. *International Journal of Online Pedagogy and Course Design*, *12*(1), 1–13. doi:10.4018/IJOPCD.2022010105

Wang, V., & Torrisi-Steele, G. (2021). The gap between theory and practice: A scenario from higher education. In K. N. LaVenia & J. J. May (Eds.), *Case studies in leadership and adult development applying theoretical perspectives to real world challenges* (pp. 167–176). Routledge.

Wasserberg, M. (2000). Creating the vision and making it happen. In H. Tomlinson, H. Gunter, & P. Smith (Eds.), *Living headship* (pp. 154–160). Paul Chapman Publishing.

Yukl, G. (2009). *Leadership in organizations*. Pearson Education.

ADDITIONAL READING

Bush, T., & Glover, D. (2003). School leadership: Concepts and evidence. *National College For School Leadership,* 1–42.

Fullan, M. (2007). *The new meaning of educational change* (4th ed.). Teachers College Press.

Harris, A., Busher, H., & Wise, C. (2003). Effective training for subject leaders. In N. Bennett, M. Crawford, & M. Cartwright (Eds.), *Effective educational leadership* (pp. 131–142). Paul Chapman.

Horner, M. (2003). Leadership theory reviewed. In N. Bennett, M. Crawford, & M. Cartwright (Eds.), *Effective educational leadership* (pp. 27–43). Paul Chapman.

Özer Ölmez, M., & Kırkgöz, Y. (2018). Investigating model proposals offered for training and appointment of Turkish school principals in terms of leadership models. *Journal of Institutional Research, 3*(2), 834–852.

Stewart, C., & Woldoko, B. (2016). University educator mindsets: How might adult constructive-developmental theory support design of adaptive learning? *Mind, Brain and Education : the Official Journal of the International Mind, Brain, and Education Society, 10*(4), 247–255. doi:10.1111/mbe.12126

Turhan, B., & Kırkgöz, Y. (2020). An analysis on mission statements of Turkish primary schools: Similarities and differences. *Research in Educational Administration & Leadership, 5*(1), 71–98. doi:10.30828/real/2020.1.3

KEY TERMS AND DEFINITIONS

Andragogy: The term Andragogy, first introduced by Malcolm Knowles, the American Educator, is a theory of learning that specifically refers to the strategies used to facilitate adult learners during the learning process. Andragogy proposes six assumptions or principles, which are adults' self-concept, learning experiences, readiness to learn, motivation, need to know, and problem-centered learning. Andragogy is a theory of learning focusing on assisting adult learners to learn (Knowles, 2005).

Leadership: Leadership involves a process of exerting intentional social influence by one person over other people or groups in order to organize the activities and relationships in an organization (Yukl, 2002). Leaders have various roles such as unifying people around key values (Wasserberg, 2000), and communicating "clear sets of personal and educational values which represent their moral purposes for the school" (Day et al., 2001, p. 53).

Management: Management refers to "maintaining efficiently and effectively current organizational arrangements" (Bush, 2008). The overall function of management is to maintain rather than bring about change, as in the case of leadership.

Online Learning Community: Online learning community is a virtual space that connects people with shared interests and educational goals. It is intended to facilitate collaboration, discussion, knowledge sharing, and support each other's learning. Members of an online learning community may be students or teachers, generally to undertake a course or a program in a particular subject. All interactions take place via the Internet, through lectures, videoconferencing, and so forth.

APPENDIX

The Leadership and Management in ELT Course Content

- What is leadership?
- Leadership and management
- Leadership preparation and development
- Leaders as change agents
- Decision-making and conflict management
- Time and stress management
- Leading with values
- Communication
- Motivation and delegation
- Working with and through teams

Chapter 8
Teaching Revolutions:
Radical Andragogical Leadership in Mexico, Russia, and the USA

Mattius W. Rischard
University of Arizona, USA

ABSTRACT

The Russian and Mexican Revolutions resulted in distinct socioeconomic programs during the 20th century; yet they produce comparable trajectories of political leadership and cultural discourse on adult educational reform. Using Soviet and Mexican educational materials from the revolutionary era, it is possible to compare the policies and practices produced in revolutionary contexts to the U.S. andragogy established through study of Mexico and Russia. The adult outcomes of schooling became a site of negotiating cultural hegemony that revolutionary groups used to design radically productive and disruptive teaching and learning theory, while at the same time indirectly modeling the progressive reforms employed in the U.S. Yet, since the dawn of the 21st century, both have struggled to refine their approach to adult learning in the context of the digital turn because of how the networked flow of information works to both support and undermine the institutionalized tenets of the revolutionary ethos.

"I felt resentment at those adherents and eulogists as well as critics and enemies, who, I felt, had misled me with constant talk and writing about Bolshevism and Communism, leaving me ignorant of the more basic fact of a revolution—one which may be hinted at, but not described, by calling it psychic and moral rather than merely political and economic, a revolution in the attitude of people toward the needs and possibilities of life." – John Dewey, Impressions of Soviet Russia and the Revolutionary World, 1929

DOI: 10.4018/978-1-6684-7832-5.ch008

INTRODUCTION

Educators in the United States know the content they plan to deliver in class, and how to assess the extent to which students make progress in learning that information. They may even have a specific learning objective or state standard in mind while determining the learning outcomes of a given lesson. But beyond measuring cognition by teaching to tests, how does one know actual progress has been made beyond the ability to regurgitate the required answers on exams? To properly gauge student progress, one must first define that progress. According to the U.S. Department of Education, there are three steps to negotiating the definition of achievement: "to determine the performance level for all students assessed by the State assessment system…to summarize that information for the school to determine the extent to which the students are attaining the proficient and advanced performance levels…[and] to determine whether the school has made adequate improvement over the previous year and the extent to which the students are attaining the proficient and advanced performance levels." (US Department of Education, 2009). The statistical nature of the State's discourse is important here, as it forces the public conception of progress into the semantics of quantitative assessment and scientism. At the same time, it establishes critical norms of school performance for locating institutions that are struggling to meet ideal government and industry standards of achievement, and measuring the extent to which students in general are acquiring the necessary abilities for available employment.

This form of centralized planning through statistics based on common learning materials and outcomes is not new—the Soviet Union had perfected such a system in primary and secondary schooling, while universities and higher education institutes generally held their own series of content-specific oral exams, essays, and interviews for admission and assessment, subsidized by the State. However, both the Russian Federation and the U.S. have moved even further towards universal standardized testing and centralized standards through the institution of the EGE (Unified State Exam) to prepare for higher education and the Common Core (No Child Left Behind) that funnels students towards college readiness in their respective nations, whilst Mexico, a State once known for its radical forms of pedagogy and andragogy in its Indigenous training programs during the early Revolutionary period, has attempted to do away with revolutionary-style trade schools since the crystallization of the revolutionary party into the modern state institution, i.e., the "Institutional Revolutionary Party"—never was there a more oxymoronically named organization.

The ultimate aim of this study is to examine, via a historiographic comparison of Russian, Mexican, and American radical educational traditions, how andragogical leadership has been exercised in unique circumstances of agency within higher education to promote not just social change, but also cultural *progress* relative to the History of prior civilizations and non-Western nations. Educational psychologists and policymakers have struggled with defining the ethos of the educational apparatus: i.e., the political authority and credibility of the institution, defined by the ultimate aims and goals of public (adult) schooling since its mass institutionalization. U.S. Progressivism, as an early 20th Century educational, movement sought to meet the educational needs of a growing proletariat increasingly made up of English Language Learners (ELLs) and first-generation immigrants from outside of Western Europe. The movement has undergone several transformations since the demand for more public-access trade schools, night schools, and apprenticeships in U.S. urban centers became a national concern in the late 19th Century. While statistical measurement and standardization were several of the techniques advocated by early reformers and fetishized by modern policymakers, these forms of analyzing outcomes for mass populations of students were only part of the Progressives' initial aims: i.e., to liberalize public education and make it accessible, instructional principles had to change beyond homogenizing admission and measurement policies.

One of the main barriers to a useful instruction for the industrialized divisions of labor was that historical andragogy traditionally implied classical preparation for a privileged few at universities, which had little utilitarian purpose for most of the new industrial and agrarian elite looking to hire wage laborers and field hands. For reform to occur, Progressives like the educational philosopher John Dewey (1859-1952) believed teachers needed to approach instruction though project-based learning, and integrate curriculum with real-world skills valued by entrepreneurs and industries. What's more, the curriculum had to prepare a growing electorate for the responsibility of civic involvement and increasingly democratized politics. Group work was thought to help with collaborative social skills and teach students to participate in the co-construction of knowledge at Dewey's famous Laboratory Schools at the University of Chicago. In "My pedagogic creed" (1897), he highlights the experiential learning observed at the laboratory schools and theorized that, through the "activity of the school," education itself could become artistic and polytechnic rather than a hyperspecialized technocracy or a romanticized badge of elitist humanism, ultimately, his "Pragmatist" education would serve as preparation for productive members of an industrial society from kindergarten through higher education, while also revolutionizing democratic participation in political life (Dewey, 1938). Generally, Dewey (1897) assumed education to be "participation of the individual in the social consciousness of the race" with a mission to prepare students holistically for the demands of present culture and society. As an early leader of the pragmatism in American philosophy, utilitarianism in education, and functionalism in psychology, his perspective was that the institution of traditional preparation for university "fails because it neglects this fundamental principle of the school as a form of community life: it conceives the school as a place where certain information is to be given, where certain lessons are to be learned, or where certain habits are to be formed." (p. 78). Because it focused on abstract concepts, traditional instruction alienated students with an imperious method meant to teach an idealized adult learner with classical training. If content was instead situated in students' lived experiences and the needs of the present community across all public institutions, learners would be empowered to participate in democratic social life at such an aggregate level that Dewey recognized mass education as one of the primary tools for social progress: "education is a regulation of the process of coming to share in the social consciousness; and that the adjustment of individual activity on the basis of this social consciousness is the only sure method of social reconstruction." (p.80). It is no surprise that Dewey would eventually spend a portion of his career touring the educational systems of revolutionary Russia and Mexico (the epigraph is drawn from his published impressions), becoming politically involved later as head of the Dewey Commission, or the "Commission of Inquiry into the Charges Made against Leon Trotsky in the Moscow Trials," in 1937 to defend Trotsky from Stalin's accusations while exiled in Mexico (Dewey, 1972).

Radical school reforms with the aim of greater democratic functionality for social justice crystallized in public discourse by transitioning from philosophical debates to scientific classifications and direct political rhetoric. These social conflicts within revolutionary nations at the beginning of the 20th Century served as the impetus for the US popularization of school reform across disciplines and classes: in revolutionary Mexico, the amended constitution of the Cardenas administration proclaimed that education was to be "socialist in nature," where the amended Article 3 of the 1934-46 Mexican Constitution mandated that, "in addition to removing all religious doctrine," socialist schools would "combat fanaticism and prejudices", "build[ing] in the youth a rational and exact concept of the universe and of social life." In the USSR, educational psychologists Anton Makarenko and Lev Vygotsky were developing the world's first Marxist theories of psychological science for the production of the "New Socialist Man" who would be most productive as part of the social collective necessary for modernizing the Soviet Union (Vygotsky, 1930).

New possibilities opened up for American educators through their conceptions of critical andragogy, based on sociocultural assumptions of cognition. Dewey travelled to Russia and Mexico in the mid-1920s to gather impressions of the new Commissariat of Education (Russia) and Secretariat of Education (Mexico), documenting observations of both major officials and student-teacher interactions in his *Impressions of Soviet Russia and the Revolutionary World* (1929). Touring the rural schools of Mexico that built literacy through community farm and industry projects "revived my [Dewey's] faith" (p. 124) in democratizing educational institutions:

In many places there is much attention to music and to design in the plastic arts, for both of which things the Indians display a remarked genius. As a rule, if what we saw may be depended upon as evidence, the designs in the small rural schools were much better, even though the work was crude, than in the industrial schools in the city, where department store art has made a lamentable invasion. If the rural schools can succeed in preserving the native arts, aesthetic traditions, and patterns, protecting them from the influence of machine-made industry, they will in that respect alone render a great service to civilization.

Dewey's review of adult education Soviet Unified Labor Schools and other public education programs in post-war Leningrad and Moscow as well as the surrounding children's colonies was also glowing, but in a more theoretical sense, qualified by some skepticism about the Bolshevik centralized state program and its ostensible connections to the Marxist-Hegelian dialectic of History. He worried that their theoretical justifications for centralizing the relations of labor to surplus capital would create new issues of inequality and injustice through the dictatorship of the proletariat in education. However, Dewey was most interested in how the educational system would affect this revolutionary transition in the psychology of the Russian people:

From the communist standpoint, the problem is not only that of replacing capitalistic by collectivistic economic institutions, but also one of substituting a collective mentality for the individualistic psychology inherited from the "bourjui" epoch—a psychology which is still ingrained in most of the peasants and most of the intellectuals as well as in the trading class itself.

While cynical about the increasing role of propaganda in the State school system, Dewey acknowledged that the RSFSR had created an institutional space for progressive experimentation in reforming adult learning for modernization, which provided United States education with a blueprint for reform. Through the experiences of American educators situated in conversation with their Mexican and Russian revolutionary contemporaries, I argue that progressive transformations in the way the United States institutionally (re)shapes its adult learners for new sociopolitical and economic demands historically depended on the revolutionary struggles of an economically and geographically connected neighbor (Mexico) and an internationally competitive political adversary (USSR). Yet, since the dawn of the 21st Century, both have revolutionary models struggled to refine their approach to adult learning in the context of the Digital Turn because of how the networked flow of information (and thus ideology) via Information Communication Technology (ICT) works to both support and undermine the historically institutionalized tenets of their revolutionary ethos.

THEORETICAL FRAMEWORK

Mass compulsory education and publicly available higher education are ambitious social experiments, laden with ideological value systems that are inextricable from the design principles guiding instruction. Hegemony in the school is balanced between various forces depending on the given State, where the Church, the government, the military, the private sector, the administration, the teachers, and the individual student or taxpayer may have fluctuating levels of control. Struggle between these factors is inevitable, but new forms of collaboration to refine educational methods (and thus producing new forms of socialization) typically occur when conflicts arise within or between social parties—whether they be political parties, warring nations, students and administrators, the government, and the Church, etc. Sociologist Georg Simmel (1904a) posited in his theorization of social struggle, "that conflict has sociological significance, since it either produces or modifies communities of interest, unifications, organizations, is in principle never contested. On the other hand, it must appear paradoxical to the ordinary mode of thinking to ask whether conflict itself, without reference to its consequences or its accompaniments, is not a form of socialization," (p. 490).

The historical pretext for conflict between the US and Bolsheviks as well as the Mexican insurgents is extant: the revolutionary hopes of both parties were met with perhaps the largest deployments of US troops outside the World Wars in the first half of the 20th Century, with General Pershing pursuing Pancho Villa with more than 10,000 men into northern Mexico while the US marines occupied Veracruz, and in Russia thousands of marines landed at Arkhangelsk, Crimea, and other ports controlled by the Whites to stop the spread of Bolshevism during the civil war. However, per Simmel the uneasy ceasefire and continued antagonism between nations is sociological proof that "conflict itself is the resolution of the tension between the contraries. That it eventuates in peace is only a single, especially obvious and evident, expression of the fact that it is a conjunction of elements." (1904b). Thus, it is through feelings of hostility that parties are drawn together, refined, and crystalized as recognizable social bodies. It is through our recognition of the parties we struggle against that we find new forms of social unity and cohesion, i.e., "the further sociological significance of struggle; namely, the influence which it exercises, not upon the relation of the parties to each other, but rather upon the inner structure of each party," (p. 673).

In an exemplary educational context, we see evidence supporting the conflict theory of social cohesion serving progressive reform most clearly after the passing of the National Defense Education Act (NDEA) in 1958 and the Elementary and the Secondary Education Act (ESEA) shortly after: an acute sense of international competition fueled both multi-million dollar investments into public primary/secondary schools and higher education following the Russian development of the Hydrogen bomb, the launching of Sputnik, and the space voyage of Yuri Gregarin—all testaments to the industrial and scientific power of Soviet education. Additionally, the Soviets had demonstrated that they could achieve mass literacy, strong employment, and public-access higher education at little cost to the student. To compete with both the quality and egalitarianism of the Soviet model during the Space Race and the War on Poverty, the US Federal Government was forced for the first time in history to spend more than 4% of its total GDP on public education ("US Education Spending", 2015). Explained via Simmel's theory, it was through the recognition of a common adversary that the inner structure of the US education system was refined. The institution needs opposing ideologies and other, competing institutions to better understand itself, and to force the resolution of its inner contradictions.

Modern Transformations of Revolutionary Struggle: Andragogy in The Digital Turn

In more recent years, digital-based humanities researchers and learners have begun to recognize how the access to knowledge through virtual means is influencing collective thinking to both undermine and reinforce trans-national competition and ideological struggle. A challenge in digital humanities studies in these countries has been capturing the impact of communication and perspectives based on cultural communities, including variables such as age, gender, and race (Balkun & Deyrup, 2020); as these institutions—despite emerging from one-party revolutionary systems of government—can no longer attempt to shape their diverse adult learners by monolithic, top-down applications of andragogical policy, when they now have access to a relatively unregulated network of social and informational communication technology.

Approximately two decades ago, social networking opportunities began to offer a wide range of sharing options across social media platforms. Through the process of encoding/decoding (Hall, 1980), consumers constructed new cognitive meanings, which influenced users' value judgments and sharing habits as they participated in making meaning of the information they would encounter online. Recent research analysis on regulated fake news (fact-checking systems) across three countries found that survey respondents were more aware of how harmful fake news is to society in Japan and South Korea, having more regulated and established political systems (Cheng et al., 2021). While implications of respondents from Thailand, "(they) are less alert to the fact that regulating fake news may harm their freedom of speech because it is already being restricted under the current regime" (Cheng et al., 2021, p.10). These studies support the need for further understanding of how cultural interpretations influence social media users and their communities.

The inherited conceptions of institutionalized revolutionary andragogy manifest in qualitatively definitive patterns of encoding/decoding the digital symbol for a meaningful message worthy of sharing with others. I agree with cultural theorist Stuart Hall that "The level of connotation of the visual sign, of its contextual reference and positioning in different discursive fields of meaning and association, is the point where already coded signs intersect with the deep semantic codes of a culture and take on additional more active ideological dimensions" (Hall, 1980, p. 129). Hall allows that "already coded" signs may take on new meanings depending on the cultural perspective of the decoding subject, due to their semantic point of reference. Despite the rhetorical goals of the producer (regardless of whether the producer is even a human), there is no inherent ideological or socio-political or even truth-value to the sign in itself when initially communicated via ICT and social media, meaning that digital educational content gains or loses cultural capital based upon how quickly and ubiquitously it is used or shared, if at all.

The cultural valuation and encoding-decoding process of communicative technologies and media, as outlined by Hall (1980), has historically been connected to the means of media's production: oral folklores, histories, and philosophies were praised for the verbal primacy of the protean poets who peddled in them, the early modern period brought about the primacy of the publisher-canon, wherein works were recognized based on their association with the aesthetics valued by the limited publishing apparatus and its patronizing classes, and the era of mechanical reproduction introduced the possibility of a primacy of the author-audience, wherein the individual and social merit of analogue ICT, such as newsreels, cinema, photography, and newspapers, could be located in their social content, context of publication and consumption, and cultural use-value. Extending Walter Benjamin's (1936) dialectic to the age of digital reproduction in which we currently live, the disembodied representations of new media in virtual ICT have cemented the primacy of the reader/user as the rubric of value, and as a primary determinant of how people choose to share information.

ICT networks do not have agency, but rather contain pretext (often in a form of power) for agency to utilize. Advantageous nodal structures and assemblages in a network garner the highest amount of power. Structural power online is thus the relation between panoptic centrality (i.e., institutions and governmental sources operating to coordinate "politically correct" publication and dissemination) and the social network it seeks to constitute: the codification of generative connections, the incessant packaging and cataloguing of variations at intersectional nodes of power (sexuality, race, gender, nationality, and other discriminatory qualia and quanta). In Mexico and Russia, there has been a significant struggles between teachers and students—especially adult learners in higher education settings—to embrace digitization of the curriculum, when to do so might compromise the centralization of knowledge production and transmission under traditionally-recognized sites of revolutionary-nationalist power, such as the PRI in Mexico, and the United Russia (Единая Россия) political bloc as the successor-in-power to the Soviet Communist Party.

The difference in their willingness to embrace the digitization of andragogy following the Digital Turn could be attributed to the relative loosening of the PRI's political grip on the Mexican State after a recent series of cartel-related scandals and failed electoral bids, while in Russia, Putin's UR party continues to maintain its centralized hold on the media and educational apparatuses. In Mexico, there is currently a movement to demand "the right to a digital education" based upon the revolutionary principles under which modern Mexican democracy was founded—i.e., the human right to education and autonomous development:

The recognition of the right to digital education comes into being in this context of profound transformation in which information and communication technology plays a decisive role in the innovation of teaching functions, as well as in the approach to research. In their design and use, information, and communication technologies (ICTs) make it possible to customize the steps to knowledge access. Alternatives, such as multimedia teaching, blended learning, or online learning, together with classroom work and distance learning, reduce or eliminate the limitations of space and time imposed by conventional teaching. The result is that learning profits from all the resources of digital technology, especially the internet, by overcoming space and time limitations. (Marcos Sanchez, 2022)

According to Mexico's Global Open University's Communications and Pedagogic Innovation Director, we should seize this post-COVID moment to recognize the fact that the very space-time continuum of learning has been collapsed into a window in which higher education learning must prioritize data literacy techniques over specific content retention. At the same time, in Russia, there is a similar discourse demanding a "transformation of the principles of relationships between the subjects of educational relations and a completely different qualitative approach to the content and orientation of the content of the educational process." (Gromova, 2020, p. 58). Yet, in order to implement the revolutionary approach to reconceptualizing the relationship between student and teacher as one of knowledge co-construction and data literacy, rather than knowledge transmission and data regurgitation, there are significant political and legal hurdles.

In order to redefine Russian andragogy by "taking into account the process of global digitalization, the concept of content quality comes to the fore," teachers and legislators must recognize that "it is not the tools themselves (new platforms and their technical capabilities) that are important, but their purpose." Thus,

A new environment can be formed only if there are five important elements of transformation (formal-legal, technical, content, personal-psychological, social-mental), which only together will provide a synergistic effect. In this case, the system must move progressively, and each element can become part of the system only after certain processes formed at the previous stage. So, without formal, legal and technical conditions, it makes no sense to move on to the development of the content part, and, even more so, the social and mental part.

Here, it is important to note that digital andragogy in Russia aims to provide legitimate techniques and training for teachers to constantly renegotiate updates to technology and content for guiding the development of college student data literacy and construction; but educators recognize that this process is rendered irrelevant unless "formal-legal" prerequisites are met.

At present, the formal-legal status of digital learning among Russia's adult learners remains a contentious issue. Russia's "Roskomnadzor" (The Federal Service for Supervision of Communications, Information Technology and Mass Media) began installing equipment to isolate Russia from 2015-19, including mobile phones, from the rest of the Internet in the event that the government directs such action, as required by a law taking effect in November 2019. The government's justification was to counteract potential cyber-attacks from the United States, but educators both within and without the country are worried that the action would create an online "iron curtain". Then, as of late February 2022, two of the world's leading social media platforms Facebook and Twitter have been restricted by Roskomnadzor as a wartime measure amid the invasion of Ukraine. Internet rights monitor NetBlocks reported that Twitter and Facebook platforms were restricted, or throttled, across multiple providers, with the bans becoming near-total by March 4th. In such an environment, it is ironic that a State, which believes itself to be the safeguard of Russia's revolutionary autonomy, would seek to restrict revolutionary activity on a national scale.

Struggle Between Classes as the Basis for The Revolutionary Educational Ethos

The first major revolutions of the 20th Century have been historically portrayed as struggles between Marxian socioeconomic classes. Antagonism between groups termed the industrial bourgeois, the agrarian elite, and the priesthood, or the *Nepmen* and *kulaks* against the proletariat and peasantry, became the source of Bolshevik legitimacy in curricular design (Fitzpatrick, 1979); *hacendados,* foreign investors, and their armies against the smallholders, the *rancheros,* and the *Indios* formed a similar kind of dialectic motivation behind the ethos of Mexican progressive rural schooling (Raby, 1978). Nikolai Bukharin (1922) wrote extensively on the guiding norms of educative behavior for communist working-class people at the dawn of the revolution, as did V.I. Lenin (1921) and his wife Nadezhda Krupskaya (1922)—a founding member of the official Commissariat of Enlightenment established after the October Revolution (abbreviated in Russian as "Narkompros"). While many Bolsheviks recognized the ideological and thus political implications of schooling, Krupskaya and her Narkompros contemporaries were especially interested in the possibilities of mass education, reformation of the family and social structures, and educating a proletarian bureaucracy and intelligentsia.

The early years of Stalinist reform during the first Five Year Plan (1928-32) produced the resulting proletarian intellectuals that would assume power after the Great Purge (Fitzpatrick, 1979, p. 242). Social engineering of class through public education became more overt in its applications to instructional materials from the onset of schooling: as a kind of pedagogical primer for parents and kindergarten teachers in the early Soviet era, Vera Fediaevskaia's *Nursery School and Parent Education in Soviet Russia* (1936) essentially promotes communist teaching practices meant to prepare children in the crèche model of childcare, rather than the nuclear family traditionally assumed in capitalist nations. Her 40 years of early childhood education experience in Russia led her to conclude that there were benefits to education through collectivization, especially for women as a class, and she mobilizes that knowledge to organize new pedagogical practice around liberating women from childcare and providing the foundations of a communist education.

The first Mexican Secretary of Education, Jose Vasconcelos, was as an official and a rector of UNAM before retiring, and an incredibly prolific writer and speaker on the topic of dislodging class privilege in education. Within the compiled works of his later philosophical career (1981), one finds many of the origins and outcomes of his andragogical theories regarding race, class, and socioeconomic power. "Qué es el comunismo (1936)" describes his understanding of communist teaching and enterprises as modeled in indigenous land collectives and labor unions already developing throughout the nation. "De Robinsón a Odiseo: Pedagogía estructurativa (1935)" articulates the philosopher's andragogical plan to introduce European literary classics to the population as a method of civilizing the peasantry. "Bolivarismo y Monroísmo: Temas iberoamericanos (1934)" discusses the various aspects of competing ideologies in Mexican education—that of American corporatism (the Monroe Doctrine) and Latin American nationalism (Bolivarismo). "Indiologia (1927)," or "Indiology" is Vasconcelos self-coined term for the study of Indian cultural methods of educating the community and understanding the means and relations of labor in a traditional pueblo, so as to translate revolutionary curriculum into a digestible format for the peasantry. Such were the revolutionary conceptions of the Secretariat's instructional principles implemented throughout the first decade of the revolutionary regime.

The Urban and the Rural

Reconciling the peasantry to a proletarian and middle-class revolution is a recurrent theme of concern for reformers. One of the first theorists on public, compulsory education in Russia, Count Lev Tolstoy (1904) gives a 19th Century account of his general considerations for a free system of adult-based elementary schooling for the peasant class. He acknowledges international models (such as the United States) to follow in establishing schools and taxes to pay for them but insists that a unique polytechnic Russian system of social learning must be applied to engage their unique population and recalls his experiences teaching and administrating peasant schools at Yasnaya Polyana to illustrate his argument. Specifically, he highlights the ability to teach rural students' literacy and collaborative skills through labor connected to the production of the peasant community from season to season. Such emphasis on experiential learning through physical activity would remain a common trait of Soviet physical education through the Stalin and Khruschev's regimes, though it was often turned towards more disciplinary and less educative purposes.

While the precepts of Leninism are grounded in a modern vanguard party instilled with proletarian consciousness, the majority of the countryside in Russia, as it was in Mexico, was still populated with largely illiterate, religious, and conservative agrarian communities at the turn of the 20th Century. To reach the majority of the population, a mass literacy and propaganda campaign was required. Perhaps the most widely read Soviet Document of the Russian Civil War period, *The ABCs of Communism* (1920), was written by two Central Committee members (Bukharin and Preobrazhensky) as a pamphlet that eventually became the basic primer for the Soviet political economy and initiation into Bolshevism. Focusing on the section "Development of Powers of Production Under Communism" (p. 60-2), the document explains that the bourgeois were actively withholding knowledge and culture from the working classes under the Imperial and Provisional Government educational systems. Thus, the goals of Communist education are to destruct the prior school apparatuses intended for the economically privileged and redesign instruction as an instrument for all Communists' socialization and enlightenment.

Lunacharsky (1920) publicly expressed his aims as the government's first Commissar for Education and Enlightenment. Therein, he communicates an attempt to reform Russian education system on Marxist principles to combat adult illiteracy and to train future generations in Communist thought among the peasants. Aside from proposing a simplified alphabet to aid with literacy instruction, he believed in a system for subsidizing the arts and removing it from the purview of bourgeois ownership and patronage. Another innovation he proposed was the unified labor school and Workers' Faculties that would provide intensive and accelerated courses to train technicians and administrators from the working classes and the peasantry, as well as night classes for the undereducated.

Writing in the post-reform period, Mexican educational philosopher Samuel Ramos (1940) looked back on the twenty years of revolutionary pedagogy since Jose Vasconcelos accepted the post of Secretary of Education. While Vasconcelos's philosophical idealism guided early reforms to the Federal funding of rural schools and the implementation of new primary and secondary institutions, Ramos actually sees the interim period of "socialist education" as an extension of *caudillismo* (warlord) ideology advocated by policymakers of the Calles administration. His concern that socialist education provided by a centralized government served to orient the rural population towards the hegemonic power of the State is a critical counterpoint to the revolutionary rhetoric in favor of rural school reform during the 1930s, echoing Dewey's earlier concerns over state control in *Democracy and Education.* Though Vasconcelos preached a relatively peaceful assimilation between indigenous and Hispanic learners in Mexico, his thought was still overtly influenced by the project of modernity and the racial supremacy of *mestizaje* professed by the elite *criollos* of Mexico, and by the broader eugenical worldview growing in popularity by the end of the 19th Century.

Yet at the height of the Great Depression, a volunteer educator from the Office of Education, Katherine Cook (1945), provided month-long account of a tour through the rural schools of Mexico, which were administered by Vasconcelos' successor, Dr. Saenz, in the 1930s. Via eyewitness accounts and interviews with Saenz, Cook found that the Mexicans were cheaply and effectively attaining the aim of rural health education and school hygiene combined with literacy and technical training to an extent impossible elsewhere in the United States during the Depression, especially in educating their own Native American populations and other marginalized groups. What's more, she remarks, it is admirable that the Mexican Federal teachers are selected for their capacity for leadership, granted high social status, and guaranteed pay by the Federal Government—something the U.S. has had historic difficulties accomplishing.

Racial/Ethnic Oppositions

Recognizing that education has been historically deployed as a tool of ideological training and the reproduction of social and ethnic class privileges for "Great Russians" while minimizing those of ethnic minorities, Lenin called for the enlistment of specialists in enhancing the general literacy and communal consciousness of the working and peasant classes across the old Empire, along with a renegotiation of power and knowledge in the institutional relationship to the individual (Lenin, 1921). Many of these ideals form the guiding principles of the first Commissariat of Enlightenment that created the early Soviet education system during the NEP and admitted the use of non-Russian languages for instruction, so long as a Russian language course was taught.

In Mexico, Vasconcelos published *La Raza Cósmica* (1925) as pseudo-prophetic and partially eugenic set of educational theses, positing the accelerated mixing of the world's races (Spaniard and Indian) would pave the way for the emergence of a new type of mestizo people, that is, the "future Cosmic Race." Vasconcelos glorifies the concept of *mestizaje* and its roots in *indigenismo* for the first time in national consciousness: as a counterpoint addressing foreign racist powers such as the United States, he argues that their distinctly Anglo-Saxon refusal to mix with other races points to their inevitable decay, as "the ultimate aim of History is to achieve a fusion of all peoples and all cultures." (p. 46). Claiming that the contradictions of racism and Darwinist ideologies, which are "scientific" theories that form exceptions to validate, explain, and justify ethnic superiority and to repress others, Vasconcelos attempts to refute these theories in an effort to improve the cultural morale of his "depressed race," offering instead a new educational model to bring about industrialization and the creation of a utopian "Universopolis" based upon his own aesthetic and racial theories of superiority.

Many of the themes present in *La Raza Cósmica* recur in the lectures Vasconcelos delivered later at the University of Chicago, following his resignation (1926). He reiterates the racial opposition between Latin and Germanic cultures and the function of Christianity as a civilizing agent in every culture, including Asiatic nations. He continues to glorify *mestizaje* through aesthetics as the mechanism of social exchange, the civilizing impetus, and genetic procreation, by correlating physical beauty with class status, while hard work and miserable living conditions are given as reasons for indigenous peoples being "decidedly plain looking," because "a race of slaves cannot be beautiful." (p. 39).

In his own travels, educator George Booth (1941) interprets more recent racial developments in the Mexican public school system through a Marxist lens, recognizing that the reforms of the later Cardenas regime sought to integrate the mestizo and the Indian peasants into educational systems in order to reconcile popular opinion with the revolutionary aims of the administration. He presents the thesis that the education system first conceived by Jose Vasconcelos was reshaping both the Mexican countryside as well as its metropolitan consciousness through a reformulation of national ideology and the development of small industry. Invited by the government and the labor unions that had been nationalized in Mexico in 1922, Tannenbaum (1933) also connected his proletarian interest in organized labor to marginalized groups such as peasants and Indians affected by revolutionary education. Based on his experience, many government officials, and teachers themselves proved that the work they described justified the social program of the Revolution. Tannenbaum's vision of Mexico is at times almost mythic in his romanticization of the Indian and the revolutionary, but through his account, the public vision of the Revolution was somewhat vindicated as a truly agrarian, patriotic, and populist movement with honest intentions.

Generational Conflicts

Komsomol, the Leninist self-governing school organization for secondary and college students became a legitimate competitor with Narkompros for control of the education system by the first Five Year Plan (Fitzpatrick, 1979). Through popular literature addressing student audiences, a transition from the Leninist and Trotskyite "Old Bolsheviks" comprising Narkompros to the industrialists of the Stalinist period facilitated a replacement of Party staff and administrative positions in prestigious institutions with Komsomol activists and younger, scientifically oriented Marxists. For example, a volume of Russian textbooks from the late Stalinist period, *I Want to Be Like Stalin* (1947) informs teachers on methods of moral instruction, Soviet patriotism, socialist humanism, collectivism, Lenin, and Stalin, from a formalized and State-approved point of view. Furthermore, there is an extensive appendix on disciplinary technique and "rules for schoolchildren" indicative of the Stalinist institutional culture. As comparative education scholar George Counts implied in the introduction to the English translation, this text is ideologically directing future generations towards the insular command economy in the name of scientific efficiency and industrial development, while largely forsaking the original Marxist project of promoting international communism among adult learners in the global proletariat. *New Russia's Primer* (1931) is another unconventional textbook that employs elements of poetry, science fiction, and political science to explain the purpose and methods of the first Five Year Plan. Written by a young engineer explaining the goals of industrialization under a planned economy, the text is immediately accessible, entertaining, and well written for a wide range of literacies. However, the motifs recurring in the book fluctuate between apocalyptic and utopian, abstracted from many of the Russian lived experiences during the civil war, the NEP, and the Five-Year Plan.

The experiences of young adults after the Russian Civil War were often those of homelessness, poverty, and crime. To remedy this situation, and to adapt the educational system to the next generation of communist youth throughout the Soviet Union during a period of state-mandated industrialization, methods of social experimentation were employed that were historically unprecedented. Perhaps the most influential Ukrainian educational theorist and sociologist studying the revolutionary generation, Anton Makarenko elaborated the theory and methodology of facilitating an upbringing in self-governing collectives and introduced the concept of "productive labor" into the educational system. Further, he was a fantastic writer with a talent for blending genres: his "pedagogical poem" *The Road to Life* (1936) is an autobiographical novel providing his account of a socialist teacher sent to work at a Ukrainian collective farm to explain the skills and principles necessary for communist education. He argues through the novel that staff must have personal boundaries and standards although they live with the residents, large amounts of youth and adults learn best in collaborative groups, and that coercion or punishment is "sometimes necessary" to integrate productive labor and education.

Soon after, Makarenko also published the *Book for Parents* (1939), an andragogical book on parenting that gives numerous examples of situations, conflicts, and misunderstandings in the family that can become teachable moments through analysis. He intended to give the reader "the useful starting points for their personal active pedagogic thinking." (p. 1). Makarenko produced an original approach to the theory of discipline used for "combating and surmounting difficulties" to build one's integrity and character, so as to resist socioeconomic class antagonism and oppression. He also gives the first detailed elaboration of the educational significance of the collective and makes use of a remarkably profound "system of perspectives," wherein an optimistic socialist vision was used to combat present adversity: "Man must have something joyful ahead of him to live for. The true stimulus in human life is the morrow's joy." (p. 1).

In Mexico, older revolutionaries like Vasconcelos were replaced by less philosophical, more pragmatic secretaries. Saenz was the most prominent Secretary of Education in Mexico after Vasconcuelos resigned, and his lectures to the Harris Foundation in the US (1927) demonstrate how crucial his conceptualization of the Mexican culture was to adapt the educational system to the entire federation of Mexican States. Assuming that the prior regimes had made little effort to include indigenous or mestizo citizens in the system of education that would prepare them for the university, and that the majority of Mexicans were actually indigenous or mestizos, his main initiatives as Secretary were aimed at the reformation of the rural school and the recruitment or mobilization of teachers to instruct communities outside the capital in methods of crop rotation, literacy, and basic hygiene to integrate the marginalized demographics of Mexico into the national economy and modern standards of living.

While Saenz attained a graduate degree from Columbia by studying with John Dewey, he recognized in his article "The School and Culture" (1932) that Western cultural concepts of education were based upon a certain level of socialization to an industrialized lifestyle that the bulk of indigenous peoples in Mexico were not accustomed to. Dewey's theories of making education functionally useful could be commandeered by the industrial elite to determine which jobs were available, and thus which courses and teachers would be valued in a state where democratic systems were not yet viable. However, in accordance with Dewey's pragmatic principles of making schoolwork culturally and socially relevant and productive, Saenz advocated adopting indigenous epistemologies and cultural methods of understanding to make knowledge of healthcare, agriculture, industry, and history relevant at the rural schools and cultural missions he had initiated to fulfill the reforms first outlined by Vasconcelos.

Priests vs. Educators

John Dewey brought attention to the roles of both the Russian Orthodox and the Roman Catholic Churches in maintaining the reproduction of class privilege through complicit educational programs for their respective 19th Century leaders, Czar Nicholas Romanov and President Porfirio Diaz (Dewey, 1929, p. 117-8). In contrast to the traditionalism of Orthodox and Catholic instruction, state-funded reforms aimed at reducing Church presence in the classroom, minimizing the use of corporal punishment, requisitioning Church assets for the sake of public schooling, and developing scientific, evidence-based methods of instruction emphasizing integrated learning experiences. Russian educational psychology developed by Iakobson (1960), Idashkin (1962), Elkonin (1961; see Ablin, 1963 for translations), and especially the early methods of Vygotsky (1925) were far more advanced in their ability to measure cognition and behavior than contemporary ideas of psychoanalysis and behaviorism in the West.

Known as the Father of Soviet Psychology and a founder of the sociocultural perspective in educational psychology, Lev Vygotsky was attempting to rethink the pedagogical methods and aims of the early Soviet educational system away from religious and philosophical scholasticism by providing a uniquely Marxist theory of psychology. In an early lecture (1925) that leads with an epigraph from Marx on materialism, Vygotsky gives Pavlovian reflexology as the basis for mental behavior before illustrating how Marxist analysis of the historical and social "reflexes" derived from cultural interaction and the inputs of other people also shape consciousness beyond any existential factors of the unconscious suggested by Western psychologists such as Sigmund Freud or William James. His argument is critical to the Soviet implementation of collaborative teacher-student learning and recognition of the sociocultural components of development, which could be potentially harnessed in the creation of the "New Socialist Man."

A primary aim of the Communist project since Chernashevsky's writing in the 19th Century was the formulation of this "New Man," whose collective consciousness represented a revolutionary new phase in human evolution at the social level. Vygotsky explains that, as the fetters of a decaying capitalist society fall away, the positive potential for establishing a new kind of productive social being emerges through a planned educational system: "New generations and new forms of their education represent the main route which history will follow whilst creating the new type of man." (1930, p. 1). His theories are predicated on a kind of dialectical materialism found in Marx, where a change in the system of human reflexes caused by relationships or technologies apparent in an environment will alter one's behavior, and thus consciousness.

Smith (1942) highlighted the tensions between the Church and the State during the Mexican Revolution that led to further difficulties in the reform period when he briefly visited Mexico and other Latin American countries to complete a study on comparative education. While the Diaz regime had allowed the Catholic Church to build up massive holdings in the countryside so long as it legitimized the government, revolutionaries advocating for land reform were eager to remove the Church from governmental activity and removing its right to property ownership. These antagonistic feelings, according to Smith's interviews, deeply affected the education system after the revolution, when many clergy were disaffected with reform and came to intense disagreements about curriculum in the rural schools as well as the secularization of public schools in the major cities.

Dislodging the Church from its home in the university was also an early aim of the Mexican revolution. Considered the founder of the secular Mexican university system, Justo Sierra (1910) gave a speech at the opening of UNAM (the Mexican Autonomous National University) to define specifically how the mechanisms of higher education would change according to revolutionary principles set in contrast to the Catholic institutions left over from the colonial administration, such as the Royal and Pontifical University. Rather than merely transmitting the officially-approved "truth" of the curriculum to a preselected, privileged stratum of society, Sierra saw the new UNAM as a place where knowledge and truth were negotiable, debatable, and revisable through research, which is to be conducted by a group selected of the people to achieve the political ideals of democracy and liberty, rather than simply for their loyalty to God and the King of Spain.

The cultural achievements of the Mexican Revolution in education were among its most important for providing a stable, multi-ethnic national consciousness. While most educators and historians would assume a break between the conservative (Diaz and Huerta) and radical (Madero and Carranza) administrations during the 1910's, Vaughan (1982) demonstrates in the accounts of Mexican teachers that pedagogy and educational policy remained fairly consistent during the transition and did not truly become "socialist" or secular in nature until the 1930's, after the Church was fully dislocated. Additionally, the dependent nature of Mexico in the capitalistic relationship with buyers of exports (the USA, mainly), and the Mexican government in pre- and post-revolutionary regimes, dictated the kind of schooling available in both contexts, regardless of early revolutionary rhetoric. During the 1920s and 30s the government sought to nationalize and modernize rural society with a series of massive investments in teacher training and construction of rural trade schools drawn from the Church's assets. As a result, according to Vaughan's later study (1997), the State's explicit program of socialist education became an arena for intense conflict between priests, students, teachers, and politicians over power, culture, knowledge, civil rights, and even gender practices.

Balancing Idealism and Utilitarianism in Later Reform

As the first of many Five-Year Plans began, educator Raymond Charques (1932) yielded the possibility that the Soviet model may have created the most educated, mobilized, and self-conscious society the world has ever seen, and may even be considered "the best educational system in the world." (p. 1). At the same time, he acknowledged the divide between the peasant collectives and the proletarian context that the Bolsheviks had struggled to reconcile since the Civil War. Setting criticisms of the model's implementation aside, the USSR's reforms to public education emphasizes the role that education could play for inspiring progressive educators in the West and for renegotiating the purpose of educational institutions in and of themselves.

As a result of growing American curiosity in these revolutionary processes, the proceedings of the first unofficial US Trade Union Delegation to the USSR were collected and organized around several different topics by specialists in their respective fields to provide a comprehensive report on Soviet industrialization. Delegate George Counts (1928) gave a different perspective on the changing purposes, structure, power dynamics, and andragogical practices of the Soviet public education system as it was developing at the beginning of the Stalinist regime. A contemporary of Dewey, he visited Russia on multiple occasions following the civil war and initially portrayed the Narkompros program as a significant "challenge to American education" (Counts, 1931) due to the Soviets' explicit aim of surpassing the US and Western Europe in economic productivity using a state-planned system of public instruction.

Darkened by his experiences with Stalinism, Counts' later description of the Soviet school system becomes more cynical and critical than records of his earlier interactions with Russia would indicate. He focuses on regimentation of the student's body, the suppression of freedom and human dignity, and the return to quantitative punishment-reward systems after the dissolution of the Commissariat of Enlightenment and Lenin's death (Counts, 1957). Yet Soviet education would still present a challenge to the US in terms of literacy rates and college graduation rates even in Stalin's and Khrushchev's administrations, given the great attainments in Soviet scientific education at the beginning of the Space Race. To discursively belittle their victories, he attributes Soviet success to coercive forms of social brainwashing inherent in the Party-State system that led to harder working and more engaged students in the USSR.

The cultural period after Stalin's death, known as the Khrushchev "Thaw" (*otepel'*), opened up new possibilities for individual expression and relaxed censorship throughout the Soviet Government. In a series of published speeches expanded into written proposals on reforming the Leninist unified labor school model for higher educational institutions and increased admissions to universities, Khrushchev (1958) sought to bring the more privileged systems associated with the intellectual class under proletarian ideological control and to make communist education as utilitarian as possible. While his attempts at liberalization allowed young new intellectuals such as Gorbachev a place in the Party, it also created many tensions with the old Stalinist elites. Counts (1959) revised his interpretation of Soviet education under Khrushchev's reforms in light of the launching of Sputnik and the true threat that Soviet education had become to the shortcomings of liberal democratic education. The Soviet leader argued in a series of speeches, which Counts translated, that the proletarians in the USSR must return to polytechnic education, which had been abandoned in favor of specialization during the Stalinist period of industrialization. It was scientists educated in the revolutionary period at polytechnic institutes who produced the Soviet satellite, cosmonauts, and the atomic bomb, rather than the party favorites raised in the Stalinist educational bureaucracy. Thus, both Counts and Khrushchev look forward to a time when Stalinist disciplinarian systems of education will be reconstructed to encourage the foundations of science, but also reintroduce polytechnic skills to make students capable of physical work and an acute sociopolitical consciousness.

Based on her extensive experience teaching in Moscow before WWII and interviewing pedagogical theorists in Russia afterwards, Levin (1963) traces the trajectory of Russian education from the point of view of the child growing up through the school system and the various obligations required of them during different educational administrations as they enter adulthood. For instance, the Narkompros attempted purely student-centered education that was compulsory for 4 years, and by the time of Khrushchev's regime, most Russians were attending at least 7 years of compulsory schooling, and curricula at state universities were far more developed. Based upon her interviews and teaching experience, Levin demonstrates the changing interpersonal and intrapersonal perspectives of students and educators throughout these periods of reform. A survey of more than 70 American educators' firsthand experiences teaching, consulting, and conducting research in the Soviet education system in 1958 demonstrate that polytechnic curriculum developed in the revolutionary period by Soviet scholars emphasizes a balance between academic and technical specialties in each student (Bereday et al., 1960). Polytechnic, socialist education showed the Americans how broadening the pedagogical approach to labor and different kinds of work through a utilitarian curriculum actually narrows the gap between mental and manual work and the achievement gaps between mass education and higher education. According to the teachers, such a society could override Capitalist divisions of labor while still preparing students technically to contribute to the growth of modernized Communist civilization.

Vasconcelos and other early Mexican educators conceived of the rural school system, *indigenismo,* and separation of Church and State as major ideological enterprises for instilling revolutionary spirit, but the movement was still fraught with racial, socioeconomic, and gendered conflict. The contradictory elements of his early Secretariat of Education would become critical points of reform in the Cardenas period, which finally established a cohesive national ideology to quell remnants of the Porfiriato (conservative, Spanish, Catholic politics) and post-revolutionary *caudillismo* (warlord-style radical populism). In an early account of the educational reforms under Cardenas, Benjamin (1935) gives a brief historical interpretation of the cultural phenomena he observes in the transformation and proliferation of rural public schools, designed specifically for the indigenous populations of Mexico. He recognizes this form of public education as a method for maintaining the hard-won freedoms of the revolution against reactionaries of the *Porfiriato* as well as ambitious and authoritarian *caudillos* like Calles and Obregon. While the new rural schools present an "atheistic" challenge to the Catholic method of catechism in indigenous villages and mestizo pueblos, Benjamin admits that this new pedagogical model suited to the mestizos and Indians of the countryside could help to establish a completely new social order in Mexico.

As liaison from the cultural mission's department to the US Department of Education in the 1930s and 40s, Bonilla y Segura was eager to afford foreign powers a greater understanding of the outstanding educational accomplishments of Mexico throughout the late 1930s and early 1940s, after the Cardenas regime had reformed rural education around indigenous schooling to provide basic polytechnic skills as well as encourage greater socioeconomic involvement in the Mexican economy. He demonstrates in his report (1945) how missionary societies, under the purview and funding of the federal government, have established normal schools for teachers, athletic facilities, improved wells and crop yields, healthcare and small industry among other aspects of Mexican peasant culture. Indeed, their successes in combatting illiteracy, low birth rates, and general education as well as hygiene revolutionized the social standards of the countryside. Furthermore, their system of public schooling, which strove to include various demographics historically rejected from education, constituted (and will continue to constitute) an inspiration and a challenge to educators of the United States.

The Contrasting Purposes of Higher Education

The Commissariat of Enlightenment attempted to establish an egalitarian system of higher education that was depleted of material and intellectual resources by the Civil War and the NEP. However, Bolshevized youth were given important new rights in admission and reviewing professors that diffused power among the bottom of the hierarchy in regulating the university system. Loyalty to the Bolshevik party and politically-correct curriculum, while upending the structures that restrained the culture of the institution, resulted in a highly centralized university system well-suited to the Stalinist regime (McClelland, 1971). Stuart Finkel (2008) gives an intellectual history based on the perspectives of almost one hundred prominent intellectuals, including professors, journalists, bourgeois engineers, writers, etc., who were expelled from the USSR for their dissenting views on the intended design of the Soviet university, finding that Lenin's policies towards the Soviet system of higher education were first and foremost aimed at eliminating perceived ideological foes from within, following the Russian Civil War. Finkel demonstrates that the Bolsheviks' main aim was to impose ideological conformity through censorship and reform, to redefine the role of the intelligentsia, and establish a Leninist public sphere. Thus the Soviet regime rigorously policed the ideological boundaries of the university and laid the institutional and discursive foundations for the specialists of the Stalinist state.

On the eve of the Mexican Revolution, Turner (1910) provided a scathing expose of the atrocities committed by the *Profiriato* (aided by American partners) that left the people of Mexico largely enslaved to land tenure and ill-educated to question their socioeconomic position in relation to the oppressive regime. He lists instances of Diaz's censorship of the press, fraudulent elections, the squelching of opposition parties, the rising number of labor strikes, and the enslavement or execution of indigenous peoples as methods of maintaining the *Porifiato's* hegemony. In the context of education, the effect, according to Turner's political theorizing, is that the intelligentsia and Mexican institutions of higher education were oriented towards corroboration and conspiracy with the government while punishing dissident opinions and criticism.

While the new constitution of Mexico under Cardenas stipulated that public education in the 1930s would be "socialist in character," intellectuals debated over what this socialism might look like in practice. Using statistics from the Mexican Secretariat of Education, Punke (1932) was able to demonstrate that Mexican universities were growing after the revolution, especially in the medical and scientific fields, due to the expansion of secondary and prep schools across Mexico. Also, he shows that the Mexican Federal Government was taking over an increasingly larger share of the cost for higher education than the local states, as the secondary and prep school system became slowly nationalized under the revolutionary program of school expansion.

In higher education specifically, Antonio Caso was a longtime philosopher in education by the time that Lombardo Toledano, Mexico's foremost Marxist intellectual and educator, proposed a curriculum for the undergraduate degree that would teach history as the dialectical development of social institutions as related to different social classes. Caso disagreed in their public debate (1933) over the future of revolutionary education, arguing that communities within the university should freely associate and disagree so as to reform new communities of thought, rather than being subjected to a universal system of socialist education. While the conclusion of the debate did not result in a decisive victory for either viewpoint, their dialogue represents a primary example of the discourse of academic freedom and socialist intent that pervades the Mexican university even in the present.

As an educator sympathetic with Mexican socialist reform in education, George Sanchez (1936) gave an account of the many weeks traversing the rural schools and cultural missions of the countryside which were previously inaccessible to many American teachers interested in Mexican revolutionary andragogy. He acknowledges that Mexico is only now beginning to confront the issues of its colonial past, and present wealth inequality, but that it was doing so with a teaching method suited to acculturating the peasant population to modern methods of self-care and basic schooling. Sanchez admits this preparation is insufficient to develop higher intellectual specialties as seen in industrialized nations like the United States but understood that this phase of general preparation for the majority of the population is necessary before their mass inclusion in higher education pursuits. In treating higher education spaces specifically, he historically traces the Spanish heritage that established Renaissance-style seminaries and universities in Mexico, complete with clerical teachings, the humanities, and history along with classical languages, until the 19th Century (Sanchez, 1944). A burgeoning liberal industrial bourgeoisie began to clamor for utilitarian reform and a curriculum that adhered to the principles of realism and pragmatism rather than idealism or humanism. While the National University and polytechnic institutes were enacted in the final years of Diaz's regime, the universities became increasingly divided ideologically between the liberal/progressive scientists at the polytechnic institutes and philosophical conservatives at the National University, creating a kind of dialectic between Right and Left seen in Mexican higher education culture in the post-revolutionary period.

CONCLUSION: THE PERMANENT STRUGGLE

Marxist andragogy transmitted through Russian and Latin American revolutionaries is once again the focus of understanding and renegotiating sociopolitical conflicts in the United States education system. Dewey's utilitarian and democratizing methods of project-based learning, polytechnic instruction, and experiential education inspired the discourses that shaped progressive education movements in revolutionary Mexico and Russia and were in turn refined through international experimentation and collaboration. Lev Vygotsky's sociocultural psychology transformed the philosophy of Marxism into a measurable developmental science of cognition. His posthumous English translations are now becoming popular as seminal works of social learning theory and constructivist pedagogy in American educational research, where his methods of analysis help researchers to understand the uses and outcomes of small group learning and co-constructing knowledge (McCaslin & Hickey, 2001). And like the Ministry of Education in revolutionary Mexico, the United States calls for progressive educational reform to accommodate the vast population of partially and completely indigenous peoples who already make up the majority of school-age children in some Southwestern States (Hussar & Bailey, 2014).

As in Mexico, reformers in the US are addressing the issue of marginalized students' retention and achievement with socialist-inspired, student-centered, experiential education where Southwestern school districts were offering Mexican American Studies curricula based upon the design principles of Paulo Freire. In *Pedagogy of the Oppressed* (1970), Freire provided a model of student-centered literacy instruction that achieved many of the consciousness-raising goals of earlier revolutionary Latin American theorists. He became a celebrated figure in Mexican and US higher education at the beginning of the Neoliberal era in global politics, when he was invited by the leading Mexican educators at UNAM to discuss and publicize his position on education and its relationship to power, real v. idealized curriculum as it confronts students' social realities, and the role of the organic, public intellectual in reconciling the divisions of labor between intellectual and manual work (Escobar et al., 1994). In Arizona, where Latinos historically dropped out of high school at nearly three times the rate of their White peers, a MAS program was integrated with core classes as legitimate options in history, literature, and social science among institutions in Tucson Unified School District, where Hispanics are more than twice as numerous as Whites. In attempting to fit the state curriculum to the student's cultural background and interests as recommended by the progressive movement, researchers from the University of Arizona found in Tucson Magnet High School that student cohorts who participated in MAS programming were between 51 and 108 percent more likely to graduate high school, and approximately twice as likely to pass Arizona's standardized exams (Cabrera et al., 2014). The research team's expert opinion would be invaluable during the ensuing court case determining the legality of a ban imposed on the program in 2012.

Even as the nation learns from the conflicts of revolutionary neighbors and competitors, struggles with the hegemonic forces inside the educational institution itself continue to impede the functioning of radical democratic reform efforts. Addressing a MAS administrator at a legislative hearing for the passing of HB2281 (the act that would withhold 10% of TUSD's funding if they did not cancel MAS courses), the then-State Senator and eventually Superintendent of Public Instruction John Huppenthal presented the State's logic for the ban:

The textbook that you use, Pedagogy of the Oppressed, my understanding is that if you go and look at the citations, you see Marx, Lenin, Mao, Che Guevara, Fidel Castro. And so, our suspicion is inside these classes, these students are being indoctrinated by people who are in power to have a certain mindset of us versus them. ... That's our suspicion. We really think we know what's going on behind those doors. People in power are doing something distasteful. (Gonzalez et al. v. ABOR, 2017, p. 11)

Already we see the beginnings of a discourse concerned with the rights of the individual over the collective (especially in regards to property and opportunity), a bourgeois appeal to "law and order," and *pathos*-invoking imagery employed to appeal to the target audience's fear of Eastern European and Latin American communists or anarchists. The Federal District Court for Tucson ruled in 2017 that the ban violated students' protections under the First and Fourteenth Amendment, and that the censorship had been politically and racially motivated; but State prohibition and the student/teacher class-action suit against the Arizona State Board of Education made national attention for years leading up to the decision (p. 42). As per Simmel's theory, conflict and conversation about it are teaching the public that the US educational system needs to become multicultural to continue calling itself democratic, and that it needs to be relevant to new kinds of people to remain useful. American education must therefore re-adapt in line with andragogy's guiding precept—that learners with adult brains need the relevance of a knowledge domain proven to them if they are to be motivated in learning the content required.

The revolutions in Mexico and Russia prompted early progressives like Dewey to think through both the possibilities and dangers of centralizing State education: on one hand, it could provide directed support to facilitate "fuller, freer, and more fruitful association and intercourse of all human beings with one another." (1938, p. 98). On the other, as Dewey comments in *Democracy and Education,* it may not be "possible for an educational system to be conducted by a national state" since "the full social ends of the educative process" are generally misguided by the nationalistic demands of the State itself (p. 97).

Arizona's Attorney General and State Superintendent recently attempting to destroy Mexican American Studies supports this theory.

Protests erupted in Russia as students and professors block exam sites for the piloting of the EGE, or Unified State Exam (Denisova-Schmidt & Leontyeva, 2014). Mexico also recently adopted standardized tests to remain internationally competitive with admissions to secondary and higher schools, which was met with teacher occupation movements in Mexico City and Oaxaca, among other regions (Botz, 2013). But Mexico and Russia taught Dewey that a state apparatus does not necessarily stand in the way of local interests and individual talent; it can actually become, momentarily, the defender of public welfare that progressives had hoped for, when it acts as a vector of revolutionary sentiment—when it insurrects against itself in a convulsion brought on by its own historical contradictions. Through funding the training and mobilization of teachers, libraries, and basic necessities, centralized networks of resources for on-the-job training and feedback could fuel the community development necessary for the "free and equal association" Dewey mythologized, especially when knowledge sharing, and co-construction is enabled by the advent of digital ICT. While it might seem impractical in present times to dislodge educational practice from statistical outcomes valued by the State, it is even less practical to continue pretending that they are designed to help students become full-rounded participants in a march towards progress for any truly democratic society.

Further, the modern struggle for andragogical digitization in these revolutionary contexts highlights that technologies for encoding and visualizing information, from the paperback novel to the digital map—far from merely critiquing culture and knowledge—provide us with superficial and indiscriminate information, such that a medium's potential for increasing democratic participation, individual autonomy, and community sovereignty is related to ways that users can format, stylize, and interpret the representations of information. In a space as complex as the globalized city, the codes used to represent it reveal how the selective visibility of marginal spaces has both individual and socio-political implications for (resisting) control over the movement of networked bodies and identities, which can work both for and against institutionalized revolutionary movements. Thus, the relative political power or sovereignty that a network agent can have over a given ICT-based culture is theoretically dependent upon their position in the network. The network theorists Galloway and Thacker begin *The Exploit: A Theory of Networks* (2013) with the phrase "we're tired of trees," revealing an obvious skepticism with the post-Digital Turn fetish for the agential freedom of rhizomatic networking made possible by digital ICT. Indeed, they demonstrate that sovereignty plays an important role in creating and controlling networks. Network sovereignty is thus rooted in the power to make unilateral decisions that can be augmented by exercising discipline over the network and accumulating knowledge about its members (i.e., what the authors term the "Foucauldian argument" of networks).

Networks should, in theory, have the revolutionary function of levelling hierarchies in social learning; indeed, they should liberate all fields of knowledge by de-centralizing them out from underneath the control of those who traditionally "posses" knowledge-power in a Marxist framework. But are these ICT functions truly liberating and revolutionary, or simply reiterating the authoritarian/institutional mechanisms of disciplinary practice and power? According to most Cultural Studies theorists and Marxists of Stuart Hall's tradition: yes, the network society is generally liberating in the knowledge-power relationship of encoding/decoding. As the network of information expands to parts of the world where knowledge is traditionally held in the hands of a privileged group, it follows logically that there would be a global revolution. A binary breakdown of the old hierarchies, as the world came into a new state of cognitive awareness.

Yet what has happened in the growth of networks in Mexico and Russia has demonstrated quite the opposite effect, in a Foucauldian sense, from a revolution. Aside from governmental censorship and control, the globalization of the capitalist firm and its economic-political relationships have penetrated the private life of the individual deeper than ever before in the age of Neoliberalism. Widening income gaps, environmental crises, wars, corporate and governmental corruption, and economic recessions are accompanied by the world's fastest growth in networking and ICT. Indeed, the disciplinary mechanisms of power are more distributed, refined, automated, and responsive than ever before. Adult learners' and teachers' appetite for "user-friendliness" is key here: the growing information and computing firms developing products for corporations and individuals understand the value of a simplistic, intuitive, yet useful technology. These products seem to liberate one from inconveniences associated with data gathering, decision-making, navigation, communication, etc. Yet these devices use digital networks dictated by formal-legal limitations to customize an experience to a subject's preferences, in the most efficient way to placate us to dominant ideologies. If systemic power is inherently architectural—that is, not a component or form of agency itself—then what happens to us in a fully automated, all-powerful network that grows increasingly independent from individual human agency? One might say that the adult learner will continue to be artfully extracted from the equation of power by the sophistication of modern technologies of discipline, if the revolution in human relations is allowed to completely ossify through institutional and governmental structure.

REFERENCES

Ablin, F. (1963). *Education in the USSR: A collection of readings from soviet journals.* International Arts & Sciences.

American Trade Union Delegation to the Soviet Union. (1928). *Soviet Russia in the second decade.* John Day Publishing.

Benjamin, H. (1935). Revolutionary education in Mexico. *The Annals of the American Academy of Political and Social Science*, *182*(1), 181–189. doi:10.1177/000271623518200121

Bereday, G. Z. F., Brickman, W., & Read, G. H. (1960). *The changing Soviet school.* Houghton Mifflin.

Bonilla y Segura, G. (1945). Report on the cultural missions of Mexico. U. S. Gov't. Printing Office.

Booth, G. C. (1941). *Mexico's school-made society.* Stanford University Press.

Bukharin, N. I., & Preobrazhensky, E. A. (1969). The ABCs of communism. Penguin. (Original publication 1920)

Cabrera, N. L., Jeffrey, M. F., Jaquette, O., & Marx, R. M. (2014). Missing the (student achievement) forest for all the (political) trees: Empiricism and the Mexican American studies controversy in Tucson. *American Educational Research Journal*, *51*(6), 1084–1118. doi:10.3102/0002831214553705

Caso, A., & Toledano, V. L. (2017). The university debate between Antonio Caso and Vicente Lombardo Toledano [1933]. In C. Sanchez & R. E. Sanchez Jr., (Eds.), *Mexican philosophy of the 20th Century: Essential readings* (pp. 73–92). Oxford University Press.

Chantrill, C. (2015). *US Education Spending History from 1900: A Century of Education Spending*. http://usgovernmentspending.com

Charques, R. (1932). *Soviet education: Some aspects of cultural revolution*. L. & V. Woolf.

Cheng, J. W., Mitomo, H., Kamplean, A., & Seo, Y. (2021). Lesser evil? Public opinion on regulating fake news in Japan, South Korea, and Thailand–A three-country comparison. *Telecommunications Policy*, *45*(9), 102185. doi:10.1016/j.telpol.2021.102185

Cook, K. (1945). *The house of the people: An account of Mexico's new schools of action*. U.S. Gov't. Printing Office.

Counts, G. (1931). *The Soviet challenge to America*. John Day Publishing.

Counts, G. (1959). *Khrushchev and the central committee speak on education*. Pittsburgh University Press.

Counts, G. (1975). *The challenge of Soviet education* (Vol. 1957). Greenwood.

Denisova-Schmidt, E., & Leontyeva, E. (2014). The unified state exam in Russia: Problems and perspectives. *Industry and Higher Education*, *76*(76), 22–23. doi:10.6017/ihe.2014.76.5530

Dewey, J. (1897, January 16). My pedagogic creed. *The School Journal*, *54*(3), 77–80.

Dewey, J. (1929). *Impressions of Soviet Russia and the revolutionary world*. New Republic.

Dewey, J. (1938). Experience and education. *Touchstone (Nashville, Tenn.)*.

Dewey, J. (1972). *Not guilty: Report of the commission of inquiry into the charges made against Leon Trotsky in the Moscow trials*. Anchor Foundation.

Escobar, M., Fernandez, A., Guevara Niebla, G., & Freire, P. (1994). *Paolo Freire on higher education: A dialogue at the National University of Mexico*. SUNY Press.

Esipov, B. P., & Goncharov, N. K. (1947). *I want to be like Stalin*. Introduction by G. Counts. John Day.

Fediaevskaia, V. (1936). *Nursery school and parent education in Soviet Russia*. Dutton.

Finkel, S. (2008). *On the ideological front: The Russian intelligentsia and the making of the Soviet public sphere*. Yale University Press.

Fitzpatrick, S. (1979). *Education and social mobility in the Soviet Union, 1921-1934*. Cambridge University Press. doi:10.1017/CBO9780511523595

Freire, P. (1970). *Pedagogy of the oppressed: Cultural action for freedom* (M. Bergman Ramos, Trans.). Continuum Press.

Galloway, A. R., & Thacker, E. (2013). *The exploit: A theory of networks*. University of Minnesota Press.

Gonzalez et al. v. Arizona State Board of Education. (2017). Case no. CV 10-623 TUC AWT. United States Arizona District Court.

Gromova, N. S. (2020). Digitalization of education in Russia: From modern technologies to an innovative model of the educational environment. *Advances in Economics, Business and Management Research, 156*, 54–59. doi:10.2991/aebmr.k.201205.009

Hall, S. (1980). Encoding/decoding. In S. Hall, D. Hobson, A. Lowe, & P. Willis (Eds.), *Culture, media, language: Working papers in cultural studies* (pp. 128–138). Hutchinson.

Hussar, W., & Bailey, T. (2014). *Projections of Education Statistics to 2022* (41st ed.). National Center for Education Statistics.

Ilin, M. (1931). New Russia's primer. Houghton Mifflin.

Khrushchev, N. (1958). *Proposals to reform Soviet Education.* Soviet News.

La Botz, D. (10 Sep 2013). Mexican teachers' massive strikes crescendo to a showdown. *Labor Notes.* http://labornotes.org/2013/09/mexican-teachers'-massive-strikes-crescendo-showdown

Lenin, V. I. (1975). *On public education* (Vol. 1921). Progress Press.

Levin, D. (1963). *Soviet education today.* Monthly Review Press.

Lunacharsky, A. V. (1981). *On education* (Vol. 1920). Progress Press.

Makarenko, A. S. (1954). *A book for parents. Foreign Languages.* (Original publication 1939)

Makarenko, A. S. (1973). *The road to life* (Vol. 1936). Oriole.

Marcos Sanchez, V. (2022). The right to a digital education. *Mexico Business News.* https://mexicobusiness.news/talent/news/right-digital-education

McCaslin, M., & Hickey, D. T. (2001). Self-regulated learning and academic achievement: A Vygotskian view. *Theoretical Perspectives, 2*, 227–252.

McClelland, J. (1971). Bolshevik approaches to higher education, 1917-1921. *Slavic Review, 4*(4), 818–831. doi:10.2307/2493850

Punke, H. H. (1932). Recent developments in Mexican secondary education. *The School Review, 40*(9), 693–706. doi:10.1086/439401

Raby, D. L. (1978). Ideology and state-building: The political function of rural education in Mexico, 1921 - 1935. *Ibero-amerikanisches Archiv, 4*(1), 21–38.

Ramos, S. (2017). Twenty years of education in Mexico [1941]. In C. Sanchez & R. E. Sanchez Jr., (Eds.), *Mexican philosophy of the 20th Century: Essential readings* (pp. 53–62). Oxford University Press. doi:10.1093/oso/9780190601294.003.0005

Rosenberg, W. (1984). *Bolshevik visions: The first phase of the cultural revolution in Soviet Russia.* Ardis Publishers.

Sáenz, M. (1927). Integrating Mexico through education. *Bulletin of the Pan-American Union, 61*, 762–777.

Sáenz, M. (1932). The School and culture. *Progressive Education, 9*(2), 99–111.

Sanchez, G. I. (1936). Mexico: A revolution by education. Viking.

Sánchez, G. I. (1944). *The development of higher education in Mexico*. King's Crown Press. doi:10.7312anc93304

Sierra, J. (2017). Discourse at the inauguration of the National University. In C. Sanchez & R. E. Sanchez Jr., (Eds.), *Mexican philosophy of the 20th Century: Essential readings* (pp. 17–26). Oxford University Press.

Simmel, G. (1904a). The sociology of conflict I. *American Journal of Sociology, 9*(4), 490–525. doi:10.1086/211234

Simmel, G. (1904b). The Sociology of Conflict II. *American Journal of Sociology, 9*(5), 672–689. doi:10.1086/211248

Smith, H. L. (1942). *Education in Mexico*. Indiana U School of Education Press.

Tannenbaum, F. (1933). *Peace by revolution: Mexico after 1910*. Columbia University Press. doi:10.7312/tann92264

Tolstoy, L. (1904). *The complete works of Count Tolstoy: Pedagogical articles*. Dana Estes & Company.

Turner, J. K. (1969). *Barbarous Mexico* (2nd ed.). University of Texas Press.

US Department of Education. (2009). *Guidance on Standards, Assessments, and Accountability*. https://www2.ed.gov/policy/elsec/guid/ standardsassessment/guidance_pg5.html

Vasconcelos, J. (1926). *Aspects of Mexican civilization*. The University of Chicago Press.

Vasconcelos, J. (1979). *The cosmic race/La raza cósmica* (Vol. 1925). Johns Hopkins University Press.

Vasconcelos, J. (1981). *Antología de textos sobre educación* [Anthology of texts on education] (A. Molina, Ed.). Fondo de Cultura Económica.

Vaughan, M. K. (1982). *The state, education, and social class in Mexico, 1880-1928*. Northern Illinois University Press.

Vaughan, M. K. (1997). *Cultural politics in revolution: Teachers, peasants, and schools in Mexico, 1930-1940* (2nd ed.). University of Arizona Press. doi:10.2307/j.ctv1d6q40z

Vygotsky, L. (1930). Socialisticheskaja peredelka cheloveka. *VARNITSO, The Journal of the All-Union Association of Workers in Science and Technics for the Furthering of the Socialist Edification in the USSR*.

Vygotsky, L. (1999). Consciousness as a problem in the psychology of behavior [1925]. In *Undiscovered Vygotsky: Etudes on the pre-history of cultural-historical psychology* (European studies in the history of science and ideas, Vol. 8). Peter Lang Publishing.

KEY TERMS AND DEFINITIONS

Dewey (John): Dewey was a major progressive educational reformer of the 20th century. He founded the University of Chicago Laboratory Schools, where he was able to apply and test his progressive ideas on teaching method. Although Dewey is known best for his publications about education, he also wrote about many other topics, including epistemology, metaphysics, aesthetics, art, logic, social theory, and ethics.

New (Socialist) Man: A template proposed by Soviet thinkers during the early days of the Revolution, for an archetypal human that would be a product of learning and physical productivity and strength; an ideal member of the Soviet state whom would no longer be Russian, Ukrainian, or any other minority, but a new kind of being whose psychology would be oriented towards the collective consciousness through radical forms of andragogy that could affect the sociopolitical awareness of the country en masse, and the generations to come.

Progressive Era: Through the turn of the 20th Century, many American middle-class activists sought to address the problems caused by rapid industrialization, urbanization, immigration, and political corruption, inflected by both a Marxian critique of industrial ownership in monopolies as well as a concern for the liberal standards of democracy. They were alarmed by the spread of slums, poverty, and mass exploitation of labor and votes. Andragogical initiatives to educate and acculturate migrants and immigrants to modern democratic participation and industrial labor meant that multiple overlapping progressive movements fought perceived social, political, and economic ills through advancing democracy, scientific methods, professionalism, and efficiency, although this work often went hand-in-hand with industrialization and the needs of the State itself. Participants included such ideologically diverse voices as both John Dewey and Teddy Roosevelt.

Vasconcelos (José): The secretary of education during the transition to institutionalized Mexican politics, he was called the "cultural caudillo" or "warlord" of the Revolution in education. Vasconcelos was an important Mexican writer, philosopher, and politician, who developed a revolutionary philosophy of the "cosmic race" that served as a template for all aspects of Mexican sociocultural, political, and economic policies, much like the "New Soviet Man" in revolutionary Russia at the same time.

Vygotsky (Lev): Early Soviet psychologist who worked on the relationship between language and thought, the development of language, and a general theory of development through actions and relationships in a socio-cultural environment. In order to a develop his "new psychology" that he foresaw as a "science of the Superman". He believed that it was critical to help build the New Soviet Man out of the current workers and peasants of the country, and then their children. Consequently, the majority of his later work involved the study of infant and child behavior, as well as the development of language acquisition, where his initial revolutionary writings centered around adult educational psychology.

Chapter 9
Contextual Influences on Leader Development Among Natural Scientists in Uganda

Julius Lukwago
Graduate School of Business and Leadership, University of KwaZulu-Natal, South Africa

Ana Martins
Graduate School of Business and Leadership, University of KwaZulu-Natal, South Africa

Orthodox Tefera
Graduate School of Business and Leadership, University of KwaZulu-Natal, South Africa

ABSTRACT

The study's aim is to ascertain the role of contextual influences on leader development among natural scientists in Uganda differ from those of non-scientist leaders in the Ugandan context. The study design, methodology, and approach include a retrospective mixed-methods multi-case comparative case study to examine the leader development lifespan experiences of technical expert scientists holding leadership positions in organizations of different sectors. Purposive sampling was used to select senior expert natural scientists providing sustainability leadership at healthcare organizations, at the Uganda Ministry of Agriculture, Animal Industry, and Fisheries and affiliate agencies. Non-scientist leaders from Uganda Revenue Authority were included as a comparator group. Findings reveal that scientists experience normative behaviors right from their undergraduate levels only focused on technical skills being devoid of soft skills. Research limitations and implications direct to the need for developing leaders through coaching and mentoring, as well as leadership succession planning.

DOI: 10.4018/978-1-6684-7832-5.ch009

INTRODUCTION

As leadership effectiveness becomes recognized as a central pillar in any organization's success, natural scientists from disciplines such as engineering, agriculture, and medicine are increasingly looking for approaches that engender the nurturing of leadership skills. Until recently, natural scientists like physicians and engineers had ignored the value of leader development and focused mostly on technical competence development (Stoller, 2009; Rice, Davis, Soistmann, Johnson, Gray, Turner-Henson, & Lynch, 2020). However, while extant literature provides some insight into what approaches work in leader development, despite the recognized influence of context, most of the research has been conducted in developed country contexts, and moreover, overly focused on non-scientist leaders (DeRue & Myers, 2014; Geerts, Goodall, & Agius, 2020). Additionally, the few studies conducted have mostly been siloed into specific sub-disciplines within the natural sciences field. For example, looking at only engineers or only physicians or nurses, thereby diminishing our understanding of the collective characteristics of the leader development phenomenon among natural scientists as a group (Collins & Holton, 2004; Blumenthal, Bernard, Bohnen, & Bohmer, 2012; Funari, Feider, & Schoneboom, 2015; Goodall & Stoller, 2017; Perry, Hunter, Currall, & Frauenheim, 2017). This study sought to identify the contextual influences on leader development unique to natural scientist leaders in Uganda. The related research question was—how do attitudes, beliefs, subjective norms, intentions, and experiences of leader development among natural scientists in Uganda differ from those of non-scientist leaders? The emerging themes are discussed in this book chapter. This study further uses a retrospective mixed-methods multi-case comparative case-study approach to examine the leader development lifespan experiences of technical expert scientists holding leadership positions in organizations at the forefront of healthcare, engineering, and agriculture in Uganda and how such experiences differ from non-scientist leaders at a comparable organization.

REVIEW OF THE LITERATURE

In this section, a review of extant literature is presented. This study highlights a critique of the contextual factors and especially, the theory of planned behavior.

Contextual Influences

Studies have shown that contextual influences such as organizational culture, societal expectations, the structure and nature of undergraduate training, and resource constraints, impact leader development. Previous research has demonstrated that omitting leadership training in medical and engineering education curricula reduces the value these scientists attach to leadership (Colcleugh, 2013; Nakanjako, Namagala, Semeere, Kigozi, Sempa, Ddamulira, Katamba, Biraro, Naikoba, Mashalla, Farquhar, Sewankambo, Afya Bora Consortium members, & Sewankambo, 2015; Perry *et al.*, 2017). No such previous studies have been reported in the area of agricultural sciences. By integrating three natural science disciplines, this study shows the impact of lack of leadership training in STEM curriculums in general. A unique contribution of this study is the demonstration that the lack of both leadership skills and soft-skills training at the undergraduate level makes scientists worse off than their non-scientist counterparts who have soft-skills training in their programs. Moreover, the non-scientists undergraduate programs are structured in such a way that students get to socialize, network, and engage in sporting activities—thereby maximizing the developmental window that a university process is for leader development.

While integrating leadership training in undergraduate curricula for engineers, doctors and agriculturalists would improve leader development and leadership effectiveness, perhaps the greatest strategy to address the contextual gaps in leader development among scientists is to structure the learning programs for these scientists to include inter-disciplinary collaborative projects, and experiential learning that makes them engage with the wider community where they are expected to apply their technical knowledge after graduation. Even more, to provide time for and encourage scientists to engage in social activities such as sports, is regarded fundamental. Participants in the current study highlighted how it was extremely difficult to engage in sports, student leadership, or other community activities at the time of their undergraduate training because, to begin with, such is counter-cultural for scientists as per subjective norms and societal expectations. Secondly, the structure of medical, engineering, or agricultural training is such that no time allowance is made for extracurricular activities. Study participants recounted their experience in experiments, ward rounds, and group discussions organized after class that may be considered optional but missing of which, one would be disadvantaged academically. This is important because previous studies have found that engaging in social activities and sports has life-long benefits in improving one's performance in leadership and other areas of life (Murphy, 2011; Kniffin, Wansink, & Shimizu, 2015; Cullen-Lester, Maupin, & Carter, 2017; Liu, Venkatesh, Murphy, & Riggio, 2020). For example, in a longitudinal study of the life-long impact of participation in sports at high school, Kniffin, Wansink, & Shimizu (2015) demonstrated that those who participated in university level sports 60-years earlier exhibited significantly more leadership than those that did not.

Previous research suggests that leader development is a multi-faceted process whose effectiveness is not limited to the content, structure and design of the program but is also affected by the broader context, timing and specifically the leader characteristics such as leader identity and leader self-efficacy (Mumford, Marks, Connelly, Zaccaro, & Reiter-Palmon, 2000; Day & Harrison, 2007; Mumford, Campion, & Morgeson, 2007; Hannah, Avolio, Luthans, & Harms, 2008; Day, Harrison, & Halpin, 2009; Harms, Spain, & Hannah, 2011; Johnson, Putter, Reichard, Hoffmeister, Cigularov, Gibbons, Chen, & Rosecrance, 2018). On the organizational side, extant literature suggests that there are subjective norms and structural processes peculiar to natural scientists like medical doctors and engineers that might impede nurturing of leadership skills (Brazier, 2005; Colcleugh, 2013). For example, the process of developing expertise in technical fields such as medicine takes years. Because of the premium attached to excellence in technical skills, little room is left for natural scientists to engage in developing social skills and other leadership competencies (Mazzoccoli & Wolf, 2016; Perry *et al.*, 2017; Geerts, Goodall, & Agius, 2020) such as "strategy, communication, persuasion, motivation, and myriad people skills" (Perry *et al.*, 2017, p. 3).

It has been a long-held view in the literature that context impacts the leader development and leadership effectiveness (Day, 2000; Dickson, Den, Hartog, & Deuling, 2003; Brazier, 2005; Ayman & Lauritsen, 2018; Oc, 2018; Ng'weno, 2020). However, researchers have approached context from a multiplicity of perspectives (such as policy, politics, learning transfer, socio-economic factors, national or industry culture) sometimes with inconsistent findings (Cheng & Hampson, 2008; Blume *et al.*, 2010; Grossman & Salas, 2011; Pollock, Jefferson, & Wick, 2015). This makes it difficult for organizations seeking to address elements within their context to enhance leader and leadership development to isolate what is critical from what is mundane (Grossman & Salas, 2011).

The current study brings additional clarity to the contextual factors that critically affect leader development among natural scientists in a developing country setting. For example, of all the factors identified as important in the model of the learning transfer process, as Grossman and Salas (2011) espouse; the work environment weighted extensively more than trainee characteristics or training design. Addressing work environment policies is, therefore, important because the current study findings identified that the practice of promoting scientists into leadership mainly on the basis of technical competence is ubiquitous. Referred to as the Peter principle, the phenomenon describes how organizational HR systems and practices promote the best performing technical personnel to the next level of management, with the responsibility to oversee their fellow technical personnel albeit taking such a person away from their area of competence into a leadership role for which they are less competent (Benson, Li, & Shue, 2018; Ghinea, Cantaragiu, & Ghinea, 2019).

Theory of Planned Behavior

This study is underpinned on the theory of planned behavior (TPB). Ajzen (1991) put forward the TPB which looks at five constructs to illuminate how expert technical leaders approach leader development practices. This theory has been proposed to guide leader development research by Carbery and Garavan (2011) as highlighted in the theory's constructs are attitude, subjective norms, perceived behavior control, intention and behavior. The attitudes of the technical experts on leader development programs deals on expectancies around personal and career growth and training system. Intrapersonal characteristics that can predict leader development, including personality (Mumford, Zaccaro, *et al.*, 2000; Zaccaro, 2007; Day, Harrison, & Halpin, 2009; Zaccaro *et al.*, 2018) cognitive capabilities (Spencer, McClelland, & Spencer, 1992; Backus *et al.*, 2010; Wai, 2014) and learning styles (Mainemelis, Boyatzis, & Kolb, 2002; Kolb & Kolb, 2005). Others include, "educational achievements, hierarchical position, past career experience, skills and competences [...] cultural values [...] career growth expectancies [...] achievement expectancies, and work-related expectancies" (Carbery & Garavan, 2011, pp. 28–34). The subjective norms denote perceived peer pressures to participate or not. For example, a characteristic of technical fields is the despise of social skills training as right-brain skills and an emphasis on learning technical skills as the premium towards performance (Stoller, 2009; Perry *et al.*, 2017). Other determinants include "development norms, perceptions of the transfer environment, contribution expectations of the organizational culture [...] achievement norms [...] and staffing norms" (Carbery & Garavan, 2011, pp. 34–37).

The theory of Perceived Behavior Control focuses on the perceptions or situations where the person may desire to participate in development activities but are held back by limitations such as resources. Other elements include self-efficacy beliefs and leader efficacy, confidence to use particular learning strategies and organizational constraints such as organizational culture, economic environment, and how much it encourages the learning of leadership (Hannah *et al.*, 2008). It also includes the HR systems, processes and practices across the employee lifecycle and how they integrate and encourage learning (Scandura & Lankau, 1996; Fairhurst, 2009). Others include "self-directness and confidence to use personal learning strategies" (Carbery & Garavan, 2011, p. 38). The Intention deals with the ambition to engage in leader development practices describes a person's willingness to exert effort in learning leadership. It includes the desire to "participate in future activities [...] felt responsibility—[the sense of obligation to participate]...and self-prediction—the expectation that one will participate in future [leader] development activities" (Carbery & Garavan, 2011, p. 40). This is sometimes demonstrated by signing up for training or taking steps to find a mentor or coach. Finally, behavior is all about the actual behavior of participating in leader development activities including the intensity of exposure to leader development practices such as formal leadership training, self-study, coaching and mentorship programs, action-learning projects, 360-degree feedback, cross-disciplinary interactions, and informal hands-on supervisory experience.

METHODOLOGY

Mixed methods were used to gather primary data thereby integrating qualitative interviews among 21 natural scientist and 11 non-scientist leaders, two focus group discussions (FGDs) and a cross-sectional survey of 221 leaders. Given that scant research exists about leadership development specifically appropriate for natural scientists (Geerts *et al.*, 2020; Perry *et al.*, 2017), particularly in Africa (Gumede, 2017; Nakanjako *et al.*, 2015), a qualitative inductive approach was essential in the first phase of the study to explore the experiences of natural scientist leaders. The current study used the lifespan model of leader development (Liu *et al.*, 2020) to explore prior university-level experiences of leaders currently at the forefront of transformational efforts in agriculture, medicine and engineering. Consequently, purposive sampling was used to select senior expert natural scientists providing sustainability leadership at healthcare organizations, at the Uganda Ministry of agriculture, animal industry and fisheries (MAAIF) and its affiliate agencies. Non-scientist leaders from Uganda Revenue Authority (URA)—a comparable government agency—were included in the study as a comparator group to ascertain any uniqueness in HEIs leadership experiences between natural scientists and non-scientists. According to Creswell (2018), purposive sampling is appropriate when the researcher needs to obtain information from experts most likely to possess the knowledge and experiences required to answer the research questions.

Data Collection and Analysis

As mixed methods approach was used, this included several primary data collection methods, the firt being the semi-structured interview with an interview protocol. The interviews lasted between 36 minutes and 108 minutes each. Collection of data via interviews was stopped after 20 scientists and 11 non-scientists were interviewed, as saturation was achieved (Creswell, 2018). Interviews and FGDs were conducted using Zoom meetings due to the COVID-19 restrictions. Recorded interviews and FGDs were then transcribed. Text analysis and thematic analysis (Creswell, 2018) were used to identify themes emerging from the data, while patterns of related ideas coalescing around concepts in the TPB theoretical framework formed sub-themes subsequently delineated as drivers or barriers to leadership development. NVivo version 12 software was used to code and organize data and conduct cross-case analyses across scientists and non-scientists. In the quantitative phase, the TPB theory was used to develop a questionnaire fielded to all 150 MAAIF and 371 URA staff holding leadership positions with a response rate of 78% and 27.8%, respectively. Due to COVID-19 restrictions prohibiting to face-to-face meetings, data collection was conducted via an online *SurveyMonkey* self-administered questionnaire. Survey data was analyzed using SPSS version 27. The Cronbach Alpha was 0.952 in relation to the questionnaire, which demonstrates strong reliability (Hulin *et al.*, 2001).

ANALYSIS AND DISCUSSION

In this subsection, various factors are analyzed and discussed. These factors include demography and the social characteristics of respondents in this study, as well as the determinants of leader development behavior.

Demography and Social Characteristics of Respondents

Table 1 shows that MAAIF had a higher proportion of scientists than URA. Most respondents were male, above 40 years, had a master's as the highest qualification, and were mostly working in middle management. Natural science disciplines were classified using the taxonomy that delineates between the natural science and humanities or social sciences (Ledoux, 2002).

Table 1. Response rate

	Leaders Invited	Respondents	Response
URA	371	103	27.8%
MAAIF	150	117	78.0%
Doctors	N/A	1	N/A
Overall	521	221	42.4%

Source: Own data

In general, the demographic results indicated that most scientist leaders are older compared to non-scientist leaders. This could explain why scientists and non-scientists were comparable in leadership effectiveness. Previous studies have indicated that age affects leadership effectiveness and leadership styles, whereby older leaders with higher legacy beliefs demonstrate empathy, collaboration and overall leadership effectiveness (Kooij *et al.*, 2011; Zacher, Rosing, & Frese, 2011; Bojadziev *et al.*, 2016). Besides the legacy beliefs, a possible explanation is that such older scientist leaders are more likely to have experienced a wide range of leader development windows to enhance their effectiveness. Liu *et al.* (2020), in a study of the lifespan model of leader development, contend that there are development opportunities at each stage of life. For example, Liu *et al.* (2020) suggested that in the older group 30-60 years, challenges at work, marriage, and parenting provide a breeding ground for leadership skills. Among those above 60 years, Liu *et al.* (2020) intimated that a relational and sensitive approach to leadership is more common.

With nearly half of the scientist leaders aged 50 years and above could explain why the current study findings depict a lack of statistical difference between the leadership effectiveness of scientists and non-scientists, despite some earlier studies intimating that scientists struggle with soft skills and leadership effectiveness in general (see for example, Nick, Reinhold, & Valerie, 2005; Colcleugh, 2013; Perry *et al.*, 2017; Nyssa, 2019). The study findings seem to reveal that there were no significant differences in many of the dimensions of leadership effectiveness among scientists and non-scientists and this could in part be explained by the scientist leaders being older and having transitioned from ineffective to effective leaders through leveraging a wider range of lifespan developmental experiences than younger leaders would (Murphy, 2011; Kjellström, Törnblom, & Stålne, 2020; Liu *et al.*, 2020). However, this is not to mean that by simply growing older, one becomes better at leadership. Maxwell (2013) has argued that some leaders simply go through life repeating the same experience year after year without reflection to learn from it. Indeed, a further look at the current study findings reveals that after controlling for age, not all scientist leaders demonstrated high leadership effectiveness; rather, the findings intimate that higher exposure to specific leader development activities leads to higher leadership effectiveness. Not surprisingly, the specific approaches include feedback and mentorship—approaches that these study findings highlighted to be impactful when combined with reflection.

The study finding tends to indicate that promotion is largely due to technical expertise confirms what Perry *et al.* (2017) found and what previous studies have established (Goodall & Stoller, 2017; Geerts, Goodall, & Agius, 2020). Nevertheless, because promotion is largely due to technical capabilities that take a long-time to accumulate, as was found in this study, the implications are that scientist leaders are likely to be older. These study findings are in line with what Abet (2021) reported, that it takes up to 12 years in university education to qualify as a consultant specialist in the Ugandan medical field. Furthermore, while the non-scientists have slightly more females in leadership positions, the scientists are skewed toward the male, with two-thirds of scientist leaders are male. This could be explained by the patriarchal, high-power distance culture inferred in the way scientist leaders lead (or 'percuss' as one respondent put it), leading to fewer females enrolling into STEM subjects in earlier life stages as intimated by one key information. A surprising finding from this study was that in all the interviews conducted, there were no notable differences between men and women in attitudes, perceived behavior control, leader-identity or how they experience leader development amidst the context and subjective norms. It had been envisaged that given the experience of career disruptions amongst young women after they start families, the perceived behavior control for leader development during that developmental window would be lower amongst women compared to men due to the work-family demands and unique needs of women in leadership development programs (Ely, Ibarra, & Kolb, 2011; Fitzsimmons & Callan, 2016, 2019). Moreover, from a diversity and inclusion standpoint, studies have shown that people bring different viewpoints to leadership (Ely, Ibarra, & Kolb, 2011). Only one respondent referred to having dropped out of her PhD program when she had three children in five years, but she quickly added that she later picked up her learning journey and went on to succeed. This could be explained by the effect of strong leader characteristics among women making such social barriers diminish their power. As Offermann *et al.* (2019) have found, women with families and those without had similar life outcomes, while those who had extended maternity experienced similar career satisfaction with those taking the standard one.

The significance of these findings is that programs that promote STEM subjects among young women are essential. However, in addition, they could benefit more from specific encouragement of females to engage in leadership in the sciences. Furthermore, by promoting a less patriarchal, power-dominated approach to leadership, the conceptualization of leadership in the sciences as more collaborative than authoritative could encourage more females to see themselves less as 'reluctant leaders. Moreover, ambition and self-efficacy were found to be a strong leader characteristic that nurtures an interest in leader development—something other studies agree with (Machida & Schaubroeck, 2011; Mesterova *et al.*, 2015).

Determinants of Leader Development Behavior

The study sought to identify the contextual influences on leader development, such as attitudes, beliefs, subjective norms, intentions, and experiences unique to natural scientist leaders in the Ugandan context.

One-Way ANOVA and the F-test were used to examine the differences in perceptions between scientists and non-scientists. Table 2 shows the mean scores on the determinants of leader development behavior based on, as per the theoretical framework.

Table 2. Effectiveness/impactfulness of leader development approaches

Development approaches	Mean score (1-5)			
	Scientist (n=93)	Non-scientist (n=96)	F	Sig.
COACHING	4.02	4.30	3.945	0.049
MENTORSHIP	4.37	4.45	0.574	0.450
FEEDBACK	4.26	4.45	3.139	0.078
EXPERIENTIAL LEARNING	4.29	4.43	1.988	0.160
E-LEARNING	3.61	3.85	3.638	0.058
FORMAL LEADERSHIP TRAINING	3.89	4.22	7.414	0.007
NETWORKING	3.98	4.32	9.361	0.003
LEADERSHIP ROLES	4.10	4.43	10.352	0.002
SELF-DIRECTED	3.89	4.26	9.498	0.002
SELF-AWARENESS	4.08	4.45	9.922	0.002

Scale 1 (Not at all impactful) -5 (Extremely impactful)

Source: own data

One-Way ANOVA and the F-test were used to examine the differences in the perceptions between scientists and non-scientists. Table 2 shows the mean scores on the determinants of leader development behavior based on, as per the theoretical framework. The mean scores were derived from Likert scales ranging from 1 to 5. The results show that within the scientists' group, *outcome expectations* were highest at 4.47, while the lowest-scoring determinant was *subjective norms* and *attitude*. The results also show that non-scientists rated higher on all determinants of leader development compared to scientists. However, only in three areas (*attitude, perceived behavior control*, and *intention*) was there a statistically significant difference in the way scientists demonstrated beliefs and perceptions that drive participation in leader development activities. For example, when it comes to the attitude, which represents the perceptions towards participation in leader development activities and the value of leadership to technical personnel, the scientists scored a mean of 3.86.

In contrast, the non-scientists scored a mean of 4.15, which was statistically significant (p=0.016). On *intention*, which represents having deliberate and active plans to engage in specific activities to grow one's leadership skills, scientists scored a mean of 3.91, which was significantly lower than non-scientists at 4.25 ($p<0.001$). Similarly, scientists scored a mean of 4.00 compared to non-scientists mean score of 4.25, a statistically significant difference ($p=0.001$). These results are similar and corroborated with those from the qualitative interviews, where it was reported by the majority of the respondents that scientists approach leader development in a lackluster manner and experience both real and perceived obstacles to leader development that are unique to their context. Specifically, medical doctors, for example, experience schedule and workload problems that make it difficult on practice to participate in leader development activities. One of the respondents summarized this context problem as follows:

Let's say you are working in a private hospital...what kind of rota system are you working on? If you're doing what I call a one-in-two, which means one night you're on, the other night you're off. Usually, the second night you are off, you're trying to sleep rather than study. But on the other hand, if you're provided with a rota which has at least a one-in-three, one-in-four, one-in-five, it means that there are certain nights that you can say "for me this night, I will read, or I can attend the Grand Round where some visiting professor is speaking. Now, these things are difficult...work is really, really busy. That extra pound of flesh that is expected from your employer can often come at the expense of your personal life. Let's say you have a young family. You've recently gotten married, and yet at the same time, you're trying to grow your career. Trying to learn something [non-technical and (mis)considered peripheral in the context] has to give. – 2SMF-Senior Leader

The potential implication of the results could be that while scientists may appreciate the benefits of leader development in comparable terms to non-scientists (for example, they scored about the same on *outcome expectations*), their attitude towards leadership and leader development activities, the limited active planning towards leader development, and the obstacles in their context that make them believe engaging in leader development is outside their control, all combine to create a disincentive to participate in leader development. For example, while promotion mainly focuses on technical skills and less on leadership skills, the technical qualification that matters most takes a very long time to attain. This phenomenon was well articulated by one of the participants in the scientists' FGD, as outlined below.

For us scientists, the things that matter to your career are not administrative. First, you get into a leadership position because of your technical expertise, and your remuneration increases based on the research you undertake and publish, and the projects you lead. Delivering on that requires you to focus on the technical work. In that case, administrative work and being bogged down by managing people becomes immensely unattractive....For example, to rise to my level, I was expected to have a PhD in crop science and to have published at least four major papers. That process takes time. Time you might not divide up to undertake a leadership course when you know that as much as such soft skills are helpful, they are not required for you to push ahead in your career. – FGD-Scientist-Male

Regarding subjective norms, the results in Figure 2 could indicate that there was no significant difference between scientists and non-scientists. This could be surprising given that the literature on social identity theory and the results from the qualitative data seem to indicate a strong influence of context and norms among scientists such as medical professionals compared to non-scientists. Each determinant of leader development is considered in greater detail below, and comparisons are made between scientists and non-scientists to identify the unique areas of interest.

With regards to the determinants of leader development behavior the following determinants of leader development are worthy of discussion.

Attitude

The findings appear to have depicted an attitude among scientists that has hampered growth in leadership skills, similar to that which Nick *et al.* (2005) found to be a barrier to adopting soft skills amongst health scientists. Yet, soft skills are an essential multiplier of the technical skills for one to be effective (Rao, 2012). Nick *et al.* (2005, p. 12) have argued that health professionals find comfort in doing the "complicated easy" tasks that require intellect, such as designing complex plans but trivialize the "simple hard" behaviors that demand courage and discipline since they depend on emotions, feelings and integrity. This attitude that belittles soft skills inadvertently gives impetus to any resistance to participation in leader development activities. In the theory of planned behavior, Ajzen (1991) and Fishbein and Ajzen (2011) have demonstrated that one's attitude predicts the performance of a behavior. Consequently, when scientists, especially the early-career scientists, believe that soft skills are for weaklings or that administrative roles and people management is not for people like them, they are unlikely to participate in leader development.

Moreover, the current study findings appear to indicate that scientists were not as involved as non-scientists in leader development activities such as sports, community, religious and school and social networks across the lifespan, something that previous studies have found to immensely add to one's learning of leadership and effectiveness (Murphy & Johnson, 2011; Cullen-Lester, Maupin, & Carter, 2017; Kjellström, Törnblom, & Stålne, 2020; Liu *et al.*, 2020). The scientists scoring significantly lower than non-scientist on the dimension of attitude aligns with the viewpoint permeating historical literature that scientists struggle with soft skills and leadership (Sapienza, 2004; Colcleugh, 2013; National Research Council, 2015; Perry *et al.*, 2017; Perry, Mobley, & Brubaker, 2017; Nyssa, 2019). For example, in a study of medical doctors in South Africa, Mdingi (2018) found that these scientists had dismissive attitudes towards leadership skills as inferior to technical skills, an element exacerbated by the lack of leadership training within the medical education curriculum. However, some studies bring to light that the attitude of technical competencies and hard skills being the only requirement for leadership in roles superintending other experts is not unique to scientists (Rao, 2012; Jain & Anjuman, 2013). For example, Irvine & Brundrett (2019) found similar attitudes among teachers and school leaders, while Martinelli & Erzikova (2017) found the same among public relations experts.

The unique contribution of the current study is that it is comparative and identifies that this problem was significantly more serious (with a lower ranking on positive attitude) amongst scientists as compared to non-scientists. Indeed, the finding that the attitude of superiority among scientists makes them desire to participate in leadership only at the senior-most ranks could be preventing them from participating as leaders at lower ranks in early career, sports, and extracurricular activities. According to the lifespan model of leader development (Liu *et al.*, 2020), this implies that many scientists miss out on nurturing their leadership skills through such opportunities and development windows. The finding that fewer scientists than non-scientist reject the notion that they are already fit for leadership and therefore do not need further training in leadership could be explained by the scientists' belief that they are already leaders by virtue of their superior technical qualifications. This, in one part, aligns to a study by Goodall & Bäker (2015) where they argued that technical expertise gives the scientist *expert signaling*, an understanding of the work that juniors do and therefore enables them to manage performance better than a non-scientist or someone without such superior industry experience. However, as Goodall & Bäker (2015) themselves highlight, expert signaling is only one part of what gives an expert leader overall effectiveness—leadership capability, being the other. Therefore, it is imperative that such attitudes that conjecture that all technical experts need to succeed and deliver superior performance is superior technical qualifications be discouraged in favor of programs that seek to improve leadership effectiveness among scientists. Requiring leadership credentials alongside technical qualifications before one is considered for promotion could be one such policy level incentive to elevate the value of leader development.

Subjective Norms

The finding that scientists experience normative behaviors right from their undergraduate levels, such as not engaging in social activities as a result of social identity and time trade-offs because of an arduous curriculum only focused on technical skills and devoid of soft skills, concurs with similar findings in the Mdingi (2018) study. Moreover, it is congruent with Hogg's (2001) theorizing that engaging in prototypical behaviors is a strong motivation partly due to the attendant influence and enhancements in self-esteem and social attraction. However, the implication is that scientists miss out on opportunities to grow their leadership. As some studies show, engaging in such activities as sports, dating, early marriage or leadership in religious or social associations contributes greatly to someone's leader development (Murphy, 2011; Kniffin, Wansink, & Shimizu, 2015; Martinelli & Erzikova, 2017; Liu *et al.*, 2020). As some participants in the study suggested, educators need to include leadership training in undergraduate programs and provide experiential learning opportunities, say with cross-disciplinary team projects that involve social scientists as some in the medical field have advocated for (Nakanjako *et al.*, 2015). Organizational development practitioners in natural science institutions should lead dialogues, mentorships and provide incentives in order for early-career scientists to value the complementary nature of leadership skills to technical skills.

Perceived Behavior Control

Building on the subjective norms discussed above, another area where scientists scored high, albeit lower than non-scientists, was on perceived behavior control. The findings seem to suggest that this is reinforced by attitudes, beliefs, and subjective norms, whereby scientists believe that sciences are hard and complex requiring one to invest fully in that—lest they do not grow in their careers (Perry *et al.*, 2017; Perry, Mobley, & Brubaker, 2017). Moreover, in Uganda, as reported by Abet (2021), the number of senior consultant specialists is so low that they are too overwhelmed to make time for learning other skills; save for the technical qualification. The result is that developing leadership skills takes a back seat. Scientists strongly believe that they could develop their leadership if they wanted to, but add that the environment makes it difficult for those who want. Leader development must, therefore, be promoted at the start of the pipeline when early-career scientists have the motivation to engage as opposed to older scientists whose legacy beliefs could have waned and who have limited time to participate (Zacher, Rosing, & Frese, 2011; Coltart *et al.*, 2012; Rice *et al.*, 2020).

Intention

The results found that intention to engage in leader development among scientist was high even though not as high as non-scientists. Nonetheless, fewer scientists had concrete plans to undertake specific leader development in the next 90 days. The results imply that leaders might have good intentions, but other influences get in the way of actualizing these intentions. One explanation is that scientist leaders are often so overwhelmed with a large span of control and overworked due to a shortage of expert staffing (Cupit *et al.*, 2019; Abet, 2021) that practical realities of work get in the way of developing leadership. This exacerbates the need to catch them early and develop leadership skills while they are still in at the start of the pipeline for technical training (Nakanjako *et al.*, 2015; Perry, Mobley, & Brubaker, 2017; Mdingi, 2018; Baas, Dewhurst, & Peyre, 2020; Manikkath *et al.*, 2020; Murugaboopathy *et al.*, 2020). Including leadership education in undergraduate curriculums and integrating experiential learning leadership programs in post-graduate and early career stages is critical not only to have a paradigm shift in the way scientists appreciate leadership skills but also how they learn them. As previous research has established, learning of leadership is cumulative, and greater impact is obtained from experiential and reflective processes (Stanton & Grant, 1999; Mumford, Campion, & Morgeson, 2007; Kjellström, Törnblom, & Stålne, 2020).

An alternative explanation, as emerged in the qualitative interviews, is that leader characteristics such as developmental readiness, learning efficacy and discipline account for why some leaders make good on their intentions to engage in leader development while others do not (Hall, 2004; Avolio & Hannah, 2008; Hannah *et al.*, 2008; Reichard *et al.*, 2017). Once again, the powerful influence of leader characteristics and self-beliefs in enabling one to navigate contextual challenges and barriers comes into play. It might therefore, be helpful for learning and development practitioners in natural science organizations, in light of budgetary constraints, to be selective in whom they invest in as a priority—not all popcorn will pop. For example, Avolio and Hannah (2008) have proposed that leader development interventions ought to "first focus on assessing and then building the developmental readiness of individual leaders, as well as the developmental readiness of the organization as prerequisite steps for accelerating positive leader development" (Avolio & Hannah, 2008, p. 331).

This study findings confirm two of the leader characteristics—self-efficacy and motivation—from what Grossman and Salas (2011) put forward as trainee characteristics that determine whether learning transfer succeeds or not. The findings also highlight that the respondents, having risen to the top of their leadership careers, were atypical compared to their counterparts. This implies that deliberate practice in leader development is something not highly adopted, and that leader development remains haphazard many years after scholars have pointed to the need to be intentional (Conger, 1993; Day, 2010). The findings underscore that leaders who are intentional, motivated and demonstrate deliberate practice are more likely to engage in leader development and possibly produce stronger learning transfer. The implication is that at the center of congruent factors that make any leader development efforts work, deliberate practice and design must be adopted (Baas, Dewhurst, & Peyre, 2020; Kjellström, Törnblom, & Stålne, 2020).

The study findings demonstrate that (a) excessive workloads, (b) the organization's policy not to require leadership qualifications as a prerequisite for promotion into leadership positions, (c) the lack of succession planning leading to promotion into incompetence, (d) the lack of an institutionalized culture that welcomes feedback and reflection, and (e) a work environment that does not support or hold trainees accountable to perform new behaviors after training, were common contextual factors impeding leader development effectiveness. The findings seem to imply that excessive workloads make it difficult for scientists to make time for socializing, networking and even attend training on leadership. This study brings light to the implications of a large workload and span of control, as was found in a study among nurse managers (Cupit *et al.*, 2019). The results also indicate that organizations employing scientists can create an environment that nurtures an appetite for leader development by benchmarking their non-scientist counterparts that have instituted strategies such as a policy requiring formal training for senior or supervisory roles. Other areas to benchmark might include a culture of feedback and after-action reviews where leaders are open to criticism; intentionality in leadership succession planning where acting in leadership roles is a deliberate practice to grow the next level of leaders; and where training transfer is improved through clear accountability, line-manager coaching and mentorship support (Day, 2010; Seifert & Yukl, 2010; Ghinea, Cantaragiu, & Ghinea, 2019; Kjellström, Törnblom, & Stålne, 2020).

CONCLUSION

This book chapter study explored how attitudes, beliefs, subjective norms, intentions, and experiences of leader development among natural scientists in Uganda differ from those of non-scientist leaders, with a view to identifying the contextual influences on leader development unique to natural scientist leaders in Uganda. Demonstrating what uniquely affects how scientists approach leader development brings additional clarity to what scientist leaders can do to make leader development more effective. The study concluded that the conceptualization of leadership as hierarchy and 'percussion' diminishes the impetus for scientists to engage in leader development. The study found evidence that leader-identity and self-view coupled with a social identity of superiority make scientists abhor soft skills. Moreover, the study concluded that compared to non-scientists, the lack of leadership training in undergraduate programs is typical for scientists. Further, not incorporating leadership in undergraduate curricula for natural sciences diminishes the appreciation of leadership skills as complementary to technical competences. Consequently, scientists miss taking advantage of a developmental window at university, which previous research has established would lead to demonstrating leadership skills in the future. This preponderance of beliefs and social identity in controlling behavior is consistent with previous literature (Bandura, 1986; Ashforth & Mael, 1989; Ajzen, 1991; Tajfel & Turner, 2004; Ellemers & Haslam, 2012; Cassar, Bezzina, & Buttigieg, 2017). Therefore, it can be concluded that without addressing the beliefs and significance that scientists attach to soft skills and leadership skills, right from their undergraduate and early career days, the effectiveness of whatever leader development approach is deployed is likely to be diminished. Furthermore, the study concludes that systemic industry and organizational practices, policies and norms make it challenging for scientists to value, let alone engage in leader development. Specifically, this study found evidence that excessive workloads make it harder for scientists to engage in leader development. It also found that the diminished acceptance of a feedback culture and a limited supportive learning transfer environment for leaders to account for and practice what they have learned, all attenuate the impact of leader development activities. Moreover, the lack of incentives and policies to require leadership skills for leadership positions pushes scientists to focus on the only sure route for career progression and better remuneration—technical skills, leaving leader development to be approached haphazardly without intentionality and deliberate practice. Additionally, compared to non-scientists, the study identified that there are few role models and mentors who demonstrate best practice in leadership, perpetuating a cycle of learning from bad practice, moreover in a community that predominantly learns through experiential learning and acting in a leadership role. Therefore, it can be concluded that benchmarking non-scientists organizations for policies and practices that elevate the value of leadership can enable organizations employing predominantly natural scientists to establish a contextual environment that cultivates and sustains effective leader development. These practices were identified as elevating the value of leadership skills through campaigns, hiring and promotion policies, managing workloads to provide time for leader development, and establishing formal leadership training programs for staff.

LIMITATIONS AND FUTURE RESEARCH DIRECTIONS

Future studies should recruit a larger sample and include junior leaders. Additionally, the organizational cultures studied in the current study may not be representative of all natural science institutions. Further studies may delve into other organizational contexts and, more broadly, other country contexts. Additionally, future research studies could use a multi-rater approach to measuring leadership effectiveness which exposes it to social desirability bias. Such an approach requires preparing the organization for such an activity and training raters to appreciate how to provide feedback. Further studies could contemplate on making use of the 360-degree multi-rater leadership assessment to obtain a strong measure of leadership effectiveness as an outcome variable.

Recommendations for practice include that Educators may consider adopting innovative pedagogical approaches that integrate collaborative projects and community-based action learning projects that enable natural scientists to interact in an interdisciplinary fashion and to learn more through problem-based case-study approaches. Educators and practitioners of organizational development should consider allocating time for social activities which include, namely, networking, sports as well as community engagement activities that pivot on and shape the scientists' emotional and social intelligence competencies instead of overburdening individuals with a disproportionate number of technical studies or technical work concentrated solely on the technical competences. Instead, determinations should be made to ensure a culture shift which enables technical personnel not to spend most of their early careers isolated and immersed in technical activities. Policymakers as well as Practitioners in organizational development, ought to consider conducting a benchmarking exercise of non-scientists organizational cultures and policies to ascertain both the importance of leadership training and the evidence of leadership skills, prior to an individual being given a leadership position, regardless of the areas of theoretical knowledge and technical expertise.

REFERENCES

Abet, T. (2021). Heart, vessel diseases leading cause of deaths, say experts. *Daily Monitor*. https://www.monitor.co.ug/uganda/special-reports/heart-vessel-diseases-leading-cause-of-deaths-say-experts-3366312

Ajzen, I. (1991). The theory of planned behavior. *Organizational Behavior and Human Decision Processes*, *50*(2), 179–211. doi:10.1016/0749-5978(91)90020-T

Ashforth, B. E., & Mael, F. (1989). Social identity theory and the organization. *Academy of Management Review*, *14*(1), 20–39. doi:10.2307/258189

Athanassoulis, N., & Wilson, J. (2009). When is deception in research ethical? *Clinical Ethics*, *4*(1), 44–49. doi:10.1258/ce.2008.008047

Avolio, B. J., & Hannah, S. T. (2008). Developmental readiness: Accelerating leader development. *Consulting Psychology Journal*, *60*(4), 331–347. doi:10.1037/1065-9293.60.4.331

Ayman, R., & Lauritsen, M. (2018). Contingencies, context, situation, and leadership. In *The nature of leadership* (3rd ed., pp. 138–166). Sage Publications. doi:10.4135/9781506395029.n6

Baas, T., Dewhurst, S., & Peyre, S. (2020). Leadership and management for scientists. In L. Infante Lara, L. Daniel, & R. Chalkley (Eds.), *BEST* (pp. 87–101). Academic Press. doi:10.1016/B978-0-12-820759-8.00006-1

Backus, C., Keegan, K., Gluck, C., & Gulick, L. (2010). Accelerating leadership development via immersive learning and cognitive apprenticeship. *International Journal of Training and Development*, *14*(2), 144–148. doi:10.1111/j.1468-2419.2010.00347.x

Bandura, A. (1986). *Social foundations of thought and action: A social cognitive theory*. Prentice-Hall, Inc.

Benson, A., Li, D., & Shue, K. (2018). Research: Do people really get promoted to their level of incompetence? *Harvard Business Review*, *8*. https://hbr.org/2018/03/research-do-people-really-get-promoted-to-their-level-of-incompetence

Blume, B., Ford, J., Baldwin, T., & Huang, J. (2010). Transfer of training: A meta-analytic review. *Journal of Management*, *36*(4), 1065–1105. doi:10.1177/0149206309352880

Blumenthal, D. M., Bernard, K., Bohnen, J., & Bohmer, R. (2012). Addressing the leadership gap in medicine: Residents need for systematic leadership development training. *Academic Medicine*, *87*(4), 513–522. doi:10.1097/ACM.0b013e31824a0c47 PMID:22361800

Bojadziev, M., Stefanovska-Petkovska, M., Handziski, V., & Barlakoska, G. (2016). Age related preferences of leadership style: Testing McGregor theory X and Y. *Journal of Management Research*, *8*(4), 187. doi:10.5296/jmr.v8i4.10088

Brazier, D. K. (2005). Influence of contextual factors on health-care leadership. *Leadership and Organization Development Journal*, *26*(2), 128–141. doi:10.1108/01437730510582563

Carbery, R., & Garavan, T. N. (2011). Conceptualizing participation in formal training and development activities: A planned behavior approach. In R. F. Poell & M. van Woerkom (Eds.), *Supporting workplace learning* (pp. 27–45). Springer Netherlands. doi:10.1007/978-90-481-9109-3_3

Cassar, V., Bezzina, F., & Buttigieg, S. C. (2017). The relationship between transformational leadership and work attitudes: Comparing mediating influences of social identity and the psychological contract. *Leadership and Organization Development Journal*, *38*(5), 646–661. doi:10.1108/LODJ-11-2015-0248

Cheng, E. W. L., & Hampson, I. (2008). Transfer of training: A review and new insights. *International Journal of Management Reviews*, *10*(4), 327–341. doi:10.1111/j.1468-2370.2007.00230.x

Colcleugh, D. (2013). *Everyone a leader: A guide to leading high-performance organizations for engineers and scientists*. University of Toronto Press.

Collins, D. B., & Holton, E. F. (2004). The effectiveness of managerial leadership development programs: A meta-analysis of studies from 1982 to 2001. *Human Resource Development Quarterly*, *15*(2), 217–248. doi:10.1002/hrdq.1099

Coltart, C. E., Cheung, R., Ardolino, A., Bray, B., Rocos, B., Bailey, A., Bethune, R., Butler, J., Docherty, M., Drysdale, K., Fayaz, A., Greaves, F., Hafferty, J., Malik, A. N., Moolla, A., Morganstein, L., Pathiraja, F., Shah, A., Sleat, G., ... Donaldson, L. (2012). Leadership development for early career doctors. *Lancet*, *379*(9828), 1847–1849. doi:10.1016/S0140-6736(12)60271-2 PMID:22531157

Conger, J. A. (1993). The brave new world of leadership training. *Organizational Dynamics*, *21*(3), 46–58. doi:10.1016/0090-2616(93)90070-H

Creswell, J. W., & Plano Clark, V. L. (2018). *Designing and conducting mixed methods research* (3rd ed.). SAGE.

Cullen-Lester, K. L., Maupin, C. K., & Carter, D. R. (2017). Incorporating social networks into leadership development: A conceptual model and evaluation of research and practice. *The Leadership Quarterly*, *28*(1), 130–152. doi:10.1016/j.leaqua.2016.10.005

Cupit, T., Stout-Aguilar, J., Cannon, L., & Norton, J. (2019). Assessing the nurse managers' span of control: A partnership between executive leadership, nurse scientists and clinicians. *Nurse Leader*, *17*(2), 103–108. doi:10.1016/j.mnl.2018.12.004

Day, D. V. (2000). Leadership development: A review in context. *The Leadership Quarterly*, *11*(4), 81–613. doi:10.1016/S1048-9843(00)00061-8

Day, D. V. (2010). The difficulties of learning from experience and the need for deliberate practice. *Industrial and Organizational Psychology, 3*(1), 41–44. . doi:10.1111/j.1754-9434.2009.01195.x

Day, D. V., Harrison, M. M., & Halpin, S. M. (2009). *An integrative approach to leader development: Connecting adult development, identity, and expertise*. Routledge.

DeRue, D. S., & Myers, C. (2014). Leadership development: a review and agenda for future research. In D. V. Day (Ed.), *The Oxford handbook of leadership and organizations* (pp. 829–852). Open University Press.

Dickson, M., Den Hartog, D., & Deuling, J. (2003). Research on leadership in a cross-cultural context: Making progress, and raising new questions. *The Leadership Quarterly*, *14*(6), 729–768. doi:10.1016/j.leaqua.2003.09.002

Ellemers, N., & Haslam, S. A. (2012). Social identity theory. In P. A. M. Van Lange, A. W. Kruglanski, & E. T. Higgins (Eds.), *Handbook of theories of social psychology* (Vol. 2, pp. 379–398). Sage Publications Ltd. doi:10.4135/9781446249222.n45

Ely, R. J., Ibarra, H., & Kolb, D. M. (2011). Taking gender into account: Theory and design for women's leadership development programs. *Academy of Management Learning & Education*, *10*(3), 474–493. doi:10.5465/amle.2010.0046

Fairhurst, G. T. (2009). Considering the context in discursive leadership research. *Human Relations*, *62*(11), 1607–1633. doi:10.1177/0018726709346379

Fishbein, M., & Ajzen, I. (2011). *Predicting and changing behavior: the reasoned action approach.* Taylor & Francis. doi:10.4324/9780203838020

Fitzsimmons, T. W., & Callan, V. J. (2016). Applying a capital perspective to explain continued gender inequality in the C-suite. *The Leadership Quarterly, 27*(3), 354–370. doi:10.1016/j.leaqua.2015.11.003

Fitzsimmons, T. W., & Callan, V. J. (2019). The diversity gap in leadership: What are we missing in current theorizing? *The Leadership Quarterly, 31*(4), 101347. doi:10.1016/j.leaqua.2019.101347

Funari, T. S., Feider, L. L., & Schoneboom, B. A. (2015). Build high-performing military nurse leaders through coaching, sponsorship, and engagement. *Nurse Leader, 13*(3), 54–59. doi:10.1016/j.mnl.2014.07.019

Geerts, J. M., Goodall, A. H., & Agius, S. (2020). Evidence-based leadership development for physicians: A systematic literature review. *Social Science & Medicine, 246*, 12709. doi:10.1016/j.socscimed.2019.112709 PMID:31887629

Ghinea, V., Cantaragiu, R., & Ghinea, M. (2019). The Peter principle and the limits of our current understanding of organizational incompetence. *Quality - Access to Success, 20*(172), 74–77.

Giber, D., & Friedman, D. (2006). Leaders of the future. *Leadership Excellence Essentials, 23*(6), 12–13.

Goodall, A., & Stoller, J. K. (2017). The future of clinical leadership: Evidence for physician leadership and the educational pathway for new leaders. *BMJ Leader, 1*(2), 8–11. doi:10.1136/leader-2017-000010

Goodall, A. H., & Bäker, A. (2015). A theory exploring how expert leaders influence performance in knowledge-intensive organizations. In I. M. Welpe, J. Wollersheim, S. Ringelhan, & M. Osterloh (Eds.), *Incentives and Performance* (pp. 49–67). Springer International Publishing. doi:10.1007/978-3-319-09785-5_4

Grossman, R., & Salas, E. (2011). The transfer of training: What really matters? *International Journal of Training and Development, 15*(2), 103–120. doi:10.1111/j.1468-2419.2011.00373.x

Gumede, V. (2017). Leadership for Africa's development. *Journal of Black Studies, 48*(1), 74–90. doi:10.1177/0021934716678392

Hall, D. T. (2004). Self-awareness, identity, and leader development. In N.J. Mahwah (Ed.), *Leader development for transforming organizations: Growing leaders for tomorrow* (pp. 153–176). Lawrence Erlbaum Associates Publishers.

Hannah, S. T., Avolio, B. J., Luthans, F., & Harms, P. D. (2008). Leadership efficacy: Review and future directions. *The Leadership Quarterly, 19*(6), 669–692. doi:10.1016/j.leaqua.2008.09.007

Harms, P. D., Spain, S. M., & Hannah, S. T. (2011). Leader development and the dark side of personality. *The Leadership Quarterly, 22*(3), 495–509. doi:10.1016/j.leaqua.2011.04.007

Hogg, M. A. (2001). A social identity theory of leadership. *Personality and Social Psychology Review, 5*(3), 184–200. doi:10.1207/S15327957PSPR0503_1

Hulin, C., Netemeyer, R., & Cudeck, R. (2001). Can a reliability coefficient be too high? *Journal of Consumer Psychology, 10*(1), 55–58. doi:10.1207/S15327663JCP1001&2_05

Irvine, P. A., & Brundrett, M. (2019). Negotiating the next step: The part that experience plays with middle leaders development as they move into their new role. *Educational Management Administration & Leadership*, *47*(1), 74–90. doi:10.1177/1741143217720457

Jain, S., & Anjuman, A. S. S. (2013). Facilitating the acquisition of soft skills through training. *The IUP Journal of Soft Skills*, *2*(2), 32–39.

Johnson, S. K., Putter, S., Reichard, R. J., Hoffmeister, K., Cigularov, K. P., Gibbons, A. M., Chen, P. Y., & Rosecrance, J. C. (2018). Mastery goal orientation and performance affect the development of leader efficacy during leader development. *Journal of Leadership & Organizational Studies*, *25*(1), 30–46. doi:10.1177/1548051817713689

Kjellström, S., Stålne, K., & Törnblom, O. (2020). Six ways of understanding leadership development: An exploration of increasing complexity. *Leadership*, *16*(4), 434–460. doi:10.1177/1742715020926731

Kjellström, S., Törnblom, O., & Stålne, K. (2020). A dialogue map of leader and leadership development methods: A communication tool. *Cogent Business & Management*, *7*(1), 1717051. doi:10.1080/23311975.2020.1717051

Kniffin, K. M., Wansink, B., & Shimizu, M. (2015). Sports at work: Anticipated and persistent correlates of participation in high school athletics. *Journal of Leadership & Organizational Studies*, *22*(2), 217–230. doi:10.1177/1548051814538099

Kolb, A. Y., & Kolb, D. A. (2005). Learning styles and learning spaces: Enhancing experiential learning in higher education. *Academy of Management Learning & Education*, *4*(2), 193–212. doi:10.5465/amle.2005.17268566

Kooij, D. T. A. M., Lange, A. H. D., Jansen, P. G. W., Kanfer, R., & Dikkers, J. S. E. (2011). Age and work-related motives: Results of a meta-analysis. *Journal of Organizational Behavior*, *32*(2), 197–225. doi:10.1002/job.665

Ledoux, S. F. (2002). Defining natural sciences. *Behaviorology Today*, *5*(1), 24–36.

Liu, Z., Venkatesh, S., Murphy, S. E., & Riggio, R. E. (2020). Leader development across the lifespan: A dynamic experiences-grounded approach. *The Leadership Quarterly*, *32*(5), 101382. doi:10.1016/j.leaqua.2020.101382

Machida, M., & Schaubroeck, J. (2011). The role of self-efficacy beliefs in leader development. *Journal of Leadership & Organizational Studies*, *18*(4), 459–468. doi:10.1177/1548051811404419

Mainemelis, C., Boyatzis, R. E., & Kolb, D. A. (2002). Learning styles and adaptive flexibility: Testing experiential learning theory. *Management Learning*, *33*(1), 5–33. doi:10.1177/1350507602331001

Manikkath, J., Matuluko, A., Leong, A., Ching, D., Dewart, C., Lim, R., Meilianti, S., & Uzman, N. (2020). Exploring young pharmacists' and pharmaceutical scientists' needs and expectations within an international pharmacy organization: Findings from FIP's needs assessment survey. *Research in Social & Administrative Pharmacy*, *16*(12), 1677–1685. Advance online publication. doi:10.1016/j.sapharm.2020.02.015 PMID:32169471

Martinelli, D., & Erzikova, E. (2017). Public relations leadership development cycle: A cross-cultural perspective. *Public Relations Review*, *43*(5), 1062–1072. doi:10.1016/j.pubrev.2017.09.002

Maxwell, J. C. (2013). *The 5 levels of leadership: proven steps to maximize your potential*. Center Street.

Mazzoccoli, A., & Wolf, G. (2016). Mentoring through the leadership journey: From novice to expert. *Nurse Leader*, *14*(4), 253–256. doi:10.1016/j.mnl.2016.05.012

Mdingi, S. L. (2018). *Investigating the levels of management and leadership development among medical doctors* [Masters Thesis]. Nelson Mandela Metropolitan University.

Mesterova, J., Prochazka, J., Vaculik, M., & Smutny, P. (2015). Relationship between self-efficacy, transformational leadership and leader effectiveness. *Journal of Advanced Management Science*, *3*(2), 109–122. doi:10.12720/joams.3.2.109-122

Mumford, M. D., Marks, M. A., Connelly, M. S., Zaccaro, S. J., & Reiter-Palmon, R. (2000). Development of leadership skills: Experience and timing. *The Leadership Quarterly*, *11*(1), 87–114. doi:10.1016/S1048-9843(99)00044-2

Mumford, T. V., Campion, M. A., & Morgeson, F. P. (2007). The leadership skills strataplex: Leadership skill requirements across organizational levels. *The Leadership Quarterly*, *18*(2), 154–166. doi:10.1016/j.leaqua.2007.01.005

Murphy, S. E. (2011). Providing a foundation for leadership development. In S.E. Murphy & R. J. Reichard (Eds.), *Early development and leadership: Building the next generation of leaders* (pp. 3–37). Routledge/Taylor & Francis Group.

Murphy, S. E., & Johnson, S. K. (2011). The benefits of a long-lens approach to leader development: Understanding the seeds of leadership. *The Leadership Quarterly*, *22*(3), 459–470. doi:10.1016/j.leaqua.2011.04.004

Murugaboopathy, V., Hebbal, M., Ankola, A., Mohandoss, S., & Sharma, R. (2020). Students' perspective of need for leadership development program in dental education. *Journal of Global Oral Health*, *2*(2), 93–98. doi:10.25259/JGOH_65_2019

Nakanjako, D., Namagala, E., Semeere, A., Kigozi, J., Sempa, J., Ddamulira, J. B., Katamba, A., Biraro, S., Naikoba, S., Mashalla, Y., Farquhar, C., & Sewankambo, N. (2015). Global health leadership training in resource-limited settings: A collaborative approach by academic institutions and local health care programs in Uganda. *Human Resources for Health*, *13*(1), 87. doi:10.118612960-015-0087-2 PMID:26581196

National Research Council. (2015). *Enhancing the effectiveness of team science*. The National Academies Press., doi:10.17226/19007

Ng'weno, B. (2020). The historical, political, and demographic context of leadership in Africa. In E. Jordans, B. Ng'weno, & H. Spencer-Oatey (Eds.), *Developing Global Leaders: Insights from African Case Studies* (pp. 59–87). Springer International Publishing. doi:10.1007/978-3-030-14606-1_4

Nick, G., Reinhold, G., & Valerie, I. (2005). *Managing health services*. McGraw-Hill Education.

Nyssa, Z. (2019). Why scientists succeed yet their organizations splinter: Historical and social network analyses of policy advocacy in conservation. *Environmental Science & Policy, 98*(2020), 8–94. . doi:10.1016/j.envsci.2019.05.007

Oc, B. (2018). Contextual leadership: A systematic review of how contextual factors shape leadership and its outcomes. *The Leadership Quarterly, 29*(1), 218–235. doi:10.1016/j.leaqua.2017.12.004

Offermann, L. R., Thomas, K. R., Lanzo, L. A., & Smith, L. N. (2019). Achieving leadership and success: A 28-year follow-up of college women leaders. *The Leadership Quarterly, 31*(4), 101345. doi:10.1016/j.leaqua.2019.101345

Perry, J., Mobley, F., & Brubaker, M. (2017). Most doctors have little or no management training, and that's a problem. *Harvard Business Review, 15*(December). https://hbr.org/2017/12/most-doctors-have-little-or-no-management-training-and-thats-a-problem

Perry, S. J., Hunter, E. M., Currall, S. C., & Frauenheim, E. (2017). Developing engineering leaders: An organized innovation approach to engineering education. *Engineering Management Journal, 29*(2), 99–107. doi:10.1080/10429247.2017.1298189

Pollock, R. V. H., Jefferson, A. McK., & Wick, C. W. (2015). *The six disciplines of breakthrough learning: how to turn training and development into business results.* John Wiley & Sons, Inc. doi:10.1002/9781119153832

Rao, M. S. (2012). Myths and truths about soft skills. *T+D*. https://www.td.org/magazines/td-magazine/myths-and-truths-about-soft-skills

Rice, M., Davis, S. L., Soistmann, H. C., Johnson, A. H., Gray, L., Turner-Henson, A., & Lynch, T. (2020). Challenges and strategies of early career nurse scientists when the traditional postdoctoral fellowship is not an option. *Journal of Professional Nursing, 36*(6), 462–468. doi:10.1016/j.profnurs.2020.03.006 PMID:33308541

Richard, R. J., Walker, D. O., Putter, S. E., Middleton, E., & Johnson, S. K. (2017). Believing is becoming: The role of leader developmental efficacy in leader self-development. *Journal of Leadership & Organizational Studies, 24*(2), 137–156. doi:10.1177/1548051816657981

Sapienza, A. M. (2004). *Managing scientists: Leadership strategies in scientific research* (2nd ed.). Wiley. doi:10.1002/047166359X

Scandura, T. A., & Lankau, M. J. (1996). Developing diverse leaders: A leader-member exchange approach. *The Leadership Quarterly, 7*(2), 243–263. doi:10.1016/S1048-9843(96)90043-0

Seifert, C. F., & Yukl, G. (2010). Effects of repeated multi-source feedback on the influence behavior and effectiveness of managers: A field experiment. *The Leadership Quarterly, 21*(5), 856–866. doi:10.1016/j.leaqua.2010.07.012

Spencer, L. M., McClelland, D. C., & Spencer, S. M. (1992). *Competency assessment methods history and state of the art* (1st ed.). Hay McBer Research Press.

Stanton, F., & Grant, J. (1999). Approaches to experiential learning, course delivery and validation in medicine. A background document. *Medical Education*, *33*(4), 282–297. doi:10.1046/j.1365-2923.1999.00272.x PMID:10336759

Stead, V., & Elliott, C. (2013). Women's leadership learning: A reflexive review of representations and leadership teaching. *Management Learning*, *44*(4), 373–394. doi:10.1177/1350507612449504

Stoller, J. K. (2009). Developing physician-leaders: A call to action. *Journal of General Internal Medicine*, *24*(7), 876–878. doi:10.100711606-009-1007-8 PMID:19455370

Tajfel, H. & Turner, J. C. (2004). *The social identity theory of intergroup behavior*. Psychology Press. . doi:10.4324/9780203505984-16

Tannenbaum, S. I., Mathieu, J. E., Salas, E., & Cohen, D. (2012). Teams are changing: Are research and practice evolving fast enough? *Industrial and Organizational Psychology: Perspectives on Science and Practice*, *5*(1), 2–24. doi:10.1111/j.1754-9434.2011.01396.x

Wai, J. (2014). Experts are born, then made: Combining prospective and retrospective longitudinal data shows that cognitive ability matters. *Intelligence*, *45*(July–August), 74–80. doi:10.1016/j.intell.2013.08.009

Zaccaro, S. J. (2007). Trait-based perspectives of leadership. *The American Psychologist*, *62*(1), 6–16. doi:10.1037/0003-066X.62.1.6 PMID:17209675

Zacher, H., Rosing, K., & Frese, M. (2011). Age and leadership: The moderating role of legacy beliefs. *The Leadership Quarterly*, *22*(1), 43–50. doi:10.1016/j.leaqua.2010.12.006

KEY TERMS AND DEFINITIONS

Leader Development: Advancing and improving the set of skills pertaining to the particular individual to fulfil the capacity as a leader through learning intervention strategies of coaching and mentoring.

Perceived Behavior Control: A form of influence based on attitudes, beliefs, subjective norms, intentions that focus on the perceptions or situations where the person may desire to participate in development activities but in reality, may feel held back by limitations such as resources.

Self-Efficacy Beliefs: Are fundamental in developing leader efficacy and these principles are embedded in the individual's confidence levels.

Subjective Norms: Rules that arise from perceived peer pressures to participate or not. For example, a characteristic of technical fields is the despise of social skills training as right-brain skills and an emphasis on learning technical skills as the premium towards performance.

Chapter 10
Transformational and Andragogical Leadership and Its Impact on Resilience and Workplace Stress

Steve Lambert
University of Chester, UK

ABSTRACT

There has been a 36.5% rise in reported cases of work-related stress between 2018/19 and 2019/20 to over 822,000 individual cases. Marcatto et al. found that in Europe stress-related absence was second only to musculoskeletal disorders while the EU-OSHA estimated that work-related stress cost European companies 20 billion euros per year and accounted for nearly 60% of lost working days. This chapter focuses on the under-researched area of stress amongst middle leaders in education. While it is widely reported that teachers face high levels of stress, as a result of organisational pressures, there is a paucity of research on middle leaders. This chapter considers whether there will be a significant negative relationship between self-reported levels of resilience and self-reported levels of stress amongst middle leaders.

INTRODUCTION

Transformational and andragogical leaders involve their staff in planning, assessing, formulating goals, carrying out activities. and evaluating results. They value their staff and trust that they are going to do a good job. However, leaders need to be confident in their own ability if they are to trust their staff. Often, leaders struggle to 'let go' to distribute meaningful activities to their staff. However, the distribution of tasks and the involvement of staff in leadership decision-making in an already busy environment have the potential to increase the amount of workplace stress that individuals experience. Grummell et al. (2015) argue that education, is already greedy of an individual's time, therefore expecting further engagement has the potential to cause increases in workplace stress and, work-life conflict. Often resilience is posited as the panacea that will solve the issues of workplace stress but this is an oversimplification. There is a

DOI: 10.4018/978-1-6684-7832-5.ch010

complex interaction among an individual's perceived levels of workplace stress, the role of resilience in moderating the impact of workplace stress and the role that the leader.

The importance of addressing workplace stress cannot be understated. The UK's Health and Safety Executive (HSE) found that in 2019 there were 17.9 million days lost to stress within the UK alone (HSE, 2022). Individuals reporting work-related stress have continued to rise since 2001 and now stand at over 800,000 in the year from April 2019 to the end of March 2020 which is now a major issue for both individuals and their employers. Occupations such as healthcare and education are particularly prone to higher levels of work-related stress. Stress-related absence is now the second highest cause of work absence, after musculoskeletal disorders (Marcatto et al., 2021). Furthermore, a report from the European Union (EU-OSHA, 2014) stated that work-related stress cost European companies an estimated 20 billion euros per year and accounted for nearly 60% of lost working days. The Chartered Institute of Personnel Development's (CIPD) (2022) report found that heavy workload was a significant contributory factor in work-related stress, resulting in record levels of *presenteeism* (working when unwell) and *leavism* (using holiday allowances to work). What isn't clear from any of these reports is whether is a result of increased prevalence or a willingness to report stress to employers.

Understanding work-related stress is not as simple as a relationship between the phenomenon causing the stress (the stressor) and the associated response to it. Instead, as Copper, Dewe and O'Driscoll (2001) propose that work-related stress should be viewed as an ongoing relationship between the individual and their working environment. This approach implies that the factor causing the stress and the subsequent response to it is dynamic and not merely a static one that will always cause the same response when an individual encounter the same factor causing the stress.

In addition to the perceived levels of workplace stress, it is important to explore how resilience interacts with stress. Resilience is often seen as the ability to bend but not break, bounce back, and perhaps even grow in the face of adverse life experiences. The American Psychological Association (APA) (2014) states that resilience is focused on developing and maintaining a healthy mental state. Yet often we see resilience as the solution which enables us to cope with the challenges life throws at us. Resilience is more nuanced than that. Indeed, overall resilience has a limited impact on work-related stress. Yet when we delve deeper into the elements which make up resilience we find that there are three aspects which are particularly important when it comes to the workplace. These include an individual's perception of their future, and whether they consider it optimistically.

In addition, the ability to interact socially, whether formally or informally is important to how we navigate the workplace relationships that we have with those around us. Given that education is a people-focused profession, this is particularly important, and this has been echoed by the section on relationships from the H&S survey. Finally, and linked to this, is the support we receive, both from family and colleagues and our ability to seek solace from them.

In this chapter, the author/s explore the role of workplace stress, resilience, and the role that leaders have in mitigating the impact of stress.

The chapter, which is based on a study conducted in in 2022, starts discussing workplace stress both the positives of stress and also the negative impact of stress. Before moving to a discussion around resilience and how resilience is often misconstrued as the means to mitigating stress. We then look in detail at the relationship between the two, based on some empirical data that we have collected to demonstrate that the role that leaders have has a direct relationship to the levels of perceived workplace stress. In addition, we explore what leaders can do, as transformational and or andragogical leaders to better support staff so that they can flourish in their roles.

STRESS, RESILIENCE AND THE TRANSFORMATIONAL LEADER

What Is Stress?

Stress is a term that is used ubiquitously and as both Quick et al., (2013) Robinson (2018) and Polizzi et al. (2020) note is one of the most widely studied topics in psychology. The aphorism '*the straw that broke the camel's back*' encapsulates the essence of stress. Nonetheless, stress is often widely used as a term relating to how individuals and organisations respond to environmental issues. Kahn (1987) states that there is little agreed-on scientific definition. However, operationally there are agreed definitions relating to the individual elements that comprise stress.

For a stress response to be experienced, a stressor must be evoked. This can be either a physical or psychological stimulus that can be related to a particular context, such as a workplace, or an individual's experience of, or perceived experience of a situation (Taylor et al. (2004). Once a stressor has been experienced, individuals are exposed to a number of possible responses These responses, according to Rees et al. (2010) result in the release of hormones, primarily adrenaline or noradrenaline. Individuals feel this response through increases in heart rate, respiration, perspiration, disturbed sleep or irregular eating. While these manifestations are the most recognisable, there are psychological responses that are less obvious such as increased levels of anxiety. These responses are typically designed to prepare an individual for a fight-or-flight response.

Not all stress leads to a fight-or-flight response. Some stress can be positive, leading to an improvement in an individual's performance (Copper, Dewe, & O'Driscoll, 2001; Robinson, 2018; Daniel, 2019). This form of 'good stress' is referred to as Eustress, and since 1908 has been expressed as the Yerkes-Dodson law (Bong, Fraser, & Oriot, 2016). This model shows how an individual's performance increases with increased stress to an optimum point after which performance levels decrease as stress levels continue to rise (Albort-Morant et al., 2020). The difficulty with this model is that the optimum level is different for each individual. This is also affected by factors such as susceptibility to stress and fatigue, as well as familiarity with tasks. Yet Copper, Dewe and O'Driscoll (2001) argue that the Yerkes-Dodson model is an oversimplification of the complex nature of stress due to not addressing any psychological responses to stress. However, Christian and Lolas (1985) argue that the model remains valid for responses to physical factors, such as heat and cold, but does not adequately explain psychological stress.

One of the major criticisms of existing models of stress is they reflect only one component of the stress process, either phenomenon causing the stress or the impact of it and say little about the process of stress. Thus, conceptualising stress as one-dimensional and negating the complexities of the phenomenon. The second criticism is the failure to consider an individual's response to stress may be moderated by individual differences. For example, personality attributes such as expectations, values and goals, or context factors, such as levels of support from families and friends as well as role and status within the organisation. Therefore, what might be perceived as stressful to one individual, may not be to another.

Copper, Dewe and O'Driscoll (2001) suggest that work-related stress is not as simple as a correlational relationship between stimulus (stressor) and associated response. Instead, they propose that work-related stress should be viewed as transactional – an ongoing relationship between the individual and their environment. This approach tries to explore the nature of the stressor-response relationship as a dynamic process and not merely a static one as in the response-based model of stress.

The HSE Management Standards also offer 6 factors which contribute to workplace stress. These commonly used descriptors allow organisations to identify areas that are causing employees particular stress.

Table 1. HSE Management standards (2022) work-related stressors contextualised for educational leaders

Six Factors of Work-Related Stress		Contextualised to Middle Leadership in Education
1	Demands *(of the role)*	These factors relate to the complexity of the job itself and the variety of tasks undertaken. Many middle leaders follow a traditional academic route from teaching into leadership, often with little training for their new role (Corbett, 2017; 2021)
2	Role *(within the organisation and how one is perceived)*	Cartwright and Cooper (1997) argue that this factor encompasses an individual's behaviours and demands associated with a role. However, Kahn et al., (1964) state that ambiguity surrounding the role and also conflicting pressures within the job are central to this element. One of the key challenges for middle leaders is managing the loyalties they have to their teams, from which they may well have progressed to balancing the demands and expectations imposed upon them by senior leaders. This has seen middle leaders struggle to balance the task-orientated and people-orientated aspects of their jobs (Lambert, 2020).
3	Relationships *(at work)*	Middle leaders in education have three dimensions of support. From their teams whom they manage, from their managers and their peers (those in equivalent roles). The role of this form of social support can be important yet Lynch, Grummell, and Devine (2012) remind us, that education is greedy of time and a lonely place. Therefore, a lack of support networks either formal or informal can have a significant psychological impact on an individual.
4	Change *(and the management of knowledge capital)*	Cartwright and Cooper (1997) suggest that this factor relates to job insecurity, for which the impact has been explored extensively in the work of Maslow (McLeod, 2007). For those who work in schools or further education, there is no career progression other than into leadership. However, there is no correlation between being a good teacher and a good leader, despite individuals often being promoted into leadership based on their performance as a teacher.
5	Support *(from leaders and managers)*	There are a plethora of factors that can act as stressors. For example, organisational culture, and approaches to leadership, particularly where individuals are not part of the decision-making process, result in either a top-down or autocratic approach to leadership. Alternatively, poor communication between senior leaders and staff can result in disengagement and poor motivation. Likewise, external factors such as external accountability measures such as government performance tables, inspections or constantly changing funding expectations can all act as stressors.
6	Control *(over working pattern)*	Lynch, Grummell, and Devine (2012) already remind us that education is time greedy. This has been exacerbated by the recent SARS-CoV-2 pandemic (commonly called Covid-19). The boundaries between work and non-work time have become increasingly blurred, particularly as working patterns have changed and more people work from home. The pandemic has resulted in changing family structures and patterns of work and schooling.

Having explored notions of stress, both in terms of prevalence and work-relatedness we now need to focus on resilience. As a potential moderating factor, it is vital to understand how it is defined and conceptualised.

Resilience

With its root found in two bodies of literature: psychological aspects of coping and the physiological aspects of stress (Tusaie & Dyer, 2004), resilience is often seen as the ability to bend but not break, bounce back, and perhaps even grow in the face of adverse life experiences (Walker, 2019 and 2020). However, the American Psychological Association (APA) (2014) states that resilience is focused on developing and maintaining a healthy mental state with Anyan et al. (2020) going on to say that this is crucial for positive development and adaptations throughout life. The study of resilience is seen positively as a way of understanding the factors that may protect or promote a healthy adaptation despite threats, chronic stressors, or adverse life situations (Reivich et al., 2013). It is important to note that being resilient does not mean that individuals will not have trouble or distress. There will be certain factors that may mitigate the level of distress someone feels, making them more resilient than others. However, resilience is not necessarily a personality trait that only some people possess. On the contrary, resilience involves behaviours, thoughts, and actions that can be learned and developed. Southwick et al., (2014) state that while useful, the definition offered by APA (2014) does not reflect the complex nature of resilience, which includes a range of biological, psychological, and social factors that interact to determine how individuals respond to distress.

Pietrzak and Southwick (2011) suggest that resilience is often viewed incorrectly, suggesting that it should not be viewed in a dichotomous manner, you either have it or not. Instead, they argue that resilience is more akin to a continuum, with an individual having differing degrees of resilience depending on the domain in which it is viewed. For example, an individual who adapts well in the workplace, thus demonstrating higher levels of resilience, may fail to adapt to the same level in their personal life. Likewise, Kim-Cohen and Turkewitz (2012) state that levels of resilience can change over time; reinforcing the notion that resilience is a learned behaviour.

Ledesma (2014) suggests the principal models of individual resilience are the compensatory model and the challenge model.

Werner and Smith (2001) define compensatory resilience as neutralising the exposure that one might experience to risk, in essence working in opposition to the risk. They argue that the risk factors and the compensatory factors independently contribute to predicting the outcome; ideally, the compensatory factors mitigate or neutralise the risks altogether. However, there is a lack of consensus on what the compensatory factors might be.

The challenge model of resilience states that assuming that the risk factors are not too extreme they can prepare an individual for their next challenge. However, Fleming and Ledogar (2008) note that exposure to both low and high levels of risk results in negative outcomes but do not define any thresholds, possibly due to levels of exposure being individual, similar to what is seen with stress.

Ahern, et al. (2006) critiqued six different approaches to measuring resilience. Of these six, two, the Adolescent Resilient Scale (ARS), and the Baruth Protective Factors Inventory (BPFI) relied on undergraduate students as their target population. This raises questions, particularly with the BPFI as to its application outside of the population group. Fletcher and Sarkar (2013) point out that most resilience theories are specific to a particular population. For example, adolescents, families, or specific occupational groups such as police officers or nurses. Both Richardson (2002) and Fletcher and Sarkar (2013) advocate for greater use of the Resilience Scale for Adults (RSA), due to its ability to be applied to different types of stressors, adversities, and life events.

As the RSA has been developed as a general model of resilience, structured around the original factors of resilience from a range of existing models. This model of resilience will be adopted for the remainder of this chapter.

Measuring Stress and Resilience

As already mentioned, there is a lot of existing data on stress within the workplace. However, there is very little in terms of stress of leaders in education. In order to start to fill this gap and act as a platform for further work, a small cross-sectional study (Haslam & McGarty, 2003) was used to measures the outcome and the exposures to stress of educational leaders at the same time. This was achieved by using an online survey which considered perceived levels of stress and resilience.

Table 2. Summary of Friborg et al. (2006) six domains of resilience

	Resilience Factor	Summary
1	Personal strength/perception of self	An individual's self-assessment of strengths and abilities
2	Personal strength/perception of future	An individual's belief about the ability to realise opportunities and future plans
3	Structured role	An individual's assessment of the importance of structure and order in achieving goals.
4	Social competence	An individual's assessment of comfort in social situations
5	Family cohesion	The extent to which family cohesion and values are important to an individual
6	Social resources	An individual's assessment of how social resources reinforce and strengthen self-coping mechanisms

Therefore, the design of this research was a correlational analytic study. The two key variables were stress and resilience and in particular, the 6 domains of resilience (see Table 2).

Sample

The sample size was 62 participants to ensure that there was a statistically valid basis for analysis. In total 66 completed the survey, based on the inclusion criteria of being in a middle leadership role in education and consenting to complete the questionnaire. Four individuals were excluded as they did not meet the aforementioned inclusion criteria. Therefore, in total 62 (*n=62*) participants were included in the analysis.

The sample had a greater proportion of females, 43 (69.40%) than male participants, 19 (30.60%). The disparity of females to males in teaching suggests that the number of males in the sample might be too low for valid statistical analysis. This sample is also reflective of the disparity in staffing generally in education.

The majority of respondents were in the age range 45 to 54 years old, 35% as shown in Table 3. Higher than that in schools with 27% of staff within the same age category. Whereas in the FE sector, the mean age of staff was 47 years old, although the distribution of staff across age brackets was similar at approximately 12% of staff within each of the aforementioned age categories.

Table 3. Distribution of participants ages

	N	%
18-24	1	1.6%
25-34	9	14.5%
35-44	16	25.8%
45-54	22	35.5%
55-64	13	21.0%
65 and over	1	1.6%

A snowball strategy (Cohen, Manion and Morrison, 2003) was used to obtain the sample of participants. The initial group of individuals were identified through professional contacts within further and higher education settings. Figure 1. Illustrates the range of educational settings that participants in the sample group worked in, with a third of individuals working within higher education, a quarter in primary schools and the remainder split across secondary schools, further education colleges and independent training providers.

Figure 1. Participant sectors

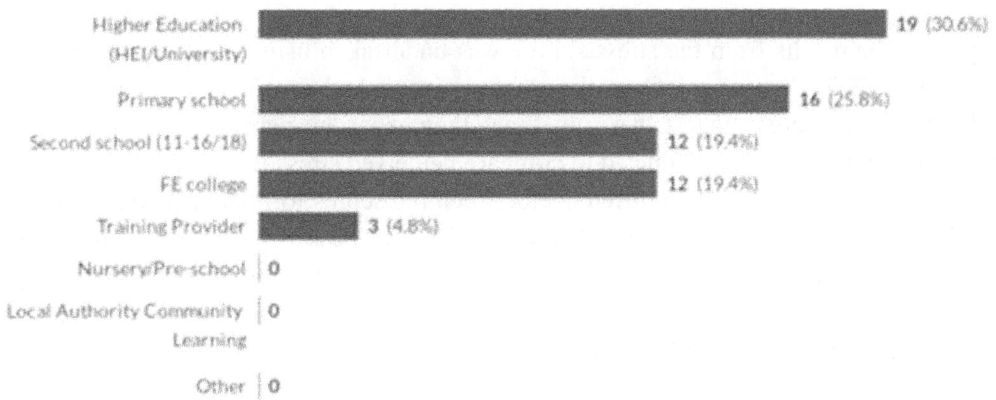

Finally, the majority of the participants, 53 (85%) reported coming from a white ethnic group or background. There were 5 participants (8%) from Asian backgrounds and 2 who were Chinese (3%). While the majority of participants were form white ethnic backgrounds, compared to government data which suggests that 78.5% of the general working age population, rising to 85.9% of the teaching population (DfE, 2020).

To collect the data, participants were sent an online questionnaire which comprised of three parts:

General Demographic Questions: A general set of questions was used to examine demographic characteristics, such as age, gender and ethnicity. These were based on the questions developed for the 2021 population census, thus ensuring their validity (ONS, 2021).

HSE Management Standards Indicator Tool (HSE-MSIT) (HSE, 2008): A questionnaire consisting of a 35-item, five-point Likert scale, from strongly disagree to strongly agree, designed to measure the six primary work-related stressors. Marcatto et al., (2014) found evidence that the HSE-MSIT offered significant levels of correlation when measuring work-related stress.

Resilience Scale for Adults (RSA): Focuses on the 6-domains of resilience (*personal strength/perception of self; personal strength/perception of future; structured style; social competence; family cohesion; social resource*). Participants were asked to use a 5-item scale to position themselves between two opposing responses to a given statement. While there is a risk of individuals opting for the middle option, this appears not to be a significant issue as tested in the validated version of the tests (Friborg, et al., 2005).

Are We Stressed?

The extent to which educational leaders are stressed and whether resilience has a positive impact on individuals' perceived level of stress in education is the principle aim of this chapter.

The results were analysed using the Statistical Package for the Social Sciences (SPSS, version 27.0.1). A linear regression was performed to examine the hypotheses. This would explore the correlation between the overall level of resilience and the level of work-related stress.

Stress Findings

The analysis of the results from the stress survey was undertaken using the analysis tool available as part of the HSE-MSIT question set. Overall male participants had a mean average of 3.31 and females of 3.42. Both sets of scores suggest that sometimes they experience stress within the workplace. An independent-samples t-test was conducted to compare perceived stress levels between male and female participants. There was a significant difference between the scores for male (M=3.31, SD= .377) and females (M=3.34, SD= .327) participants; t(30)= -1.120, p=0.0272. These results suggest that gender does affect the perceived level of stress that individuals experience within the workplace.

However, further analysis was undertaken to visualise the data. Figure 2 provides the averages for each of the elements associated with stress.

Figure 2. Averages by group vs. stress domains

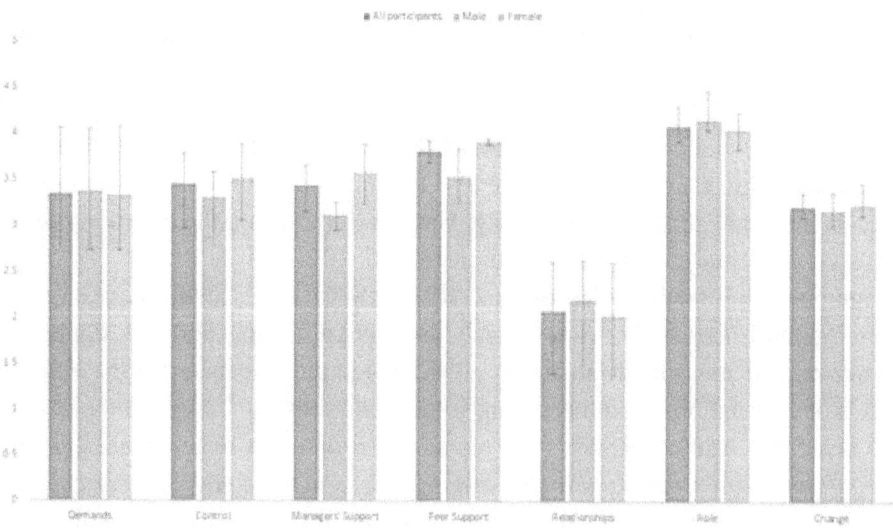

The results presented are on a scale from 1 (Never), 2 (Seldom), 3 (Sometimes), 4 (Often) and 5 (Always). What this chart shows is that individuals (middle leaders) are clear about their role with an average score of 4.09 (*often*), although male respondents report a slightly higher score of 4.16. Within the domain of work relationships, participants scored 2.08, and female participants scored 2.03, which has questions on personal harassment, friction or anger between colleagues, bullying and relationships being strained, score lower. This suggests that participants *Seldom* experience any form of bullying or harassment. This is possible as a result of the emphasis employers place on requiring staff to complete equality and diversity training, which covers the legislation pertaining to protected characteristics and forms of harassment.

The domain of demands focuses on conflicting workplace demands, unachievable deadlines, having to neglect tasks due to work pressures and an assumption that long hours are the norm. Participants scored on average 3.35, within the *Sometimes* range, however, males scored slightly more at 3.38 and females less at 3.33. This suggests that sometimes middle leaders' work causes demands on them, but looking at the scores to individual questions, staff report that they *often* must work very intensively to complete their tasks.

What was interesting was the perceived level of support, both from their manager and their peers. All participants said that they sometimes received support from their managers and their peers. Female participants scored higher in this section, still within the boundary of *Sometimes*, while peer support was notably higher than that from managers. Suggesting that support was sought from peers more so than managers. This raises an issue as andragogical or transformational leaders do we have the balance correct. Are we at risk of slipping towards being autocratic in our approach and simply telling staff what they need to be doing and how, rather than co-creating and leading our teams?

Interestingly, the scores relating to change showed that on average participants scored 3.29 (scores ranged from 3.12 to 3.50), which is in the lower half of the *Sometimes* range. In particular, the questions on being consulted about changes scored particularly low (3.12), however, respondents did feel that they can challenge managers about potential changes. Yet the findings of the change element portray the change in a single dimension. What is neglected is that change is multi-dimensional, in so much as the phenomena being changed, which the questions in the HSE-MSIT relate to, but there is also the dimension of how change is to be implemented. In addition, the impact the change is likely to have on individuals needs to be factored in and is usually considered through Kübler-Ross's model of change (see Buller, 2015).

The above scores provide greater insights into the experiences and challenges that middle leaders face in their role. Particularly around their understanding of the role and the expectations placed on them. Despite saying the in paragraph above we may be at risk of becoming too autocratic, we are also at risk of being too laissez-faire and abdicating our responsibilities as a leader. What is emerging is that being a transformational or andragogical leader is akin to walking a tightrope. Where we are balanced between either being too hands-off, resulting in little clear direction, or too directing and controlling. This is important as De Nobile (2018) states that middle-leadership is under-theorised and ambiguities persist in relation to who and what they do. Additionally, identified that the and supports studies Bennett et al., (2007) that state the middle leadership role in education is one of the most demanding due to opposing pressures from senior leaders and teaching staff.

While the above section presented the scores from the survey, it does not tell us whether any of the individual factors were significant in relation to respondents overall perceived stress level. The multiple regression showed that some of the perceived stress factors such as how much control and individual had, or the relationship with colleagues had a greater impact on stress compared to the demands of the role which did not have a significant impact on stress.

A multiple regression was used to determine perceived levels of stress based on individual stressors. The model explained a statistically significant amount of variance in stress, $F(7, 54) = 1602.20$, $p = .001$, $R^2 = .99$, $R^2_{adjusted} = .99$. What the findings tell us is that there are key factors, some of which are in an individuals' control, others less so, that are strong predictors of the perceived level of stress an individual will encounter.

For example, the element of control that an individual had over their job was a significant predictor of overall perceived stress level, $\beta = .36$, $t(-.95) = 4.79$, $p = .001$. In addition, the level of support received from their peers was also a significant predictor to overall perceived stress level, $\beta = -.29$, $t(-3.21) = -4.22$, $p = .001$. An increase in one in the score of peer support corresponds to a decrease, by -.257, in the perceived levels of stress, $B = -.257$, 95% CI [-.379, -.135]. Alongside this, the level of support from a manager was a significant predictor on perceived stress level $\beta = .29$, $t(1.50) = 3.62$, $p = .001$. However, an increase in one in the score of manager's support corresponds to a slight increase, by .251, in the perceived levels of stress, $B = .251$, 95% CI [.112, .390]. An individual's relationship again like the other elements has an impact on perceived levels of stress, $\beta = .57$, $t(14.87) = 5.29$, $p = .001$. Finally understanding of an individual's role also has a significant impact on stress $\beta = .15$, $t(4.83) = 6.53$, $p = .001$.

Interestingly, while the demands of the role element were not significant $p = .001$ it did have a positive impact on the perceived levels of stress. With an increase of one in this score corresponding to a decrease by -.133 in stress, $B = -.133$, 95% CI [-.210, -.056]. This suggests that while the role may be demanding, clarity of expectations within the role does reduce the level of stress and individual experiences. On the other hand, issues relating to change did not have a significant impact on perceived stress levels, $p = .582$.

The significance of the aforementioned factors (demands of the role, control over their own, perceived support, responses to change) have not been considered in relation to gender, as the sample population is skewed towards females and as discussed earlier the number of males in the sample is too small to generalise the results relating to male leaders. While this is a limitation of this study, future work could be done to explore these findings in respect of gender. In addition, as already stated the perceived level of overall stress experienced does not appear to be significantly different based on gender.

Therefore, it could be suggested that the level of control an individual has over their work, and the interactions individuals have with their peers and managers all have a significant contribution to the level of perceived stress they experience.

Resilience Findings

A Pearson's correlation was performed to assess the relationship between perceived workplace stress and resilience. The results showed that there was a negative relationship between levels of resilience and perceived stress.

The dataset was inspected to determine whether there were any missing data or extreme values due to input errors. There was one participant with missing data. Therefore, data for this participant were excluded from the analyses, meaning that the total dataset was n=61.

As predicted, there was a significant negative relationship between these two variables: r -.377, $N = 61$, $p < .001$ (one-tailed). Having found support for the research hypothesis, that resilience does help to reduce the perceived level of stress, the next step was to investigate the individual contributions of the six factors of resilience in terms of how they relate to perceived workplace stress.

The Cronbach Alpha test is used to determine the reliability of the findings. In this each how reliable was the data for each of the six domains of resilience. A score of 0.7 or above is good, with 0.9 being a very good score in terms of how reliable the findings were. What we found was that most of the scores were reliable however, perception of an individual's self was not a reliable indicator of resilience.

Internal consistency was satisfactory for six of the seven measures – ranging from .719 for the measure of Perception of Self to .935 for Perceived Workplace Stress. However, Cronbach's Alpha for the measure of Structured Style was .574. and the removal of any of the items for this measure reduced the internal consistency, therefore, this measure of resilience was not included in the regression model. In addition, the measure of Perception of Self, which evidenced a low corrected item-total correlation of .284, was removed from this measure. For the measure of the dependent variable of Perceived Workplace Stress, two items also had corrected item-total correlations below .3 (Field, 2013, p. 713). Similarly, these items were removed for the regression analysis. The means and standard deviations for each of the measures were computed and are presented in Table 4.

Table 4. Minimum and maximum scores, means and standard deviations (SD) for perceived workplace stress and the six factors of resilience (N = 61 for all measures)

Variable	Min	Max	Mean	SD
Perceived Workplace Stress	1.03	3.82	2.55	0.61
Perception of Self	2.60	4.60	3.78	0.58
Perception of the Future	1.75	4.50	3.57	0.71
Structured Style	2.25	4.50	3.69	0.55
Social Competence	1.50	4.50	3.55	0.69
Family Cohesion	2.33	4.67	3.93	0.59
Social Resources	2.29	4.43	3.89	0.54

This data is important as it provides a level of reliability, in so much that across all respondents, their scores did not deviate significantly from the mean average. Therefore, we can infer that the responses gained were representative of the group.

CORRELATIONS

The seven measures of resilience under investigation were then correlated to assess the relationships between them and to the perceived levels of stress. Pearson's correlations and associated p-values are presented in Table 5 below.

Aside from how the six predictor variables related to the criterion variable of workplace stress, one wanted to assess possible instances where there was a strong correlation between predictor variables which might be indicative of excessive multicollinearity. As can be seen in Table 5 the relationship between family cohesion and social resources was .717. A preliminary multiple regression analysis revealed that Variance Inflation Factor values for these two variables were greater than 2.0; indicating a problem with the multicollinearity of predictor variables. Given this issue, the fact that these two variables were highly correlated, and they were both concerned with support, it was decided to create a composite variable by combining the items from these two factors. Reliability analysis of this composite yielded a Cronbach's alpha of .888 and corrected item-total correlations were all above .40. This new variable was given the label 'Family and Social Support'. The participant mean for this variable was 3.91 (SD = 0.52) with minimum and maximum scores of 2.38 – 4.54. The relationships between this variable and the remaining original variables are also presented in Table 5 (item 7).

It can also be seen in Table 5 that the variable of Structured Style (which also suffered from poor internal consistency) evidenced a weak relationship with both the criterion variable and the other measures of resilience. These findings gave additional support for the decision not to include this variable in the final regression model.

Table 5. Pearson's correlation coefficients and associated p-values for the variable under investigation (N = 61, all tests two-tailed)

	1	2	3	4	5	6	7
Perceived Workplace Stress	$r = -.202$ $p = .118$	$r = -.447$ $p < .001$	$r = .009$ $p = .946$	$r = -.283$ $p = .027$	$r = -.205$ $p = .114$	$r = -.329$ $p = .010$	$r = -.290$ $p = .023$
1 Perception of Self	-						
2 Perception of the Future	$r = .450$ $p < .001$	-					
3 Structured Style	$r = .132$ $p = .309$	$r = .035$ $p = .787$	-				
4 Social Competence	$r = .555$ $p < .001$	$r = .313$ $p = .014$	$r = -.111$ $p = .395$	-			
5 Family Cohesion	$r = .226$ $p = .080$	$r = .309$ $p = .015$	$r = .148$ $p = .246$	$r = .372$ $p = .003$	-		
6 Social Resources	$r = .382$ $p = .002$	$r = .496$ $p < .001$	$r = .136$ $p = .297$	$r = .497$ $p < .001$	$r = .717$ $p < .001$	-	
7 Family and Social Support	$r = .331$ $p = .009$	$r = .438$ $p < .001$	$r = .153$ $p = .240$	$r = .471$ $p < .001$	-	-	-

Multiple Regression

A multiple regression was used to determine the strength of the effect between the elements of resilience (as shown in Table 5) with the perceived level of stress. As the data will demonstrate only an individual's perception of themselves had a significantly strong effect on their stress level. However, there are other factors which will be demonstrated that strongly influence stress levels. What senior leaders need to be aware of, is how they interact with individuals will directly contributed to the notions of perceived stress.

The data were screened to determine whether they satisfied the assumptions for multiple regression analysis. Cook's D was computed and indicated that there were no multivariate outliers. A histogram indicated that residuals were normal and a scatterplot indicated that there was a linear relationship between the predictor variables and the criterion variable. Finally, Variance Inflation Factor values were satisfactory, with the highest value being 1.67, indicating that there was no excessive multicollinearity between the predictor variables.

The multiple correlations between the four remaining predictor variables measuring resilience and the criterion variable of Perceived workplace stress were .482. $R^2 = .232$ and $R^2_{adjusted} = .177$; indicating that 17.7% of the variance in Perceived Workplace Stress was accounted for by these four variables. The standard error of the estimate was 0.55. There was a significant model fit: $F_{(4, 56)} = 4.23, p = .005$.

The individual contributions of the four predictor variables revealed that only Perception of the Future gave a significant contribution to the model fit: $\beta = -.411, t_{(56)} = 2.94, p = .005$. The next strongest predictor was Social Competence ($\beta = -.186, t_{(56)} = 1.23, p = .226$), then Perception of Self ($\beta = .105, t_{(56)} = 0.70, p = .488$), and finally Social Support ($\beta = -.057, t_{(56)} = 0.40, p = .688$). Therefore, when taken individually through Pearson's correlations, Perception of the Future, Social Competence, and Family and Social Support were significantly related to Perceived Workplace Stress. However, the unique contributions of these variables revealed through multiple regression analysis revealed that only Perception of the Future remained a significant predictor.

The domain of Perception of the Future, which focuses on the extent to which a middle leader has a medium or long-term plan, or do they take every day as it comes, was a significant predictor in relation to perceived workplace stress. This agrees with the HSE-MSIT (2008) and Cartwright and Cooper's (1997) identification of work-related stressors, both of whom identify perceptions of an individual's role within the organisation and factors relating to personal career development are key factors in determining levels of workplace stress.

Marelich and Piercy (2020) suggest that positive expectancies of the future are a form of dispositional trait based on a model of behavioural self-regulation. They go on to state that having a positive perception of the future is associated with various positive outcomes, including enhanced mental and physical health, survival, mortality, and well-being. Cervera-Torres, et al. (2021) found in their recent study that resilience and levels of optimism about the future can have a positive impact on how individuals manage stress. However, they caveated this by suggesting that resilience had a greater impact, in their study, on individuals who experienced low-level of stress, whereas, optimism had a greater positive impact on how those experiencing high levels of stress coped. Yet the authors did not define any form of a threshold for low- or high levels of stress. This notion of a positive outlook on the future links well with Werner and Smith's (2001) compensatory model of resilience which has already been discussed in this chapter.

The next most prominent element was that of Social Competence. This is how well individuals find engaging with new people, and whether they need to be with other people or are comfortable with their own company. As before, this element aligns well with that of the HSE-MSIT (2008) and Cartwright and Cooper (1997) who both state that relationships at work are important. Although the findings in this study suggest that relationships with peers are more important than those with managers. This was closely followed by Perception of Self. Interestingly, Cervera-Torres et al. (2021) found that how individuals engage with others depends on the levels of stress experienced. They observed that in general, women who reported higher levels of stress were more likely to engage with those around them. Men, however, showed the opposite pattern, preferring to isolate themselves from others. An important observation in Cervera-Torres et al. (2021) study to note, is the intention to self-isolate and the evidence is not sufficient to establish a causal relationship.

The combined domain of family and social support did have a particularly high impact on the stress score. This is particularly interesting given that resilience is not an innate feature of an individual's personality (UCL, 2014) and therefore has to be developed primarily within the family, but also through society's structures, notably schools. The notion of social support refers to the 'verbal and nonverbal communication between recipients' and 'that reduces uncertainty about the situation, the self, the other, or the relationship' (Albrecht & Adelman, 1987, p. 19).

The results here pose several questions to follow up, regarding the extent to which families are involved in the development of participants' resilience. An alternative hypothesis could be that individuals do not communicate about work-related issues with sufficient meaning to their partners. Carr and Kella (2017) argue that communicated support from a marital partner was particularly important in coping with work-related stress and for developing higher levels of resilience.

The data suggests that it is our internal perceptions of ourselves and our future that are the strongest contributors to how one manages workplace stress. This aligns with Bem's (1967) literature on self-perception theory, where certain situations lead to self-described attitudes that result from an individual's observations of their behaviour. Holland et al., (2002) state that an individual's attitude moderates both the extent to which their attitude predicts future behaviours and the extent to which it is impacted by past behaviour, both of which appear to have an impact on the level of workplace stress one experiences.

CONCLUSION

The findings of this research demonstrated that there is a relationship between levels of resilience and perceived levels of stress of middle leaders in education. This agrees with previous studies (Friborg et al., 2006; Lara-Cabrera et al., 2021) that state there is an inverse correlation between high levels of resilience and levels of work-related stress. Although little has been published in relation to leadership in education.

The study is a holistic view of middle leadership and the relationship between resilience and stress. Yet it is more nuanced than this. The findings of this research demonstrated that an individual's:

- Perception of the future,
- Ability to engage in social situations, either formally or informally (making casual conversation)
- Support from family and peers

are all key attributes in managing work-related stress. Interestingly, the support from peers featured amongst the work-related stress scores, compared to support from managers. This implies that peer support is possibly valued more than support from senior managers.

Therefore, for those in leadership positions within education the following should be considered as way of recommendations:

1. Provide staff with clear direction when it comes to asking for actions to be undertaken. Ensuring that you avoid any form of ambiguity.
2. Leaders need to be valued and they need to understand how their contribution helps to fulfil the vision of the organisation.
3. Recognise that staff will work outside of normal hours in order to get tasks completed. Recognise this and if time off cannot be provided, then at least thank them for the work they have done.
4. Allow staff, where possible the flexibility to have an element of control over their work. This will allow them to feel trusted and valued within the organisation.

As andragogical leaders we walk a tightrope between being too directing and autocratic at one end of the continuum or too hands-off at the other end of a continuum. At times, we will move more towards one end or the other, as the situation dictates. Yet the findings of this chapter imply that maybe we are a little to hands-off, resulting in middles perceiving a lack of clarity when they are provided with direction from more senior staff.

It is worthwhile noting that this study, and thus a limitation of it, was that it was completed towards the end of the Covid-19 pandemic, which resulted in imposed government restrictions on population movement and adaptations to working patterns. The effects of these environmental and organisational restrictions would undoubtably have tested an individuals' resilience and stress. This means that there is a strong argument for extending this study, ensuring that there is a more gender balanced sample in order to test the validity of the findings.

While the data is statistically valid, further research is needed to determine the extent to which the findings are correlated by sector. Given that most of the participants were from further and higher education, are there any notable differences between these two sectors, or indeed the school sector? As before, this is a limitation of the study which builds and argument that more work is needed to determine what the stress-causing factors are. While resilience can help to reduce workplace stress, it is a coping mechanism and it is more important to alleviate the causes rather than treat the symptoms.

REFERENCES

Albort-Morant, G., Ariza-Montes, A., Leal-Rodríguez, A., & Giorgi, G. (2020). How does positive work-related stress affect the degree of innovation development? *International Journal of Environmental Research and Public Health, 17*(2), 520. doi:10.3390/ijerph17020520 PMID:31947589

Albrecht, T. L., & Adelman, M. B. (1987). Communicating social support: A theoretical perspective. In T. L. Albrecht & M. B. Adelman (Eds.), *Communicating social support* (pp. 18–38). Sage.

Allen, T. D., Herst, D. E., Bruck, C. S., & Sutton, M. (2000). Consequences associated with work-to-family conflict: A review and agenda for future research. *Journal of Occupational Health Psychology, 5*(2), 278–308. doi:10.1037/1076-8998.5.2.278 PMID:10784291

American Psychological Association. (2014). *The road to resilience. American Psychological Association*. Retrieved from http://www.apa.org/helpcenter/road-resilience.aspx

Anyan, F., Hjemdal, O., Bizumic, B., & Friborg, O. (2020). Measuring resilience across Australia and Norway. *European Journal of Psychological Assessment, 36*(2), 280–288. doi:10.1027/1015-5759/a000509

Bem, D. J. (1967). Self-perception: An alternative interpretation of cognitive dissonance phenomena. *Psychological Review, 74*(3), 183–200. doi:10.1037/h0024835 PMID:5342882

Bennett, N., Woods, P., Wise, C., & Newton, W. (2007). Understandings of middle leadership in secondary schools: A review of empirical research. *School Leadership & Management, 27*(5), 453–470. doi:10.1080/13632430701606137

Bong, C. L., Fraser, K., & Oriot, D. (2016). Cognitive Load and Stress in Simulation. In V. J. Grant & A. Cheng (Eds.), *Comprehensive healthcare simulation: Paediatrics* (pp. 3–17). Springer International Publishing. doi:10.1007/978-3-319-24187-6_1

Buller, J. L. (2015). The only thing we have to change Is—change itself. *Change Leadership in Higher Education: A Practical Guide to Academic Transformation*, 1-27.

Cervera-Torres, S., Ruiz-Fernández, S., Godbersen, H., Massó, L., Martínez-Rubio, D., Pintado-Cucarella, S., & Baños, R. M. (2021). Influence of resilience and optimism on distress and intention to self-isolate: Contrasting lower and higher COVID-19 illness risk samples from an extended health belief model. *Frontiers in Psychology*, *12*, 662395. doi:10.3389/fpsyg.2021.662395 PMID:34108919

Chartered Institute of Personnel and Development. (2020). *Health and wellbeing at work 2022*. CIPD.

Christian, P., & Lolas, F. (1985). The stress concept as a problem for a theoretical pathology. *Social Science & Medicine*, *2*(12), 363–365. doi:10.1016/0277-9536(85)90444-7 PMID:4095592

Cohen, L., Manion, L., & Morrison, K. (2003). *Research methods in education*. Routledge Farmer.

Cooper, C. L., Dewe, P. J., & O'Driscoll, M. P. (2001). What is stress? In *Organizational stress: A review and critique of theory, research, and applications*. SAGE Publications, Inc., https://www.doi.org/10.4135/9781452231235.n1

Daniel, C. O. (2019). Effects of job stress on employee's performance. *International Journal of Business. Management and Social Research*, *6*(2), 375–382.

De Nobile, J. (2018). Towards a theoretical model of middle leadership in schools. *School Leadership & Management*, *38*(94), 395–416. doi:10.1080/13632434.2017.1411902

EU-OSHA. (2014). *Calculating the cost of work-related stress and psychosocial risks*. Publications Office of the European Union., doi:10.2802/20493

Field, A. (2013). *Discovering statistics using IBM SPSS statistics: And sex and drugs and rock "N" roll* (4th ed.). Sage.

Fleming, J., & Ledogar, R. J. (2008). Resilience, an evolving concept: A review of literature relevant to aboriginal research. *Pimatisiwin*, *6*(2), 7–23. PMID:20963184

Fletcher, D., & Sarkar, M. (2013). Psychological resilience: A review and critique of definitions, concepts, and theory. *European Psychologist*, *18*(1), 12–23. doi:10.1027/1016-9040/a000124

Friborg, O., Hjemdal, O., Rosenvinge, J. H., Martinussen, M., Aslaksen, P. M., & Flaten, M. A. (2006). Resilience as a moderator of pain and stress. *Journal of Psychosomatic Research*, *61*(2), 213–219. doi:10.1016/j.jpsychores.2005.12.007 PMID:16880024

Fribory, O., Barlaug, D., Martinussen, M., Rosenvinge, J. H., & Hjemdal, O. (2005). Resilience in relation to personality and intelligence. *International Journal of Methods in Psychiatric Research*, *14*(1), 29–42. doi:10.1002/mpr.15 PMID:16097398

Greenhaus, J. H., & Beutell, N. J. (1985). Sources of conflict between work and family roles. *Academy of Management Journal*, *10*, 76–88.

Haslam, S. A., & McGarty, C. (2003). *Research method and statistics in psychology*. Sage (Atlanta, Ga.).

Health and Safety Executive (HSE). (2008). *HSE Management Standards Indicator Tool*. HSE, Bootle.

Health and Safety Executive (HSE). (2022). *Working days lost in Great Britain*. https://www.hse.gov.uk/statistics/dayslost.htm

Holland, R. W., Verplanken, B., & van Knippenberg, A. (2002). On the nature of attitude-behavior relations: The strong guide, the weak follow. *European Journal of Social Psychology*, *32*(6), 869–876. doi:10.1002/ejsp.135

Kim-Cohen, J., & Turkewitz, R. (2012). Resilience and measured gene-environment interactions. *Development and Psychopathology*, *24*(4), 1297–1306. doi:10.1017/S0954579412000715 PMID:23062298

Krouse, S. S., & Afifi, T. D. (2007). Family-to-work spillover stress: Coping communicatively in the workplace. *Journal of Family Communication*, *7*(2), 85–122. doi:10.1080/15267430701221537

Lara-Cabrera, M. L., Betancort, M., Muñoz-Rubilar, C. A., Rodríguez Novo, N., & De las Cuevas, C. (2021). The mediating role of resilience in the relationship between perceived stress and mental health. *International Journal of Environmental Research and Public Health*, *18*(18), 9762. doi:10.3390/ijerph18189762 PMID:34574684

Lynch, K., Grummell, B., & Devine, D. (2012). *New managerialism in education commercialization, carelessness, and gender*. Palgrave Macmillan. doi:10.1057/9781137007230

Marcatto, F., Di Blas, L., Luis, O., Festa, S., & Ferrante, D. (2021). The perceived occupational stress scale: A brief tool for measuring workers' perceptions of stress at work. *European Journal of Psychological Assessment*. Advance online publication. doi:10.1027/1015-5759/a000677

Mauno, S., & Kinnunen, U. (1999). The effects of job stressors on marital satisfaction in Finnish dual-earner couples. *Journal of Organizational Behavior*, *20*(6), 279–895. doi:10.1002/(SICI)1099-1379(199911)20:6<879::AID-JOB982>3.0.CO;2-2

Office for National Statistics (ONS). (2021). *Demography question development for Census 2021*. https://www.ons.gov.uk/census/censustransformationprogramme/questiondevelopment/demographyquestiondevelopmentforcensus2021

Pietrzak, R. H., & Southwick, S. M. (2011). Psychological resilience in OEF-OIF Veterans: Application of a novel classification approach and examination of demographic and psychosocial correlates. *Journal of Affective Disorders*, *133*(3), 560–568. doi:10.1016/j.jad.2011.04.028 PMID:21624683

Polizzi, C., Lynn, S. J., & Perry, A. (2020). Stress and coping in the time of COVID-19: Pathways to resilience and recovery. *Clinical Neuropsychiatry, 17*(2), 59.

Quick, J. C., Wright, T. A., Adkins, J. A., Nelson, D. L., & Quick, J. D. (2013). *Preventive stress management in organizations* (2nd ed.). American Psychological Association. doi:10.1037/13942-000

Rees, T., Mitchell, I., Evans, L., & Hardy, L. (2010). Stressors, social support and psychological responses to sport injury in high-and low-performance standard participants. *Psychology of Sport and Exercise*, *11*(6), 505–512. doi:10.1016/j.psychsport.2010.07.002

Reivich, K., Gillham, J. E., Chaplin, T. M., & Seligman, M. E. (2013). From helplessness to optimism: The role of resilience in treating and preventing depression in youth. In S. Goldstein & R. B. Brooks (Eds.), *Handbook of resilience in children* (pp. 201–214). Springer. doi:10.1007/978-1-4614-3661-4_12

Robinson, A. M. (2018). Let's talk about stress: History of stress research. *Review of General Psychology, 22*(3), 334–342. doi:10.1037/gpr0000137

Roehling, P. V., Roehling, M. V., & Moen, P. (2001). The relationship between work-life policies and practices and employee loyalty: A life course perspective. *Journal of Family and Economic Issues, 22*(2), 141–170. doi:10.1023/A:1016630229628

Southwick, S. M., Douglas-Palumberi, H., & Pietrzak, R. H. (2014). Resilience. In M. J. Friedman P. A. Resick, & T. M. Keane (Eds.), *Handbook of PTSD: Science and practice* (2nd ed., pp. 590-606). Guilford Press.

Taylor, S. E., Lerner, J. S., Sage, R. M., Lehman, B. J., & Seeman, T. E. (2004). Early environment, emotions, responses to stress, and health. *Journal of Personality, 72*(6), 1365–1394. doi:10.1111/j.1467-6494.2004.00300.x PMID:15509286

Tusaie, K., & Dyer, J. (2004). Resilience: A historical review of the construct. *Holistic Nursing Practice, 18*(1), 3–8. doi:10.1097/00004650-200401000-00002 PMID:14765686

University College London (UCL). (2014). *Local action on health inequalities: Building children and young people's resilience in schools*. Public Health England.

Walker, B. (2019). *Finding resilience: Change and uncertainty in nature and society*. CSIRO PUBLISHING.

Walker, B. (2020). Resilience: What it is and is not. *Ecology and Society, 25*(2), 11. doi:10.5751/ES-11647-250211

KEY TERMS AND DEFINITIONS

Correlations: Correlation is a statistical measure that expresses the extent to which two variables (stress and resilience) are related (meaning they change together). It is a measure that is used to assess the relationship between, in this case stress and resilience, without making a definitive statement about the cause and effect.

Middle Leaders: Middle leaders are those individuals who report to the executive leadership team. Sometimes called tier 3 leaders, with tier 1 being the CEO, tier 2 being the senior or executive management or leadership team and tier 3 being the middle or operational leaders. They will often lead a function area of the organisation.

Resilience: Resilience is the ability to adapt to life's misfortunes and setbacks. Resilience helps you rebound from a setback or challenge, such as work-related stress, or a form of loss, such as a job loss or a bereavement. If you lack resilience, you might dwell on problems, feel victimized, become overwhelmed. Resilience won't make your problems go away—but resilience can give you the ability to see past them to better handle stress.

Stress: Stress is how we feel when under pressure or threatened. It happens when we are in a situation that we don't feel we can manage or control. When we experience stress, we can experience physiological responses, such as increased heartrate, or psychological responses such as feeling more worried or anxious.

Work-Related Stress: This is stress that is specifically experienced as a result of the workplace or the job role.

Chapter 11
IT Students as Design Leaders

Geraldine Torrisi-Steele
Griffith University, Australia

ABSTRACT

Graduates of computer science and related degrees represent the capability to continue the evolution of technologies and drive the future in emerging technologies. Some of these graduates will be technological leaders, innovating and pushing the boundaries of technological possibilities throughout their careers. Even though it is a popular opinion that 'the technical' and 'design' do not mix, the author's proposition is that graduates with technological degrees must also be 'design leaders. Design is imperative to meet the needs and desires of users; the design provides some solution to a problem in a user or human-centered manner. Equipping technological students with capabilities for both technological leadership and design leadership will help ensure a human-centered vision for our technological future.

INTRODUCTION

Undeniably, digital technologies are on the center stage of human society. For millions of people around the world, technology is present in daily existence. Digital technologies are everywhere, being used in a multitude of contexts from business to medicine, education, and industry. Artificial intelligence, machine learning, robotics, the internet of things – the future of technology is exciting and there are promises that technologies will reshape human activity. Not surprisingly then, many students envisage success and exciting technology careers. Hence, information technology and computer science degrees are popular choices for many students entering university. These degrees usually deliver the required technical knowledge in key areas such as programming, software development, databases, networks, security, etc. With a focus on technical knowledge, mention of 'design' is scarce. At best, the term design in the technical domains is associated with areas such as human-computer interaction, user interface design, and user experience. Within the author's experiences of teaching computers in technical degrees, the perception is that 'design' is almost in the 'opposite corner' to the technical world. The concept of design for most technical students is associated with aesthetics and for many of the technical student's world view is that design is not relevant to them. The authors' proposition is that design is highly relevant to degree

DOI: 10.4018/978-1-6684-7832-5.ch011

programs in information technology and science. In the present chapter, the author explores the idea of design and crafts an argument that, in the context of emerging technologies, graduates with information technology and computer science degrees must also be technology leaders but also design leaders.

DESIGN AND ANTICIPATORY DESIGN

While the topic of leadership in the business world has received a great deal of attention over the years and resulted in substantial literature, design leadership is only recently emerging as a dimension of leadership. Unfortunately, the concept of "design leadership" is muddied by the common use of the word 'design'. For many, the term design simply refers to the aesthetic appearance of something and conjures images of artworks and galleries. For others, the concept of design is more mature - "design is seen as the manipulation of visual or tangible aspects of physical matter or information at the point of output" (Celi & Rudkin, 2015, p. 61). At the core of the design is the imperative to meet the needs and desires of users; the design provides some solution to a problem in a user-centered way such that users can effectively and efficiently meet their goals.

Design meets technology in the field of Human-Computer Interaction (HCI)and its derivative fields. The advent and subsequent ubiquitousness of the personal computer and the accompanying graphical user interfaces provoked the emergence of the field of digital design. From a historical perspective, the field of digital design can be seen to have evolved from the field of HCI as well as the more traditional pen-and-paper field of graphic design.

The beginnings of HCI may be traced back to the National Bureau of Standards conference called Human Factors in Computing Systems which took place in 1982 (Carroll, 2001). This conference, together with the foundation laid by earlier work in cognitive science in the 1960s and 1970s (which related to areas such as software engineering, prototyping, and user interfaces) inspired the subsequent growth and popularization of the field of HCI (Carroll, 2001). HCI is concerned with usability (i.e. the ease with which humans are able to interact with a system in order to achieve their objectives) and has subsequently spawned the fields such as user interface design (UID), user interaction design (UxD), human-centered design, and most recently User experience design (UX).

Reconsider now the definition of design given earlier as "manipulation" of "aspects of physical matter or information at the point of output" (Celi & Rudkin, 2015, p. 61). Implicit in such a conceptualization of design is the notion that design is about problem-solving in relation to some already identified problem. Hence, it may be considered that the designer's task occurs only at the endpoint of a process. Indeed, a dominating perception of design has been that designers are responsible for delivering products appealing to consumers, including packaging, graphics, and logos (Han & Bromilow, 2010). In literature, though, we find that some authors encourage a redefinition of the design discipline in a way that highlights the capacity of design to be future-facing rather than simply being the solution or endpoint to a pre-determined problem. Futurists such as Zamenopoulos and Alexiou (2007) propose what they call an anticipatory view of design. From the anticipatory perspective, the discipline of design must go beyond merely generating a design solution for an apparent problem to communicating a vision of the future. No longer designed simply about output and an end-point, it is also about envisioning the future and being a catalyst for change and innovation (Celi & Rudkin, 2015).

Now more than ever, the anticipatory view of design is important. To understand why this is so, consider the present digital age. It is a time when technologies are infiltrating almost every aspect of existence, and the exponential growth of technology is in full swing, to such an extent that the term 'emerging technologies' has come into being and is a catchphrase of the present time.

LOOKING AT THE FUTURE: EMERGING TECHNOLOGIES

While exactly what is meant by the term "Emerging technologies" is open to interpretation, the definition adopted in the present paper is based on the World Economic Forum's Summit on the Global Agenda, 2010 held in Dubai in which emerging technologies are defined as either new technologies or innovative applications of existing technologies, and the essence of emerging technology is that it leads to the development of new capabilities and has some long-lasting impact on individuals, society, political and economic systems. Emerging technologies can simplify life or enable people to do things they've not been able to do before and have the potential to change behaviors. Emerging technologies are disruptive – meaning that the new technology or new application of the technology will fully replace the old one.

As the world continues along the curve of exponential technology growth and emerging technology is becoming the catchphrase of the present time, and likely to be in decades to come. Information technology and computer science graduates provide capabilities for industry, organizations, and technology research. Emerging technologies arise largely because someone has identified a problem or challenge which needs to be addressed. In their efforts to solve society's and industry's problems with technology, technology graduates inevitably shape the landscape of emerging technologies. When the latter statement is juxtaposed with the statement about anticipatory design: "No longer is design simply about output and an endpoint, it is also about envisioning the future and being a catalyst for change and innovation (Celi & Rudkin, 2015)", the rift between information technologists, computer scientists, and design narrows considerably.

Knemeyer (2015, p. 249) refers to a shift from "designers as creators", to "designers as facilitators". In the context of emerging technologies, the role of the designer is one of a facilitator of innovation, and as in Knemeyer's (2015, p. 249) view this will "accelerate in the emerging technologies fuelled decades to come". If programmers and other technology graduates are considered from the perspective put forward by Knemeyer (2015), then every programmer, networking and security specialist, database programmer, software developer, and so on, must also be considered as a designer. Technology graduates can be facilitators of technological innovation. Writing an API (application programming interface) is a design endeavor – it is coming up with a solution to a problem, it is something that will ultimately be used by a human being and it must be human-centered – the central premise of design.

The idea that programmers and digital technology developers are designers is not a commonly held view. It appears to be absent in formal literature – at least as far as a brief literature search on keywords related to design and programmers is conducted. However, the search for literature did uncover an article in Medium, under UX (user experience design). The article, written by Mikael Ainalem (2020) is titled: "Programmers are not designers, or are they?". Mikael questions: "Are these two roles so different that we have to believe?" (2020, para 1). He points out that historically, art and science were not so distinct from each other, with famous people like Da Vinci being both artists and engineers so that design and technical endeavor could not only co-exist but were "inseparable" (Ainalem, 2020). Mikael (2020) points out that in more modern times design and science have drifted far apart with claims that designers think differently from technical folk. To make the point, Mikael compares the daily task of a programmer, a designer, and UX designer Table 1)

Table 1. Similarities among a day in a life of a programmer, visual designer, and UX designer (Ainalem, 2020, para.12)

Programmer	Visual Designer	UX Designer
Starts with an empty text file	Starts with an empty canvas	Starts with an empty document/canvas/video
Code, change, refine, tweak.	Draw, change, refine, tweak, redo, rework, iterate.	Write, illustrate, change, refine, tweak, redo, rework, iterate, read, interview
Seek inspiration from external sources (e.g. Github)	Seek inspiration from external sources, e.g., Dribbble, Behance	Seek inspiration from external sources, e.g., Medium, blogs
Discusses with colleagues about trends, tools, algorithms, patterns	Discusses with colleagues about trends, tools, styles, patterns	Discusses with colleagues about trends, tools, interactions, UX patterns
Works with stakeholders to define requirements, estimates, go-to-market	Works with stakeholders to define requirements, estimates, go-to-market	Works with stakeholders to define requirements, estimates, go-to-market strategies, asset drops, handovers

In the context of emerging technologies, graduates of computer science and other technology degrees can and should be positioned as designers alongside being technologists. They should be encouraged to become visionary, and inspiring, and set a direction for technological change and innovation by designing technologies in a human-centered way. 'Vision', 'inspiration', and 'change' are, as Kotter (1999) explains essential elements of leadership.

Design Leadership

Organizations are increasingly valuing 'designers' in organizations. Since innovation in the present fast-paced world can give 'the edge' (Lapierre & Giroux, 2003), organizations are now beginning to view design as strategic and imperative to the survival of the organization (Han & Bromilow, 2010). Hence, growing capability in design leadership in technological graduates is highly desirable, if not necessary.

Historically, design leadership is about leading visual and product design and perhaps user experiences, and not apparently related directly to programming (although it is acknowledged that some digital designers also have skills in 'hard-core' technical areas such as programming). The notion of design leadership began to really take a foothold around the late 1990s, and early 2000 with the advent of the internet and the World Wide Web. It was at this time that designers began to position themselves more strategically within organizations and began seeking more input into the strategic direction of the organization they worked for (Joziasse, 2011a). In the present context where organizations are challenged by increased global competition and unstable economies, design is being recognized as essential for obtaining and maintaining competitive advantage, and thus design leadership is highly sought after and valued (Banfield, 2015; Turner, 2013; Han & Bromilow, 2010). For example, large organizations such as Phillips, 3M, Barclays, Pepsi have appointed chief design officers (Michlewski, 2015). Gloppen (2009) refers to design leadership as being a strategic undertaking by which design leverages the achievement of corporate goals. Given the rising importance of design leadership, it is worth noting that although much has been written about design in terms of aesthetics, usability, user experience, etc. Banfield (2015, p. vii) observes 'there's a lot of misunderstanding about what design leadership means" and subsequently leading design in an organization is at the present time both "challenging and confusing".

Seeking to define design leadership, a few authors have identified the skill set of effective design leaders. For example, Turner and Topalian (2002) along with Jozaisse (2011b) and Han and Bromilow (2010) maintain that design leaders have a vision for the future, develop strategic intent, direct design investment, make a positive impact on corporate reputation, nurture an environment supporting innovation, embed design awareness into the organization, and are inspirational.

In an effort to identify the skills of design leaders, Miller and Moultrie (2013) explored design leaders' perceptions of skills important to design leadership by conducting semi-structured interviews with practitioners from across seven large organizations based around the UK international fashion retail organizations. The results were organized into five categories: Design (inspire, imagine, envision, design, edit), Cognitive (draw, listen, verbalize, observe, learn, write), Interpersonal (motivate, nurture, persuade, negotiate), Business (structure, analyze, synthesize) and Strategic (identity, plan, connect, translate and maneuver).

Of these skills design leaders perceived as most important the design skills and closely aligned interpersonal skills (Miller & Moultrie, 2013). Not surprisingly, literature-based research conducted by Han and Bromilow (2010, p. 29) reveal the most demanded skills among graphic design leaders as being communication skills alongside" the ability to influence and manage interpersonal relationships between people and disciplines". Broadly speaking then, it seems that in the literature on design leadership, the identified skills and qualities of design leaders do not deviate greatly from the generic functions of leadership which are: establishing vision and direction, motivating and inspiring, and aligning people (Ren & Fangfang, 2014). It is desirable to consider what may be perhaps more specific leadership skills and qualities for design leaders in technological domains. Follett (2015) in his edited book Designing for Emerging Technologies, identifies eight tenets for designing for emerging technologies. Whilst Follett's tenets are not referred to specifically as leadership skills, they do however emerge from Follett's encouragement of designers to take up the opportunity to be drivers of change. urges designers "to help define the parameters of and sculpt the interactions between man and technology" (p. 8) and to become the arbitrators as well drivers of creation. Design leadership is thus implicit in Follett's view of what designers must be. To help designers fulfill their role Follett identifies eight key design tenets for emerging technologies:

1. Identify the problems correctly vision.
2. Learn constantly.
3. Think systematically.
4. Work at a variety of scales.
5. Connect people and technology.
6. Provoke and facilitate change.
7. Work effectively in cross-disciplinary teams.
8. Take risks responsibly.

It is now instructive to examine Follett's design tenets and consider how these tenets align with leadership to provide a useful framework for articulating desirable attributes of design leaders. Each of the above tenets is thus now discussed in light of the generic functions of leadership and of the principles of design leadership identified by various authors.

In the first tenet, Follett (2015) encourages designers to identify problems correctly. Designers must be problem identifiers, of significant and valuable problems, not just problem solvers, and designers must seek to "guide the technology" not merely allow it to "guide us" (Follett, 2015, p. 10). Follett's first tenet aligns clearly with the notion of vision in leadership, in fact, vision is generally considered the foundation of the leadership undertaking. In relation to vision, Chopra's (2010) states that defining a vision begins with looking and listening. In juxtaposing Follett's first tenet and Chopra's views, it may be said that design leaders should formulate a vision by looking and listening for significant and valuable problems, especially those that are significant to humanity. Not only will this facilitate the creation of vision but will more easily inspire and align others to the cause because there is a clear benefit to humanity. Extending on the theme of benefit to humanity, Follett's fifth tenet is connecting people and technology. As Follett (2015) so eloquently puts it "designers should provide the connective tissue between people and technology" (p. 19). Just as Chopra (2010) views awareness and consciousness as essential to leadership, Follett's fifth tenet requires that designers ask "why" this is being done, not just the "what" of what technology can

Design Leadership in the Context of Emerging Technologies

A clear vision embodies a direction and establishing direction, and navigating to a direction requires an understanding of the issues and challenges. In the context of emerging technologies, this means that designers must understand not only the technology but also the implications of technology for people. As Joziasse (2011) advises, design leaders, must" be aware of the questions that underlie every challenge". To do this, occurring, design leaders must acquire the capacity to learn constantly (Follett's second tenet) if they are to gain such understanding and awareness that is required to set vision and direction in a dynamic technological context. Furthermore, to set a direction design leaders must be capable of both thinkings systematically (tenet 3), and working at a variety of scales (tenet 4). Design leaders will need to become system thinkers to truly understand and gain awareness of context. This is because emerging technology is increasingly being embedded and connected into ecosystems; it is being integrated into both our daily activities and into the physical objects, both natural and artificial, which surround us. Furthermore, design in emerging technologies is happening at both macro and micro levels. Design leaders must fully understand and be able to work comfortably at system levels, and in micro and macro domains be able to set viable direction.

The value of Follett's sixth tenet, provoke and facilitate change, firmly places design leadership within the sphere of transformational leadership. Transformational leadership is about progress and development. Transformational leaders tend to lead by inspiring others to follow. Successful transformational leaders work towards gaining trust and fostering a safe non-threatening environment wherein visions can be shared. Avolio et al. (1991) identify the characteristic of transformational leadership as an idealized influence (role models admired and respected), inspirational motivation (providing meaning to the work, creating enthusiasm and optimism), intellectual stimulation (encouraging innovation and creativity), and individualized consideration (acceptance of individual differences).

Applying Follett's sixth tenet to design leadership, the advice is sound: "recognize when things are not working and rather than smooth things over promote wholesale change" (Follett, 2015, p. 20). Follett's sixth tenet also has implications for the leadership function of aligning people to a vision. Emerging technologies are usually disruptive, and people may be uncomfortable in the face of 'seemingly unchecked advances' (Follett, 2015, p. 20). The design leader must understand this, and be empathetic to it as they begin discussion and lead the transformation.

Follett (2015) maintains that designing for emerging technologies necessitates working effectively in cross-disciplinary teams (tenet 7). The complexity and challenge of working with emerging technologies are simply too great for any one person to master. Emerging technologies draw on a few disciplines from engineering and electronics to the sciences and such a broad knowledge base is simply beyond the capacity of any one person. From a design leadership viewpoint, the implication of tenet 7 is that collaborating and leading with a cross-disciplinary team is of paramount importance. The ability to align team members from diverse disciplinary backgrounds will be an obvious challenge since the diversity of backgrounds, terminologies and paradigms of different team members result in conflict. The challenge faced by leaders of multi-disciplinary teams is to espouse positive relationships, especially using interpersonal skills and communication skills to enable maximum comprehension and understanding among all team members (Reuveni & Vashdi, 2015). It is important to create a shared mental model of the tasks at hand if conflicts and misunderstandings are to be avoided. (Reuveni & Vashdi, 2015).

Finally, Follett (2015) in the eighth tenet urges designers to Take risks responsibly. The disruption caused by emerging technologies means that moving forward is inherently risky. Thus, design leaders must be risk-takers but they must do so after carefully evaluating the benefits and consequences.

Evidently, Follett's (2105) eight tenets for design in emerging technologies offer a valuable framework for better understanding the desirable capacities of a design leader. Taking all of the eight tenets together, Follett's tenets, with their emphasis on understanding and empowerment, appear conducive to what Turner (2010) calls the transformational synergy that is necessary to realize the full potential of design, that is, a balance between design pushers, who use design to make strategy tangible, and design pullers, who create an understanding of the possibilities and demonstrate the value of the design.

Hand-in-hand with bringing design and design leadership to bear on students' learning experiences in technological degrees is the design thinking mindset. Design thinking is the combining of "empathy, creativity, and rationality to analyze and fit solutions to particular contexts" (Wrigley & Straker, 2015); It is a "sharing of empathetic insights" such that all participants are immersed in the design process (Gasparini, 2015). Design thinking pivots on empathy – on understanding the worldview of others. While design thinking is extensively applied to product design, design thinking is gaining momentum as a leadership philosophy, especially in leadership for digital transformation (Gusakov, 2019) (Himsworth, 2018) (Appleyard, Enders, & Velazquez, 2020). In parallel to the call for design thinking as a leadership philosophy, is the recognition that a design thinking mindset is useful in teaching computer science because 1) design thinking is a valuable platform for developing the problem-solving capabilities needed by programmers and 2) problem solving in computer science and its related discipline requires an element of creativity and innovation (Schmalfeldt, 2022). In the context of emerging technologies, it is important to cultivate leaders who will drive innovation, address meaningful problems, and do so with a human-centered mindset. In addition to Follett's (2015) eight tenets as a framework for design leadership in technological disciplines, the design thinking mindset should be cultivated as a guiding philosophy for technological leaders who are also developed design leaders.

CONCLUSION

Emerging technologies are infiltrating almost every aspect of our activities and technologies are being embedded in an array of objects in our natural and built environment and are disrupting well-established. ways of doing things as well as creating the ability to do things that could never have been done before.

Graduates of computer science and related degrees represent the capability to continue the evolution of technologies and drive the future in emerging technologies. These graduates must possess the knowledge and leadership skills to move digital technologies. They must be technology leaders but, they must also be design leaders because they must propel technology forward by identifying the problems of human society and developing ethical, human-centered technologies to address these problems. Recognizing that programmers and other technologists are designers, and embedding this view into technical programs of study is seen as essential to supporting the development of emerging technologies that will serve the needs of humanity in a positive way because, design, by its definition, is an endeavor to solve a problem to meet a human need. This helps to humanize technology and helps every programmer and technology specialist that although their knowledge is focused on technology, they are working to create products for human use. By virtue of the functionalities and affordances that coders and other technologists build they are, in some way, either constraining or enabling human activity. When technology leaders are also design leaders, the techno-centric view is mediated by the human-centered view and technologies lose the illusion of neutrality.

Unfortunately, not much has been written about design leadership and the topic of design leadership remains relatively unexplored (Turner, 2013). Follett's eight principles of design for emerging technologies (identify the problems correctly, learn constantly, think systematically, work at a variety of scales, connect people and technology, provoke and facilitate change, work effectively in cross-disciplinary teams, take risks responsibly) provide a useful framework for identifying the qualities and capacities of effective design leaders. The eight principles are applicable to designers, in the traditional sense of the word, and applicable to designers that are programmers and technologists. However, what little is written about design leadership tends to be generic, referring to the general qualities of leaders with little to guide the development of design leadership. If technology is serving the good of society then there is a need for discourse around design leadership – what it means and how it might be enacted by technologists.

REFERENCES

Ainalem, M. (2020). *Programmers are not designers, or are they?* https://uxdesign.cc/programmers-are-not-designers-or-are-they-7c0cbe5fca7

Appleyard, M., Enders, A., & Velazquez, H. (2020). Regaining R&D leadership: The role of design thinking and creative forebearance. *California Management Review*, *62*(2), 12–29. doi:10.1177/0008125619897395

Avolio, B., Waldman, D., & Yammarino, F. (1991). Leading in the 1990s: The four Is of transformational leadership. *Journal of European Industrial Training*, *15*(4), 9–16. doi:10.1108/03090599110143366

Banfield, R. (2015). *Design leadership*. O'Reilly Media, Inc.

Carroll, J. M. (2001). *Human-computer Interaction in the new millennium*. Addison-Wesley.

Celi, M., & Rudkin, J. (2015). Towards user involvement in envisioning practices: Opening the use of the future. In D. Bihanic (Ed.), *Empowering users through design: Interdisciplinary studies and combined approaches for technological products and services* (pp. 59–78). Springer., doi:10.1007/978-3-319-13018-7_4

Chopra, D. (2010). *The soul of leadership: Unlocking your potential for greatness*. Harmony Books.

Follett, J. (2015). Designing for emerging technologies. In J. Follett (Ed.), *Designing for rmerging technologies: UX for genomics, robotis, and the iInternet of things* (pp. 1–26). O'Reilly Media, Inc.

Gasparini, A. (2015). Perspective and Use of Empathy in Design Thinking. In *ACHI 2015: The Eighth International Conference on Advances in Computer-Human Interactions*. IARIA.

Gloppen, J. (2009). Perspectives on design leadership and design thinking and how they relate to European. service industries. *Design Management Journal, 4*(1), 33–47. doi:.00005.x doi:10.1111/j.1942-5074.2009

Gusakov, A. (2019). The role of design thinking for achieving leaders in the digital transformation of business. *Advances in Social Science, Education and Humanities Research, 441*, 356–362.

Han, Y. J., & Bromilow, D. (2010). Graphic designers win the leadership game. *Design Management Journal, 5*(1), 20–31. doi:.00011.x doi:10.1111/j.1948-7177.2010

Himsworth, J. (2018, June 19). *Why design thinking should also serve as a leadership philosophy*. https://www.forbes.com/sites/forbesagencycouncil/2018/07/19/why-design-thinking-should-also-serve-as-a-leadership-philosophy/?sh=861c66a5a90f

Jozaisse, F. (2011b). Design leadership: Current limits and future opportunities. In R. Cooper, S. Jung-inger, & T. Lockwood (Eds.), *The handbook of design management* (pp. 398–413). Berg. doi:10.5040/9781474294126.ch-025

Joziasse, F. (2011a). The soul of design leadership. *Issue Design Management Review Design Management Review, 22*(3), 34–42. doi:.00138.x doi:10.1111/j.1948-7169.2011

Knemeyer, D. (2015). The changing role of design. In J. Follett (Ed.), *Designing for emerging technologygies* (pp. 427–438). O'Reilly Media Inc.

Kotter, J. P. (1999). *John Kotter on what leaders really do*. Harvard Business School Press.

Lapierre, J., & Giroux, V. (2003). Creativity and work environment in a high-tech context. *Creativity and Innovation Management, 12*(1), 11–23. doi:10.1111/1467-8691.00263

Michlewski, K. (2015). *Design attitude*. Gower.

Schmalfeldt, T. (2022). Design thinking in computer science edcuation. In I. Schmidberger, S. Wippermann, S. Stricker, & U. Müller (Eds.), *Design thinking im bildungsmanagement* (pp. 147–159). Springer. doi:10.1007/978-3-658-36951-4_9

Sperka, M., & Stolár, A. (2005). Graphic Design in the Age of Interactive Media. In *3rd International Symposium of Interactive Media Design*. Yeditepe University.

Turner, R., & Topalian, A. (2002). *Core responsibilities of design leaders in commercially demanding environments*. Design Leadership Forum, organized by Alto Design.

Wrigley, C., & Straker, K. (2015). Design Thinking pedagogy: The Educational Design Ladder. *Innovations in Education and Teaching International*.

Zaleznik, A. (1992). *The Harvard business review on leadership: ''Managers and leaders: Are they? different?* Harvard Business School Press.

Zamenopoulos, T., & Alexiou, K. (2007). Toward an anticipatory view of design. *Design Studies*, *28*(4), 411–436. doi:10.1016/j.destud.2007.04.001

KEY TERMS AND DEFINITIONS

Anticipatory Design: Zamenopoulos and Alexiou (2007) propose what they call an anticipatory view of design. From the anticipatory perspective the discipline of design must go beyond merely generating a design solution for an apparent problem to communicating a vision of the future. No longer is design simply about output and an endpoint, it is also about envisioning the future and being a catalyst for change and innovation (Celi & Rudkin, 2015).

Digital Designer: The role title of the digital designer is being used to refer to design roles that are broad and multidisciplinary and encompass both the elements of visual communication design and human-computer interaction. Alongside these design skills, digital designers are also adept in technical skills including coding for websites and mobile apps.

Emerging Technology: A new technology or innovative use of an existing technology that makes some lasting impact or change on how a task is performed.

Graphic Design: The use of visual language of images (i.e., drawing) can be used to communicate to individuals, communities, and societies. Traditionally graphic designers worked with pen and paper to produce images designed for effective communication of ideas, whether it was for educational, industrial, entertainment, or business purposes. The role of the graphic designer has expanded with the advent of digital technologies to include technical skills in areas such as web development and digital image manipulation, and the user experience of the interface and interaction with technology.

Human-Computer Interaction (HCI): HCI is concerned with usability (i.e., the ease with which humans are able to interact with a system in order to achieve their objectives) and has subsequently spawned the field of user interface design (UID), user interaction design (UxD), human-centered design, and most recently User experience design (UX).

Chapter 12

Team Creativity by Integrating Knowledge Management: A Guide for Andragogical Leadership

Xiuxiu Wu
Hong Kong Polytechnic University, Hong Kong

Kin Wai Michael Siu
The Hong Kong Polytechnic University, Hong Kong

Jörn Bühring
The Hong Kong Polytechnic University, Hong Kong

Caterina Villani
The Hong Kong Polytechnic University, Hong Kong

ABSTRACT

Knowledge management (KM) closely relates to team creativity and is a critical strategy for businesses to maintain competitiveness. The synergy between KM and team creativity benefits and empower andragogical leadership by increasing team engagement, which results in organizational success. Prior studies examined the KM models, processes, systems, and components that influence team creativity. However, the integration of KM into facilitating team creativity is yet to be fully understood. Therefore, this study provides a rationale for the relationships between KM and team creativity. Based on this interplay, the authors create a novel framework that serves as a practice guideline that aggregates KM models and team creativity components. The framework aims to assist andragogical leadership in effectively and purposefully managing teams while considering the major challenges of practice. The study goes beyond the boundaries to theoretically connect KM and team creativity, which benefits both academic research and the application of KM in business.

DOI: 10.4018/978-1-6684-7832-5.ch012

INTRODUCTION

Andragogical leadership is a significant research area in education and in business (Jeanes, 2021; Wang & Torrisi-Steele, 2022). Andragogy refers to the art and science of assisting adults in their learning (Knowles et al., 1998). Andragogical leadership requires a growth mindset to learn and solve problems actively. Andragogical leaders believe that employees can handle problems and achieve self-realization (Jeanes, 2021).

In today's competitive context, where technology and knowledge are changing rapidly every day, organizations must be innovative, creative, and capable of making changes (Wang & Torrisi-Steele, 2022). To empower andragogical leaders in this context, knowledge management (KM) is the starting point to turning the company and the employees into learning organizations and individuals, which is the primary goal of andragogical leaders. KM entails the creation, storage, transmission, and use of knowledge in organizations (Alavi & Leidner, 2001). More practically, KM manages knowledge generation and use, including identifying knowledge, managing it, and designing strategies to harness knowledge successfully in any company (Vaio et al., 2021; Crane, 2016).

Regarding andragogical leadership, KM procedures, and practices may help stakeholders to be more forward-looking and more actively engaged in innovation. In practice, KM has also been emphasized through its implementation of three creative activities to maximize or exploit knowledge (Kazanjian & Drazin, 2012): product line extensions, new platform development, and new business creation. Product line extensions are widely observed in software and hardware companies. For example, IBM, Apple, Dell, and others started to sell the first generation of hardware products (laptops), then gradually expanded their product line based on the accumulated knowledge and user feedback. This process heavily relies on knowledge sharing among team members. When businesses target a new industry or technological area, they create a new product platform to serve as the framework for product line extensions by leveraging related advantages of economical scope and resource sharing. New business creation is when a firm embarks on an entirely new and unrelated market with existing or new technology.

Overall, leveraging and utilizing knowledge in the business process and applying it directly to new applications has distinct advantages for improving efficiency (Ode & Ayavoo, 2020). In fact, it is impossible to keep organizational competitiveness that depends on unsystematic knowledge accumulation or completely relies on the human brain. However, if KM applies to the whole organization, it is challenging to meet different department requirements. Hence, KM could be conducted and used in teams to improve creativity and efficiency by developing a standard protocol (Rong & Liu, 2021; Johnson,2016).

On the same line of research, a comprehensive framework for leadership is provided by Price (2004), cited by Callahan and Grunberg (2018). In this framework, leadership functions at four levels: personal, interpersonal, team, and organizational. According to the principles of this framework, managing a team and understanding the team's dynamics is a core competence for leadership, as one individual is not equipped with all the knowledge and experience needed to solve the issues (Wang & Torrisi-Steele, 2022). As a result, the attention to teams has increased in psychology and management (Tang & Schmidt, 2021; Kratzer & Mrożewski, 2021). Research reveals that team creativity is a collection of fresh and practical ideas developed at the team level (Dong et al., 2017). Team creativity is essential for andragogical leadership to consider, as it facilitates value contribution and the competence to solve uncertain problems (Peng et al., 2019). Moreover, the success of an enterprise is fundamentally dependent on team creativity (Kratzer & Mroewski, 2021). Undoubtedly, significant moves can be made by seeing the importance of team creativity for andragogical leadership, as the advantage of team creativity has surpassed individual creativity (Rong & Liu, 2021).

The benefit of KM in teams is the management of complexity, which improves team resilience and creativity. Resilience is not just an individual trait but also a characteristic of situations that allows employees to react quickly when faced with problems or difficulties (Cristancho, 2016). Notably, knowledge is the main input for creativity in a collaborative setting since it enables the recombination, restructuring, and heuristic search for new knowledge and information (Ohlsson, 2011). Accordingly, team creativity is greatly stimulated by KM (Rong & Liu, 2021), and KM is the predictor of creativity in teams (Johnson,2016).

Irrespective of the company requirements for KM concerning team creativity, it also has a tenuous connection with the recent developments in machine learning and artificial intelligence, such as ChatGPT and LaMDA. As more technologies are accessible to solve problems, this tendency may impact team creativity. Yet, AI presents benefits and concerns, as Noam Chomsky (1928) noted in a New York Times piece published in 2023. On the one hand, it solves daily problems. Another side is that it brings false information into technical development and human cognition. ChatGPT or other AI tools are programmed to filter data and offer descriptions and predictions. They can answer what is, what was, and what will be. But true intelligence offers what is not possible, what might be conceivable, and what is impossible. This connects with human minds because they are adaptable, elegant, and efficient at processing fewer data to produce connections and explanations. More importantly, andragogical leadership requires moral judgment, but AI can merely provide generalized, objectionable, apathetic reactions by blindly following the code without taking a stance. Also, team creativity has a deep foundation in the humanities, where feeling, emotions, and social interaction play crucial roles that AI finds difficult to match. Considerably, ChatGPT and relevant tools are unable to balance the demand between creativity, which is flexible, divergent, dynamic, and the constraints, which are programs, codes, and algorithms.

For this reason, KM assists andragogical leaders or teams to capture, share and reflect their knowledge which forms the foundation for creativity. Furthermore, leaders must learn to anticipate the innovation challenges and opportunities emerging over the medium and longer term while concurrently, building sustainability-minded, futures literate, highly collaborative, and creative capabilities needed to deal with the demands of rapid social, technological, and environmental change to remain feasible in the long-term (Bühring & Liedtka, 2018, pp. 134-135). In this context, maximizing team creativity by exploiting their knowledge is a pressing matter, as the knowledge generated by teams is crucial implicit capital. Although, andragogical leaders trust and support their subordinates completely. They are companions when employees need help (Wang & Torrisi-Steele, 2022). Regardless, this trust should not be blind but based on evidence. Implementing KM in a team is an effective way to give more autonomy through knowledge sharing, storage, and transformation. The finding supports this debate that leaders' crucial responsibility is to create a learning environment that empowers workers to exploit their unlimited potential for learning (Nuckles, 2000). KM generally provides structure and direction to the team tasks, while andragogical leaders can act as a coach (Wang & Torrisi-Steele, 2022).

This chapter investigates the relationship between KM and team creativity in the context of andragogical leadership. First, we briefly discuss the KM cycle, models, and team creativity, identifying the sole research area. Second, we identify components that have been suggested to influence team creativity involved in KM. Third, we develop the framework for andragogical leaders to conduct KM that facilitates team creativity. Meanwhile, the challenge is also addressed.

KNOWLEDGE MANAGEMENT

Knowledge encompasses cognition, judgment, or skill acquired via study, practice, or investigation (Vaio et al., 2021). The two most common types of knowledge are tacit and explicit (Polanyi, 1966). Tacit knowledge is hard to verbalize and transfer to others, such as memories, beliefs, and motivations. This knowledge can be diffused by imitation, practice, or observation. Moreover, tacit knowledge is challenging to articulate as it involves personal feelings. While it is a valuable intangible asset for organizations, it is hard to share, learn and transmit. Explicit knowledge is structured and can be easily shared with others by using documents, meetings, or symbols.

KM is the management of knowledge creation and application to make knowledge more sustainable by facilitating decision-making (Vaio et al., 2021). Therefore, developing a quantitative and qualitative knowledge system in the organization would enable it to update and add to its knowledge in an ongoing manner, thereby accumulating and integrating both personal and organizational knowledge continuously. As a result, enterprises can make the right decisions and adapt to the changes in the market (Adaileh et al., 2020; cited by Rong & Liu, 2021). In addition, knowledge management is an essential component of enterprise management and a crucial skill for managers (Ode & Ayavoo, 2020).

Knowledge Management Cycle

The knowledge management cycle (KMC) identifies the core phases: knowledge sharing and distribution, capture and storage, creation, application, learning, and update (Pereira et al., 2021). The Evans et al. (2014) model was analyzed as an example to illustrate the KM cycle: knowledge identification and creation, storage, sharing, application, learning, and improvement. The five phases comprise a non-linear process to guide knowledge creation and conversion, summarized below.

1. Identify or create valuable tacit and explicit knowledge by scrutinizing documents, observations, and meetings according to organizational goals, cultures, and metrics. For example, when a team meets a new request to solve a problem or make a decision, team members must search for the relevant knowledge that is held. In case knowledge is missing, the knowledge will be created or acquired from external resources. In this phase, knowledge diversity may affect the team's creativity as various resources are needed to solve complicated problems, which informs the team members to generate more ideas or offer special perspectives. According to Hou et al. (2021), knowledge-based diversity is closely related to team creativity. It predicts positive team performance compared with other types of diversity, such as the diversity of demographic and ventures.
2. To store the valuable knowledge based on the assessment made in the last stage. The process involves codifying the knowledge so employees can efficiently approach, edit, update, and share it. Usually, the storage is achieved by an IT-based system that enables retrieval and update. The challenge is that tacit knowledge can hardly be codified, but explicit knowledge can be codified as written documents and procedures for later tasks. However, the stored knowledge must be used in practice to improve productivity. As Zaim et al. (2019) reported, knowledge storage mediates the influence of knowledge creation and sharing on knowledge utilization. In addition, knowledge storage is the foundation for later knowledge utilization.

3. Knowledge sharing refers to the process of exchanging and transferring knowledge to solve problems among teams. It can be significantly affected by intrinsic motivation and organizational knowledge-sharing practices (Natu & Aparicio, 2022). Intrinsic motivation refers to team members who perform well because their interests and satisfaction drive them to do so without needing external rewards. Members who are driven by intrinsic motivation intend to share their knowledge enthusiastically. Thus, members can learn from each other. Organizational knowledge-sharing practices indicate how organizations manage and facilitate their knowledge-sharing. Knowledge-sharing behavior varies depending on the context. Managers must consider their practice and strategy in light of their unique cultures and departments. To some degree, the opposite of knowledge sharing is knowledge hiding, which refers to team members intentionally concealing their relevant knowledge when colleagues seek support (Connelly et al., 2012). Organizations suffer several negative consequences as a result of knowledge hiding. It increases distrust during team collaboration and deters creativity (Arain et al., 2020). Moreover, Chatterjee et al. (2021) further indicated that knowledge hiding decreases the performance of the whole organization. Fortunately, leadership is a profound factor in preventing knowledge hiding and promoting knowledge sharing, such as the manager's honesty, patient listening, and positive assumptions about subordinates. Moreover, unnecessary conflicts could be decreased, and creative outcomes could be generated by team members' knowledge of diversity (Hou et al., 2021).

Figure 1. Knowledge management cycle (KMC) model
Source: Evans et al. (2014).

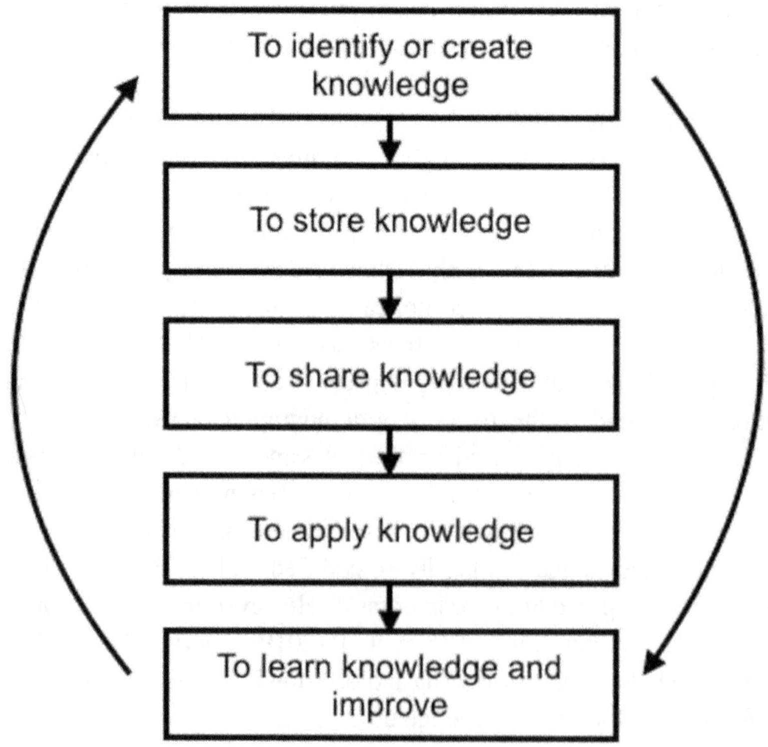

4. Knowledge activation and application is the most crucial step in facilitating the employee's daily work. If the knowledge is not utilized properly, it may lead to misleading guidance. As Alavi and Leidner (2001) suggest, knowledge sharing and creation do not guarantee improved organizational performance, but effective knowledge application does. Zaim et al. (2019) also examined how knowledge utilization greatly influences organizational KM performance. To some degree, the success of KM depends on the organization's capacity to utilize and leverage the knowledge resources that are captured and stored. In addition, knowledge utilization partly mediates the connection between knowledge creation and KM performance. As a result, knowledge utilization is the primary mechanism for the rest of knowledge creation, sharing, and capture, which impacts KM's effectiveness (Ode & Ayavoo, 2020).
5. Learn knowledge and improve: once the knowledge is applied, it is a good chance for employees to reflect on and refine the existing knowledge by reconstructing, integrating, connecting, internalizing, and externalizing knowledge (Evans et al., 2014). In the future, identified valuable knowledge can be refined and leveraged, but irrelevant knowledge must be excluded. The common approaches that assist in this stage include debriefing, reflection, and meetings. Knowledge reflection will be triggered by instructions, failure, dissatisfaction, and break-up routines (Schley & Woerkom, 2014). For example, when routine knowledge is inadequate to cope with new problems, employees reflect on what they know to transfer subconscious tacit learning into conscious awareness (Rausch, 2012). Therefore, andragogical leaders offers challenging tasks as a plausible way to encourage knowledge reflection.

Knowledge Management Models

To understand the process of knowledge creation, storage, and application, four KM models have been summarized that are highly relevant to this research: the SECI model (socialization, externalization, combination, and internalization) by Nonaka (1995), Radermacher's knowledge model (1996), the I-space model (Boisot & Cox (1999), and the knowledge maturity model (Pereira et al., 2021). These modes are fundamental inspirations for this study, which aims to generate KM guidance for the andragogical leaders practice that could facilitate the team's creativity.

SECI model. Nonaka and Takeuchi (1995, p. 72) proposed the SECI model, a spiral framework to show knowledge conversions based on tacit and explicit knowledge: socialization, externalization, combination, and internalization. SECI model is the most influential theory for studying KM, creativity, and innovation.

1. Socialization refers to the synthesis of tacit knowledge between individuals' social interactions, a process of sharing experiences to create new tacit knowledge from tacit knowledge. It is a way to get tacit knowledge without using language but by observation, imitation, and practice. "Sympathized knowledge" will be generated in this type, such as shared mental models and technical skills.
2. Externalization transforms tacit knowledge into explicit knowledge; it articulates knowledge into explicit concepts through metaphors, analogies, collective reflection, and hypotheses.
3. The combination is the explicit knowledge generated from synthesizing multiple explicit sources of knowledge. Combination systemizes the explicit knowledge into a knowledge system using documents, meetings, and other media. It creates systematic knowledge in areas such as prototypes and ergonomics.

Figure 2. Knowledge spiral
Source: Adopted from Nonaka and Takeuchi (1995, p. 72).

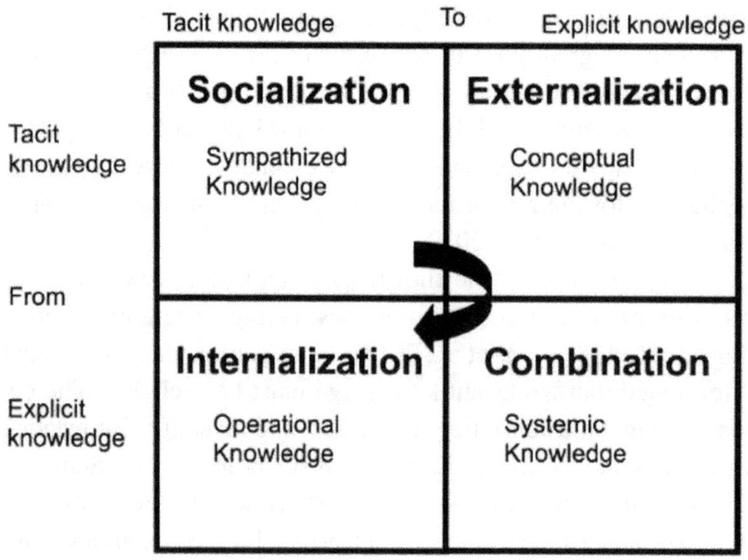

4. Internalization is the transfer of explicit knowledge into tacit knowledge related to learning by doing. At this stage, "operational knowledge" is generated about project management, new product development, and policy implementation.

Radermacher's knowledge model. Radermacher's knowledge model (1996) suggests that knowledge is a broad concept that contains all human behaviors and actions. The model includes four levels of knowledge based on general knowledge processing and inspired by the biological system. The knowledge processes consist of four levels and three transformations between the four levels. The A level is physical or embodied knowledge that contains the physical or signal resources expressed through the physical formation. Level B includes the tacit or neuronal knowledge that processes input using dynamic neural networks or holistic methods. All knowledge that facilitates human body movement is at this level. Level C is symbolic knowledge expressed by language, logic, and symbolic materials. Level D is the knowledge that mathematical or scientific models can express. Four levels are evolving and dependent on each other.

Knowledge representation can be transformed from level A to level B by adjusting the selected signals. From level B to level C, the transaction is possible by understanding the clear concepts of the situation and mental states. Tacit knowledge is codified and externalized through moving from B to C. Moving from C to B, explicit knowledge becomes unconscious and tacit. Lastly, the structured and systematical model will be generated from moving C to D through accumulated knowledge.

I-Space model. Boisot and Cox (1999) developed the I-Space model, a three-dimensional dynamic framework for analyzing knowledge that combines structuring and sharing and divides it into three dimensions:

Figure 3. Four-level view on knowledge
Source: Based on Radermacher (1996).

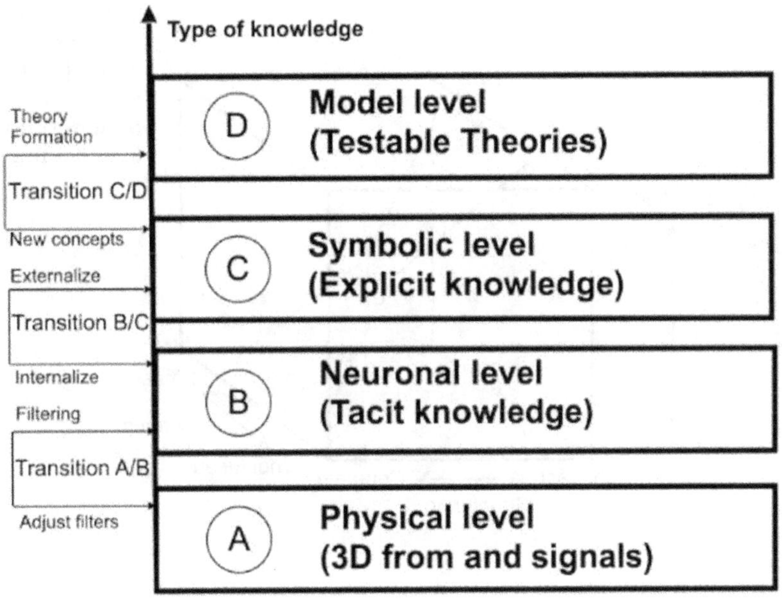

1. Codification is the quickness and simplicity with which a phenomenon or experience object can be assigned to a set of perceptual or mental categories. The degree of the categories' codification is indicated by how quickly and easily assignments may be made.
2. Abstract measures the amount of perceptual or conceptual categories to describe or capture a phenomenon. Greater generality and higher abstraction of the phenomenon depend on having fewer categories.
3. Diffusion refers to a specific group that has timely access to relevant knowledge. The proportion of the population with access to knowledge increases the degree of diffusion.

Each dimension includes two types of knowledge: uncodified (tacit) and codified (explicit), abstract and concrete, and diffused and undiffused. New knowledge emerges from the social learning process through knowledge structure and sharing.

Phases:

1. Threats and opportunities are identified by scanning for relevant and valuable knowledge.
2. Codification gives structure and a definite frame for problems.
3. Abstraction is the process of broadening the scope of codified knowledge.
4. Diffusion is sharing codified and newly created insights with a wider population.
5. Absorption applies knowledge to different situations in a learning-by-doing format.
6. Impacting refers to embodying knowledge into concrete practices and physical artifacts.

Figure 4. I-Space model
Source: Based on Boisot and Cox (1999) and Thoring et al. (2022).

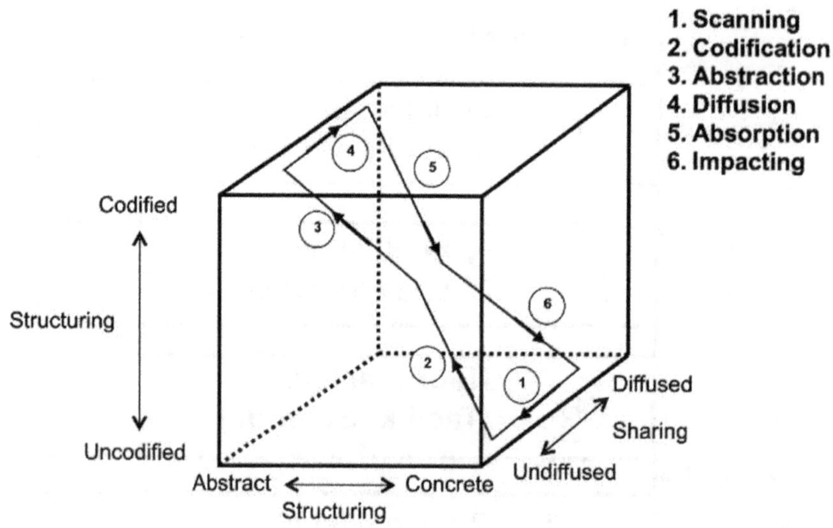

Knowledge Maturity Model

The models reviewed above help andragogical leaders understand the definition and dimensions of knowledge. Managers in different industries can choose the model that suits them to deploy knowledge management. However, the important note for practice is to measure the current KM maturity before implementation. Pereira et al. (2021) proposed KM stages to assess KM maturity and identify the stages and main steps to develop highly efficient KM in project-oriented organizations, which andragogical can use as a reference to evaluate their maturity stage and thus implement appropriate strategies.

1. The starting point is when employees understand the concept of KM, but procedures must still be applied.
2. The second level is realizing the value and applicability of KM to the organization.
3. Third, the organization has set up some fundamental infrastructure.
4. Fourth, KM processes receive systematic support, are monitored, and are evaluated as a component of organizational culture.
5. Lastly, these procedures are continuously enhanced and included in the external network.

In summary, all four models explicitly address KM research and practice. In the following section, we look at the relationship between KM and team creativity.

Figure 5. Knowledge management maturity models
Source: Pereira et al. (2021).

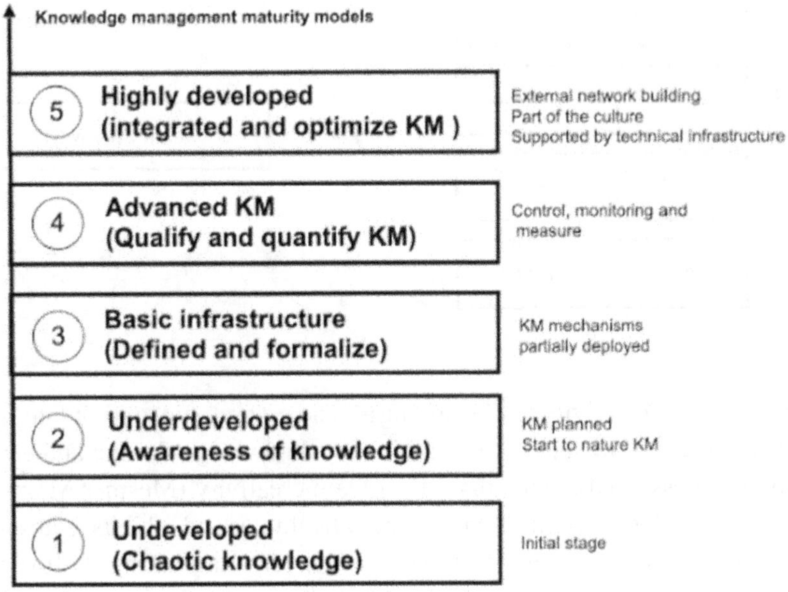

TEAMS AND TEAM CREATIVITY

As evidenced by the growing importance of teams within organizations (McKay et al., 2021), teams are the new expected trend in creative and innovative work. Organizations and institutions depend on teams or groups to solve rapid challenges (Tang & Schmidt, 2021; Mathier et al., 2017). Teams strongly emphasize individual commitment and cooperation to achieve common objectives. According to Kozlowski and Ilgen (2006), a team has the following characteristics: "(a) it is formed of two or more individuals who (b) communicate and interact; (c) share one or more goals; (d) complete organizationally relevant tasks; € display interdependencies regarding workflow, goals, and outcomes; (f) have various roles and responsibilities; and (g) are embedded collectively in the organizational system with boundaries to the broader context. As a result of their strong commitment, shared duties, interdependence, and mix of abilities, teams are able to support the pursuit of creative and innovative goals. Teams are also able to make more deliberate efforts to enable the functionality of diversity, which is essential for team creativity (Hülsheger et al., 2009). Team creativity is an inventive outcome produced by the group's decision-making process after carefully weighing all available information (Dong et al., 2017).

KNOWLEDGE MANAGEMENT AND TEAM CREATIVITY

KM is the catalyst toward creative problem-solving through developing new products and services within the organization (Kazanjian & Drazin, 2012). Creativity emerges during a knowledge exchange, internalization, and externalization at the individual and team levels, which empowers organizations to cope with challenging opportunities.

Figure 6. Knowledge management and team creativity

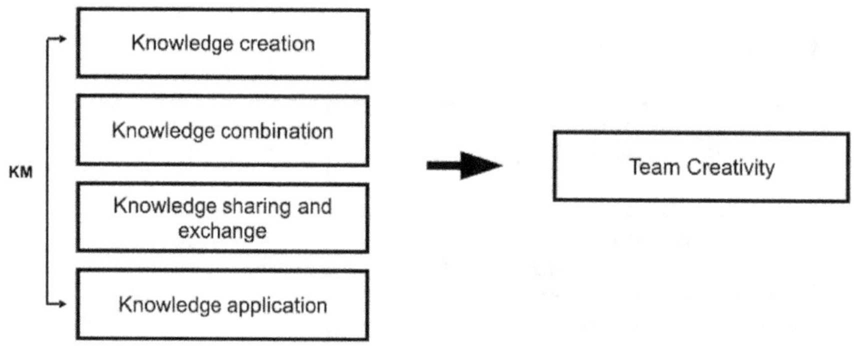

KM could improve teams' creativity by developing a standard protocol to stimulate knowledge sharing or management (Kazanjian & Drazin, 2012). Additionally, knowledge sharing and exchange have been consistently identified as essential factors that affect team creativity (Mesmer-Magnus & De Church, 2009; Sung & Choi, 2012). Notably, knowledge diversity facilitates and inhibits teams' creativity, caused by the contradictory relationship between divergent and convergent thinking (Pollok et al., 2021).

The challenge is that KM's definition and relevant analysis vary significantly in different contexts. Knowledge is more meaningful in a specific context. Context shapes and influences our perception and interpretation of the world (Crane, 2016). For instance, Jin Tong and colleagues (2009) argue that KM needs to consider the different cultures represented by various organizational departments to maintain the circulation of knowledge. A "one size fits all" approach will never work as organizations vary in structure, scope, and culture. Hence, a team-based strategy is more practical and efficient.

Components for Team Creativity in The Knowledge Management Process

Several studies have been published to illustrate the factors impacting a team's creativity (Tang & Schmidt, 2021; Pollok et al.,2021; Schley & Woerkom, 2014). Six components were chosen as the commonly used factor affecting team creativity, which the KM needs to consider (Wu et al.,2022a; Wu et al.,2022b).

Shared Mental Models (SMM)

SMM is a common understanding, such as the shared purpose, mission, and vision, that aids teams' collaboration to unleash members' full potential when working in a challenging and uncertain environment (Klimoski & Mohammed, 1994). As a result, SMM allows team members to comprehend tasks precisely and integrate their participation with another team member's behavioral patterns, which are related to the team's performance (Cannon-Bowers et al., 1993). Task-based models and team-based models are the two types that constitute SMM (Mathieu et al., 2000). Task models emphasize relevant knowledge, such as project schedules and success measurements. In contrast, team models include understanding team members and their interaction patterns. This is a critical factor for teams to perform better because it allows members to collaborate more efficiently. SMM facilitates idea generation and evaluation by making the information clear and easy to progress, which could be related to team creativity as the

Figure 7. Six components affecting team creativity

SMM improves the effectiveness of knowledge exchange (Harvey & Chia-yu, 2017). In addition to the advantages of SMM, it may result in inadequate adjustment when facing challenges.

Psychological Safety and Trust

Psychological safety and trust refer to team members feeling safe to interact with other individuals in the workplace without worrying about judging or rejecting their opinions and having positive attitudes when confronting conflict (Edmondson & Lei, 2014). Consequently, psychological safety and trust encourage employees to express their concerns confidently and seek support or feedback from colleagues. This consists of team climate and results in positive interpersonal communication, improving team creativity, and knowledge sharing among team members (Newman et al., 2017). Andragogical leaders believes in positive assumptions about employees, which are critical to fostering psychological safety and trust in teams. In turn, team creativity will be unleashed. A significant relationship occurs between supportive leadership and psychological safety based on the social learning theory. More specifically, a leader may demonstrate to subordinates that taking risks and being honest with one another is safe by listening to them, offering support, and providing clear, consistent directions (Newman et al., 2017; Bandura, 1977). Team psychological safety has been strongly linked to the innovative manufacturing process (Lee et al., 2011) and knowledge generation at the team level (Choo et al., 2004).

Team Motivation

Motivation explains why individuals show particular behaviors and desires. Individuals will persevere in an activity more if their motivation is higher (Graham & Weiner 1996). In team collaboration, members' work performance may be encouraged by different motivations. But, among them, the internal motivation to achieve success positively correlates with and predicts team members' beliefs about whether they can be creative (Wu et al., 2022a), as the motivation to pursue success involves confidence and persistence, which share similar features with creativity. Motivation is a critical factor for knowledge-sharing willingness and behavior, which can be positively affected by providing feedback to give teams recognition and appreciation (Liu & Fang, 2010; Gagné & Deci, 2005).

Team Intuition

Team intuition is how team members use their intuition collectively to generate ideas and make decisions without rationality or deliberative thinking. It is a tacit and subconsciously shared feeling that guides team members based on previous teamwork (Samba et al., 2022). The team enables the context in which individual intuition can be encouraged, applied, filtered, verified, and examined. The collaboration forms the team's intuition to benefit the decision-making process, creating a shared goal to increase effectiveness and interpret the information collected (Samba et al., 2022). In early research, Eisenhardt (1999) stated that repeated and timely internal team interactions create a deep collective intuition that empowers managers to maintain their competitiveness. Team intuition plays a significant role in team collaboration. It refers to how team members leverage their intuition to collaborate with individuals to enable a high level of team creativity (Akgün & Keskin, 2021). Seven different types of team institutions have been investigated that influence team creativity during new project development (Akgün & Keskin, 2021): holistic intuition, inferential intuition, affective intuition, attitudinal intuition, free intuition, social intuition, and moral intuition. These intuitions can be referenced by andragogical leaders to analyze and cultivate according to team composition.

Team Empathy

Empathy is associated with team creativity in negative and positive ways, particularly in generating more ideas, as it allows individuals to interact intelligently (Wu et al., 2022b; Alzayed et al., 2021). Empathy is the ability to put oneself in another person's shoes and feel their feelings and experiences, which is critical for resolving team problems. It is a key component of emotional and social intelligence that has a strong relationship with the firm's innovative performance by allowing the team to address problems in a creative, compassionate, and empathic manner (Bühring and Moore, 2018). Two common ways of showing empathy are the mirroring and reconstructive routes (Bloom, 2017). The mirroring route is spontaneous and happens automatically without processing much information. For example, when one team member acts anxious and aggressive, the other may react similarly by empathizing with others emotionally and subconsciously. The reconstructive route requires deliberative efforts to replicate others' experiences. Self-awareness and conscious understanding are necessary for this route by analyzing vast amounts of information.

Nonetheless, both routes should be revised to emphasize collaboration with others while avoiding misleading understanding. Therefore, managing empathy is necessary to benefit long-term goals, as emotions drive empathy, but deliberative thinking is more critical in the workplace. Furthermore, emphasizing with other members can easily reach the agreed discussion because of the revealed motivation and thinking style. Then a homogeneous team will be formed, which deters the team's creativity.

Team Reflectivity

Team reflectivity refers to the degree to which teams collectively reflect their knowledge and experience accumulated from practice; it is a valuable predictor of innovative team performance across time (Konradt & Eckardt, 2016; Schippers et al., 2012). Therefore, how teams reflect consciously on their performance will be the catalyst for creatively and effectively handling further tasks. Despite the fixed rules in daily work, team members often solve temporary tasks or problems depending on the context, interaction with colleagues, and newly emerging technology, which indicates that learning and practice will happen continuously. In addition, human beings are used to adapting the established behavior patterns for future operations, it is a natural tendency to save cognitive resources, but the context changes every day, and these patterns will become obsolete without reflecting what we knew and what we could do better (Schley & Woerkom, 2014). Hence, reflecting on challenging tasks or problematic aspects support teams in reimagining imposed boundaries and effectively transferring the knowledge to other projects.

Regarding andragogical leadership, reflectivity assists teams in examining their working process or performance, thus learning from mistakes or conflicts and assisting the team in monitoring and regulating their work without managerial intervention (Tjosvold, 1991). It's also a way to foster team intrinsic motivation, which refers to the team members being motivated to share knowledge and be engaged in the discussion. Team reflectivity will not happen spontaneously. Instruments and tools are needed to guide the process, such as debriefings and briefings in the format of seminars and formal or informal meetings (Schley & Woerkom, 2014). One notable point is that team performance improves greatly when members reflect less frequently but deeply on content quality (Otte et al., 2018).

HOW ANDRAGOGICAL LEADERSHIP LEVERAGES KNOWLEDGE MANAGEMENT AND SUPPORTS TEAM CREATIVITY

The above discussion links KM (knowledge creation, sharing, implications, reflectivity) and team creativity, which in return requires andragogical leaders to avoid isolated perspectives and promote a holistic picture that includes the interplay between KM and components of team creativity into the practice. In the same line of research, Ode and Ayavoo (2020) prove that knowledge creation, storage, and application positively influence firm innovation as KM protects organizational intangible assets and makes the knowledge more sustainable to apply it more efficiently in the long term. If organizations start to deploy KM, a clear strategy or model is needed to address the possible positive results and potential negative consequences in terms of the effectiveness of team creativity.

Moreover, context is related to what type of knowledge andragogical leaders adopts, whether it is tacit or explicit, abstract or concrete, domain-specific based, or generalized knowledge. Besides the context, the current pattern of how team members interact with others is the foundation for further KM deployment. The strategy should be deployed step by step to avoid potential resistance.

In practice, incorporating individual KM applications into business practices aid individuals in managing their knowledge by leveraging algorithms and metacognitive theories to act much like a second brain, helping individuals make better decisions. Moreover, collaboration platforms are great solutions that assist teams in collaborating, visualizing, and creating faster, enabling team members to be more confident in their decisions as they all strive for the same goal. IBM, for instance, uses Watson Discovery which uses natural language processing and AI-based search to uncover answers and gather insights from documents. Disney implemented for KM approach to improve their waiting line process. This led to the development of FASTPASS, a creative solution that helped bolster profits and customer satisfaction (Cope et al.,2011). Ultimately, team-based KM frameworks and tools should be invented to match specific contexts more precisely. Thus, andragogical leaders could leverage KM and support teams to respond creatively, profoundly, and successfully to the ever-changing world.

However, the challenge must be noticed. To stimulate team creativity, andragogical leaders must anticipate individual-level performance to predict team-level performance (Hunter et al., 2018), which is difficult as many mediating variables affect the transition from individual creativity to team creativity. Ployhart (2004) highlights that the goal of team composition is to transfer individual attributes, such as knowledge and skills, to improve team performance. To enable a successful transition, the individual predictors have been identified by Hunter et al. (2018): developing trust at the early phases of team collaboration, a sense of psychological safety, and shared mental models are all mediators that alleviate conflicts and solve problems effectively.

Figure 8. Individual predictors and team-level outcomes

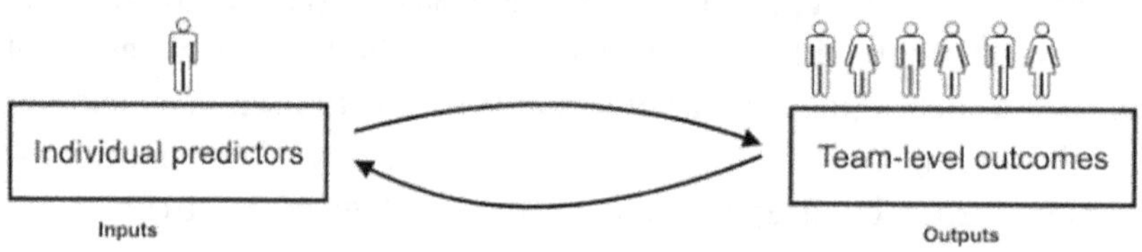

By contrast, Kozlowski and Ilgen (2006) categorized the mediating variables into two dimensions. Team cognition includes idea generation, shared mental models, and team reflectivity. The social process includes collaboration, communication, psychological safety, support, conflicts, cohesion, and team efficacy (Carmeli et al., 2013; Hunter et al., 2018). Instead, Hou et al. (2021) propose cultivating heterogeneous knowledge compositions by integrating members' diverse backgrounds. This could promote the team's creativity by allowing for a broader perspective on problems and minimizing disagreements. Understanding the present team member's knowledge composition is another transition prerequisite. For example, a product design team in the medical industry may require a medical expert, even if they are unfamiliar with product design. Nonetheless, the project could be aided by the use of interdisciplinary expertise.

Another potential challenge is that, regardless of the benefits of employing KM, employees may be reluctant to shift knowledge from tacit to explicit (Serenko & Bontis, 2016). This can happen either because of technical issues or a lack of time. However, it can also be caused by an undervaluation of KM or an unwillingness to transfer ownership of any individual knowledge (Friedrich et al., 2020). As an incentive, gamification design could be advised, which uses game design approaches to impact behavior and provide a pleasurable experience that encourages particular behaviors (Deterding et al., 2011).

Competitions, challenges, status, instant and long-term feedback, and performance graphs can provide individuals with the internal and extrinsic motivation to share their expertise (Friedrich et al., 2020).

A MODEL

These reflections and discussions above indicated that KM is closely related to team creativity, and several components affect team creativity. In this section, we contribute a comprehensive framework for KM based on the previous significant studies. Our goal is for andragogical leaders to incorporate all aspects of team creativity at various stages of KM.

Figure 9. Team creativity by integrating knowledge management

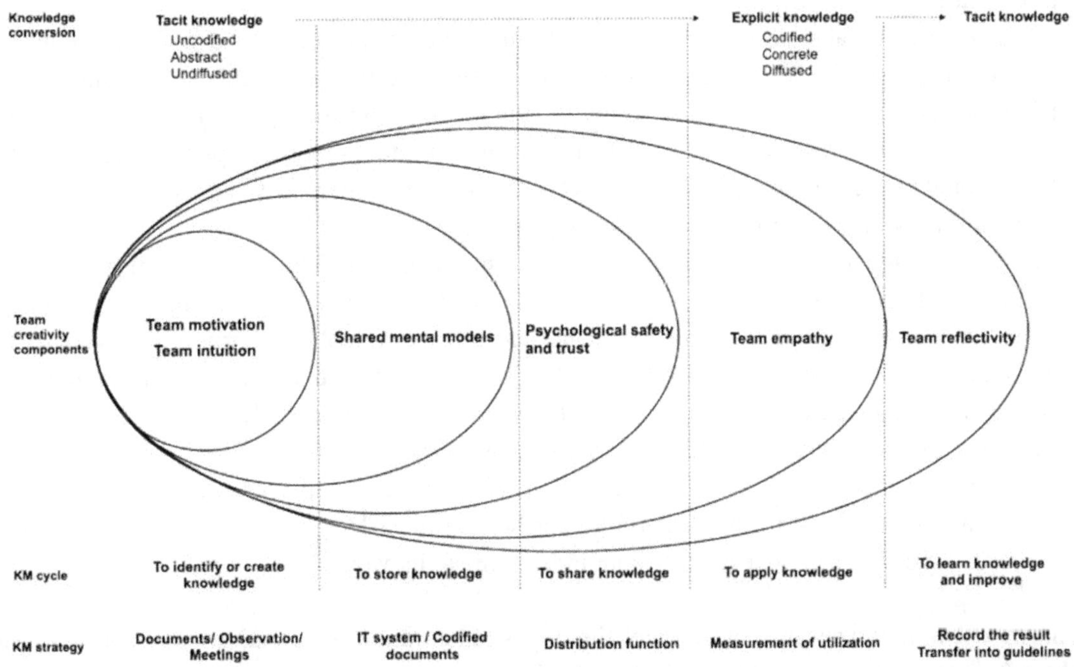

We integrate team creative components into different KM stages, where the KM strategy should promote the team's creativity, according to the KM cycle (Evans et al., 2014). At the beginning of the KM cycle, the three major forms of knowledge are tacit, abstract, and undiffused knowledge. Nonetheless, tacit information is translated into explicit format during the stages of knowledge utilization through tangible outputs produced by teams. Notably, the knowledge appears tacit again during the team reflection phase. The knowledge conversion process and the KM cycle are non-linear processes that repeat over time to be monetized. In this case, the value and effectiveness of knowledge management are determined and judged by the level of team creativity based on the satisfaction of components across KM stages, regardless of whether the knowledge is tacit or explicit.

1. In the knowledge identification and creation stage, analyzing documents, conducting meetings, and observing could be used to discuss what knowledge is required for tasks. Meanwhile, the components that affect team creativity should be involved. The strategy should improve the team's motivation and intuition, as internal motivation to achieve success positively correlates with and predicts team members' confidence in creativity (Wu et al.,2022a). In contrast, team intuition enables creativity (Akgün & Keskin, 2021). Furthermore, a newly built problem is required for knowledge identification or production. Teams must assess their previous experience, domain-relevant knowledge, and required expertise to evaluate whether it is adequate for new problems encountered. Thus, from an andragogical standpoint, the "new problem" might be considered a suitable moment to examine knowledge.
2. During the knowledge storage stage, an IT system or codified documents can be utilized to create shared mental models for members to comprehend the team's purpose and mission. Knowing the history of projects and how the team solved challenges might be useful resources for similar tasks in the future.
3. The knowledge-sharing stage often occurs during the team ideation stage as a divergent process focused on producing ideas that could be used as future solutions. In this stage, knowledge distribution is critical to ensuring that the individual receives the appropriate knowledge at the right moment. However, regardless of whether the setting is digital or physical, psychological safety and trust should be the primary goal at this point so that team members may share their ideas fearlessly and honestly without fear of negative consequences.
4. In the knowledge application stage, utilization measurement could be applied. Team empathy should be examined internally and externally, as it may negatively affect whether team members fully express their stance or have been substantially influenced by other members (Form & Kaernbach, 2018), and externally, whether the applied knowledge is ethically desirable to the targeted stakeholders.
5. In the knowledge learning and improving stage, accumulated knowledge should be recorded and transferred into guidelines that facilitate team reflectivity. Team reflectivity is crucial to make the accumulation more sustainable, avoid repeated mistakes, and empower knowledge identification and creation. Although team reflectivity is critical in the KM cycle, it is frequently overlooked because organizations move quickly and underestimate its value.

This model summarizes and consolidates the stages of KM and team creativity components, extending them further by including andragogical leadership in management.

CONCLUSION AND FUTURE RESEARCH

The analysis shows that KM effectively provides organizations with more competitiveness by improving team creativity. To fully realize the potential of KM, a framework is developed to guide andragogical leadership in considering its application. The challenge, however, is determining how to successfully translate individual predictors into team-level performance and reconcile them with the context-specific requirements that the KM strategy can be best performed. Theoretically, KM research as a core competitive strategy should focus on team creativity to better understand how KM tools and resources foster collaboration, innovation, and creative problem-solving. Future research could involve different dimensions and components of team creativity that influence KM effectiveness. Additionally, andragogical leadership on team creativity and KM could be further studied in a more specific context.

ACKNOWLEDGMENT

The authors would like to thank the research fund support of the RPg studentship and the Eric C. Yim Endowed Professorship of The Hong Kong Polytechnic University.

REFERENCES

Adaileh, M. J., Alrawashdeh, M., Elrehail, H., & Aladayleh, K. J. (2020). Assessing the nexus between knowledge management and firm performance: A data article. *Data in Brief*, *32*(10), 1–8. doi:10.1016/j.dib.2020.106283 PMID:32984475

Akgün, A. E., & Keskin, H. (2021). Team intuition and creativity in new product development projects: A multi-faceted perspective. *Journal of Engineering and Technology Management*, *62*, 101660. doi:10.1016/j.jengtecman.2021.101660

Alavi, M., & Leidner, D. E. (2001). Knowledge management and knowledge management systems: Conceptual foundations and research issues. *Management Information Systems Quarterly*, *25*(1), 107. doi:10.2307/3250961

Alzayed, M. A., Miller, S. R., & McComb, C. (2021). Empathic creativity: Can trait empathy predict creative concept generation and selection? *Artificial Intelligence for Engineering Design, Analysis and Manufacturing*, *35*(4), 369–383. doi:10.1017/S0890060421000196

Arain, G. A., Bhatti, Z. A., Ashraf, N., & Fang, Y. H. (2020). Topdown knowledge hiding in organizations: An empirical study of the consequences of supervisor knowledge hiding among local and foreign workers in the Middle East. *Journal of Business Ethics*, *164*(11), 611–625. doi:10.100710551-018-4056-2

Bloom, P. (2017). *Against empathy: The case for rational compassion*. Random House.

Boisot, M., & Cox, B. (1999). The I-Space: A framework for analyzing the evolution of social computing. *Technovation*, *19*(9), 525–536. doi:10.1016/S0166-4972(99)00049-8

Bühring, J., & Liedtka, J. (2018). Embracing systematic futures thinking at the intersection of Strategic Planning, Foresight and Design. *Journal of Innovation Management*, *6*(3), 134–152. doi:10.24840/2183-0606_006-003_0006

Bühring, J., & Moore, P. (2018). Emotional and Social Intelligence as 'Magic Key' in Innovation: A Designer's call toward inclusivity for all - Letter from Academia. *Journal of Innovation Management*, *6*(2), 6–12. doi:10.24840/2183-0606_006.002_0002

Callahan, C., & Grunberg, N. E. (2018). *Military medical leadership*. Academic Press.

Cannon-Bowers, J. A., Salas, E., & Converse, S. A. (1993). Shared mental models in expert team decision making. In N. J. Castellan (Ed.), *Individual and group decision making* (p. 221246). Lawrence Erlbaum Associates.

Carmeli, A., Gelbard, R., & Reiter-Palmon, R. (2013). Leadership, creative problem-solving capacity, and creative performance: The importance of knowledge sharing. *Human Resource Management, 52*(1), 95–121. doi:10.1002/hrm.21514

Chatterjee, S., Chaudhuri, R., Thrassou, A., & Vrontis, D. (2021). Antecedents and consequences of knowledge hiding: The moderating role of knowledge hiders and knowledge seekers in organizations. *Journal of Business Research, 128*(5), 303–313. doi:10.1016/j.jbusres.2021.02.033

Chomsky, N., Roberts, I., & Watumull, J. (2023, March 8). Opinion | Noam Chomsky: The False Promise of ChatGPT. *The New York Times.* https://www.nytimes.com/2023/03/08/opinion/noam-chomsky-chatgpt-ai.html?smid=tw-share

Choo, A. S., Linderman, K., & Schroeder, R. G. (2004). Social and method effects on learning behaviors and knowledge creation in six sigma projects. *Academy of Management Proceedings, 2004*(1), C1–C6. doi:10.5465/ambpp.2004.13857620

Connelly, C. E., Zweig, D., Webster, J., & Trougakos, J. P. (2012). Knowledge hiding in organizations. *Journal of Organizational Behavior, 33*(1), 64–88. doi:10.1002/job.737

Cope, R. F., Cope, R. F. III, Bass, A. N., & Syrdal, H. A. (2011). Innovative knowledge management at Disney: Human capital and queuing solutions for services. *Journal of Service Science, 4*(1). Advance online publication. doi:10.19030/jss.v4i1.4268

Crane, L. (2016). *Knowledge and discourse matters: Relocating knowledge management's sphere of interest onto language* (1st ed.). Wiley.

Cristancho, S. (2016). Lessons on resilience: Learning to manage complexity. *Perspectives on Medical Education, 5*(3), 133–135. doi:10.1007/S40037-016-0277-1 PMID:27256287

Deterding, S., Khaled, R., Nacke, L. E., & Dixon, D. (2011, May). Gamification: Toward a definition. In *CHI 2011 gamification workshop proceedings* (Vol. 12, p. 15). ACM.

Di Vaio, A., Palladino, R., Pezzi, A., & Kalisz, D. E. (2021). The role of digital innovation in knowledge management systems: A systematic literature review. *Journal of Business Research, 123*, 220–231. doi:10.1016/j.jbusres.2020.09.042

Dong, Y. T., Bartol, K. M., Zhang, Z. X., & Li, C. W. (2017). Enhancing employee creativity via individual skill development and team knowledge sharing: Influences of dual-focused transformational leadership. *Journal of Organizational Behavior, 38*(3), 439–458. doi:10.1002/job.2134

Edmondson, A. C., & Lei, Z. (2014). Psychological Safety: The History, Renaissance, and Future of an Interpersonal Construct. *Annual Review of Organizational Psychology and Organizational Behavior, 1*(1), 23–43. doi:10.1146/annurev-orgpsych-031413-091305

Eisenhardt, K. M. (1999c). Strategy as Strategic Decision Making. *Sloan Management Review, 40*(3), 65–72. https://dialnet.unirioja.es/servlet/articulo?codigo=2485499

Evans, M., Dalkir, K., & Bidian, C. (2014). A holistic view of the knowledge life cycle: The knowledge management cycle (KMC) model. *Electronic Journal of Knowledge Management, 12*(2), 85–97. https://academic-publishing.org/index.php/ejkm/article/view/1015

Form, S., & Kaernbach, C. (2018). More is not always better: The differentiated influence of empathy on different magnitudes of creativity. *Europe's Journal of Psychology, 14*(1), 54–65. doi:10.5964/ejop.v14i1.1432 PMID:29899798

Friedrich, J., Becker, M., Kramer, F., Wirth, M., & Schneider, M. (2020). Incentive design and gamification for knowledge management. *Journal of Business Research, 106*, 341–352. doi:10.1016/j.jbusres.2019.02.009

Gagné, M., & Deci, E. L. (2005). Self-determination theory and work motivation. *Journal of Organizational Behavior, 26*(4), 331–362. doi:10.1002/job.322

Graham, S., & Weiner, B. (1996). Theories and principles of motivation. *Handbook of Educational Psychology, 4*(1), 63-84.

Harvey, S., & Chia-yu, K. (2017). Social Processes and Team Creativity. In R. Reiter-Palmon (Ed.), *Team Creativity and Innovation* (pp. 87–128). Oxford University Press. doi:10.1093/oso/9780190222093.003.0005

Hou, F., Su, Y., Qi, M., Wang, L., & Wang, Q. (2021). Entrepreneurial team knowledge diversity and creativity: A multilevel analysis of knowledge sharing, individual creativity, and team creativity. *Frontiers in Psychology, 12*, 717756. doi:10.3389/fpsyg.2021.717756 PMID:34539518

Hülsheger, U. R., Anderson, N., & Salgado, J. F. (2009). Team-level predictors of innovation at work: A comprehensive meta-analysis spanning three decades of research. *The Journal of Applied Psychology, 94*(5), 1128–1145. doi:10.1037/a0015978 PMID:19702361

Hunter, S. T., Neely, B. H. Jr, & Gutworth, M. B. (2018). Selection and team creativity: Meeting unique challenges through diversity and flexibility. In R. Reiter-Palmon (Ed.), *Team creativity and innovation* (pp. 259–281). Oxford University Press.

Jeanes, E. (2021). A meeting of mind (sets). Integrating the pedagogy and andragogy of mindsets for leadership development. *Thinking Skills and Creativity, 39*, 100758. doi:10.1016/j.tsc.2020.100758

Johnson, T. C. (2016). *Diversity, knowledge management, and team creativity in virtual healthcare teams: A predictive study*. ProQuest Dissertations & Theses.

Kazanjian, R. K., & Drazin, R. (2012). Organizational learning, knowledge management, and creativity. In *Handbook of organizational creativity* (pp. 547–568). Academic Press. doi:10.1016/B978-0-12-374714-3.00021-5

Klimoski, R., & Mohammed, S. (1994). Team mental model: Construct or metaphor? *Journal of Management, 20*(2), 403–437. doi:10.1177/014920639402000206

Knowles, M. S., Holton, E., & Swanson, A. (1998). *The adult learner*. Gulf Publishing Company.

Konradt, U., & Eckardt, G. (2016). Short-term and long-term relationships between reflection and performance in teams: Evidence from a four-wave longitudinal study. *European Journal of Work and Organizational Psychology*, *25*(6), 804–818. doi:10.1080/1359432X.2016.1160058

Kozlowski, S. W., & Ilgen, D. R. (2006). Enhancing the effectiveness of work groups and teams. *Psychological Science in the Public Interest*, *7*(3), 77–124. doi:10.1111/j.1529-1006.2006.00030.x PMID:26158912

Kratzer, J., & Mrożewski, M. (2021). Factors fostering creativity in start-up teams. In *Creative Success in Teams* (pp. 123–144). Academic Press. doi:10.1016/B978-0-12-819993-0.00007-2

Lee, J. Y., Swink, M., & Pandejpong, T. (2011). The roles of worker expertise, information sharing quality, and psychological safety in manufacturing process innovation: An intellectual capital perspective. *Production and Operations Management*, *20*(4), 556–570. doi:10.1111/j.1937-5956.2010.01172.x

Liu, W. C., & Fang, C. L. (2010). The Effect of Different Motivation Factors on Knowledge-Sharing Willingness and Behavior. *Social Behavior and Personality*, *38*(6), 753–758. doi:10.2224bp.2010.38.6.753

Mathieu, J. E., Heffner, T. S., Goodwin, G. F., Salas, E., & Cannon-Bowers, J. A. (2000). The influence of shared mental models on team process and performance. *The Journal of Applied Psychology*, *85*(2), 273–283. doi:10.1037/0021-9010.85.2.273 PMID:10783543

McKay, A. S., Reiter-Palmon, R., & Kaufman, J. C. (2021). An interdisciplinary view on team creativity: Toward integration across fields. In *Creative Success in Teams* (pp. 231–241). Academic Press. doi:10.1016/B978-0-12-819993-0.00012-6

Mesmer-Magnus, J. R., & Dechurch, L. A. (2009). Information sharing and team performance: A meta-analysis. *The Journal of Applied Psychology*, *94*(2), 535–546. doi:10.1037/a0013773 PMID:19271807

Natu, S., & Aparicio, M. (2022). Analyzing knowledge sharing behaviors in virtual teams: Practical evidence from digitalized workplaces. *Journal of Innovation &Amp. Knowledge (Beverly Hills, Calif.)*, *7*(4), 100248. doi:10.1016/j.jik.2022.100248

Newman, A., Donohue, R., & Eva, N. (2017). Psychological safety: A systematic review of the literature. *Human Resource Management Review*, *27*(3), 521–535. doi:10.1016/j.hrmr.2017.01.001

Nonaka, I., & Takeuchi, H. (1995). *The Knowledge-Creating Company: How Japanese Companies Create the Dynamics of Innovation* (1st ed.). Oxford University Press.

Nuckles, C. R. (2000). Student-Centered Teaching: Making It Work. *Adult Learning*, *11*(4), 5–6. doi:10.1177/104515959901100403

Ode, E., & Ayavoo, R. (2020). The mediating role of knowledge application in the relationship between knowledge management practices and firm innovation. *Journal of Innovation & Amp. Knowledge (Beverly Hills, Calif.)*, *5*(3), 210–218. doi:10.1016/j.jik.2019.08.002

Ohlsson, S. (2011). Deep learning: How the mind overrides experience. *Choice Reviews Online*, *49*(03), 49–1770. doi:10.5860/CHOICE.49-1770

Otte, K. P., Konradt, U., & Oldeweme, M. (2018). Effective team reflection: The role of quality and quantity. *Small Group Research*, *49*(6), 739–766. doi:10.1177/1046496418804898

Peng, J., Wang, Z., & Chen, X. (2018). Does self-serving leadership hinder team creativity? A moderated dual-path model. *Journal of Business Ethics, 159*(2), 419–433. doi:10.100710551-018-3799-0

Pereira, L., Fernandes, A., Sempiterno, M., Dias, L., Lopes Da Costa, R., & António, N. (2021). Knowledge management maturity contributes to project-based companies in an open innovation era. *Journal of Open Innovation, 7*(2), 126. doi:10.3390/joitmc7020126

Ployhart, R. E. (2004). Organizational staffing: A multilevel review, synthesis, and model. *Research in Personnel and Human Resources Management, 23*, 121–176. doi:10.1016/S0742-7301(04)23003-1

Polanyi, M. (1966). *The tacit dimension*. P. Smith.

Pollok, P., Amft, A., Diener, K., Lüttgens, D., & Piller, F. T. (2021). Knowledge diversity and team creativity: How hobbyists beat professional designers in creating novel board games. *Research Policy, 50*(8), 104174. doi:10.1016/j.respol.2020.104174

Price, P. A. (2004). *Genesis and evolution of the United States Air Force academy's officer development system*. http://www.dtic.mil/dtic/tr/fulltext/u2/a428315.pdf

Radermacher, F. J. (1996). Cognition in systems. *Cybernetics and Systems, 27*(1), 1–42. doi:10.1080/019697296126651

Rausch, A. (2012). Task characteristics and learning potentials—Empirical results of three diary studies on workplace learning. *Vocations and Learning, 6*(1), 55–79. doi:10.100712186-012-9086-9

Rong, P., & Liu, S. (2021). Top management team knowledge hiding and enterprise innovation performance: A moderated mediation model. *Frontiers in Psychology, 12*, 783147. Advance online publication. doi:10.3389/fpsyg.2021.783147 PMID:34987451

Samba, C., Williams, D. W., & Fuller, R. M. (2022). The forms and use of intuition in top management teams. *The Leadership Quarterly, 33*(3), 101349. doi:10.1016/j.leaqua.2019.101349

Schippers, M. C., West, M. A., & Dawson, J. F. (2012). Team reflexivity and innovation: The moderating role of team context. *Journal of Management, 41*(3), 769–788. doi:10.1177/0149206312441210

Schley, T., & van Woerkom, M. (2014). Reflection and reflective behavior in work teams. *Discourses on professional learning: On the boundary between learning and working*, 113-139.

Serenko, A., & Bontis, N. (2016). Understanding counterproductive knowledge behavior: Antecedents and consequences of intra-organizational knowledge hiding. *Journal of Knowledge Management, 20*(6), 1199–1224. doi:10.1108/JKM-05-2016-0203

Sung, S. Y., & Choi, J. N. (2012). Effects of team knowledge management on the creativity and financial performance of organizational teams. *Organizational Behavior and Human Decision Processes, 118*(1), 4–13. doi:10.1016/j.obhdp.2012.01.001

Tang, M., & Schmidt, J. T. (2021). Fostering creativity in interdisciplinary and intercultural teams: The importance of team building and team management. *Creative Success in Teams*, 55–79. doi:10.1016/B978-0-12-819993-0.00004-7

Thoring, K., Mueller, R. M., Desmet, P., & Badke-Schaub, P. (2022). Toward a unified model of design knowledge. *Design Issues, 38*(2), 17–32. doi:10.1162/desi_a_00679

Tjosvold, D. (1991). *Team organization: An enduring competitive advantage.* John Wiley & Sons.

Tong, J., & Mitra, A. (2009). Chinese cultural influences on knowledge management practice. *Journal of Knowledge Management, 13*(2), 49–62. doi:10.1108/13673270910942691

Wang, V., & Torrisi-Steele, G. (2022). Leading adult learning in organizations. In V. Wang (Ed.), *Handbook of Research on Educational Leadership and Research Methodology* (pp. 1–15). IGI Global. doi:10.4018/978-1-6684-4144-2.ch001

Wu, X., Siu, K. W. M., & Bühring, J. (2022b). Empathy between designers in the design synthesis stage. In C. Stephanidis, M. Antona, S. Ntoa, & G. Salvendy (Eds.), *HCI International 2022 – Late Breaking Posters. HCII 2022. Communications in Computer and Information Science* (Vol. 1654). Springer. doi:10.1007/978-3-031-19679-9_27

Wu, X., Siu, K. W. M., Bühring, J., & Villani, C. (2022a). The relationship between creative self-efficacy, achievement motivation, and job burnout among designers in China's e-market. *Social Sciences, 11*(11), 509. doi:10.3390ocsci11110509

Zaim, H., Muhammed, S., & Tarim, M. (2018). Relationship between knowledge management processes and performance: Critical role of knowledge utilization in organizations. *Knowledge Management Research &Amp. Practice, 17*(1), 24–38. doi:10.1080/14778238.2018.1538669

KEY TERMS AND DEFINITIONS

Andragogical Leadership: Leaders who hold positive assumptions about their employees and encourage them to solve problems creatively through self-learning and practice and foster a collaborative and supportive learning environment inside organizations.

Knowledge Management: It refers to the management of knowledge creation and application to make knowledge more sustainable by facilitating decision-making (Vaio et al., 2021).

Knowledge Management Cycle: The knowledge management process includes knowledge sharing and distribution, capture and storage, creation, application, learning, and updating (Pereira et al., 2021).

Team Creativity: Is an inventive outcome produced by the group's decision-making process after carefully weighing all available information (Dong et al., 2017).

Chapter 13
Toward Social Justice:
Leadership's Crucial Role in an Increasingly Expansive Technology-Dependent Context

Karen Escalante
California State University, San Bernardino, USA

Melissa Meetze-Hall
Riverside and San Bernardino County Offices of Education, USA

ABSTRACT

This chapter asserts that transformative leadership is essential to educational organizations. The authors identify andragogical leadership as synonymous with transformative leadership, with critical attributes hinged upon activism, praxis, technology, and social justice supporting adult learning. The transformative leader can engage in dialogue with their learners/employees to facilitate critical consciousness through a constructivist framework and provide evidenced-based feedback. The orientation of this chapter focuses on what social justice is and why it is central to leadership, access and use of technology as a form of social justice, the essential role of dialogue and inquiry for the andragogical leader, where learning takes place, and the necessity of transformative leadership to move organizations further along the social justice continuum to enact change for the modern world. This chapter contributes to the knowledge base of transformative leadership as a promising vehicle to address the growing need for humane (expansive/accessible) learning spaces.

BACKGROUND

Educational technology use continues to expand; it thereby impacts learning by both increasing access to knowledge and altering the way we learn. At both the K-12 level and the post-secondary level, some students and instructors/employees may not yet see technology as an opportunity. Leaders must recognize technology as a tool and refrain from being reactionary. For example, adopting all new technology

DOI: 10.4018/978-1-6684-7832-5.ch013

every year is rarely the right option without examining outcomes from extant tools and practices. If we consider technology a tool, we must understand learning; learners; curricular decisions based on learning theory; and people. Considering all of these facets, leaders must be able to "...promote an environment of professional learning and innovation that empowers educators to enhance student learning through the infusion of contemporary technologies and digital resources" (ISTE, 2023).

As an advocate for citizenship and equity and given that technology can expand access for all, leaders must be able to envision technology uses for social justice. Holding this vision while embracing the potentiality of contemporary technologies and digital resources, the leader's ability to facilitate the use of technology will be crucial to moving employees, learners, and staff forward. To do so, leaders must (in part): know about whom they are supporting; be courageous; encourage creativity; and support others by making decisions based on learning theory.

This chapter was written to support the educational leader in promoting professional learning and innovation. To do so, the chapter provides theoretical understanding so leaders may consider and form the rationale for their leadership decisions. The chapter also guides leaders who are navigating new technologies as well as the management of transitions for learners/employees. Some examples of these decisions are how to allocate budgets for hardware and teacher training (Scghaffhauser, 2016). The leader must consider what they know about themselves as a learner, what they believe about learning, and the leadership practices they will enact. These practices include creating a culture learners/employees are empowered to use technology in innovative ways to enrich learning.

The authors' purpose for writing this chapter is to support leaders committed to social justice. These leaders realize the crucial role of technology in expanding access, recognize the importance of supporting the growth of others, and are concurrently developing their own capacities to engage, connect, and lead (Drago-Severson et al., 2023). Expansive, liberating use of technology must include the leaders' ability to move beyond technology management toward a transformative leadership style. In contrast to transformational or transactional leadership, transformative leadership begins with the end in mind and fiercely advocates for justice and democracy (Shields, 2010).

INTRODUCTION

Leadership is known to impact educational organizations; those with dynamic skill sets can lead to an organization's success (Bass, 2008; Collins, 2001). The authors assert that dynamic skill sets for the modern world reflect the beliefs and praxis of a transformative leader. Transformative leadership is grounded in social justice; such leaders demonstrate the potential to engage in equitable and inclusive democratic practices. "Transformative educational leadership begins by challenging inappropriate uses of power and privilege that create or perpetuate inequity and injustice" (Shields, 2010). The modern world also reflects increased access to and use of technology as the educational landscape shifts and changes to adapt to the assets and needs of learners/employees. The transformative leader uses technology as a tool to support adult learners and the growth of the organization.

This chapter will provide insights and practical strategies for individuals participating in graduate leadership programs, graduate faculty, students, researchers, and librarians in leadership programs, and any other leader interested in learning more about transformative leadership and the integration of technology. The focus is to present the authors' perspectives and ideas, encourage the reader to think critically and broaden their views on learners, learning, and leadership. As the reader moves through the chapter, reflect on your beliefs and perspectives on leadership and technology integration.

Leaders in various organizations will find the content of this chapter relevant and informative. Whether you seek to improve your leadership abilities or support the development of others, this chapter is an essential resource. It will support leaders in advancing their understanding of transformative leadership's role in increasingly technology-dependent contexts and moving an organization forward on the social justice continuum. As one moves their organization forward, the function of this chapter is also to move this conversation forward.

In this chapter, the authors conceptualize social justice using a definition from Furman (2012), understanding this to be a key feature of educational organizations despite often differing perceptions. This is followed by Vygotsky's (1934; 1986) theory of social constructivism and how this theory provides opportunities for technology to be used as a pathway for knowledge acquisition and as a promoter of social justice. Freire's (2000) understanding of praxis is then introduced, as well as humanizing spaces by way of the transformative leader. Learning and motivation theories are introduced to provide a timeline and progression for the reader, and adult learning theory is then expanded upon. Next, leadership styles are introduced, with detailed insight into transformative leadership, including beliefs and praxis. Finally, evidence-based feedback is discussed in addition to innovative technology integration. In addition to the written content and reference tables, the authors present a conceptual framework for further consideration.

Two key questions are provided to guide one's thinking during this chapter:

- What beliefs and behaviors of the transformative leader create space for evidence-based reflection and digital transformation in education?
- How can increased expertise in technology implementation impact social justice?

CONCEPTUALIZING SOCIAL JUSTICE

Educational literature is replete with definitions for social justice (Bertrand & Rodela, 2018; Hytten & Bettez, 2011; North, 2008; Shields, 2004). Increasingly, one is hard-pressed to engage with an educational program that omits "social justice" from its mission, vision, or diversity statements. While the term "social justice" may be used or even promulgated in organizational mission statements, faculty, teachers, leaders, and support staff often do not share a common understanding of social justice (Hytten & Bettez, 2011). Towards the goal of creating a consistent vernacular, and to move the discussion ahead throughout this chapter, we will rely on Furman's (2012) definition to support our mutual understanding. Therefore, in this chapter, social justice centers on "the experiences of marginalized groups and addresses inequities in educational opportunities and outcomes." Marginalized groups represent the most underserved, underrepresented, and undereducated; addressing these inequities demands purposeful, complete, and equal participation of all societal groups. The authors believe educational leaders can impact change when they facilitate these efforts by mutually shaping society to meet their needs (Bell, 2016). To that end, educational leaders for social justice must create and demand spaces and opportunities for all voices to be heard (Bogotch, 2002).

Reviews of research and journal publications provide evidence that the past century has shown an increased effort on the part of educators and educational leaders to increase their understanding of social justice related to power, identity, and social construction (Quin, 2009). A leader grounded in social justice believes in respect, care, recognition, and empathy (Theoharis, 2007). They actively work to create safe and affirming spaces for individuals who have been "othered" (Kumashiro, 2000). Consequently, social justice educational leaders empower other educators, learners, and employees to act in anti-oppressive ways (Quin, 2009). However, leadership programs can be hesitant to engage in these efforts given the broad definition (Cambron-McCabe & McCarthy, 2005). Social justice work requires ongoing investigative dialogue and continuous growth; it is not a fixed point.

The principles and practices of social justice include ongoing discourse and actions for both leaders and aspiring leaders. Thus, advocacy must be part of leadership preparation programs (Cambron-McCabe & McCarthy, 2005; Theoharis, 2007); it cannot be separated from the role of educational leadership. This concern includes the omnipresent role of systemic racism and marginalization (Bertrand & Rodela, 2018). Leaders must be intentionally taught to look within the margins and identify who is being ignored and how to address what is being missed (Kumashiro, 2000).

EXPANDED ACCESS TO EDUCATION: ONE FORM OF SOCIAL JUSTICE

This chapter argues that transformative leadership is essential to educational organizations. To unpack meaningful, actionable practices, the authors begin by reviewing learning theories in order to then move forward to discuss the expanded use, management, and access to technology toward increased social justice. And thus, we begin with how learners/employees' access and build knowledge, which is often grounded in social constructivism. Vygotsky's (1934; 1986) cultural-historical social activity theory emerged in the 1920s and 30s and provided a system and process by which we understand the world. In part, Vygotsky suggested that knowledge construction and organization are generative. The development occurs first on a social level and later on an individual level; therefore, the potential for development occurs when children participate in social behavior. This understanding is labeled as the Zone of Proximal Development (ZPD) and has been a foundational building block for classroom instruction with a focus on language development. Vygotsky proposed three aspects that are significant to cognitive development. First, knowledge is a historical process; our unique social and cultural backgrounds shape our cognitive development. Second, knowledge cognition is social; as such we rely on discourse, dialogue, and processing with others to support understanding and meaning-making. Finally, language, numbers, and tools support the mediation of knowledge acquisition (Aubrey & Riley, 2015).

Moving beyond Vygotsky's work in the early 20th Century, with an adjusted focus on adults, Mezirow (1991) conceptualized a different understanding of adult learners and how they build and shape meaning; he also suggests three factors to cognitive development. The first is an epistemic perspective which acknowledges patterns, abstract thinking, and scope of awareness, to name a few. Second is a sociolinguistic perspective which includes, but is not limited to, cultural and language codes, socialization, and differing philosophies. The third perspective is psychological; this includes self-concept, inhibitions, neurotic needs, including others. The theory and process of learning and acquiring knowledge are complex.

Adult development is seen as an adult's progression toward an enhanced capacity to validate prior learning through reflective discourse and to act upon the resulting insights. Anything that moves the learner/employee toward a more "inclusive, differentiated, permeable, and integrated meaning perspective, the validity of which has been established through rational discourse, aids an adult's development" (Mezirow, 1991, p. 7). Engaging in discourse through technology allows learners to engage in a (virtual) community of practice. Lave and Wenger (1991) suggest that members within a community of practice learn to gradually take up space as they engage in knowledge generation, application, and reproduction. The theories of Mezirow and Lave and Wenger are explored more in subsequent sections of this chapter to provide increased understanding.

Increasingly, technology has become a tool to facilitate meaning-making and process values and beliefs during learning construction. Growth-focused technology leaders explore their meaning-making systems considering how they affect leaders' understandings of diversity, equity, and social justice (Drago-Severson et al., 2023). Technology allows for increased access to continued education as we consider ways individuals engage in discourse to support knowledge acquisition and development. In its many and varied forms, technology can promote social justice. Examples of positive impact can be traced back to the fourteenth century with the invention of book printing, to more recent technologies, including sewage treatment for clean water. The invention and use of technology is a form of social justice (Dyson, 1997). Online capabilities enable all individuals to access, connect, and gain knowledge of events worldwide (Dyson, 1997). Online access is now embedded into our daily lives, providing flexible opportunities to learners/employees (Neiman & Wang, 2017). Use and access to technology address inequities in educational outcomes and opportunities, allowing equal participation to all groups in society. The use of technology facilitates knowledge organization through individual and community discourse (Lowyck, 2014) and active inquiry (Weis et al., 2002). Technology has shifted learning from instructor or program controlled to learner controlled (Lowyck, 2014). A shift from instructor to learner-centered is what educational theorist John Dewey argues results in a more democratic, community foundation of learning (Aubrey & Riley, 2015)

With advances in the availability of technology, increased media speed, and deep media catalogs, learners can increasingly engage in authentic, real-world problems (Reiser, 2001). Learning at one's own pace and under conditions more favorable to the learner allows the learner/employee a safer learning environment and the opportunity to take breaks as needed (Ladson-Billings, 2021). Previous generations of media allowed for minimal interpersonal engagement; television, phone, and radio (Schejter & Tirosh, 2015). Current iterations are interactive, mobile, abundant, and multi-media, facilitating immediate and personal connections and allowing for increased access and distribution of knowledge and information (Schejter & Tirosh, 2014). Recent work in the field has also produced the cognitive theory of multimedia learning; virtual reality experiences have demonstrated promising learning outcomes (Kyaw et al., 2019).

One such example of increased access is digital connectivism via online learning. Over time, online learning opportunities have progressed from reliance on recording lessons with companion worksheet packets to the ever-growing demand for synchronous (real-time interactions with instructors) and asynchronous (learning on their own schedules) instruction. As online learning has continued to expand, researchers interested in the topic have increased in numbers; so too have the number of studies focused on technology use and, more specifically, online learning. The term 'ubiquitous' is often used now to describe technology. Identifying technology as 'being everywhere' may not be incorrect. Following the acknowledgment that technology is ubiquitous, how do leaders process this? How do leaders move forward from this noticing to making sense of online learning and its potentiality? Again, exploration and shared definitions can aid learners and leaders alike. Corbett and Spinello (2020) identify four online learning and leadership principles, which they associate as the fundamentals of digital connectivism. Digital connectivism is described as the axis between knowledge and the technological environment. The ability to access facts and information, answer questions, and solve problems, which technology allows individuals to do, provides for more significant social capital (Saunders, 2017), a form of social justice.

Four Principles of Digital Connectivism (as adapted from Corbett & Spinello, 2020):

- **Autonomy:** Learners/employees are given freedom and choice of technological tools; no sole person is the source of knowledge in the learning process.
- **Connectedness:** Connections and contributions are encouraged via digital platforms, including blogs, social media, Youtube, Ted Talks, Coursera, and substacks.
- **Diversity:** Learners/employees are encouraged to seek the opinions of others; there are many sources of knowledge and expertise.
- **Openness:** Such as Massive Online Open Communities (MOOCs) attract thousands of participants worldwide. Massive Online Open Communities are courses available for free or minimal charge for anyone to enroll in to advance or deepen their skills, understanding, and knowledge.

Accessing and acquiring new skills via technology is available to all learners worldwide. Technology is a cornerstone of our daily lives; individuals access it for work, play, and socialization. "Today, more than at any other time in history, thanks to Web-based learning platforms, such as social media tools, TED Talks, YouTube, open courseware, blogs, and various phone Apps that interact with the Internet, it is possible for anyone to become knowledgeable and expert in a broad range of content topics" (Neiman & Wang, 2017, p. 39). Therefore, it can be argued that educational leaders for social justice must provide access to technological platforms and opportunities for their learners/employees, intentionally evaluate access and address inequities to personal and professional growth and consider learning and motivation theories when examining their leadership styles. All of these combined wills result in actions that positively move organizations.

HUMANIZING PRAXIS: LEADERSHIP, DIALOGUE, AND INQUIRY

In the opening paragraphs of this chapter, the authors bring forth the importance of a shared understanding of terms. Beginning with defining social justice, the educational leader must also support their learners/employees to move beyond understanding terms to making new connections that are only sometimes consciously formed (Costa & Garmston, 2016). Experienced practitioners are both thinkers and observers. They create a sense of possibility, build on existing expertise, and build learning and reflection into all they do. To be precise, this is not about micromanaging but rather about "uplifting leaders" who are both leaders and players. While it may seem that this notion is oxymoronic, the dual roles work interdependently in an educational setting. To humanize the profession and, by extension, leaders, individuals who can model and create supportive organizations are sought for educative leadership roles.

As leaders emphasize and model the practices of social justice, the theoretical framework of Paulo Freire provides conceptual understandings to move concept development toward praxis. Praxis, in short, is an action or conduct based on theory and the practical application of a theory. Freire (2000) defined praxis as "critical reflection which leads to action." Most often, praxis is applied to an art, science, or skill. Certainly, it can be argued that teaching includes all three; art, science, and skill. Essential to the development of meaningful praxis is dialogue. The dialogue should center on a genuine, authentic discourse between individuals (such as leader and learner/employee) for conscious reflection to occur. Such conscious reflection is the catalyst for creation and action (Freire, 2000). As noted previously, technology expands the allowance for dialogue and reflection and functions as an impetus for social justice when promoted and utilized by educational leaders. At the same time, it is also true that educational leaders are necessary to their organizations in order to bring the through-line of humanizing spaces and places, dialogue, connections, and communities. Therefore, it is not an overstatement to assert: leaders are crucial to a learner/employee's growth toward social justice.

What does it mean to humanize education or the educational setting? A humanized space is a space in which less of us is hidden, and more of us are known. Students feel valued, seen, respected, and safe in humanized school classrooms. The same applies to a humanized school campus or an online learning environment. A learner's/employee's identity and intersectionalities influence how one shows up on campus or in shared spaces. Identity and intersectionalities embody beliefs and values; it is crucial to understand our identity and that of others, given that perceptions and critical consciousness of ourselves, others, and the environment flow from this sense of identity (del Carmen Salazar, 2013). This principle applies to learners, employees, and leaders. Because values and beliefs are evident to others through our actions, a self-exploration of our universal human needs is essential and is how we interpret the actions of others. There is value in exploring and embracing emotions and needs, even when these can be messy and unpredictable. Bettez (2008) outlines seven humanizing practices grounded in critical reflection, dialogue, and activism:

- Promoting a mind/body connection
- Conducting artful facilitation that promotes critical thinking
- Engaging in explicit discussions of power, privilege, and oppression
- Maintaining compassion for students (learners/employees)
- Believing that change toward social justice is possible
- Exercising self-care
- Building critical communities

The role of an educational leader is demanding and complex; it is highly contextualized and personalized (Khilji, 2022). A leader who authentically knows their learners/employees provides a more humanizing space for promoting social justice. The social justice leader engages in praxis, supporting fellowship through purposeful and critical dialogue (Freire, 2000). The following section provides an overview of learning and motivation theories, leading to a comprehensive review of adult learning, central to the knowledge base of educational leaders.

LEARNING AND MOTIVATION THEORIES

This section begins with Table 1 to organize the reader's review of learning and motivation theory. Upon initial reading, the reader may identify the omission of a known or favorite theorist. The authors acknowledge at the onset that this table is not exhaustively comprehensive. Rather, the timeline of learning and motivation theorists is meant to serve as an organizational tool for educational leaders so that they may think about their leadership style and how best to move their learners/employees toward social justice. Researchers recognize the broad number of those who have contributed and those who continue to contribute to the refinement of our understanding. To all, the field remains ever appreciative.

Table 1. Timeline of learning and motivation theories progression of theory development

Theory	Theorists	Timeframe	Central Concepts
Behaviorism	Watson Skinner	1900- 1930 1940 - 1980	Objective behavior - Learning is a *response* to negative or positive stimuli
Cognitive Constructivism	Freud Erickson Piaget Bruner Bloom Lave and Wenger	1880 1959 1920 1950 1965 1990	Stages of life cycle Took Freud's work which ended with adolescence and extended into late adulthood Four stages of development - knowledge is *constructed* through individual experiences Found that learners construct ideas based on previous knowledge Learning occurs both cognitively and affectively (beliefs, values, attitudes) Communities of Practice Focused on situated learning for adults with engagement in communities of practice
Social Constructivism	Vygotsky and Social Cognition Bandura	1920 1960	Development of the individual occurs first on a social level and later on an individual level. The potential for development occurs when children participate in social behavior. The Zone of Proximal Development (ZPD) Four steps to an operant view of learning
Humanism	Maslow	1930	Experiential learning with an emphasis on choice, creativity, values, self-realization, and dignity
Motivation	Dewey Brown Gilligan	1930 1980 1980	Experiential learning leads us to more learning. Internal and external LOC factors that influence how we view ourselves and others. The Feminist voice and perspective are added to adult learning and cognition
Adult Learning	Knowles Cronbach and Snow Friere	1970 1970 1970	Founding father of adult learning views learning as cyclical: Experience leads to reflection, which leads to action, which leads to concrete reflection, and so on. Andragogy refers to adult learning while pedagogy focuses on children. Importance of strategies geared to learner's specific abilities. The critical analysis of experience and acting on that analysis lead to more learning.
Adult Learning in Professional/ Educative Settings	Joyce and Showers Schön Feiman-Nemser	1980 1983 1992	The importance of coaches to support teacher's implementation, included the importance of feedback. The reflective practitioner engages in reflection both during and after practice Professionals support and guide the enhancement of the new teacher through reflection, collaboration and shared inquiry

Note. Based on a model by Papa (2010). This version by Meetze-Hall includes the addition of theories and practice in educative settings.

ADULT LEARNING THEORY

The understanding of adult learning theory/theorists, whether it be based upon long held beliefs, or something recently learned, focuses on how adults learn. This focus on *how adults learn* then informs how we engage adult learners. One might further surmise that adult learning theory also impacts or sets the stage for managing and using technology. Sounds simple enough: what one believes impacts management and leading decisions.

If this sounds so simple, why then, are leaders so often challenged in regard to providing meaningful professional development or guiding groups to explore their ability to embrace new learning? The dilemma may be, as Papa (2010) states, "The maze of learning theories provides no single answer to define how one learns but does permit a substantial perspective to the process of learning." As leaders, learners, and authors, we concur; there are significant numbers of learning theories that educational leaders should consider. Therefore, this section provides an overview of learning and motivation theories for consideration, including questions regarding attention and motivation.

Based on a leader's observations and analysis, they may determine that they are working with powerful examples of highly functional teams. Contrastively, the leader may realize that they are not there, yet. The path forward will require continued observation, analysis, and intentional instructional decisions. Our instructional decisions are complex and may be multi-faceted. Most assuredly, there are social-political and learning theory perspectives that may both support or confound potential solutions.

Many of the theories reviewed in the chapter thus far rely on learners' motivations as the central theme when considering teaching and learning approaches and strategies. This includes first the notion of intrinsic and extrinsic motivation. It may be important to affirm that it is also possible for a learner to be concurrently both. There may be times when, what a person wants most at a given point in time, is external recognition. That same person may experience a psychological re-orientation; this transition is internal as they come to terms with change and could result in an increased focus on intrinsic motivators. As educational leaders, we may imagine that our learners/employees are intrinsically motivated and naturally curious. If so, then, as the leader, we might decide to present information by providing resources and allowing for time for personal exploration. Contrastively, we may consider delivering the content with a focus on external motivation.

In this section we provide a review and connection between seminal theories. By definition, these are the theories that keenly influenced later developments. The reader will not find the latest research in this section, rather, this is a reminder of the progression of theory development; all to support understanding of learning. The newest research and theories are included in subsequent discussion points regarding technology.

Knowles (1970) is often identified as the "founding father" of adult learning theory, a facet of which views learning as cyclical. This cycle includes the starting point of experience, leading to reflection and action, which more precisely informs the next round of evidence-based reflection. Cronbach and Snow's *Attitude Treatment Interaction* (1970) added a learner-focused perspective and then proffered that learning is best achieved when strategies are geared directly to the learner's specific abilities. We might think of this as similar to Vygotsky's Zone of Proximal Development (ZPD) or Krashen's i+1. While similar, the difference between these two concepts is that the ZPD focuses on what learners can accomplish with the help of others, and Krashen's i+1 seeks to describe a learner's language improvement with a change of input slightly more advanced than the existing or current level.

Based on this admittedly brief review of learning theory, it may now be apparent that adult learning theory is complex and continues to evolve. While adult learning theory is grounded in both cognitive and social constructivism (Papa, 2010) much weight is placed on the decisions that the leader will make based on their understandings. Leaders must acknowledge that education is part of a complex set of systems while also recognizing the reality of their context-dependent organization. Considering that adult learners accept responsibility for their learning, engage in critical analysis of experience, and acting upon that analysis leads to more learning; the leader must model the pursuit of knowledge and can affect change by providing space for conversations. These conversations may be organized or structured to include the acknowledgment that we are navigating a changing system. For some, this may be disorienting. At the same time, it will be essential to consider that as we grow older, engaging in communities of practice increases our ability to analyze our experiences (Lave and Wenger, 1991). Lave and Wenger call this *intentional reflection*, which is becoming an increasingly enacted practice in adult learning.

Following this review of learning theory and theorists, this subsequent section will briefly compare three leadership styles and explore which style aligns with a leader grounded in activism, humanity, and social justice, the type of leadership needed for the modern world.

LEADERSHIP STYLES

Leadership plays a crucial role in shaping the success of organizations and their learners/employees (Bass, 2008; Collins, 2001). Three widely recognized leadership styles include transactional, transformational, and transformative. Table 2 (adapted from Shields, 2009; 2010) summarizes these leadership styles and highlights their focus, operational process, key values, and leadership dispositions. As noted in the table, transformative leadership prioritizes "deep and equitable change in social conditions."

Table 2. Leadership styles distinctions among three leadership styles

	Transactional Leadership	Transformational Leadership	Transformative Leadership
Emphasis of Focus	Means	Organization	Deep & equitable change in social conditions
Process	Immediate cooperation through mutual agreement and benefit	Understanding of organizational culture; setting directions, developing people, redesigning the organization, and managing the instructional program.	Deconstruction and reconstruction of social/cultural knowledge frameworks that generate inequity; acknowledgement of power & privilege; dialogue between individual & social
Key Values	Honesty, responsibility, fairness and honoring commitments	Liberty, justice, equality	Liberation, emancipation, democracy, equity, justice
Leader	Ensures smooth and efficient organizational operation through transactions	Looks for motive, develops commons purpose, focuses on organizations goals	Lives with tension & challenge; requires moral courage, activism

Note. Adapted from Shields (2009; 2010).

TRANSFORMATIVE LEADERSHIP

Transformative leadership, not to be confused with transformational leadership, focuses on adult learners/employees, and provides a conceptual framework for best supporting learning outcomes (Dirkx, 1998). Critical attributes of transformative leadership include: a combination of critique and promise, attempts to effect profound and equitable change, deconstruction and reconstruction of knowledge inquiry, acknowledgment of power and privilege, emphasis on individual and community achievements, focus on liberation, equity and justice, and evidence of moral courage (Shields, 2010; Shields & Hesbol, 2020). Transformative leadership is grounded in an activist agenda and has come into being through the work of multiple scholars; in this section, the authors examine two notable scholars' throughlines: Paulo Freire and Jack Mezirow.

Freire

As noted above, the work of Freire (2000) is central to the role of an educational leader, social justice, and humanizing the modern world. The transformative leader's role is to dialogue with the learners/employees to develop mutual respect and better understand their backgrounds, cultures, and lived experiences. Through these dialogues, greater awareness emerges; individuals can move forward with a deeper conscientiousness regarding knowledge, critique, community, activism, and self-direction. In Freire's (2000) words:

As women and men, simultaneously reflecting on themselves and on the world, increase the scope of their perception, they begin to direct their observations toward previously inconspicuous phenomena... That which had existed objectively but had not been perceived in its deeper implications (if indeed it was perceived at all) begins to "stand out," assuming the character of a problem and, therefore, a challenge. Thus, men and women begin to single out elements from their "background awareness" and to reflect upon them. These elements are now objects of their consideration and, as such, objects of their action and cognition. (pp. 82-83, italics added for emphasis)

Core to Freire's work is the understanding of praxis; this process consists of reflection followed by action rooted in changing the world for the better. Dialogic relationships allow for critical reflection of our spaces and organizational communities, followed by the ability to remake them more humane and equitable; the transformative leader stimulates these relationships. An educational leader must clearly understand power and equity at the intellectual level to lead with and model transformative practices (Shields, 2010).

Mezirow

Mezirow's (1990, 1991,1997, 2008) contributions to transformative leadership are well-established. His early understanding of this theory demonstrated the importance of dialogue with others to identify different perspectives and provide empathy and support as individuals navigated and negotiated new interpretations (Mezirow & Associates, 1990). Critical to his work are reflexivity and meaning-making, the ability of a learner/employee to reflect on problem-solving and problem-posing to make meaning; Mezirow has identified ways in which leaders can support growth within their organizations.

Transformative learning involves an enhanced level of awareness of the context of one's beliefs and feelings, a critique of their assumptions and particularly premises, an assessment of alternative perspectives, a decision to negate an older perspective in favor of a new one or to make a synthesis of old and new, an ability to take action based upon the new perspective, and a desire to fit the new perspective into the broader context of one's life. (Mezirow, 1991, p. 161)

One's frames of reference influence meaning-making. These are cultural and linguistic backgrounds or funds of knowledge from which learners/employees attribute coherence and significance to an experience. Mezirow's (2008) five structures for transformative learning include:

- Reflecting critically on the source, nature, and consequences of relevant assumptions - our own and those of others;
- In instrumental learning, determining that something is true (is as it is purported to be) by using empirical research methods;
- In Communicative learning, arriving at more justified beliefs by participating freely in an informed continued discourse;
- Taking action on our transformed perspective – we make a decision and live what we have come to believe until we encounter new evidence, argument or a perspective that renders this orientation problematic and requires reassessment;
- Acquiring a disposition – to become more critically reflective of our own assumptions and those of others, to seek validation or our transformative insights through more freely participating in discourse and to follow through on our decision to act upon transformed insight.

It is important to highlight again that Freire and Mezirow have made significant contributions to the educational leadership field with their theoretical stance of transformative leadership. Their work focuses on supporting individuals and communities through praxis, empathy, reflection, dialogue, and meaning making. These collaborative efforts provide space and opportunity for promoting social justice and engaging in advocacy. The social justice leader assumes an activist role for educational organizations and social change (Cambron-McCabe & McCarthy, 2005); the works of Freire and Mezirow provide language, beliefs, and actions to support these intentions. In the following section, the authors highlight the conditions and locations for learning.

CONDITIONS AND LOCATIONS FOR LEARNING

Technology-driven learning opportunities are created or thwarted based on both access and use. Particularly when considering dimensions related to technology use, the transformative leader must consider locations for learning. The notion of learning being impacted by contexts and perspectives is connected to researchers such as Mezirow and Merriam. Mezirow (1991) identified the importance of understanding the conditions and locations for learning; these contexts shape the learner's/employee's frames of reference. Frames of reference are influenced by a range of perspectives, including epistemic, sociolinguistic, and psychological. These frames of reference were identified earlier in the chapter and will now be expanded upon. The transformative leader needs to understand the relationship between these perspectives and the learning process, as they can profoundly impact the outcome of the learning experience.

Epistemic perspectives refer to a learner's/employee's beliefs and attitudes toward knowledge and how knowledge is acquired, processed, and used. Individuals with a social constructivist epistemic perspective may believe that knowledge is subjective and constructed through personal experiences, interactions, and dialogue with others (Mezirow, 1991).

Sociolinguistic perspectives refer to the social and cultural context in which language and communication occur. A learner/employee who is a native English speaker may have a different sociolinguistic perspective compared to an individual who is new to the English language (Mezirow, 1991). Considering an individual's background, language proficiency, and communication style allows the transformative leader the opportunity and access points to design relevant learning experiences.

Psychological perspectives refer to the learner's/employee's cognitive, emotional, and behavioral characteristics that impact the learning process (Mezirow, 1991). Dweck's (1999) research is especially illuminating regarding cognitive-motivational patterns and how these patterns originate in people's self-theories. An individual with a growth mindset may be more apt to take risks and try new things than an individual with a fixed mindset who may resist change (Dweck, 2006). From this work, we learn that when an educational leader is aware of an individual's needs and preferences, it is easier to provide support and encouragement, allowing the learner/employee to reach their full potential.

Additionally, the learner's/employee's position in the sociocultural, physical, and individual sense should also be considered. As Merriam (2008) proposed, a holistic approach to learning recognizes the significance of all experiences and learning, including storytelling, music, dance, folklore, and others. In this space, technology allows for increased access to learning through varied perspectives. These approaches consider the learner's/employee's body, emotions, mind, and spirit and incorporate all aspects of the individual into the learning experience. In this way, the transformative leader ensures a safe, accessible, and humane opportunity for learning.

Mezirow's (1991) and Merriam's (2008) ideas underscore the importance of understanding the contexts and perspectives that shape the learning experience. By considering these elements, the transformative leader can create an inclusive, effective, and meaningful environment. The following section provides the reader with context about the beliefs and praxis of the transformative leader.

THE TRANSFORMATIVE LEADER: BELIEFS AND PRAXIS

Leadership for social justice is action-oriented and transformative, committed and persistent, inclusive and democratic, relational and caring, reflective, and oriented toward a socially-just pedagogy/ (Furman, 2012)

Transformative leaders are characterized by several beliefs and practices that drive their leadership approach and style. Mountouri & Donnelly (2018) identify four beliefs or orientations that guide the transformative leader: ways of being, relating, knowing, and doing. Central to their belief system is recognizing that learners/employees are creative and always creating (Montuori & Donnelly, 2018), particularly through their understandings based upon interpretations of worldviews, knowledge, and frames of reference (Mezirow, 1991).

With leadership traditionally viewed as a top-down approach, Mountouri & Donnelly (2018) argue that relating to others is a key belief. This style requires following a democratic process and engaging in creativity; the end game is to benefit the community, learners/employees, and the organization as a whole. Winners and losers are not part of the transformative leader's belief system.

Transformative leaders believe in people first. Their beliefs about knowing center on sense-making, humanity, self-reflection, and relationships. Transformative leaders recognize their humanity and are open to uncertainty while focusing on contextualizing, connecting, and engaging in complex thought (Montuori & Donnelly, 2018). While centering on humans and sense-making, the transformative leader believes adults can critique their own assumptions, assess alternative perspectives, take action on a new perspective, and hold a desire to fit a new perspective into a broader understanding (Mezirow, 1991).

The transformative leader believes in identifying multiple ways to achieve desired outcomes; there are no prescribed rules. This includes being open to a broad worldview and drawing on the assets and skills of the learners/employees. A steady balance remains between listening, collective inquiry, and dialogue (Montuori & Donnelly, 2018). Transformative leaders believe in others taking action; action is integral to transformative learning. Because of this, the transformative leader believes in reflective and transformative learning practices. Reflective learning supports the confirmation, addition, or transformation of ways of interpreting the learner's/employee's experience (Mezirow, 1991).

Lastly, the transformative leader holds firm beliefs about creating a community of practice and is a community builder (Dewey, 1927). Creating a community of practice requires developing an emotionally safe space where learning, growth, and diverse perspectives flourish and are appreciated (Khilji, 2022).

Transformative leaders are people-centered and prioritize relationships, sense-making, and self-reflection. Their beliefs and practices keep social justice at the forefront. They engage in democratic processes, creativity, and collective inquiry to benefit the community, learners/employees, and their organization. Transformative leaders also focus on creating a community of practice that fosters emotional safety and encourages learning, growth, and opportunity. A practice that supports learning and growth is the use of evidence-based feedback.

EVIDENCE-BASED FEEDBACK TO SUPPORT GROWTH

Once again, the specificity of terminology is crucial to move the conversation and practice forward. For example, the term 'feedback' can be understood as information about reactions to a product, a person's performance of a task, etc., which is used as a basis for improvement. With other audiences, the term 'feedback' may indicate the screeching or humming sound from an amplifier. One would hope that when a leader provides feedback to an employee or staff member, the feedback is usable and not received as noise. Here the authors rely on the term to describe information for the basis of improvement.

Multiple theorists and resultant research support the importance of feedback; this frequently includes the contention that such exchanges between learner and coach should include both support and challenge. Other researchers, such as Bruner, add the notion of scaffolding as an essential facet of structured learning. This section asks the leader to consider various ways they may enter into a learning-focused conversation with their learners and the fulcrum of instructional decisions.

Bruner (1960) built upon the theory of active learning with the development of discovery learning and suggestions for scaffolding. One of the guiding principles was that learning takes place "in situ" (p. 28). To support a learner, Bruner suggested the concept of a scaffold, where. supports are in place until they can be removed for greater autonomy. The author argued that educators should consider the difference between learning and thinking and defined thinking as the "operation of utilizing information to go beyond the information" (p. 29). Bruner added to the field the inclusion and importance of dialogue in the learner's discovery and use of reflection.

Bandura's (1977) social learning theory sought to explain the phenomena of how individuals process via observational learning. Bandura contended that behavior modeling could include students observing students for social clues and norms as well as how to function in the school environment. This theory has also been applied to mentor and mentee roles, where the mentor provides the model, and the mentee is the observer. In either of these relationships, reality is reinforced, and the observer can be acculturated to the context. The combined impact of Dewey's constructivism, Bruner's discovery learning, and Bandura's social learning support the important role that observation, feedback, and reflection contribute to successful learning experiences.

Wenger (1998) began describing community of practice (COP) theory by stating the first underlying assumption that humans are social beings. Wenger then contended, "The primary focus of this theory (COP) is on learning as social participation" (p. 4). The claim further contended that communities of practice are groups of people who share a concern or a passion. They develop a shared repertoire of resources: experiences, stories, tools, and ways of addressing recurring problems—in short, a shared practice. This takes time and sustained interaction.

The work of Mezirow (1991) was based on a more generalized population of adult learners, while Schön (1983) was interested in reflective learning by professional practitioners, particularly in the medical field. The field of education quickly adopted the importance of reflection in developing the skills and knowledge of teacher practice. What has perhaps been lost is the distinction that Schön (1983) made between the structure for reflection in action versus reflection on action. Schön argued that professionals learn while doing when they may need to improvise in the moment. For Schön, learning from reflecting on the action of their own professional experiences after an event is especially important in the iterative nature of learning cycles and the resultant application of experience-based learning.

Mezirow's (1991) theoretical distinction centered on knowledge learning versus perspective learning. According to Mezirow, transformational learning (TL) is a change in perspective or beliefs (a paradigm shift). The first step in TL requires a disorienting dilemma and a resulting exploration and action plan. Knowledge and perspective learning are necessary for supporting new teacher development. Without a change in paradigm, educators might not consider the necessity of reflecting on knowledge learning, which represents the how and what of their professional practice.

Built upon these seminal theorists' work, the educator development field has been dramatically altered by the debate about learning as a socially constructed activity and the importance of reflection. The elements of teacher preparation and the attributes of teacher induction have been studied extensively (Cherubini, 2009; Darling-Hammond, 2000; Delaney, 2012; Feiman-Nemser, 2001; Ingersoll, 2014). With constructivism, discovery learning, and social learning as their foundation, studies have used the theories of reflective practice, transformative learning, and communities of practice to understand the perspectives of experienced educators as they engage with teachers new to the profession. Collectively, these learning theories inform leader and coaching development and provide a framework for understanding educative, evidence-based feedback.

While the notion of learning resulting from reflection is connected to theorists such as Mezirow, Dewey, and Schön, the concept of reflective practitioner also extends beyond education to Schön's work in the area of professional knowledge development. Indeed, a broad study of developing educators might be similar to the medical example labeled "situations of practice" (Schön, 1983, p. 16) in which the professional must be knowledgeable and analytical of technical components and be skillful in synthesis and creative in solutions. Markie (1994), too, made the connection to teacher analysis: "The first criterion is intellectual competence . . . but he should have some capacity for analysis. Without this capacity, he cannot develop it in his students" (p. 90). This focus on the analysis of classroom practice will move a teaching practice forward.

As they created their instructional guide for mentors, Lipton, and Wellman (2013) acknowledged and identified challenges for new teachers: "new teachers often have a mistaken belief in the existence of a readily available package that can transform their classes. Lipton and Wellman further claimed that mentoring relationships are central to the success of ongoing learning within induction programs. The necessity of ongoing learning is increasingly true as the educational landscape expands beyond brick and mortar's building-bound constraints into cyber communities.

With an anomalous shift from these learning theorists and their concepts, we focus now on both the technology leader and technology-supported learning. It could be imagined that the leader may be tasked with taking an institution's delivery from a face-to-face delivery model to a hybrid learning model or, on a smaller but no less challenging scale, they may be tasked with helping others to embrace Bring Your Own Device (BYOD) models. As is often the case, leaders face significant challenges as they balance between leadership and management, concepts that are not mutually exclusive. Papa (2010) reminded readers that the leader will need to be reasonably fluent in technology; it is no longer a new frontier but a necessity to bring about systemic improvements to learning experiences. Leaders must consider the strain on learners/employees when adopting technology.

So, what are some of the points a leader may need to consider? Some of the initial studies of technology in education began by identifying technology tools. Built initially for advanced research projects in the 1970s, networks for electronic communication are now commonplace. Originally, the term Computer-Supported Collaborative Learning (CSCL) focused on issues related to learning through collaboration and promoting productive collaborative discourse with the help of the computer and other communications technologies. The term CSCL is not as widely used, but the technology that it brought forth is central to most of today's learning management systems (LMS). This includes relying on such components as discussion boards, electronic whiteboards, and application add-ons for workforce productivity. It also contended that the newest versions experienced today are crucial tools for social justice. While the technology may have changed over time, CSCL studies have continued to inform technology use by educative mentors. One such study is by Baglione and Nastanski (2007), where the research findings provide important information regarding faculty perceptions of technology use by the educational leader.

In addition to Baglione and Nastanski, Klecka et al. (2004) explored the potential of electronic mentoring as distributed communities of practice when they explored asynchronous communication within an Illinois novice teacher program. Relying on open-ended surveys, focus group interviews, field notes, and reviewing posted messages and user login data, after three years of data collection, they found participation depends on much more than providing access to workshops or electronic mentors. With a new medium, such as electronic mentoring, the barriers that prevent or incentives that encourage one to log in and engage in conversations help to define the nature of participation. The nature of participation is an important contextual element of any community.

Moving from these early studies, the field has now moved far beyond the exploration of potential electronic mentoring. As practitioners and the research field have progressed, it is now possible to find a plethora of studies based on the cognitive theory of multimedia learning. This may be an area that a technology leader will need to draw upon, whether leading technology enthusiasts or technology skeptics (Collins & Halverson, 2018).

When reviewing the significant number of technology studies, it becomes evident that significant numbers of skeptics still discount technology as a viable tool for learning and communication. Some of these skeptics include university instructors and school boards (Allen & Seaman, 2013). Many of the studies and arguments by Dixson (2015), Koutropoulos (2011), Surrette and Johnson (2013) and Lineweaver (2010) were positioned to either support or refute the effectiveness of online learning, and they often included measures of engagement. Still, other studies, such as Duncan-Howell (2010), Noroozi et al. (2011), and Quintana and Zambrano (2014) began with the perspective that technology and online mentoring may be helpful where geography creates challenges of access. Collins and Halverson (2018) add useful clarifying language to help with identifying ways that we may need to rethink learning, motivation, and educational leadership (among other topics). Certainly, these considerations are congruent with the argument contentions laid out in the chapter and remind us to continue to observe and consider how to lead in schooling, learning, and technology.

The nature of participation is an important contextual element of any community and is often addressed in educational studies. The contextual elements are also essential considerations for the educational leader. Of significant consideration to the technology leader are the instances where technology supports access to collaboration and evidence-based instructional practices to support all participants' goals. In a study by Zwart et al. (2009), the analysis indicated that there were characteristics of a mentoring relationship that had a greater impact and caused teachers to try out new teaching strategies while being observed. The characteristics included partnered observation, which reduced the potential embarrassment of "trying something new in front of another when partnered with a peer" (p. 254).

A recent addition to educational terminology is that of a reciprocal coach. The early literature outlines behaviors and skill sets that are congruent with other coaching models. Alger and Kopcha (2011) examined the clinical experiences of preservice teachers supported with technology tools. They found a type of non-reflective coaching described by study participants who were in the preservice level of teacher preparation. Examples of the coded language in the study included the following behaviors grouped under coaching: "experts make comments on lessons; expert feedback on videotape lessons; experts made comments on lessons; feedback provided via templates; triad members shared advice and solutions." From these responses, the study found a preferred differentiation between what the more novice practitioners expected and wanted compared to those more mature in their practice. The more novice practitioners wanted 'advice giving,' which is to a degree contrastive to other adult learners. For more advanced practitioners, the leader or coach facilitates the adult's sense-making and co-creates the generation of solutions.

How transformative leaders guide the adult learner will be impacted by contextual realities, including the strength of existing systems. What do the studies and learning theorists then tell leaders about planning and their actions? In short, what is the starting point and the way forward? A question to consider might be: how does the leader create a space for professional exchanges that support, challenge, and provide evidence-based feedback?

Based on Knowles' work (1990), leaders must acknowledge that relevance to their jobs (or personal life) is as important to adult learners as the need to plan and evaluate their instruction. Because adult learning requires an analysis based on critical reflection, the transformative leader provides experiences and space for reflection so that the adult learner can capitalize on the experience. This also includes creating safety for individuals to be open and honest. The leader will want to include professional growth opportunities centered on positive, evidence-based conversations by including the elements of psychological safety, clear focus, and differentiated support.

When considering the notion of differentiated support, the educational leader will analyze what they know about the adult learners and use their own theory of action and understanding of learning and social justice to make leadership and learning decisions. For example, the stance and sentence stems of the leader night include coaching, with the use of facilitative language, wherein the leader begins with questions to deepen the learner's thinking. These include examples that are beyond information queries. Rather, facilitative questioning may start with sentence stems, such as:

- As you examine.
- What might it look like if…
- Let's examine…

- On the basis of what you know about your students…
- What do the multiple sources of data…
- What does it mean when…
- What do you see and hear when your students…
- What does it look like when your students…

As the transformative leader and learner/employee enter further into the conversation, psychological safety continues to be provided by using data as the "third point" (Grinder, 1997). In selecting impactful data points for analysis and discussion, the leader and learner will need to determine those data which will be most productive. With careful selection of data, the conversation focuses not on the teaching practice but puts the discussion evidence to the side to explore it together.

Differentiation with learners also depends on the leader's knowledge and awareness of the learner's expertise. As leaders work to move employees toward expanded use and access to technology, the aim is also to develop efficacy and enhance capacities. When needed, the transformative leader may be called to provide information directly. Such might be the case when responding to changes in preferences to learning modalities. Helping learners/employees envision a change of practice that includes a broad range of offerings and flexible learning modalities will necessitate leaders monitoring their approaches carefully. An example of which is the MOOC Pivot (Reich, 2019). Unpacking or making their thinking transparent is helpful to both the leader and learner and can assist the learner in making connections to personal contexts.

As depicted in Figure 1, multi-directional interactions occur during an evidence-based reflective conversation. As previously mentioned, there should be an agreed-upon selection of data points. The third point (for example, video recordings and/or student work samples) may be used as evidence during the conversation. The determination of the topic of focus and data points may take place in a community of practice or between the individuals interested in improvement. As with a community of practice, all parties bring forth evidence as the agreed-upon focus. The communities of practice, which must include conversations that provide both support and challenge, contribute to optimum conditions to propel growth. Said in another manner, evidence-based reflection is at the center of growth-focused conversations.

In addition to the studies and considerations above, there has also been increased understanding and research about "the understandings, capacities, and dispositions need to make sense of information" (Carless & Boud, 2018). This schema includes the epistemological dimension (engagement of learners) as well as the ontological dimension (investment related to identity). Again, this is a reinforcement and reminder of the learner/employee's self-identity and how the learner/employee may be poised to be an active participant in growth-focused conversations.

THE TRANSFORMATIVE LEADER: ACCELERATING SOCIAL JUSTICE

Transformative leadership, as with other forms of leadership, is dependent upon the individual and their beliefs and practices. As the authors have set forth in this chapter through research, claims, and a conceptual framework (Figure 1), this work depends upon many connecting and interweaving ideas and practices. Headlining the conceptual framework is the understanding that technology-aided dialogue, collaboration, and communities of practice will consistently be in play within an organization. Moving then to the bottom of the conceptual framework, we work up. As noted by Freire (2000) and Mezirow

(1991), within communities of practice, dialogue is central, as this allows for feedback to occur. Feedback aids in supporting and challenging learners/employees, which leads to evidence-based reflection. Through this reflective practice, knowledge acquisition and cognition occur in addition to raised levels of consciousness. As leaders, learners, and employees continue to elevate their stance of growth, humanity, and self-concept, one reaches a level of intentional praxis. Through praxis, purposeful reflection results in action for social justice and equity. Moreover, we move the needle to a more equitable and just world in this space.

Figure 1. Technology-aided coaching model for social justice

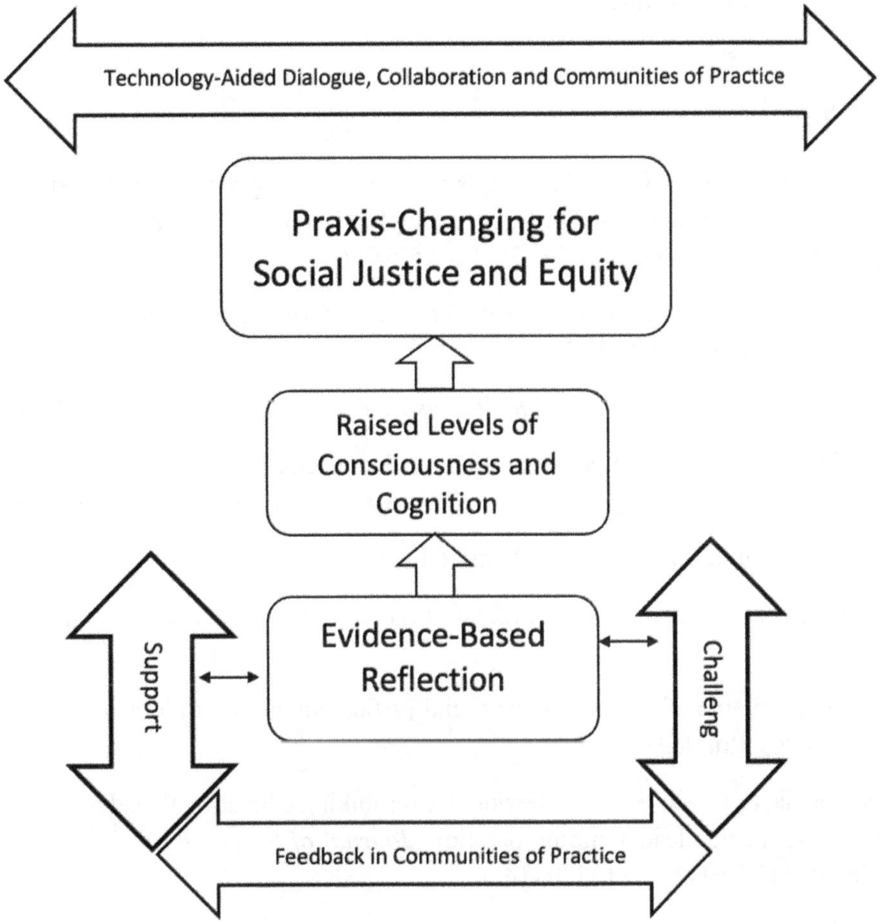

CONCLUSION

Progressing from thinking about learners, learning, and leadership, the authors underscore the importance of perspective and reflection while holding fast to foundational underpinnings of transformative change through individualized and community dialogue. To carry out and support the implementation of new technologies, the leader must be willing to examine and alter existing leadership practices.

A leader who can balance the needs of both learners/employees and the organization is key to promoting growth. This balance between the learner's support and engagement and the leader's experience, inclination, and willingness to experiment has the potential to create a safe, trusting environment where there are opportunities to coach and be coached. While implementing new practices, the leader will be mindful of ways that interactions are increased but also mindful of pitfalls that could further marginalize learners and employees.

This chapter has asserted that transformative leadership is essential to educational organizations. While management is necessary when supporting growth, transformative leadership allows new models to be developed. As the modern world increases in connectivity through shared dialogue and technology, the transformative leader is situated to address the expanding need for humane, socially just learning spaces to move organizations forward.

REFERENCES

Alger, C., & Kopcha, T. J. (2011). Technology supported cognitive apprenticeship transforms the student teaching field experience: Improving the student teaching field experience for all triad members. *Teacher Educator*, *46*(1), 71–88. doi:10.1080/08878730.2010.529986

Allen, I., & Seaman, J. (2013). *Changing course: Ten years of tracking education in the United States.* https://files.eric.ed.gov/fulltext/ED541571.pdf

Aubrey, K., & Riley, A. (2015). *Understanding and using educational theories.* SAGE Publications.

Baglione, S., & Nastanski, M. (2007). Superiority of online discussion: Faculty perceptions. *Quarterly Review of Distance Education*, *8*(2), 139–150. https://www.infoagepub.com/index.php?id=89&i=4

Bandura, A. (1977). *Social learning theory.* Prentice Hall.

Bass, B. M. (2008). *The Bass Handbook of Leadership: Theory, Research, and Managerial Applications* (4th ed.). Free Press.

Bell, L. A. (2016). Theoretical foundations for social justice education. In *Teaching for diversity and social justice* (pp. 3–26). Routledge.

Bertrand, M., & Rodela, K. C. (2018). A framework for rethinking educational leadership in the margins: Implications for social justice leadership preparation. *Journal of Research on Leadership Education*, *13*(1), 10–37. doi:10.1177/1942775117739414

Bettez, S. C. (2008). Social justice activist teaching in the university classroom. *Unsettling beliefs: Teaching theory to teachers*, 279-296.

Bogotch, I. E. (2002). Educational leadership and social justice: Practice into theory. *Journal of School Leadership*, *12*(2), 138–156. doi:10.1177/105268460201200203

Bruner, J. (1960). *The process of education.* Harvard College Press.

Cambron-McCabe, N., & McCarthy, M. M. (2005). Educating school leaders for social justice. *Educational Policy*, *19*(1), 201–222. doi:10.1177/0895904804271609

Carless, D., & Boud, D. (2018). The Development of Student Feedback Literacy: Enabling Uptake of Feedback. *Assessment & Evaluation in Higher Education, 43*(8), 1315–1325. doi:10.1080/02602938.2018.1463354

Cherubini, L. (2009). New teachers' perceptions of induction: Insights into principles practices. *The Alberta Journal of Educational Research, 55*(2), 185–198. http://www.ajer.ca/

Collins, A., & Halverson, R. (2018). *Rethinking education in the age of technology: The digital revolution and schooling in America*. Teachers College Press.

Collins, J. (2001). *Good to great*. Random House Business Books.

Corbett, F., & Spinello, E. (2020). Connectivism and leadership: Harnessing a learning theory for the digital age to redefine leadership in the twenty-first century. *Heliyon, 6*(1), e03250. doi:10.1016/j.heliyon.2020.e03250 PMID:31993523

Costa, A., & Garmston, R. (2002). *Cognitive coaching: A foundation for Renaissance schools*. Academic Press.

Darling-Hammond, L. (2000). *Teacher quality and student achievement: A review of state policy evidence*. doi:10.14507/epaa.v8n1.2000

del Carmen Salazar, M. (2013). A humanizing pedagogy: Reinventing the principles and practice of education as a journey toward liberation. *Review of Research in Education, 37*(1), 121–148. doi:10.3102/0091732X12464032

Delaney, Y. A. (2012). Research on mentoring language teachers: Its role in language education. *Foreign Language Annals, 45*(1), 184–202. doi:10.1111/j.1944-9720.2011.01185.x

Dewey, J. (1927). *The public and its problems*. Holt and Company.

Dirkx, J. M. (1998). Transformative learning theory in the practice of adult education: An overview. *PAACE Journal of Lifelong Learning, 7*, 1–14.

Dixson, M. (2015). Measuring student engagement in the online course: The online student engagement scale. *Online Learning, 19*(4), 1–14. http://www.online learningconsortium.org

Drago-Seversen, E., Blum-DeStefano, J., & Brooks Lawrence, D. (2023). *Growing for justice: A developmental continuum of leadership capacities and practices*. Corwin.

Duncan-Howell, J. (2010). Teachers making connections: Online communities as a source of professional learning. *British Journal of Educational Technology, 41*(2), 324–340. doi:10.1111/j.1467-8535.2009.00953.x

Dweck, C. (2006). *Mindset: The new psychology of success: How we can learn to fulfill us potential*. Ballantine Books.

Dweck, C. S. (1999). *Self-theories: Their role in motivation, personality, and development*. Psychology Press.

Dyson, F. J., Hansma, P. K., Roof, W. C., & Proctor, J. D. (2001). *Technology and social justice*. UCTV, Instructional Resources, University of California.

Feiman-Nemser, S. (2001). Helping novices learn to teach. *Journal of Teacher Education, 52*(1), 17–26. doi:10.1177/0022487101052001003

Freire, P. (2000). *Pedagogy of the oppressed.* Continuum.

Freire, P. (2020). Pedagogy of the oppressed. In P. Kuppers & G. Robertson (Eds.), *Toward a sociology of education* (pp. 374-386). Routledge. doi:10.4324/9781003060635-5

Furman, G. (2012). Social justice leadership as praxis: Developing capacities through. preparation programs. *Educational Administration Quarterly, 48*(2), 191–229. doi:10.1177/0013161X11427394

Grinder, M. (1997). *The science of nonverbal communication.* Michael Grinder and Associates.

Hytten, K., & Bettez, S. C. (2011). *Understanding education for social justice*. Educational.

Ingersoll, R. (2014). *Seven trends: The transformation of the teaching force.* Consortium for Policy Research in Education.

International Society for Technology in Education. (2023). *Standards for educational leaders.* https://www.iste.org/standards/iste-standards-for-education-leaders

Khilji, S. E. (2022). An approach for humanizing leadership education: Building learning. community & stakeholder engagement. *Journal of Management Education, 46*(3), 439–471. doi:10.1177/10525629211041355

Klecka, C. L., Cheng, Y.-M., & Clift, R. T. (2004). Exploring the potential of electronic mentoring. *Action in Teacher Education, 26*(3), 2–9. doi:10.1080/01626620.2004.10463327

Knowles, M. (1996). Andragogy: An emerging technology for adult learning. *Boundaries of adult learning*, 82-98.

Koutropoulos, A. (2011). Digital natives: Ten years after. *Journal of Online Learning.*

Kumashiro, K. K. (2000). Toward a theory of anti-oppressive education. *Review of Educational Research, 70*(1), 25–53. doi:10.3102/00346543070001025

Kyaw, B., Saxena, N., Posadzki, P., Vseteckova, J., Nikolaou, C. K., George, P. P., Divakar, U., Masiello, I., Kononowicz, A. A., Zary, N., & Tudor Car, L. (2019). Virtual reality for health professions education: Systematic review and meta-analysis by the digital health education collaboration. *Journal of Medical Internet Research, 20*(1), e12959. doi:10.2196/12959 PMID:30668519

Ladson-Billings, G. (2021). I'm here for the hard re-set: Post pandemic pedagogy to preserve. Our culture. *Equity & Excellence in Education, 54*(1), 68–78. doi:10.1080/10665684.2020.1863883

Lave, J., & Wenger, E. (1991). *Situated learning: Legitimate peripheral participation*. University Press. doi:10.1017/CBO9780511815355

Lineweaver, T. T. (2010). Online discussion assignments improve students' class preparation. *Teaching of Psychology, 37*(3), 204–209. doi:10.1080/00986283.2010.488546

Lipton, L., & Wellman, B. (2013). *Learning focused supervision: Developing professional expertise in standards-driven systems*. Mira Via.

Lowyck, J. (2014). Bridging learning theories and technology-enhanced environments: A Critical appraisal of its history. In *Handbook of research on educational communications and technology* (pp. 3-20). Springer. doi:10.1007/978-1-4614-3185-5_1

Markie, P. J. (1994). *A professor's duties: Ethical issues in college teaching*. Rowman & Littlefield.

Merriam, S. B. (2008). Adult learning theory for the twenty-first century. *New Directions for Adult and Continuing Education, 119*(119), 93–98. doi:10.1002/ace.309

Mezirow, J. (1990). *Fostering critical reflection in adulthood*. Jossey-Bass Publishers.

Mezirow, J. (1991). *Transformative dimensions of adult learning*. Jossey-Bass Publishers.

Mezirow, J. (1997). Transformative learning: Theory to practice. *New Directions for Adult and Continuing Education, 74*(74), 5–12. doi:10.1002/ace.7401

Mezirow, J. (2008). An overview on transformative learning. *Lifelong Learning,* 40-54.

Montuori, A., & Donnelly, G. (2018). Transformative leadership. In J. Neal (Ed.), *Handbook of personal and organizational transformation* (pp. 319–350). Springer. doi:10.1007/978-3-319-66893-2_59

Neimann, T., & Wang, V. (2017). Deep Learning and Online Education as an Informal Learning Process: Is There a Relationship between Deep Learning and Online Education as an Informal Learning Process? In V. Wang (Ed.), *Adult education and vocational training in the digital age* (pp. 37–57). IGI Global. doi:10.4018/978-1-5225-0929-5.ch003

Noroozi, O., Busstra, M. C., Mulder, M., Biemans, H. J. A., Tobi, H., Geelen, A., van't Veer, P., & Chizari, M. (2011). Online discussion compensates for suboptimal timing of supportive information presentation in a digitally supported learning environment. *Educational Technology Research and Development, 60*(2), 193–221. doi:10.100711423-011-9217-2

North, C. E. (2008). What is all this talk about "social justice"? Mapping the terrain of education's latest catchphrase. *Teachers College Record, 110*(6), 1182–1206. doi:10.1177/016146810811000607

Papa, R. (Ed.). (2010). *Technology leadership for school improvement*. Sage.

Papa, R. (Ed.). (2014). *Media rich instruction: Connecting curriculum to all learners*. Springer.

Quin, J. (2009). Growing social justice educators: A pedagogical framework for social justice education. *Intercultural Education, 20*(2), 109–125. doi:10.1080/14675980902922192

Quintana, M. G. B., & Zambrano, E. P. (2014). E-mentoring: The effects on pedagogical training of rural teachers with complex geographical accesses. *Computers in Human Behavior, 30*, 629–636. doi:10.1016/j.chb.2013.07.042

Reich, J., & Ruipérez-Valiente, J. A. (2019). The MOOC pivots. *Science, 363*(6423), aav7958. Advance online publication. doi:10.1126cience.aav7958 PMID:30630920

Reiser, R. A. (2001). A history of instructional design and technology: Part I: A history of instructional media. *Educational Technology Research and Development*, *49*(1), 53–64. doi:10.1007/BF02504506

Saunders, L. (2017). Connecting information literacy and social justice: Why and how. *Communications in Information Literacy*, *11*(1), 15. doi:10.15760/comminfolit.2017.11.1.47

Schaffhauser, D. (2016). Report: Education Tech Spending on the Rise. *The Journal*. https://thejournal.com/articles/2016/01/19/report-education-tech-spending-on-the-rise.aspx?m=1

Schejter, A., & Tirosh, N. (2014). New Media Policy: The Redistribution of Voice 1. In Y. Liu & R. Picard (Eds.), *Policy and marketing strategies for digital media* (pp. 73–86). Routledge. doi:10.4324/9781315794303-6

Schejter, A. M., & Tirosh, N. (2015). "Seek the meek, seek the just": Social media and social justice. *Telecommunications Policy*, *39*(9), 796–803. doi:10.1016/j.telpol.2015.08.002

Schön, D. (1983). *The reflective practitioner: How professionals think in action*. Basic Books.

Shields, C. M. (2004). Dialogic Leadership for Social Justice: Overcoming Pathologies of Silence. *Educational Administration Quarterly*, *40*(1), 109–132. doi:10.1177/0013161X03258963

Shields, C. M. (2009). Transformative Leadership: A Call for Difficult Dialogue and Courageous Action in Racialised Contexts. *International Studies in Educational Administration, 37*(3).

Shields, C. M. (2010). Transformative leadership: Working for equity in diverse contexts. *Educational Administration Quarterly*, *46*(4), 558–589. doi:10.1177/0013161X10375609

Shields, C. M., & Hesbol, K. A. (2020). Transformative leadership approaches to inclusion, equity, and social justice. *Journal of School Leadership*, *30*(1), 3–22. doi:10.1177/1052684619873343

Surrette, T., & Johnson, C. (2013). Assessing the ability of an online environment. *School Science and Mathematics*, *115*(6), 260–270. doi:10.1111sm.12132

Theoharis, G. (2007). Social justice educational leaders and resistance: Toward a theory of social justice leadership. *Educational Administration Quarterly*, *43*(2), 221–258. doi:10.1177/0013161X06293717

Vygotsky, L. (1986). *Thought and Language*. Massachusetts Institute of Technology Press. (Original work published 1934)

Weis, T. M., Benmayor, R., O'Leary, C., & Eynon, B. (2002). Digital technologies and pedagogies. *Social Justice, 29*(4), 153-167.

Wenger, E. (1998). *Communities of practice: Learning, meaning, and identity*. Cambridge University Press. doi:10.1017/CBO9780511803932

Zwart, R., Wubbels, T., Bergen, T., & Bolhuis, S. (2009). Which characteristics of a reciprocal peer coaching context affect teacher learning as perceived by teachers and their students? *Journal of Teacher Education*, *60*(3), 243–257. doi:10.1177/0022487109336968

KEY TERMS AND DEFINITIONS

Dialogue: Skillful, intentional, and authentic discussion between individuals that produces and facilitates critical consciousness and meaning-making.

Digital Connectivism: Axis between learning and the technological environment. A form of social justice, it allows for autonomy, connectedness, diversity, and openness. Digital connectivism increases one's social capital.

Frames of Reference: Influence an individual's meaning-making. These are cultural and linguistic backgrounds or funds of knowledge from which learners/employees attribute coherence and significance to an experience.

Praxis: An action or conduct which is based on theory. Praxis is the practical application of a theory. Resulting from an understanding of/from reflection turned into action.

Reflective Practice: Refers to an inquiry-based approach to teaching inclusive of critical thinking and personal commitment to continuous learning and improvement.

Social Justice: Centers on the experiences of those who have been "othered," often identified as marginalized groups. Social justice addresses inequities in educational opportunities and outcomes. Marginalized groups represent the most underserved, underrepresented, and undereducated.

Technology: Use of devices, software, and machines that allow individuals and communities to communicate, access information, and acquire/build new knowledge. Available at all times and easily accessible.

Third Point: Refers to the three-point interaction between the facilitator, learner, and item of focus. The third point can be an external artifact, such as student work or videotape.

Transformative Leadership: Synonymous with andragogical leadership; a leader who is action-oriented, technology-centered, and dialogue-focused to move their learners/employees and organization as a whole to a more humane, socially just space in the modern world.

Chapter 14
Andragogical Leadership and Technology in the Future of Adult Education

Yuan Zhang
Auburn University, USA

Jonathan E. Taylor
Auburn University, USA

ABSTRACT

The purpose of this chapter is to foster a deeper understanding of andragogy, explore the features of andragogical leadership and investigate the impact of technology in supporting adult learning. With an interest in investing and carving out new territory in andragogy, the authors discuss andragogical leadership and technology in promoting collaborative learning in adults. In this chapter, they revisit andragogy from a different angle: andragogy's relation with leadership. With the teaching practice drawn from the integration of technology into adult English as a second or foreign language (ESL/EFL) online learning, they investigate the role of andragogical leadership in constructing the community of inquiry.

INTRODUCTION

With an "andragogical tradition" (Knowles, 2005, p. 237), the use of technology is changing all forms of interactions and is also affecting the educational context immensely (Bonfield et al., 2020). Technology is a critical pedagogical element in educational evolution and calls for a new line of research (Núñez-Canal et al., 2022). In addition, the role of the educator is critical for any technology (Caena & Redecker, 2019). Leadership plays a vital role in the structural and cultural dynamics designed for education improvement (Movahedazarhouligh et al., 2022).

The literature distinguishes many leadership styles – from traditional (e.g., transactional leadership) (Bass,1990) to modern concepts (e.g., agile leadership) (Cinnioglu, 2020). Leadership theory emphasizes various outcomes (Day, 2000; Ehrich et al., 2016; Owens & Hekman, 2012; Rego et al., 2017) and

DOI: 10.4018/978-1-6684-7832-5.ch014

spans across multiple levels (Braun et al., 2016; Dinh et al., 2014). Different models of leadership have certain behaviors in common, such as open communication (Purcell & Chahine, 2019; Seemiller, 2021); collaboration (Thakhathi et al.,2019); innovation (Braun et al., 2016; Iqbal & Piwowar-Sulej, 2021); and connecting (Sa & Serpa, 2020). Leadership is a process (Komives et al., 2009) by which leadership is produced and enabled (Nhlapo & Hlalele, 2023). Leadership is about participation and relationships (Elbayadi, 2014), as well as engagement and action (Carbone et al., 2017; Jones et al., 2017; Simmons & Taylor, 2019). Both the current functioning and future development of an organization depend upon the characteristics of its leaders (Piwowar-Sulej & Iqbal, 2023). But there is no consensus in terms of the essence of leadership or the means by which it may be identified (Hassan et al., 2018).

With a focus on "investing and carving out new territory in andragogy" (Henschke, 2020, p. xviii), we will discuss andragogical leadership and technology in promoting collaborative learning in adults through the "Community of Inquiry'' framework (e.g., Cleveland-Innes, 2019; Garrison, 2022; Garrison et al., 2000). The aim of this chapter is threefold: (1) to explore the characteristics of andragogical leadership; (2) to identify the role of andragogical leadership in building a community of inquiry in distance learning; and (3) to investigate the impact of andragogical leadership and technology in supporting adult learning. The chapter begins with a glimpse of the current leadership focus. In an attempt to explore the characteristics of andragogical leadership, we will examine andragogical assumptions and their relationship with educational leadership by revisiting andragogy. The chapter goes on to investigate the role of andragogical leadership in constructing a community of inquiry. By exploring the meaning of learning and teaching, we will discuss the strength of the combination of andragogical leadership and technology in creating an inquiring, collaborative, and reflective learning community to support lifelong learning.

The third part of the chapter reviews the current technology usages in adult ESL/ EFL education, through these examples, we are going to discuss how andragogical leadership plays its role in a community of inquiry in terms of (1) building a connection through communication; (2) learning from and with others in collaboration; and (3) walking the journey together with adult learners. Further, challenges and inspirations for the integration of technology in education will also be discussed.

Finally, the last part of the chapter reviews significant philosophical discussions about adult education. By internalizing andragogical assumptions and processes put forth by Malcolm Knowles and the andragogical scholarship that has followed, we strive to explore "the future of andragogy" (Henschke, 2011) and shed light on the future of adult education.

TENETS OF ANDRAGOGY AND THE CHARACTERISTICS OF ANDRAGOGICAL LEADERSHIP

Literature based on educational leadership is significant (Antonakis et al., 2004; Gmelch & Wolverton, 2002). In the context of higher education, some scholars emphasize the contribution of transformational leadership (Burawat, 2019; Du et al., 2013; Jiang et al., 2017; Widisatria & Nawangsari, 2021), some explore the positive effects of servant leadership (Litano et al., 2016; Sendjaya et al., 2017; Siddiquei et al., 2021; Ying et al., 2020), some address the advantage of responsible leadership (Correia, 2019; Liao & Zhang, 2020; Ur Rehman et al., 2023), some focus on the impact of sustainable leadership (Avery & Bergsteiner, 2011; Iqbal & Ahmad, 2020; Iqbal et al., 2021; Zhang & Bartol, 2010), some advocate authentic leadership (Avolio & Gardner, 2005), and some extensively discuss the effectiveness of distributed leadership (Avissar et al., 2018; Bolden, 2011; Gronn, 2008; Spillane, 2012). As Kjellstrom et al. (2020) have pointed out that leadership is a multifaceted phenomenon that needs closer exploration. In addition, there is little framing research on what leadership characteristics are practiced by adult educators. As Fleming and Caffarella (2000) observed in their study, "leadership in adult education is an ambiguous concept" (p. 119). Since leader development is embedded in ongoing adult development, theories of adult learning seem critical to leadership in the context of higher education (Day & Liu, 2019; Day et al., 2009). Therefore, with the interest of understanding the characteristics of andragogical leadership, we ground our investigation with andragogy as a framework and focus on an andragogical leader's characteristic behaviors when directing, guiding, motivating and influencing behaviors of others (Aboramadan & Dahleez, 2020; Armstrong, 2012).

Andragogy, influenced by Houle's classic "The Inquiring Mind" (1961), and Tough's seminal publication "Learning Without a Teacher" (1967), was "a new label and a new technology" of adult learning proposed by Malcolm Knowles (1968). Contrasting with pedagogy, the art and science of helping children learn, the concept of andragogy was defined as" the art and science of helping adults learn" (Knowles, 1980, p.43). In his classic text "The Adult Learner", Knowles (2005) articulated six core andragogical assumptions as "the learner's need to know, self-directed learning, prior experience of the learner, readiness to learn, orientation to learning and problem-solving, and motivation to learn" (p. 183). (See Figure 1 which outlines the principles of andragogy).

Figure 1. Malcolm Knowles' principles of andragogy
Source: Adapted from Knowles (1984).

1. The need to know	Why they need to learn
2. The learner's self-concept	Autonomy (e.g., being responsible for their own decisions and lives)
3. The role of experience	A great volume and variety of prior experience
4. Readiness to learn	To cope with real-life situations
5. Orientation to learning	life-centered/ task-centered/ problem-centered
6. Motivation	Responsiveness to external motivators (e.g., promotions, higher salaries) is driven by internal pressures (e.g., the desire to increase job satisfaction, and quality of life)

Andragogy has become a code word for identifying the education and learning of adults (Cooper & Henschke, 2001). This "visibility of andragogy has sharpened our awareness and understanding of adult learning" (Holmes & Abington-Cooper, 2000, p.54). Learning in adulthood is viewed as a highly personal endeavor. Teaching adult learners involves principles such as featuring the student at the center of the learning process, focusing on the transformation that takes place from their experience, centering learning on a co-constructivist approach, and promoting opportunities for students to become active learners. Therefore, the role of the teacher shifts from the "one who transmits knowledge" to "one who primarily helps students to inquire" (Knowles, 1962).

With the emphasis on the characteristics of adult learners and the shift in responsibility of adult educators, Knowles (1990) promoted the idea that creative leaders are those who: (1) make a different set of assumptions (essentially positive) about human nature from the assumptions (essentially negative) made by controlling leaders; (2) accept as a law of human nature that people feel a commitment to a decision in proportion to the extent that they feel they have participated in making it; (3) believe in and use the power of self-fulfilling prophecy; (4) highly value individuality; (5) stimulate and reward creativity; (6) are committed to a process of continuous change and are skillful in managing change; (7) emphasize internal motivators over external motivators; and (8) encourage people to be self-directing (pp. 183-190).

When zooming in on these descriptions of creative leaders, it is not hard to identify the characteristics of a creative leader, such as discernment, inclusiveness, encouraging, enlightening, adaptability, reflexivity, supporting, and progressive. A leader with these characteristics will emphasize adult learners' autonomy, individuality, self-drivenness, creativity, goal-orientedness, self-fulfillment, and independence. This description of creative leaders echoes "the art and science of helping adults learn" (Knowles, 1980, p.43); it also sheds light on our understanding of what an andragogical leader might look like.

To deepen our understanding of the role of andragogical leadership, in the following section, we will discuss how these characteristics play their role in building a community of inquiry to facilitate adults to learn, particularly, in the context of the integration of technology in teaching and learning.

ANDRAGOGICAL LEADERSHIP IN BUILDING A COMMUNITY OF INQUIRY

As early as half a century ago, Knowles (1962) projected that the purpose of education will shift from focusing primarily on the transmission of soon-to-be-outdated knowledge to the development of the person's capacity to learn. Teaching should connect to the learner's needs, goals, and rich life/work experiences in order to generate meaningful effects (Knowles, 1980; 1984). As the provision of andragogical instructional methods described by Knowles et al. (1998):

A different set of functions that required a different set of skills. Instead of performing the function of content planner and transmitter, which required primarily presentation skills, I was performing the function of process designer and manager, which required relationship building, needs assessment, involvement of students in planning, linking students to learning resources, and encouraging student initiative. (p. 201).

This andragogical teaching model echoes the existential conception advanced by Dwayne Huebner, as cited by Macqurrie (1972):

Teaching is seen as a mode of being-with, a positive mode of solicitude in which one leaps ahead of the other so as to open his possibilities for him, but never leaps in for the other, for this would be really to deprive him of his possibilities. (p. 208)

Andragogical instructional methods emphasizing the importance of contextualizing learning, active exploration, and application opportunities situated in real-life scenarios (Brown, 2006; Scott & Webber, 2008), are deeply rooted in the sociocultural perspective on learning, which encourages learners with diverse levels of competence to transform their understandings through socially shared activities and to learn from one another as well as from their instructors (Vytotsky, 1978). These learner-centered and co-learning-oriented concepts provide the opportunity for learners to negotiate meaning, diagnose misconceptions, and challenge accepted beliefs, which leads to deep and meaningful educational experiences (Ramsden, 1988).

An additional characteristic of adult learners, compared to younger students enrolled in a traditional campus-based degree program, is that they often have to manage both work and family responsibilities alongside their study commitments (Wang & Torrisi-Steele, 2015; Wang, 2006, 2008). They usually take courses part-time due to their work constraints (Hansen et al., 2019), and their learning perseverance is often associated with acquiring new knowledge and skills for their professional growth (Stein & Glazer 2003; Waits & Lewis 2003). Because of their own physical, social, and cognitive characteristics toward learning, adult learners are identified as a unique learner group (Gold, 2005). They want the convenience of learning anytime, anywhere (Mariani, 2001; Green, 2000).

As Knowles (1970, 1975, 1984) foretold, many adult learners are keen to seize the learning opportunities that technology can provide. Consistent with adult learners' self-directedness (Jones, 2019), technology is in the "andragogical tradition" (Knowles, 2005, p. 237). Those already in the labor force may find distance learning the only option for upgrading skills, finishing a degree, or pursuing another degree. Distance learning helps many adult learners balance the demands of work and family with their pursuit of more education (Mariani, 2001). These assumptions of adult learners are concordant in recent studies. For example, Liaw, Huang, and Chen (2007) found that perceptions of distance learning as an effective method of education correlate positively with perceptions of online learning as a self-governing, self-paced environment, an instructor-led learning environment, and/or an effective multimedia learning environment.

Technology frees learning from the limits of time or space. The number of students joining online courses has been rising dramatically and steadily since 2000 (Smith et al., 2019). According to the National Center for Educational Statistics (NCES), the percentage of undergraduates receiving a part or all of their education via online platforms increased from 8% to 29% between 2000 and 2015 (Radford, 2011). The flexibility and/or accessibility of online programs make education available to more people (Ilgaz & Gulbahar, 2017).

In the context of the distance online learning environment, Knowles, Holton, and Swanson (2005) argued that adults have distinctive features as learners. Both the needs of adult online learners (Wolfson et al., 2014) along with a sociocultural learning theories foundation, provide a rationale for building a community of inquiry facilitating adults in discovering content knowledge, solving practical problems, thinking critically, forming identity, and collaborating with diverse learners across their endeavors in adult education.

Community of inquiry was grounded in Dewey's (1959) philosophy that education is a collaborative reconstruction of experience. The concept of community inquiry is consistent with Raven's (1982) Action Learning methodology, Wenger's (1998) communities of practice model, and Senge's (2006) concept of the learning organization. Lipman (1991) notes the importance of higher-order thinking in the community of inquiry and sees a community of inquiry as an invaluable context for an educational experience where critical thinking is promoted and deep learning is facilitated. Later, Lipman (2003) describes the features of a community of inquiry as "questioning, reasoning, connecting, deliberating, challenging, and developing problem-solving techniques" (p. 12).

There are three components in the community of inquiry:

- **Social Presence.** Beyond an emotional sense of belonging, social presence can be defined as" the collaborative nature of the community and its activities" in which the participants are "identifying with the community, communicating purposefully in a trusting environment, and developing interpersonal relationships" (Garrison et al., 2009, p.7-8).
- **Cognitive Presence.** Cognitive presence was operationalized through the practical inquiry model (Garrison et al., 2000). It was defined by Garrison (2006) as the process of creating meaning with collaborative inquiry. From the perspective of the practical inquiry model, cognitive presence can be defined as a research process that involves determining/defining an issue, challenge, or problem (triggering event), then conducting a thorough investigation on information associated with this issue (exploration), combining ideas to develop a meaningful structure or for obtaining a solution (integration), and then testing the usefulness or validity of the solution (resolution) (Garrison, 2006; Garrison et al., 2000).
- **Teaching Presence.** Teaching presence has been defined as "the design, facilitation, and direction of cognitive and social processes for the purpose of realizing personally meaningful and educationally worthwhile learning outcomes" (Anderson et al., 2001, p. 5). This begins prior to the start of a course (e.g., in the preparation and planning by an instructor of a subject related to a course) and continues throughout the course (e.g., an instructor facilitating discussions and activities) (Anderson et al., 2001; Ke, 2010).

These characteristics of a community of inquiry such as creativity, self-directedness, problem orientation, and fulfillment parallel most features of principles of andragogy which includes its emphasis on self-directed learning, transformative learning, and co-constructions of teachers and students. In a similar manner, the emphasis on the three components of a community of inquiry relies closely on andragogical leadership with characteristics such as: genuine, supportive, encouraging, inclusive, and inspiring. In addition, as a framework for both blended and online courses (Akyol et al., 2009; Wicks et al., 2015; Zhang, 2020), the community of inquiry is one of the most frequently used models for online learning in higher education (Boston et al., 2009; Harrell & Wendt, 2019). Thus, both the historical background and conceptual framework of the community of inquiry seem to perfectly match our investigation of adult learning (e.g., in the context of ESL/EFL online learning), as well as our exploration of the future of adult education regarding the integration of technology in teaching and learning.

In the following discussion, we will use adult ESL/EFL online learning practice and research as examples to further our understanding of adult learners, and andragogical scholarship in the context of adult distance learning. Inspirations gained from both success and challenges that we have encountered will be discussed as guidance walking towards the further of adult education.

USE OF TECHNOLOGY WITH ADULT ESL/EFL PRACTICE IN THE COMMUNITY OF INQUIRY

Adult English as second/ foreign language learners are widely diverse (Esterline & Batalova, 2022) and identified as a unique population of students (Floyd, 2022) —they usually come from a variety of cultural backgrounds and are with low literacy. Many of them are facing challenges such as difficult financial situations, long working hours, and immigration experience (Wilson, 2014). In 2015, nearly 26 million adults in the United States needed English literacy (Batalova & Zong, 2016). As of 2019, nearly 43 million U.S. adults lack the basic English literacy skills required to succeed in the workforce and achieve economic self-sufficiency (U.S. Department of Education, National Center for Education Statistics, 2019). Thus, these data illustrate the needs for strong ESL programs and rigorous research on instructional strategies to support the learning needs of these adults (Borradaile et al., 2021).

Many of the studies in literature involved foreign language learning. In the context of the adult learners online experience in learning English as a Second/ Foreign language, collaborative language learning builds on collaboration between learners (Mazzoni & Benvenuti, 2015; Mubin et al., 2013), between teachers and learners (Garrison, et al., 2000, 2009), or between a media used (e.g., robot) and a learner (Khalifa et al., 2017; Mazzoni & Benvenuti, 2015). Technology is used to enable learning experiences in the 'real world', especially through the formation of a community of inquiry, within which students are able to engage in authentic exchanges about things that are of meaning to them personally.

Based on our discussion on the features of adult online learners and the concept of a community of inquiry, we are going to discuss how andragogical leadership plays its role in constructing a community of inquiry in terms of (1) building a connection through communication; (2) learning from and with others in collaboration; and (3) walking the journey together. Each aspect is related to the social presence, cognitive presence, and teaching presence components of the community of inquiry, respectively. The practice and research in adult ESL/EFL online learning will be used as a lens to gain a better understanding of the concept of community of inquiry in practice.

Building the Connection Through Communication

Education largely depends on social interaction and effective communication (Lowenthal & Dunlap, 2018). Interactions in working groups carry out not only functions concerning information and the initiation of relationships but also those of mutual relief and assistance. Interactivity and a positive culture create a sense of cooperation where group learning and productivity becomes more significant than the total sum of individual students' learning and productivity (Covey, 1989).

As adults enter higher learning, many of them are seeking a kind of 'academic redemption, re-constructing their lives and their sense of self". Their experience affirms their sense that they are limited by lack of knowledge and/or lack of ability (Dirkx et al., 2001). In his book Democracy and Education, John Dewey (1916) highlighted the importance of generating a supportive teaching environment that promotes students' persistence in learning rather than the pure acquisition of knowledge and subject matter (Benson, 2001). The community of inquiry, first, will provide a space "where everyone has an opportunity to bring their needs to the table and participate in understanding the curriculum of their lives" (McMahon & Portelli, 2004, p.70). Since we not only have different experiences but also interpret our experiences differently, we should always challenge our own prior understandings or assumptions to obtain understanding and perceptions in new and undiscovered ways (Heron, 1967; Husserl, 1931;

Marton, 1986; Sadler, 1969; Spigelberg, 1982; Wolff, 1984). Before attempting to make use of learners' prior experience in content and instructional design, andragogical leaders should try to understand adult learners and their lived experiences from their perspective of themselves in the first place. As Horton (1998) eloquently stated:" It is very important that you understand the difference between your perceptions of people's problems and their perceptions of them. You shouldn't be trying to discover your perception of their perception. You must find a way to determine what their perception is" (pp. 70-71, also cited by Taylor, 2022, p.188)

"[E]ach group of adult learners . . . will be heterogeneous in nature and every individual in the group will be a complex mixture of style and ability" (Mackeracher, 2005, p. 79). From an interpretive point of view– which is also termed as social construction, constructivist, or constructionist– "both reality and knowledge are constructed and reproduced through communication, interaction, and practice" (Tracy, 2020, p.51). Authentic communication invites strong peer connections that limit isolation in distance learning and therefore may address persistence in online students (Faulconer et al., 2018; Sinclair, 2017). In interactive communication, learners respond to and engage with others. Both learners and instructors project their personalities into the community where cohesive responses encourage interaction (Swan & Shih, 2005).

Interaction has been stated as one of the critical aspects of the online learning environment (Artino 2007; Bolliger & Martindale, 2004; Bray et al., 2008; Chejlyk, 2006; Garrison & Shale, 1990; Keeler, 2006; Robles, 2006). Current research is consistent with this finding. Kuo & Belland (2016) investigated online adult learners' perceptions of interaction, satisfaction, and performance within an online course using the Blackboard platform by studying 167 minority students enrolled in six online undergraduate-level courses from a university in the southeastern United States. Results indicated that learner–content interaction and learner–instructor interaction were significant predictors for student satisfaction in online settings.

In the virtual learning environment, learners' sense of belonging involves feelings of acceptance, respect, inclusion, and support. In the context of an English language classroom, promoting learners' cultures and providing opportunities for social interaction can foster a feeling of belonging (Chung, 2022). In Park and Kim's (2011) study, they set up a collaborative online learning environment by creating an online English study group on Ning (an online social network site) to facilitate adult learners' online reading tasks. The English study group members can freely create social networks such as post written messages, photos, audio streams, and videos and participate in asynchronous online discussions.

As Watzlawick, Beavin, and Jackson (1967) argued, "every communication has a content and a relationship aspect" (p. 54). Fenwick suggests that "people's sense of their own knowledge in work and the knowledge valued by the group to which they see themselves belonging form a critical element of their sense of identity" (p. 22). As participants of communication, we constantly interpret the relationship with statements such as "This is how I see myself… this is how I see you… this is how I see you seeing me...and so forth in theoretically infinite regress" (p.51). Thus, andragogical leaders are observant of this emotional process and sensitive to the dynamic within the group. A deeper understanding of self and one another will help understand expectations and needs, as well as co-construct a mutual plan and knowledge with partners to achieve joint goals (King, 2017).

Learning From and With Others in Collaboration

Knowles's view of adult learners is very close to Maslow's model of the 'self-actualizing' person, or Rogers' fully functioning' person, which all emphasize 'becoming' rather than 'being'. According to Knowles, adults feel a need to learn, participate actively in learning, perceive the goals of the learning experience to be theirs, and accept a share of the responsibility for learning. In such a sense, individuals' construction and development of their experience constitute a distinctive and authentic existence.

Collaboration means that "people labor together in order to construct something that did not exist before the collaboration, something that does not and cannot fully exist in the lives of individual collaborators" (Peters & Armstrong, 1998, p. 75). Collaborative learning provides an opportunity for the learner to take responsibility for direct personal learning, engaged in knowledge creation and sustainability. In his book, Schools That Learn, Senge (2000) discusses the importance of creating within the culture a learning community that is cooperative. "Through such techniques as dialogue and skillful discussion, small groups of people transform their collective thinking, learning to mobilize their energies and actions to achieve common goals and drawing forth an intelligence and ability greater than the sum of individual members' talents." (pp. 7-8).

Indeed, collaborative learning and influence amongst peers can change practice and ways of seeing the world, which aligns with Mezirow's work on transformative learning theory (Mezirow, 2000) where adult learners change their frame of reference leading to a deep shift in perspective (Mezirow, 1978).

With an emphasis on individual and group learning (Peters & Armstrong, 1998; Peters et al., 2011), adult learners share responsibility for group learning, which forces everyone to develop new connections and find ways to work together. By working as both facilitator and learner, each individual is involved in a much more active learning experience. Learning with and from others encourages participants to listen actively to gain a fuller understanding of the topic and to think critically about the way of organizing their thoughts and presenting them to their peers. Collaborative learning also provides one with an outstanding opportunity to observe others' behavior and to reflect on his/her own within the group, which strengthens adult learners' management and leadership skills. By learning with and from people with varied backgrounds and being exposed to diverse viewpoints, adult learners are likely to develop a more comprehensive range of knowledge on the topic discussed.

To create an atmosphere in the online learning environment where adult learners express themselves freely in appropriate ways, share their ideas, and ask questions (Porter, 1997), andragogical leaders integrate positive relationships with educational intentions and combine intentional instruction with warm and respectful interactions (Kingston & Siraj, 2017)."In learning with technology, the connections to content, context, and community can be powerful" (Milliron & Miles, 1998). Asynchronous online discussions are among the most widely used instructional techniques to support students' online learning (Gao et al., 2013; Koszalka et al., 2021). Through sharing thoughts, asking questions, and providing feedback, students are provided with ample possibilities that promote interaction and communication, thus building community in the online learning environment (Yang et al., 2010).

In Bohinski and Mule's (2016) qualitative study, researchers found asynchronous telecollaboration and synchronous discussion boards allowed language learners to utilize a variety of strategies to negotiate meaning in their target languages. Further, discussion board communications offered more space for participants for more authentic and engaged meaning-making.

Similarly, Hilliker (2020) investigated ESL/EFL teacher education students paired with EFL students from Mexico. Through the data from recorded real-time interactions via Zoom and weekly journal entries from the education students, they found that these experiences resulted in participants' learning intercultural communication skills through their ability to make connections between culture and language, as well as, interacting with their language partners.

Walking the Journey Together

The role and responsibility of adult educators are extremely demanding in coordinating the whole educational and learning process and its success (Papanaoum, 1994). In preparing educators of adults, "andragogy becomes a way of being or an attitude of mind and needs to be modeled/exemplified by the professor. Otherwise, if we are not modeling what we are teaching, we are teaching something else" (Cooper & Henschke, 2001).

The involvement of an educator in a collaborative community of inquiry depends on how the educator is seen by students, and how the educator sees oneself. This aligns with the andragogical leaders' characterization of reflexivity. Reflexivity is defined as "People's careful consideration of the ways in which their past experiences, points of view, and roles impact their interactions with, and interpretation of, any particular interaction or context" (Tracy, 2020, p.2). Reflective practice is understanding based on metacognitions being awarded and being able to monitor the development of one's learning and to put its application into practice (Parsons & Stephenson, 2005). Through interaction with others, we learn the basic questions of reflective inquiry: How did I arrive at this point? How can I explain it? As we internalize these questions and develop the habit of asking them of ourselves, we can gradually complete many tasks that we are unable to complete on our own at first.

People give meaning to things they perceive in the world (Merleau-Ponty, 1962). Different adults have different journeys and different visions. Thus, andragogical leaders often use methods adapted from Piaget's "clinical method." They begin by telling students, "Let me hear you think," and then follow up with prompts, challenges, variations on the initial problems, and questions about the general usefulness of the methods students have chosen (Tracy, 2020). A key role of andragogical leaders is creating visions and showing adult learners how far we have come. As learning facilitators, adult educators provide empathy, motivation, support, and the development of learning skills. They foster self-knowledge and enhance the self-assessment and self-confidence of the adult learner. They motivate adult learners internally and externally to participate in formal, non-formal, and informal learning, resolve personal disjuncture (Jarvis, 2006), institutional hindrances, social issues, and promote social change (Merriam & Baumgartner, 2020).

In a virtual environment, instructor feedback especially plays a significant role in establishing the instructor's teaching presence in facilitating critical thinking, and increased engagement with the learning content (Holzweiss et al., 2014; Hosler & Arend, 2012). In Chung's (2022) study investigating the best practice for teaching adults English as a second language, she argued for using constructive feedback to reinforce the teaching presence in an ESL classroom. Instructor feedback should not merely focus on error correction but also on building a relationship with the learners and providing directions that they see as relevant to their effort. She encouraged educators to" ask the 'why' question behind the learners" in order to better understand learners' thought processes. This approach allowed for the facilitation of more individualized interactions and fostered learner-centeredness and agency in their learning process (Carver et al., 2007).

Another element of the teaching presence involves designing of the educational experience (e.g., selection, organization, and presentation of course contents), as well as developing learning activities and assessments. Simonds and Brock (2014) listed several main online teaching and learning activities such as conducting live chats; commenting on online discussion boards; exploring web links and online materials; watching videos or listening to audio content; reading lecture notes, course texts, and articles; communicating through emails or conferencing meetings; and participating in online small group projects. Techniques such as committee-building, lesson study, breakout rooms, micro-teaching, and guest speakers are ways to enable critical reflection, and engagement in dialogue from a variety of perspectives, to enable shifts in thinking, and action-taking for new frames of understanding (Veletsianos, 2011).

Also, teaching presence is embedded in shared responsibility between the teacher and students. Building on the principle that language learning is facilitated when learners are cooperatively involved in working on projects or tasks, Knutson (2003) proposed using the four-phase experiential learning process in the ESL classroom to encourage students' responsibility for their own learning, such as exposure (activating past experiences and previous knowledge about the subject), participation (the actual activity or experience), internalization (skillful reflection and questioning), and dissemination (transferring classroom experiences into learners' day-to-day contexts) (p. 56). In accordance with Dickinson's (1993) five autonomous learner characteristics which encourages the use of autonomous learning and the individual's contribution to his/her own learning, Po-ying (2007) used a strategy, which she named exploratory practice (EP), for helping her Chinese students to learn English.

"The creation of a learning community supports and encourages knowledge acquisition. It creates a sense of excitement about learning together and renews the passion involved with exploring new realms in education" (Palloff & Pratt, 1999, p. 163). In the community of inquiry, both adult educators and adult learners can appreciate new topics, dialogue with a variety of people, renew a sense of childlike wonder and curiosity (Tracy, 2020, p. 59), allowing andragogical leaders to develop and refine new models thereby moving the field of adult education forward.

The learner's needs, aims and motives are at the center of the learning process (Lytovchenko et al., 2020). Quality education is about helping students make meaningful connections with the content as well as with others in the learning environment (Cross, 1999). Teachers must begin with the purpose of their students, steer them into potentially rich experiences, and watch carefully for signs of growth (Dewey, 1959). Andragogical leaders should draw connections between learners and content (e.g., planning the content of teaching), connections among learners (e.g., selecting instructional strategies), and connections between learners and community (e.g., offering learning opportunities beyond the classroom). Taylor (2011) said it well: Connection is "the underlying mechanism by which a teacher is able to take the intended content of the teaching situation and match it up to the distinct needs, preferences, and applications of each learner in their course– to meet them where they are and walk the journey with them" (p.55).

Effective adult education involves not only information dissemination but communication and collaboration among its participants, online learning activities, and entirely online courses that influence how we view the field itself. The utilization of technology has benefited individuals who are unable to attend classes due to illness, mobility, or other limitations (Hansen et al., 2020). It has provided adults the opportunity to socialize with peers (Talmage et al., 2019), reduced social isolation, increased the social well-being of older adults (Berkowsky et al., 2013), and overcome some of the barriers to participation by older adults in lifelong learning programs (Hansen et al., 2019).

Without doubt, the integration of technology has brought forth new markets for higher education and, along with it, questions regarding educational quality (Kilburn et al., 2017). From the best practice of the integration of technology in adult ESL/EFL teaching and learning, it is obvious that the value of technology for learning is attributed to its flexibility, ease of access, efficiency, and solving 'instructional problems' (Torrisi-Steele, 2014; Veletsianos, 2011). But 'teaching' and 'learning' are not activities that take place only at specified times, places, and ages: 'technology invites a tools-first emphasis, but technology is only as good as our knowledge of how to use it to enhance learning' (Olgren, 2000, p. 7).

In terms of the integration of technology in teaching, most of the students felt comfortable in distance learning and expressed general positive attitudes towards it (Chirca & Pânişoară, 2020; Işık et al., 2010; Pekel, 2002; Lytovchenko & Voronina, 2020; Mukan & Lavrysh, 2020). But students' negative experiences can also be explained by factors which are perceived in literature as challenges in distance learning.

For instance, there is the lack of technological competencies on the part of teacher or student (Yusuf, 2020). As is shown in literature, perceived online interaction may be dependent on users' ability to use the Internet (e.g., locate information and navigate websites) or other technologies (Kuo et al., 2014). Learners' confidence in their ability to use the Internet effectively and their knowledge about the Internet may affect their interaction level in an online course (Liang & Tsai, 2008; Kuo & Belland, 2016; Robles, 2006). In addition, poorly designed mediated communication may fail to accommodate effective human coordination (Pracana & Wang, 2018).

Also, online learning requires a lot of self-discipline and initiative from learners (Smith et al., 2019). Multiple previous studies have identified some difficulties experienced by adult learners in distance learning, such as "the lack of teacher presence (Kirtman, 2009), limitations on interaction with teacher and peers (Smith et al., 2019), poor self-management and independent learning skills (Xu & Jaggars, 2013), and technical issues (Smidt et al., 2014).

In "the new, modern practice of adult education" (Donavant, 2009, p. 227), Herrington (2010) suggests that "technologies need to be used as cognitive tools for learning rather than as simply alternative delivery platforms" (p. 3). The 'proper' use of technology is through the adoption of a learning-centric rather than a technology-centric perspective (Wang, 2010). As Jody Donovan (2015) argued, "taking an online course should be more than sitting in front of a computer – real engagement involves becoming a part of the community of learners." An appropriate pedagogical design and proper exploitation of social technologies are crucial in fostering the processes of communication and interaction (Saadatmand et al., 2017).

The community of inquiry aims to increase comfort, communication, and collaboration among students and instructors. Online collaboration tools and social media can be incorporated into online communities to promote learners' sense of belonging and increase the knowledge flow between students, thereby facilitating social negotiation of meaning as learners construct their own understandings (Bliss & Lawrence, 2009; Dawson, 2018; Kumi-Yeboah, 2018). By engaging with each other, we experience growth in self-understanding.

With an interest in understanding and promoting the transformation of learning (Mertens, 2007), andragogical leaders should combine their theoretical knowledge and practical experience to promote co-learning with adult learners by creating an active, collaborative, and reflective learning environment that allows transformation for adults. Some suggestions are: respect for personality; participation in decision-making; freedom of expression and availability of information; mutuality of responsibility in defining goals, planning, conducive activities, and evaluating (Knowles,1980, 1990). Therefore, the adult educator can fulfill their role as a facilitator for greater student self-directed learning approaches (Henschke, 2011).

THE FUTURE

"The core principle and requirement of a fulfilled human being is the ability to inquire and create constructively, independently, without external controls." – Wilhelm von Humboldt, German humanist

Across the literature, it was clear that virtual, technology-enhanced learning can engage students across academic disciplines in interactive exchanges that help them to compare their own cultural perspectives with those of others; to learn to negotiate meaning with others who have different native languages, cultural interaction styles, and expectations; and to analyze, problem-solve, and collaborate within virtual learning environments.

Advances in technology will continue to play a significant role in the future. We may see that the purpose of education will shift from focusing primarily on the transmission of soon-to-be-outdated knowledge to the development of the person's capacity to learn. This educational change will change the role of the teacher from the "one who transmits knowledge" to "one who primarily helps students to inquire" (Knowles, 1984)

To envision the future of adult education, Knowles (1984) concluded his book Andragogy in Action by noting that:

We are nearing the end of the era of our edifice complex and its basic belief that respectable learning takes place only in buildings and on campuses. Adults are beginning to demand that their learning take place at a time, place, and pace convenient to them. I feel confident that most educational services by the end of this (20th) century (if not the decade) will be delivered electronically…Our great challenge now is to find ways to maintain the human touch as we learn to use the media in new ways. (p. 7)

Education is in the context of human relationships (Tolstova & Levasheva, 2019). Adult education advocates an approach of gaining empathic insight into others' viewpoints, beliefs, and attitudes without judgment: to see the world as others see it; to understand another's feelings; to communicate the understanding (Wiseman, 1996). This 'human touch' is about "standing in your participants' shoes and attempting to understand, from a first-person perspective their role, experience, history, and culture" (Tracy, 2020, p. 51). This "human touch "is a way to understand adult learners' points of view, their goals, their struggles, their hopes, and their ways of being by extension ourselves.

Envisioning the future of adult education, Merriam (2008) felt that there have been two main shifts in the focus of adult learning. One shift is from the individual learner to the learner within the various contexts in which learning takes place. This new perspective considers "learning as part of the system's cultural and historical norms… (and) how physical space and spatiality encourages or inhibits learning" (p.94). ... and the second shift is from learning as a purely cognitive activity to a multidimensional phenomenon where "learning is construed as a much broader activity involving the body, the emotions, and the spirit as well as the mind" (p.95).

In practice, adult educators strive to become the Skillful Teacher (Brookfield, 2015). It is much more meaningful for people to have the skills to discover and continue the pursuit of knowledge as opposed to just being fed information to remember and regurgitate. Andragogical leaders are those that are concerned with challenging students and pushing them to the next level. They are creative and resourceful, and they balance content knowledge with the ability to situate theory with practice. Learning "involves not just human change but interconnections of humans and their actions with rules, tools and texts, cultural, and material environments" (Fenwick, 2008, p. 19). "Guiding students towards greater learning autonomy for social and self-directed learning is imperative for continuous lifelong learning post-graduation" (Merriam & Bierema, 2014, p. 73).

In "Potential futures: Implications for adult educators", Tough (1987) states the importance for adult educators to "regain their sense of social mission" (p.11), in particular, he argues:

We must accept and treasure the fact that we can make a difference to the future of our community and even our civilization as well as to the lives of individual men and women...any successful path to a positive human future will require enormous changes in the attitudes, understanding, priorities, and behavior of adults and youth. Surely we adult education will play a key role in fostering and facilitating these changes, insights, and skills in individuals. (p. 11)

Action without philosophical reflection leads to mindless activism (Elias and Merriam, 1995, p. 4). As Holmes & Abington-Cooper (2000) advised "A major key for educators is to be aware of their personal philosophies for working with adult learners' (p.54). Adult educators need theoretical and practical knowledge and the capacity to adapt their expertise to different situations both individually and collectively, and the capability to justify their actions. Andragogical leaders should critically reflect on a philosophical adult education position by "constantly examining, evaluating, and perhaps rejecting or modifying what has been received from the past." (Elias and Merriam, 1995, p. 251) to look at the role of the teacher, a teacher's responsibility to teach, ways of teaching and learning, etc. As we move towards the next exciting scene of adult education, adult education's future will benefit from adult educators who have a strong commitment to making a positive difference (Tough, 1987).

In the discussion on what being truly educated is, Noam Chomsky profoundly stated: "Know where to look, how to look, how to question, how to challenge, how to proceed independently, to deal with the challenges that the world presents to you and that you develop in the course of yourself, education, inquiry, investigations, cooperation and solidarity with others. That's what an educational system should cultivate from kindergarten to graduate school" (2015). As we envision the future, education is viewed as going beyond what needs to be learned and focuses on how we learn and why we learn.

We hope, in this chapter, concepts extracted from theories and philosophies related to andragogical leadership and technology will provide a strong base of knowledge to foster the evolution of educational leadership and adult education. Further, we believe that discussions related to the innovative research in promoting transformative learning in adult ESL/EFL education will offer diverse perspectives on adult learners' learning experience. In addition, the challenges we are facing and influencing factors we learned from research can help adult educators tailor their instructions to contribute to learners' collaborative learning.

By exploring the concept of andragogical leadership, we intend to understand more deeply the qualities, characteristics, and practices of the future of adult education. We hope the lens of technology learning and community of inquiry and their connection with andragogical leadership will throw great light on the academic enterprise.

Throughout history, the adult education movement in the United States has allowed for diversity, flexibility, innovation, and imaginative responses to a wide range of changing educational needs (Andrews, 1980; Harrington, 1977). We believe the combination of andragogical leadership and technology will contribute to adult educators' ability in creating a supportive teaching and learning environment. We also believe this evoking conversation will spark inspiration for scholars, researchers, administrators, practitioners, and leaders at all levels of education and in various organizations in the exploration of the future of adult education.

In the end, in the quest for answers to "What is the future of adult education", we invite you to join us in continuously examining the old responses, generating new ones induced by changing conditions, and reflecting on current responses in the interest of making adult education as good as it can be.

REFERENCES

Aboramadan, M., & Dahleez, K. A. (2020). Leadership styles and employees' work outcomes in nonprofit organizations: The role of work engagement. *Journal of Management Development, 39*(7/8), 869–893. doi:10.1108/JMD-12-2019-0499

Akyol, Z., Garrison, D. R., & Ozden, M. Y. (2009). Online and blended communities of inquiry: Exploring the developmental and perceptional differences. *International Review of Research in Open and Distance Learning, 10*(6), 65–83. doi:10.19173/irrodl.v10i6.765

Anderson, T., Rourke, L., Garrison, D. R., & Archer, W. (2001). Assessing teaching presence in a computer conferencing context. *Journal of Asynchronous Learning Networks, 5*(2), 1–17. doi:10.24059/olj.v5i2.1875

Anderson, T. D., & Garrison, D. R. (1995). Critical thinking in distance education: Developing critical communities in an audio teleconference context. *Higher Education, 29*(2), 183–199. doi:10.1007/BF01383838

Andrew, G. J. (1980). *Continuing education accreditation. Power and conflict in continuing education.* Wadsworth Publishing Co.

Antonakis, J., Cianciolo, A. T., & Sternberg, R. J. (2004). *The nature of leadership.* Sage Publications.

Armstrong, M. (2012). *Armstrong's handbook of management and leadership: Developing effective people skills for better leadership and management.* Kogan Page.

Artino, A. R. (2007). Online military training: Using a social cognitive view of motivation and self-regulation to understand students' satisfaction, perceived learning, and choice. *Quarterly Review of Distance Education, 8*(3), 191–202.

Avery, G. C., & Bergsteiner, H. (2011). Sustainable leadership practices for enhancing business resilience and performance. *Strat Leader, 39*(3), 5–15. doi:10.1108/10878571111128766

Avissar, I., Alkaher, I., & Gan, D. (2018). The role of distributed leadership in mainstreaming environmental sustainability into campus life in an Israeli teaching college: A case study. *International Journal of Sustainability in Higher Education, 19*(3), 518–546. doi:10.1108/IJSHE-07-2017-0105

Avolio, B. J., & Gardner, W. L. (2005). Authentic leadership development: Getting to the root of positive forms of leadership. *The Leadership Quarterly, 16*(3), 315–338. doi:10.1016/j.leaqua.2005.03.001

Benson, P. (2001). *Teaching and researching autonomy in language learning.* Longman.

Berkowsky, R. W., Cotton, S. R., Yost, E. A., & Winstead, V. P. (2013). Attitudes towards and limitations to ICT use in assisted and independent living communities: Findings from a specially designed technological intervention. *Educational Gerontology*, *39*(11), 797–811. doi:10.1080/03601277.2012.734162 PMID:24244065

Bliss & Lawrence. (2009). Is the whole greater than the sum of its parts? A comparison of small group and whole class discussion board activity in online courses. *Journal of Asynchronous Learning Networks*, *13*(4), 25–39.

Bohinski, C. A., & Mulé, N. (2016). Telecollaboration: Participation and negotiation of meaning in synchronous and asynchronous activities. *MEXTESOL Journal*, *40*(3).

Bolden, R. (2011). Distributed leadership in organizations: A review of theory and research. *International Journal of Management Reviews*, *13*(3), 251–269. doi:10.1111/j.1468-2370.2011.00306.x

Bolliger, D. U., & Martindale, T. (2004). Key factors for determining student satisfaction in online courses. *International Journal on E-Learning*, *3*(1), 61–67.

Bonfield, C. A., Salter, M., Longmuir, A., Benson, M., & Adachi, C. (2020). Transformation or evolution?: Education 4.0, teaching and learning in the digital age. *Higher Education Pedagogies*, *5*(1), 223–246. doi:10.1080/23752696.2020.1816847

Borradaile, K., Martinez, A., & Schochet, P. (2021). *Adult education strategies: Identifying and building evidence of effectiveness. Study Snapshot. NCEE 2021-007*. Distributed by ERIC Clearinghouse.

Boston, W., Diaz, S. R., Gibson, A. M., Ice, P., Richardson, J., & Swan, K. (2009). An exploration of the relationship between indicators of the community of inquiry framework and retention in online programs. *Journal of Asynchronous Learning Networks*, *13*(3), 67–83. doi:10.24059/olj.v13i3.1657

Braun, S., Peus, C., Frey, D., & Knipfer, K. (2016). Leadership in academia: Individual and collective approaches to the quest for creativity and innovation. *Leadership Lessons from Compelling Contexts*, *8*, 349–365. doi:10.1108/S1479-357120160000008013

Bray, E., Aoki, K., & Dlugosh, L. (2008). Predictors of learning satisfaction in Japanese online distance learners. *International Review of Research in Open and Distance Learning*, *9*(3). Advance online publication. doi:10.19173/irrodl.v9i3.525

Brookfield, S. (2015). *The skillful teacher*. Jossey-Bass.

Brown, K. M. (2006). Leadership for social justice and equity: Evaluating a transformative framework and andragogy. *Educational Administration Quarterly*, *42*(5), 700–745. doi:10.1177/0013161X06290650

Burawat, P. (2019). The relationships among transformational leadership, sustainable leadership, lean manufacturing, and sustainability performance in Thai SMEs manufacturing industry. *International Journal of Quality & Reliability Management*, *36*(6), 1014–1036. doi:10.1108/IJQRM-09-2017-0178

Caena, F., & Redecker, C. (2019). Aligning teacher competence frameworks to 21st century challenges: The case for the European digital competence framework for educators (DIGCOMPEDU). *European Journal of Education*, *54*(3), 356–369. doi:10.1111/ejed.12345

Carbone, A., Evans, J., Ross, B., Drew, S., Phelan, L., Lindsay, K., & Ye, J. (2017). Assessing distributed leadership for learning and teaching quality: A multi-institutional study. *Journal of Higher Education Policy and Management, 39*(2), 183–196. doi:10.1080/1360080X.2017.1276629

Carver, R., King, R., Hannum, W., & Fowler, B. (2007). Toward a model of experiential e-learning. *Journal of Online Learning and Teaching, 3*(3), 247–256.

Chejlyk, S. (2006). *The effects of online course format and three components of student perceived interactions on overall course satisfaction.* https://www.editlib.org/p/117746/

Chirca, R. C., & Pânişoară, I.-O. (2020). Study on Romanian students' digital level and learning styles. *Journal of Educational Sciences and Psychology, 10*(1), 42–52.

Chomsky, N. (2015, May 26). *On being truly educated* [Interview]. The Brainwaves Video Anthology. https://youtu.be/eYHQcXVp4F4

Chung, H. (2022). Best practices teaching adult English as a second language in the virtual classroom. *New Directions for Adult and Continuing Education, 2022*(175/176), 45–57. doi:10.1002/ace.20467

Clark, R. E. (1994). Media will never influence learning. *Educational Technology Research and Development, 42*(2), 21–29. doi:10.1007/BF02299088

Cleveland-Innes, M. (2019). The community of inquiry theoretical framework. Designing collaborative online and blended learning. In H. Beetham & R. Sharpe (Eds.), *Rethinking pedagogy for a digital age* (pp. 85–101). Routledge. doi:10.4324/9781351252805-6

Cooper, M. K., & Henschke, J. A. (2001, Sept. 26-28). Andragogy: Its research value for practice. In W.C. Hine (Ed.), *Annual Midwest Research-to-Practice Conference in Adult, Continuing & Community Education*. Academic Press.

Correia, M. S. (2019). Sustainability: An overview of the triple bottom line and sustainability implementation. *International Journal of Strategic Engineering, 2*(1), 29–38. doi:10.4018/IJoSE.2019010103

Covey, S. R. (1989). *The seven habits of highly effective people: Restoring the character ethic.* Simon and Schuster.

Cross, K. P. (1999). *Learning is about making connections. The Cross Papers, 3.*

Dawson, V. R. (2018). Fans, friends, advocates, ambassadors, and haters: Social media communities and the communicative constitution of organizational identity. *Social Media + Society, 4*(1). Advance online publication. doi:10.1177/2056305117746356

Day, D. (2000). Leadership development: A review in context. *The Leadership Quarterly, 11*(4), 581–613. doi:10.1016/S1048-9843(00)00061-8

Day, D. V., Harrison, M. M., & Halpin, S. M. (2009). *An integrative approach to leader development: Connecting adult development, identity, and expertise.* Routledge. doi:10.4324/9780203809525

Day, D. V., & Liu, Z. (2019). What is wrong with leadership development and what might be done about it? In R. E. Riggio (Ed.), *What's wrong with leadership: Improving leadership research and practice* (pp. 226–240). Routledge. doi:10.4324/9781315163604-13

Dewey, J. (1916). *Democracy and Education.* Free Press.

Dewey, J. (1959). My pedagogic creed. In J. Dewey (Ed.), *Dewey on education* (pp. 19–32). Teachers College, Columbia University. (Original work published 1897)

Dickinson, L. (1993). Talking shop: Aspects of autonomous learning. *ELT Journal, 47*(4), 330–336. doi:10.1093/elt/47.4.330

Dinh, J. E., Lord, R. G., Gardner, W. L., Meuser, J. D., Liden, R. C., & Hu, J. (2014). Leadership theory and research in the new millennium: Current theoretical trends and changing perspectives. *The Leadership Quarterly, 25*(1), 36–62. doi:10.1016/j.leaqua.2013.11.005

Dirkx, J. M., Smith, R., & Amey, M. (2001, Sept. 26-28). Being a "college student": The community college experience of under-prepared adult learners. In W. C. Hine (Ed.), *Annual Midwest Research-to-Practice Conference Adult Continuing Community Education.* Academic Press.

Donavant, B. W. (2009). The new, modern practice of adult education online instruction in a continuing professional education setting. *Adult Education Quarterly, 59*(3), 227–245. doi:10.1177/0741713609331546

Donovan, J. (2015). *The importance of building online learning communities.* https://blog.online.colostate.edu/blog/uncategorized/the-importance-of-building-online-learning-communities/

Du, S., Swaen, V., Lindgreen, A., & Sen, S. (2013). The roles of leadership styles in corporate social responsibility. *Journal of Business Ethics, 114*(1), 155–169. doi:10.100710551-012-1333-3

Ehrich, L., Kimber, M., & Ehrich, J. (2016). The micropolitics of management in universities: Challenges and opportunities. In J. Ryan & D. Armstrong (Eds.), *Working without the system* (pp. 187–204). Information Age Publishing.

Elbayadi, M. E. (2014). *Relational leadership, DevOps, and the Post-PC era: Toward a practical theory for 21st century technology leaders* [Unpublished doctoral thesis]. Antioch University.

Elias, J. L., & Merriam, S. B. (1995). *Philosophical foundations of adult education.* Krieger Publishing Company.

Faulconer, E. K., Griffith, J., Wood, B., Acharyya, S., & Roberts, D. (2018). A comparison of online, video synchronous, and traditional learning modes for an introductory undergraduate physics course. *Journal of Science Education and Technology, 27*(5), 404–411. doi:10.100710956-018-9732-6

Fenwick, T. (2008). Workplace learning: Emerging trends and new perspectives. *New Directions for Adult and Continuing Education*, 119, 17–26. doi:10.1002/ace.302

Fleming, E. A., & Caffarella, R. S. (2000). *Leadership for adult and continuing education. In Adult Education Research Conference, Vancouver, BC, Canada, 2000.* New Prairie Press.

Gao, F., Zhang, T., & Franklin, T. (2013). Designing asynchronous online discussion environments: Recent progress and possible future directions. *British Journal of Educational Technology, 44*(3), 469–483. doi:10.1111/j.1467-8535.2012.01330.x

Garrison, D. R. (2006). Online collaboration principles. *Journal of Asynchronous Learning Networks, 10*(1), 25–34. doi:10.24059/olj.v10i1.1768

Garrison, D. R. (2022). Shared metacognition in a community of inquiry. *Online Learning*, *26*(1), 6–18. doi:10.24059/olj.v26i1.3023

Garrison, D. R., & Shale, D. (1990). *A new framework and perspective. In Education at a distance: From issues to practice*. Kreiger.

Garrison, R., Anderson, T., & Archer, W. (2000). Critical inquiry in a text-based environment: Computer conferencing in higher education. *The Internet and Higher Education*, *2*(2-3), 87–105. doi:10.1016/S1096-7516(00)00016-6

Garrison, R., Anderson, T., & Archer, W. (2009). The first decade of the community inquiry framework: A retrospective. *The Internet and Higher Education*, *13*(1), 5–19.

Gmelch, W., & Wolverton, M. (2002). *College deans: Leading from within*. American Council on Education and Oryx Press.

Gold, H. E. (2005). Engaging the adult learner: Creating effective library instruction. *Portal (Baltimore, Md.)*, *5*(4), 467–481. doi:10.1353/pla.2005.0051

Green, K. C. (2000, February 17-19). *Technology and instruction: Compelling, competing, and complementary visions for the instructional role of technology in higher education*. Paper prepared for a conference on the changing American University: The future of the City of Intellect, convened by the Center for Ideas & Society at the University of California, Riverside, CA, United States.

Gronn, P. (2008). The future of distributed leadership. *Journal of Educational Administration*, *46*(2), 141–158. doi:10.1108/09578230810863235

Hansen, R. J., Talmage, C. A., Thaxton, S. P., & Knopf, R. C. (2019, March 26). Barriers to age-friendly universities: Lessons from Osher lifelong learning institute demographics and perceptions. *Gerontology & Geriatrics Education*, *40*(2), 1–23. doi:10.1080/02701960.2019.1572003 PMID:30688166

Hansen, R. J., Talmage, C. A., Thaxton, S. P., & Knopf, R. C. (2020). Enhancing older adult access to lifelong learning institutes through technology-based instruction: A brief report. *Gerontology & Geriatrics Education*, *41*(3), 342–351. doi:10.1080/02701960.2019.1618852 PMID:31116688

Harrell, K. B., & Wendt, J. L. (2019). The impact of blended learning on community of inquiry and perceived learning among high school learners enrolled in a Public Charter School. *Journal of Research on Technology in Education*, *51*(3), 259–272. doi:10.1080/15391523.2019.1590167

Harrington, F. H. (1977). *The future of adult education: New responsibilities of colleges and universities*. Jossey-Bass Publishers.

Hassan, A., Gallear, D., & Sivarajah, U. (2018). Critical factors affecting leadership: a higher education context. *Transforming Government: People. Process and Policy*, *12*(1), 110–130. doi:10.1108/TG-12-2017-0075

Henschke, J. A. (2011). Considerations regarding the future of andragogy. *Adult Learning*, *22*(1-2), 34–37. doi:10.1177/104515951102200109

Henschke, J. A. (2020). *Facilitating adult and organizational learning through andragogy: A history, philosophy, and major themes*. Information Science Reference.

Heron, J. (1967). *Feeling and personhood: Psychology in another key*. Sage Publications.

Herrington, J., Reevcs, T., & Oliver, R. (2010). *A guide to authentic e-learning*. Routledge.

Hilliker, S. (2020). Virtual exchange as a study abroad alternative to foster language and culture exchange in TESOL teacher education. *TESL-EJ, 23*(4).

Holmes, G., & Abington-Cooper, M. (2000). Pedagogy vs andragogy: A false dichotomy? *The Journal of Technology Studies*, *26*(2), 50–55. doi:10.21061/jots.v26i2.a.8

Holzweiss, P. C., Joyner, S. A., Fuller, M. B., Henderson, S., & Young, R. (2014). Online graduate students' perceptions of best learning experiences. *Distance Education*, *35*(3), 311–323. doi:10.1080/01587919.2015.955262

Horton, M., Kohl, J., & Kohl, H. (1998). *The long haul: An autobiography*. Teachers College Press.

Hosler, K. A., & Arend, B. D. (2012). The importance of course design, feedback, and facilitation: Student perceptions of the relationship between teaching presence and cognitive presence. *Educational Media International*, *49*(3), 217–229. doi:10.1080/09523987.2012.738014

Houle, C. O. (1961). *The inquiring mind*. University of Wisconsin Press.

Husserl, E. (1931). *Ideas: General introduction to pure phenomenology*. Allen & Unwin.

Igaz, H., & Gulbahar, Y. (2017, July 20-22). *Why do learners choose online learning: The learners' voices*. International Association for Development of the Information Society. 11th International Conference on e-Learning, Lisbon, Portugal.

Iqbal, Q., & Ahmad, N. H. (2020). Sustainable development: The colors of sustainable leadership in learning organization. *Sustainable Development*, *29*(1), 108–119. doi:10.1002d.2135

Iqbal, Q., Ahmad, N. H., & Li, Z. (2021). Frugal-based innovation model for sustainable development: Technological and market turbulence. *Leadership and Organization Development Journal*, *42*(3), 396–407. doi:10.1108/LODJ-06-2020-0256

Işık, A. H., Karakış, R., & Güler, İ. (2010). Postgraduate students' attitudes towards distance learning. *Procedia: Social and Behavioral Sciences*, *9*, 218–222. doi:10.1016/j.sbspro.2010.12.139

Jarvis, P. (2006). *Towards a comprehensive theory of human learning* (Vol. 1). Routledge.

Jiang, W., Zhao, X., & Ni, J. (2017). The impact of transformational leadership on employee sustainable performance: The mediating role of organizational citizenship behavior. *Sustainability*, *9*(9), 1567. doi:10.3390u9091567

Jones, C., Penaluna, K., & Penaluna, A. (2019). The promise of andragogy, heutagogy and academagogy to enterprise and entrepreneurship education pedagogy. *Education + Training*, *61*(9), 1170–1186. doi:10.1108/ET-10-2018-0211

Jones, S., Harvey, M., Hamilton, J., Bevacqua, J., Egea, K., & McKenzie, J. (2017). Demonstrating the impact of a distributed leadership approach in higher education. *Journal of Higher Education Policy and Management, 39*(2), 197–211. doi:10.1080/1360080X.2017.1276567

Ke, F. (2010). Examining online teaching, cognitive, and social presence for adult students. *Computers & Education, 55*(2), 808–820. doi:10.1016/j.compedu.2010.03.013

Keeler, L. C. (2006). Student satisfaction and types of interaction in distance education courses. *Dissertation Abstracts International, 67*(9). (UMI No. 3233345)

Khalifa, A., Kato, T., & Yamamoto, S. (2017). *Measuring the effect of repetitive queries and implicit learning with joining-in type robot-assisted language learning system.* Paper presented at ISCA workshop on Speech and Language Technology in Education, Stockholm, Sweden. 10.21437/SLaTE.2017-3

Kilburn, A., Kilburn, B., & Hammond, K. (2017). The role of quality in online higher education. *Journal of Higher Education Theory and Practice, 17*(7), 80–86.

King, K. (2017). *Technology and innovation in adult learning*. Jossey-Bass.

Kingston, D., & Siraj, I. (2017). Supporting the implementation of the foundation phase through effective professional development. *Wales J. Educ., 19*(1), 39–68. doi:10.16922/wje.19.1.3

Kirtman, L. (2009). Online versus in-class courses: An examination of differences in learning outcomes. *Issues in Teacher Education, 18*(2), 103–116. https://eric.ed.gov/?id=EJ858508

Kjellstrom, S., Stålne, K., & T¨ornblom, O. (2020). Six ways of understanding leadership development: An exploration of increasing complexity. *Leadership, 16*(4), 434–460. doi:10.1177/1742715020926731

Knowles, M. S. (1962). *The adult education movement in the United States*. Holt, Rinehart, and Winston.

Knowles, M.S. (1968). *Andragogy, not pedagogy*. Adult Leadership.

Knowles, M. S. (1970). *The modern practice of adult education: Andragogy versus pedagogy*. Association Press.

Knowles, M. S. (1975). *Self-directed learning: A guide for learners and teachers*. Association Press.

Knowles, M. S. (1980). *The modern practice of adult education: From pedagogy to andragogy* (2nd ed.). Cambridge Books.

Knowles, M. S. (1984). Introduction: The art and science of helping adults learn. In *Andragogy in action* (pp. 1–21). Jossey-Bass.

Knowles, M. S. (1984). *The adult learner: A neglected species* (3rd ed.). Gulf.

Knowles, M. S. (1990). *The adult learner: A neglected species* (4th ed.). Gulf Publishing Company.

Knowles, M. S., Holton, E., & Swanson, A. (1998). *The adult learner*. Gulf Publishing Company.

Knowles, M. S., Holton, E. F., & Swanson, R. A. (2005). *The adult learner: the definitive classic in adult education and human resource development* (6th ed.). Elsevier, Inc., doi:10.4324/9780080481913

Knutson, S. (2003). Experiential learning in second-language classrooms. *TESL Canada Journal, 20*(2), 52–64. doi:10.18806/tesl.v20i2.948

Komives, S. R., Lucas, N., & McMahon, T. R. (2009). *Exploring leadership: For college students who want to make a difference*. John Wiley & Sons.

Kumi-Yeboah, A. (2018). Designing a cross-cultural collaborative online learning framework for online instructors. *Online Learning, 22*(4), 181–201.

Kuo, Y. C., & Belland, B. (2016). An exploratory study of adult learners' perceptions of online learning: Minority students in continuing education. *Educational Technology Research and Development, 64*(4), 661–680. doi:10.100711423-016-9442-9

Kuo, Y. C., Walker, A., Schroder, K. E. E., & Belland, B. R. (2014). Interaction, Internet self-efficacy, and self-regulated learning as predictors of student satisfaction in online education courses. *The Internet and Higher Education, 20*, 35–50. doi:10.1016/j.iheduc.2013.10.001

Liang, J.-C., & Tsai, C.-C. (2008). Internet self-efficacy and preferences toward constructivist internet-based learning environments: A study of pre-school teachers in Taiwan. *Journal of Educational Technology & Society, 11*(1), 226–237.

Liao, Z., & Zhang, M. (2020). The influence of responsible leadership on environmental innovation and environmental performance: The moderating role of managerial discretion. *Corporate Social Responsibility and Environmental Management, 27*(5), 2016–2027. doi:10.1002/csr.1942

Liaw, S., Huang, H. M., & Chen, G. D. (2007). Surveying instructor and learner attitudes toward e-learning. *Computers & Education, 49*(4), 1066–1080. doi:10.1016/j.compedu.2006.01.001

Lipman, M. (1991). *Thinking in education*. Cambridge University Press.

Lipman, M. (2003). *Thinking in education* (2nd ed.). Cambridge University Press. doi:10.1017/CBO9780511840272

Lowenthal, P. R., & Dunlap, J. C. (2018). Investigating students' perceptions of instructional strategies to establish social presence. *Distance Education, 39*(3), 281–298. doi:10.1080/01587919.2018.1476844

Lytovchenko, I., Lavrysh, Y., Lukianenko, V., & Ogienko, O. (2020). How to teach grammar to adult ESP learners at technical university more communicatively: Task-based approach. *Multidisciplinary Journal for Education. Social and Technological Sciences, 7*(1), 54–71. doi:10.4995/muse.2020.12419

Lytovchenko, I., & Voronina, H. (2020). MOOC as remote ESP learning tool at university in quarantine: Focus on Students' Attitudes. *Revista Românească pentru Educaţie Multidimensională, 12*(2), 70–76. doi:10.18662/rrem/12.2Sup1/291

Mackeracher, D. (2005). *Making sense of adult learning*. University of Toronto Press.

Mariani, M. (2001). Distance learning in postsecondary education: Learning whenever, wherever. *Occupational Outlook Quarterly, 45*(2), 2–10.

Marton, F. (1986). Phenomenography: A research approach to investigating different understandings of reality. *Journal of Thought, 21*(3), 28-49.

Mazzoni, E., & Benvenuti, M. (2015). A robot partner for preschool children learning English using socio-cognitive conflict. *Journal of Educational Technology & Society*, *18*(4), 474–485.

McMahon, B., & Portelli, J. P. (2004). Engagement for what? Beyond popular discourses of student engagement. *Leadership and Policy in Schools*, *3*(1), 59–76. doi:10.1076/lpos.3.1.59.27841

Merleau-Ponty, M. (1962). *Phenomenology of perception* (C. Smith, Trans.). Routledge & Kegan Paul.

Merriam, S., Caffarella, R., & Baumgartner, L. (2020). *Learning in adulthood: A comprehensive guide* (4th ed.). John Wiley & Sons.

Merriam, S. B. (2008, Fall). Adult learning theory for the twenty-first century. *New Directions for Adult and Continuing Education*, *2008*(119), 93–98. doi:10.1002/ace.309

Merriam, S. B., & Bierema, L. L. (2014). *Adult learning: Linking theory with practice*. Jossey-Bass.

Mertens, D. M. (2007). Transformative paradigm: Mixed methods and social justice. *Journal of Mixed Methods Research*, *1*(3), 212–225. doi:10.1177/1558689807302811

Mezirow, J. (1978). Perspective transformation. *Adult Education Quarterly*, *28*(2), 100–110. doi:10.1177/074171367802800202

Mezirow, J. (2000). *Learning as transformation: critical perspectives on a theory in progress* (1st ed.). Jossey-Bass.

Miller, M. T., & Grover, K. S. (2014). Characteristics and training of adult education program leaders in community colleges. *International Journal of Educational Studies*, *1*(2), 83–90.

Milliron, M. D., & Miles, C. L. (1998). *Technology, learning, and community (TLC)*. League for Innovation in the Community College.

Movahedazarhouligh, S., Banerjee, R., & Luckner, J. (2022). An examination of current leadership practices in early childhood and early childhood special education: A mixed methods study. *Early Education and Development*, *33*(4), 700–722. doi:10.1080/10409289.2021.1909937

Mubin, O., Shahid, S., & Bartneck, C. (2013). Robot-assisted language learning through games: A comparison of two case studies. *Australian Journal of Intelligent Information Processing Systems*, *13*(3), 9–14.

Mukan, N., & Lavrysh, Y. (2020). Video conferencing integration: Challenges and opportunities at universities. *Revista Românească Pentru Educație Multidimensională*, *12*(1Sup2), 108–114. doi:10.18662/rrem/12.1sup2/253

Nhlapo, M. D., & Hlalele, D. J. (2023). Successes and failures of the university-community partnership: A case study of Imbali semi-rural settlement. *Education and Urban Society*, *55*(2), 222–237. doi:10.1177/00131245211027368

Núñez-Canal, M., de Obesso, M. de las M., & Pérez-Rivero, C. A. (2022). New challenges in higher education: A study of the digital competence of educators in Covid times. *Technological Forecasting and Social Change*, *174*, 121270. Advance online publication. doi:10.1016/j.techfore.2021.121270

Olga, T., & Yuliya, L. (2019). Humanistic trend in education in a global context. *SHS Web of Conferences, 69*. 10.1051hsconf/20196900121

Olgren, C. H. (2000). Learning strategies for learning technologies. In E. J. Burge (Ed.), The strategic use of learning technologies (pp. 7–16). Jossey-Bass. doi:10.1002/ace.8801

Owens, B. P., & Hekman, D. R. (2012). Modeling how to grow: An inductive examination of humble leader behaviors, contingencies, and outcomes. *Academy of Management Journal, 55*(4), 787–818. doi:10.5465/amj.2010.0441

Palloff, R., & Pratt, K. (1999). *Building learning communities in cyberspace*. Jossey-Bass.

Papanaoum, Z. (1994). Adult educators: Issues of professionalization. In R. Benn & R. Fieldhouse (Eds.), *Training and professional development in Adult and Continuing Education*. University of Exeter, Center for Research in Continuing Education.

Park, H.-R., & Kim, D. (2011). Reading-strategy use by English as a second language learners in online reading tasks. *Computers & Education, 57*(3), 2156–2166. doi:10.1016/j.compedu.2011.05.014

Parsons, M., & Stephenson, M. (2005). Developing reflective practice in student teachers: Collaboration and critical partnerships. *Teachers and Teaching, 11*(1), 95–116. doi:10.1080/1354060042000337110

Pekel, N. (2002). *Students' attitudes towards web-based independent learning at Bilkent university school of English language* [Master's thesis]. Bilkent University.

Peters, J., Taylor, J., & Doi, M. (2011). Reflections on three types of teaching and learning in a model-based, blended online course. *International Journal of Learning, 18*(2), 357–366. doi:10.18848/1447-9494/CGP/v18i02/47465

Peters, J. M., & Armstrong, J. L. (1998). *Facilitating Collaborative Learning*. Jossey-Bass.

Piaget, J. (1951). *Play, dreams and imitation in childhood.* Norton.

Piwowar-Sulej, K., & Iqbal, Q. (2023). Leadership styles and sustainable performance: A systematic literature review. *Journal of Cleaner Production, 382*, 134600. Advance online publication. doi:10.1016/j.jclepro.2022.134600

Po-ying, C. (2007). How students react to the power and responsibility of being decision-makers in their own learning. *Language Teaching Research, 11*(2), 225–241. doi:10.1177/1362168806074613

Porter, L. R. (1997). *Creating the virtual classroom distance learning with the internet.* John Wiley &Sons.

Pracana, C., & Wang, M. (2018). *Psychological applications and trends 2018.* Distributed by ERIC Clearinghouse.

Purcell, W. M., & Chahine, T. (2019). Leadership and governance frameworks driving transformational change in an entrepreneurial UK university. *Leadership and Organization Development Journal, 40*(5), 612–623. doi:10.1108/LODJ-07-2018-0280

Radford, A. W. (2011). *Learning at a distance: Undergraduate enrollment in distance education courses and degree programs. Stats in Brief. NCES 2012-154.* Distributed by ERIC Clearinghouse.

Ramsden, P. (1988). *Improving learning: new perspectives.* Kogan Page.

Ravens, R. (1982). *The origin and growth of action learning.* Chartwell Bratt.

Rego, A., Owens, B., Leal, S., Melo, A. I., Cunha, M. P., Gonçalves, L., & Ribeiro, P. (2017). How a leader's humility helps teams to be humbler, psychologically stronger, and more effective: A moderated mediation model. *The Leadership Quarterly, 28*(5), 639–658. doi:10.1016/j.leaqua.2017.02.002

Resnick, L. B. (1991). Shared cognition: Thinking as social practice. In L. B. Resnick, J. M. Levine, & S. D. Teasley (Eds.), *Perspectives on socially shared cognition* (pp. 1–20). American Psychological Association. doi:10.1037/10096-018

Robles, F. M. R. (2006). *Learner characteristic, interaction and support service variables as predictors of satisfaction in Web-based distance education.* The University of New Mexico.

Rogers, E. (2003). *Diffusion of innovations.* The Free Press.

Sa, M. J., & Serpa, S. (2020). The COVID-19 pandemic as an opportunity to foster the sustainable development of teaching in higher education. *Sustainability, 12*(20), 8525. Advance online publication. doi:10.3390u12208525

Saadatmand, M., Uhlin, L., Hedberg, M., Åbjörnsson, L., & Kvarnström, M. (2017). Examining learners' interaction in an open online course through the community of inquiry framework. *The European Journal of Open and Distance Learning.* doi:10.1515/eurodl-2017-0004

Sadler, W. A. (1969). *Existence and love: A new approach in existential phenomenology.* Charles Scribner's Sons.

Scott, S., & Webber, C. F. (2008). Evidence-based leadership development: The 4L framework. *Journal of Educational Administration, 46*(6), 762–776. doi:10.1108/09578230810908343

Seemiller, C. (2021). Preparing leaders of tomorrow: An analysis of leadership competencies within accredited academic program learning outcomes. *The Journal of Leadership Studies, 15*(1), 6–27. doi:10.1002/jls.21737

Sendjaya, S., Eva, N., Butar, I., Robin, M., & Castles, S. (2017). SLBS-6: Validation of a short form of the servant leadership behavior scale. *Journal of Business Ethics, 156*(4), 941–956. doi:10.100710551-017-3594-3

Senge, P. M. (2006). *The fifth discipline: The art and practice of the learning organization.* Doubleday Currency.

Senge, P. M., Cambron-McCabe, N., Lucas, T., Smith, B., Dutton, J. & Kleiner, A. (2000). *Schools that learn: A fifth discipline field book for educators, parents, and everyone who cares about education.* Doubleday/Currency.

Siddiquei, A., Asmi, F., Asadullah, M. A., & Mir, F. (2021). Environmental-specific servant leadership as a strategic tool to accomplish environmental performance: A case of China. *International Journal of Manpower, 42*(7), 1161–1182. doi:10.1108/IJM-07-2020-0350

Simmons, N., & Taylor, K. (2019). Leadership for the scholarship of teaching and learning: Understanding bridges and gaps in practice. *The Canadian Journal for the Scholarship of Teaching and Learning, 10*(1), 1. doi:10.5206/cjsotl-rcacea.2019.1.7995

Simonds, T. A., & Brock, B. L. (2014). Relationship between age, experience, and student preference for types of learning activities in online courses. *Journal of Educators Online, 11*(1), 1. doi:10.9743/JEO.2014.1.3

Sinclair, E. (2017). A case study on the importance of peer support for e-learners. *Proceedings of the 9th International Conference on Computer Supported Education*, 280–284. 10.5220/0006263602800284

Smidt, E., Bunk, J., McGrory, B., Li, R., & Gatenby, T. (2014). Student attitudes about distance education: Focusing on context and effective practices. *The IAFOR Journal of Education, 2*(1), 40–64. doi:10.22492/ije.2.1.02

Smith, Ch., Hoderi, M., & McDermott, W. (2019). A preliminary study of students' perception and learning from different delivery methods. *Academy of Educational Leadership Journal, 23*(2). https://www.abacademies.org/articles/a-preliminary-studyof-students-perception-andlearning-from-different-delivery-methods-8235.html

Spigelberg, H. (1982). *The phenomenological movement: A historical introduction* (3rd ed.). Martinus Nijhoff.

Spillane, J. (2012). *Distributed leadership*. Jossey-Bass.

Stein, D., & Glazer, H. R. (2003). Mentoring the adult learner in academic midlife at a distance education university. *American Journal of Distance Education, 17*(1), 7–23. doi:10.1207/S15389286AJDE1701_2

Sun, P.-C., Tsai, R. J., Finger, G., Chen, Y.-Y., & Yeh, D. (2008). What drives successful e-Learning? An empirical investigation of the critical factors influencing learner satisfaction. *Computers & Education, 50*(4), 1183–1202. doi:10.1016/j.compedu.2006.11.007

Swan, K., & Shih, L. F. (2005). On the nature and development of social presence in online course discussions. *Journal of Asynchronous Learning Networks, 9*(3), 115–136.

Talmage, C. A., Hansen, R. J., Knopf, R. C., Thaxton, S. P., McTague, R., & Moore, D. B. (2019). *Unleashing the value of lifelong learning institutes: Research and practice*. Academic Press.

Taylor, J. E. (2022). Motivational immediacy. [electronic resource]: fostering engagement in adult learners. Stylus Publishing, LLC.

Thakhathi, A., Le Roux, C., & Davis, A. (2019). Sustainability leaders' influencing strategies for institutionalizing organizational change towards corporate sustainability: A strategy-as-practice perspective. *Journal of Change Management, 19*(4), 246–265. doi:10.1080/14697017.2019.1578985

Torrisi-Steele, G. (2014). Factors predisposing academics towards the use of blended strategies: A model. *International Journal of Adult Vocational Education and Technology, 5*(3), 1–20. doi:10.4018/ijavet.2014070101

Tough, A. (1967). *Learning without a teacher*. Institute for Studies in Education.

Tough, A. (1987). Potential futures: Implications for adult educators. *Spons Agency*, 13.

Tracy, S. J. (2020). *Qualitative research methods: collecting evidence, crafting analysis, communicating impact* (2nd ed.). Wiley Blackwell.

Ur Rehman, Z., Shafique, I., Khawaja, K. F., Saeed, M., & Kalyar, M. N. (2023). Linking responsible leadership with financial and environmental performance: Determining mediation and moderation. *International Journal of Productivity and Performance Management*, *72*(1), 24–46. doi:10.1108/IJPPM-12-2020-0626

U.S. Department of Education, National Center for Education Statistics. (2019). *Adult literacy in the United States*. https://nces.ed.gov/datapoints/2019179.asp

Veletsianos, G. (2011). Designing opportunities for transformation with emerging technologies. *Educational Technology*, *51*(2), 41–46.

Vygotsky, L. S. (1978). *Mind in society*. Harvard University Press.

Waits, T., & Lewis, L. (2003). *Distance education at degree-granting postsecondary institutions: 2000–2001*. U.S. Department of Education, National Center for Education Statistics.

Wang, V. (2006). *Essential elements for andragogical styles and methods: How to create andragogical modes in adult education*. Pearson Education.

Wang, V. (2008). *Facilitating adult learning: A comprehensive guide for successful instruction* (Rev. ed.). Pearson Education.

Wang, V., & King, K. (2010). Transformative learning and ancient Asian educational perspectives. In V. Wang (Ed.), *Assessing and evaluating adult learning in career and technical education* (pp. 13–26). IGI Global. doi:10.4018/978-1-61520-745-9.ch002

Wang, V., & Torrisi-Steele, G. (2015). Online teaching, change, and critical theory. *New Horizons in Adult Education and Human Resource Development*, *27*(3), 18–26. doi:10.1002/nha3.20108

Watzlawick, P., Beavin, J. H., & Jackson, D. D. (1967). *Pragmatics of Human Communication*. Norton.

Wenger, E. (1998). *Communities of practice: Learning, meaning, and identity*. Cambridge University Press. doi:10.1017/CBO9780511803932

Wicks, D. A., Craft, B. B., Mason, G. N., Gritter, K., & Bolding, K. (2015). An investigation into the community of inquiry of blended classrooms by a Faculty Learning Community. *The Internet and Higher Education*, *25*, 53–62. doi:10.1016/j.iheduc.2014.12.001

Widisatria, D., & Nawangsari, L. C. (2021). The influence of green transformational leadership and motivation to sustainable corporate performance with organizational citizenship behavior for the environment as a mediating: Case study at PT Karya Mandiri Sukses Sentosa. *European Journal of Business and Management Research*, *6*(3), 118–123. doi:10.24018/ejbmr.2021.6.3.876

Wilson, J. H. (2014). *Investing in English skills: The limited English proficient workforce in US metropolitan areas*. The Brookings Institution.

Wiseman, T. (1996). A concept analysis of empathy. *Journal of Advanced Nursing*, *23*(6), 1162–1167. doi:10.1046/j.1365-2648.1996.12213.x PMID:8796464

Wolff, K. H. (1984). Surrender-and-catch and phenomenology. *Human Studies*, *7*(2), 191–210. doi:10.1007/BF00144932

Wolfson, N. E., Cavanagh, T. M., & Kraiger, K. (2014). Older adults and technology-based instruction: Optimizing learning outcomes and transfer. *Academy of Management Learning & Education*, *13*(1), 26–44. doi:10.5465/amle.2012.0056

Xu, D., & Jaggars, S. S. (2013). The impact of online learning on students' course outcomes: Evidence from a large community and technical college system. *Economics of Education Review*, *37*, 46–57. doi:10.1016/j.econedurev.2013.08.001

Yang, Y., Yeh, H., & Wong, W. (2010). The influence of social interaction on meaning construction in a virtual community. *British Journal of Educational Technology*, *41*(2), 287–306. doi:10.1111/j.1467-8535.2009.00934.x

Ying, M., Faraz, N. A., Ahmed, F., & Raza, A. (2020). How does servant leadership foster employees' voluntary green behavior? A sequential mediation model. *International Journal of Environmental Research and Public Health*, *17*(5), 1792. doi:10.3390/ijerph17051792 PMID:32164222

Yusuf, B. N. (2020). Are we prepared enough? A case study of challenges in online learning in a private higher learning institution during the Covid-19 outbreaks. *Advances in Social Sciences Research Journal*, *7*(5), 205–212. doi:10.14738/assrj.75.8211

Zhang, R. (2020). Exploring blended learning experiences through the community of inquiry framework. *Language Learning & Technology*, *24*(1), 38–53.

Zhang, X., & Bartol, K. M. (2010). Linking empowering leadership and employee creativity: The influence of psychological empowerment, intrinsic motivation, and creative process engagement. *Academy of Management Journal*, *53*(1), 107–128. doi:10.5465/amj.2010.48037118

KEY TERMS AND DEFINITIONS

Cognitive Presence: The process of creating meaning with collaborative inquiry.

EFL: English as a Foreign Language.

ESL: English as a Second Language.

Social Presence: The collaborative nature of the community and its activities in which the participants identify with the community, communicate purposefully in a trusting environment, and develop interpersonal relationships.

Teaching Presence: The design, facilitation, and direction of cognitive and social processes for the purpose of realizing personally meaningful and educationally worthwhile learning outcomes.

Chapter 15
Leading Through Digital Andragogy:
On the Cutting Edge Educationally

Linda Marie Ellington
Southern New Hampshire University, USA

ABSTRACT

The tsunami of technological change in higher education is causing disruption throughout academia and is exerting influence on leading in formal education. In much the same way that industries experienced disruption from the digital revolution, andragogy is on the verge of what could be called a learning-tech culture revolution. This chapter focuses on two critical aspects that present potential impact in education, the leader, and the adult learner. The purpose of this literature review is to identify challenges leaders face in the digital landscape and strategies relevant to innovative techniques for digital adult learners.

INTRODUCTION

The objectives of this chapter are to interpret the progress, direction and current research on the impact technology has on andragogical leadership and to increase the conversations around adult students who are immersed in digital technology, who are accustomed to twitch-speed, multitasking, random access, graphics-first, active, connected, fun, fantasy, and a payoff world (Gisbert & Bullen, 2015). The aim of the chapter is an exploration of literature that examines key breakthroughs in the immersing of technology and how this digital disruption is influencing andragogical leadership. The goal is to lend a written voice to considering different perspectives in leading andragogical organizations.

Digital students are asking for new approaches to education with a better approach that affects how they think, interact, and make sense of their world (Greene & Larsen, 2018). The technologies and practices that have developed, and are still being developed, are heralded as opportunities to transform both online and traditional education systems. While proponents of these new ideas often postulate that they have the potential to address the educational problems facing both students and institutions, they also provide an opportunity to rethink the ways that education is organized.

DOI: 10.4018/978-1-6684-7832-5.ch015

Adult learners bring vast experience to learning opportunities and should have the maturity to stay motivated, focused, and engaged. Yet, in an online environment, many of the andragogical ideals laid out by Malcolm Knowles can get lost. Lessons learned from the experiences of crafting curriculum and leading adult students are analyzed by Green and Larsen (2018) who offer an examination of strategies that can empower instructors, instructional designers, and administrators, allowing the presence of the instructor, fidelity of implementation, and emotional/cognitive support for the student.

Digital tools and instruments are "*rapidly becoming as infrastructural as electricity*" (Cascio & Montealegre, 2016, p. 350). Barley (2015) suggests that although digital technology is rapidly becoming as infrastructural as electricity, there is surprisingly little research on how it is altering work systems or the work that people do. The next wave of technological innovation will feature the emerging technology paradigm known as ubiquitous computing. This concept is not about one technology. Rather, it reflects information and communication environments in which computer sensors are all unified with various objects, people, information, and computers as well as the physical environment. The combination of these developments is a hyperconnected and data-saturated world, where an Internet of everyone is linked to an Internet of Everything (Wooldridge, 2015).

BACKGROUND

Given the rapid advances and the increased reliance on technology, the question of how it is changing leadership is highly salient for scholars and practitioners. Even if and when technology has been applied to digital andragogical practices, technology will not fully be exploited unless methods of leading and teaching also change.

According to Gisbert and Bullen (2015, p.5) there are several reasons for changes in leading and teaching to occur:

- To achieve existing academic goals in a more productive way.
- To achieve new academic goals that are required in a digital society, such as knowledge. management, independent learning, collaborative learning, and IT skills embedded within a subject domain.
- To use the technology to make learning more engaging and motivating for students, already immersed in digital technologies.
- To help students develop digital literacy - an essential skill in a digital world.

The impact of being immersed in a digital world is at the heart of the digital discourse for institutions. The challenge is knowing that any enterprise that runs forward on only new technology, ignoring the landscape of existing knowledge, will be sub-optimal and likely will fail. To build a strong future of digital learning in the academy, we must first take stock of what we know and what has been researched (Siemens, GaSevil, & Dawson, 2015). Clearly it is not enough for leaders to be digitally literate, they must also improve their digital competence and fluency to a high level.

Digital competence has been defined in the principal reports of The Organization for Economic Cooperation and Development (OECD) as one of the key competencies of the 21st century (Gisbert & Bullen, 2015). Leaders need to speak in terms of environments, or spaces, rather than classrooms. One of the main didactic challenges educational professionals face is to integrate all the elements into an educational process that takes into account the points of view of both educators and adult learners. Although leaders do not need to be technology wizards, they must understand what can be accomplished at the intersection of higher education and technology. They should also be prepared to lead the way in conceptualizing how technology can transform the institution. Does the institution have confidence in its leadership's digital fluency (Kane, et al., 2015)? Educators' digital fluency has been noted as one of the most important skills required for effective curriculum delivery and a necessary competence for effective leading.

Digital Fluency

Digital fluencies or literacies are required because digital technologies have amplified the intensity and complexity of literate environments. Educators need to possess a wide range of digital abilities and competencies to be able to navigate and teach effectively within the connected environment (NCTE 2013). Thus, digital fluency is to be able to navigate, communicate and access opportunities in a technology intensity and complex environment.

Bell and Gilliam (2011) define digital fluency as "the aptitude to effectively and ethically interpret information, discover meaning, design content, construct knowledge, and communicate ideas in a digitally connected world". Considering extrinsic factors that can impact one's fluency in a digital environment, Briggs and Makice (2011) refer to "digital fluency as an ability to reliably achieve desired outcomes through the use of digital technology. This ability is helped or hindered by the situational forces and the digital fluency of others". This means, to be digitally fluent is more than having technological ability whereby digital tools could be used to create and communicate complex ideas and interpretations of communications. Niessen (2013) concluded from the definitions stating that "digital fluency is an emerging aptitude that involves knowing when and why we use the digital media that we choose and using it with ease to communicate and/or retrieve information". This is an improvement from traditional literacy which meant an individual had the understanding of what to do, and how to do it but had no capacity to explain when and why communication and the mode would be appropriate (Briggs & Makice 2011).

NCTE (2013) has argued that for learners and educators in the digital age to be able to engage with each other and the curriculum delivery effectively, they need to:

- Be proficient and fluent in the digital technologies
- Be able to manage, analyze and synthesize vast information on the digital platforms
- Create, critique, and evaluate multimedia texts
- Display ethical responsibilities required by the multimedia environments.

The nature of communication in the 21st Century requires a shift from traditional literacy to digital literacy and digital fluency. Leaders and educators have no choice but to acquire digital fluency to effectively assist and communicate with the 21st Century learners. This entails that leaders, educators and learners should have the ability to effectively use technology, to locate, and critically evaluate information if they are to achieve the teaching and learning goals in the digitally networked environments of the 21st century (Duke & Ward 2009).

Higher Education as a Business

Viewing higher educational institutions as a business, digital technology platforms have become the foundation for an increasing share of economic activity resulting in a changing business environment. Digital transformation—the reinvention of an organization's vision and strategy, organizational structure, processes, capabilities, and culture to match the evolving digital business context—is not only changing the institutions but also redefining educational formal and informal markets. In a research conducted by Gurbaxani & Dunkle (2019) they identified six dimensions of digital transformation for a successful competitive stance due to digital transformation. They are an organization's strategic vision, alignment of the visions and its investments in digital transformation, the suitability of the culture for innovation, possession of know-how, the strength of its digital capabilities, and its use of digital technologies. Leaders basically lead at two levels, including the proximal level, referring to the context which is the closest to the leader and follower, and the distal level, which concerns the entire institutional environment and culture (Avolio et al., 2014; Mihelj, et al., 2019).

In the last few decades, scholars have been trying to monitor the effects of digitalization processes. Part of the academic debate has been focused on the role of the leader's ability to integrate digital transformation into their institutions and, at the same time, inspire stakeholders to gain an in-depth understanding of who the digital university learners are. To bring clarity to this debate, the construct of e-leader was introduced to describe a new profile of a leader who constantly interacts with technology (Avolio et al., 2014).

To foster clarity and alignment in the academic debate, Cortellazzo et al. (2019) provided a comprehensive analysis of the contribution of studies on leadership and digitalization, identifying patterns of thought and findings across various disciplines in connection to learners. Their study identified categories at the macro level such as e-leadership and organization, digital tools, ethical issues, and social movements (Agarwal, et al., 2014). At the micro level they examined the role of C-level leaders, leaders' skills in the digital age, and practices for leading virtual teams. The *e-learner* is clearly connected to the characteristics and behaviors of those who lead in such institutions.

The academic debate continues within the relationship between digital transformation and leadership, organizing and structuring the main emerging themes at the macro and micro levels of leading. Micro and macro approaches underscore that information technology and strategic management need greater alignment. Digital transformation is successful in the long term when the overall organizational objectives match the need to adopt new digital tools or instruments. In a similar vein, individuals embrace technological advancement only when they perceive it is relevant to their tasks. It is an important responsibility of the leader, particularly of C-level leaders, to steer this strategic alignment and the proliferation of a digital culture. In a networked economy, digital transformation has led organizations to open their boundaries, and connect with other industries, stakeholders, and customers, to generate innovation. From a micro perspective, this openness is also required by leaders who need to invest in networking. This means to being *out there* (Grafström & Falkman, 2017) and being willing to communicate with different types of stakeholders, through digital tools and social media. To overcome the challenges derived from the digital transformation, leaders are required to develop a combination of digital and human skills, mainly related to the ability to communicate effectively in a digitalized context, create cohesion between geographically distant followers, foster initiative and change attitudes, and deal with complex and fast problem-solving. Basically, leading into digital ubiquity.

DIGITAL UBIQUITY

The notion of digital ubiquity (Gerth & Peppard, 2016; Schwarzmüller et al., 2018), describes the pervasive proliferation of technology that influences leadership (Roman et al., 2018). According to Lu et al. (2014), it cannot be assumed that leadership skills identified in an offline context should be transferred to virtual leadership without any adjustment. However, some authors make this assumption tacitly (e.g., Schwarzmüller et al., 2018), without explicitly addressing any related theoretical framework. Bolden and O'Regan report that "there is no one approach to leadership," since leadership is context specific and must be adapted to the needs of the day (2016, p.439). It is stated in *Guide to Administering Online Learning* that the issues and needs that digital learning administrators currently face are barriers to the adoption of emerging technologies, policies, legalities, ethics, strategic planning, design of professional development, management of the course development process, quality assurance, student support, and recruitment and marketing (Cifuentes, 2021; Latchem 2012). The studies on digitalization and leadership can be classified into four different categories, according to whether they focus on: (1) The relationship between e-leaders and organizations; (2) How leaders adopt technology to solve complex organizational problems; (3) The impact of digital technologies on ethical leadership; or (4) The leader's use of digital technologies to influence social movements (Cortellazzo, 2019).

Leaders have increasingly become the key players in driving positive results from investments in digital tools and technologies and yet, despite the call to understand how organizations and e-leaders are intertwined, few studies provide an empirical explanation of the new organizational configurations emerging from the interaction between technology and the human/social system. Berman and Korsten (2014) suggest future leaders should base their organizations on three pillars: (1) Assuring a highly connected and open working environment at any hierarchical levels and units in organizations; (2) Engaging customers by gathering knowledge about the whole person; and (3) Establishing more integrated and networked relationships with partners and competitors. Berman and Korsten posit these three pillars transform organizations at all levels (2014). This implies organizations are becoming boundaryless, at both the internal and external levels. Further, the organizational structure is no longer a static feature, but an ongoing process – where organizational boundaries are loose and permeable that will enrich innovation (Van Outvorst et al., 2017).

Leaders at any level can foster innovation if they go beyond the formal organization, to create real or virtual networks among internal and/or external communities of practice. These communities of practice enable a more agile response to change, promoting the free flow of information and breaking down information silos (Petrucci and Rivera, 2018), thereby empowering both leaders and stakeholders to integrate, transform and stimulate knowledge that fosters innovation. This way, information and communication technology enable the creation of shared information pools wherein diverse staff across the organization contribute to a collaborative and dynamic process of idea generation (Laal & Ghodsi, 2011). By breaking the organizational boundaries within and between internal and external stakeholders, the traditional leader-centered information and decision-making process is giving way to novel processes that democratize access to information and share decision power among all involved.

Although the changes made by today's technology may be impressive, and digital innovations will continue for the foreseeable future, technology by itself is not enough. Fulfilling this potential will require leaders to recreate the way their organizations operate in the world of digital ubiquity (Torre & Sarti, 2014).

ACADEMIC C-LEADERS IN THE DIGITAL AGE

According to Gerth and Peppard, (2016) the topic of leadership and digital technology can be classified into three different categories, depending on whether they focus on: (1) The increased complexity of C-level roles; (2) The skills e-leaders need; and (3) The practices for leading virtual teams effectively. The evolution of C-Level roles has an impact that digitalization has had in the competitive academic environment which has made it necessary for organizations to adopt a high-level strategic view on digital transformation. New responsibilities on the selection of digital technologies that will drive an organization's ability to remain competitive in a highly digitized world, are given mainly to its top-level leaders.

Higher education (HE) CEOs in the digital age assume the additional role of digital change agents and digital enablers, implying that they should recognize the opportunities offered by new technologies, and also push for their implementation. They are required to instill a digital culture into the top management team, involving it in actively sustaining a digital change inside the organization (Gerth and Peppard, 2016). For this matter, a greater interaction is needed between the CEO and the Chief Information Officer (CIO), who will increasingly become a key player in the digital strategy definition and implementation, rather stay confined to an IT-is-a-mess-now-fix-it flavor of a role (Gerth and Peppard, 2016; Bekkhus and Hallikainen, 2017). Bekkhus and Hallikainen (2017) acknowledge an increased ambidexterity in the role of CIOs and developed a toolbox related to their role as gatekeepers and contributors. In order to reach their goals successfully, CIOs need to have a clear picture of both the characteristics of the digital strategy and the organizational needs it is supposed to satisfy. They should also carefully evaluate the readiness of the organization in every step of the changing process in order to adopt the proper pace. To avoid IT project failures, CEOs need to facilitate the recognition of the CIO's role, as well as promote collaboration between the CIO and other top managers. As described before, digital technologies are not only used to support internal processes but are also a way to build relationships with different actors in the external environment. Social media platforms, in particular, are powerful tools that C-level executives use to build communications channels with their followers (Obschonka et al., 2017). In a study analyzing the rhetoric of CEOs in social media, Grafström and Falkman (2017) suggest that C-Level leaders' willingness and ability to construct a continuative dialogue through digital channels is a powerful way not only to manage the organizational crisis but also to sustain the reputation and the image of the organization, positioning the brand and communicating the organizational values. Thus, as Tsai and Men (2017) unveil, by properly using social media, organizational leaders can build trust, satisfaction, and advocacy among the stakeholders. According to Tsai & Men, digital technologies, and social media in particular, support CEOs in becoming "Chief Engagement Officers, who develop meaningful interpersonal interactions and relationships with today's media savvy public" (2017, p. 1859). Even if CEOs have always been considered the personification of the organization, the rising need for transparency and authenticity has led CEOs to embrace the task of being visible, approachable, and social leaders who actively contribute to the engagement of followers and customers (Tsai and Men, 2017).

THE PERSPECTIVE OF ANDRAGOGY

The term andragogy has a history that spans over 180 years. It was first coined by Alexander Kapp, a German high school teacher, in his 1833 book on methods for teaching adults (Reischmann, 2004). The term did not surface again until Eugen Rosenstock-Huessy, discussed his views on the implementation of his ideas about andragogy starting in the early 1920s through the 1970s (Reischmann, 2004). Eduard Linderman, a theorist in the field of andragogy, credited Rosenstock-Huessy with defining the term as separate from pedagogy and demagogy. Andragogik e.g., andragogy represents the learning process as one in which theory and practice become one - a process according to which theoretical knowledge and practical affairs become resolved in creative experience (Linderman, 1926 as cited in Draper, 1998). Lindeman published a book about andragogy, *The Meaning of Adult Education* in 1926, but the term "was not popularized until over four decades later (Draper, 1998).

Malcolm Knowles, who is considered the father of andragogy, first wrote about the term in 1968 in an article, *Andragogy, Not Pedagogy* (Reischmann, 2004). He viewed adult learners as self-directed and autonomous and that the role of the instructor was to facilitate the attainment and construction of knowledge. The idea of andragogy is a core foundational element for transformational adult learning has been noted by multiple researchers (Arghode et al., 2017; Giannoukos, et al., 2015; Huang, 2002). Draper posits that, "being familiar with the evolution of andragogy is the foundation for professionalism in adult education," (1998, pp. 3-26).

THE DIGITAL ADULT LEARNER AND PRACTICES

Digital andragogical learning is one of the fastest-growing segments of the educational world. One of the greatest challenges to a thorough exploration of the theories, tools, and practices that should be identified as ideal or most effective for serving the adult student is the lack of a single framework that juxtaposes cogent knowledge with the affordances of digital tools and virtual educational environments. Such a framework would enable designers and instructors to make informed decisions about how to best meet the needs of both the individual adult student and the learning objectives (Greene & Larsen, 2018).

With much online, or distance, learning taking place via a learning management system (e.g., Blackboard, Canvas, or Moodle), a formal educational experience for the adult learner can take place with little interaction among students or between students and instructor (McDonough, 2014). This issue has caused a great deal of concern for many colleges and universities as digital learning courses are not designed to ensure substantive interaction between faculty and students (Office of Inspector General, 2016). Rather than trying to modify pedagogical truisms, Greene & Larsen (2018) suggest leaders and faculty utilize a learner-focused, andragogical model of teaching. The model acknowledges the self-directedness, motivation, and rich knowledge base that adults bring to their learning environments. An andragogical model of teaching also incorporates relevant learning experiences and problem-solving performance tasks throughout the instructional design (Knowles, Holton III, & Swanson, 2015).

According to Knowles, Holton III, and Swanson (2015), optimal learning among adult learners requires attention to the following adult learning principles:

- Adults are intrinsically motivated to learn.
- Adults are inclined to learn when there is an obvious connection to their lives.
- Adults have current and past experiences as rich resources for learning.
- Adults prefer real-world learning contexts that incorporate a subject-centered, problem- solving approach.
- Adults learn best when they are empowered as autonomous, self-regulated learners.
- Adults require advanced information about a topic under study in order to determine its relevance.

Blackley and Sheffield (2015) explored the use of "andragogical practices within a digitally expanded educational context" (p. 407). They recognized that the current and prominent discourse related to 21st -century knowledge and skills provided a compelling reason to reevaluate andragogy and related adult learning principles within a digital context. The efficient use of digital learning environments in higher education is an important research topic from both a scientific and a practical perspective. Learning in digital environments is characterized by the provision of learning materials that are independent of time and location and by broad access to learning materials. Moreover, digital learning environments also support educational opportunities for all types of learners and provide digitally enhanced instruction (Simonson, et al., 2011; McDonough, 2014; Office of Inspector General, 2016).

SUCCESS FACTORS OF LEARNING

Educational researchers from diverse disciplines have been trying to identify the success factors of learning with digital media in higher education for about two decades (Green & Larsen, 2018). One central aim of higher education is to foster students' potential for high-quality accomplishments and support them in applying their knowledge to future challenges in their professional lives (Forehand, 2017). Students should be encouraged and empowered to use digital media for communication and collaboration as well as for learning and knowledge exchange in an appropriate way to become competent and proficient members of a knowledge society. Additionally, digital learning environments need to make learning more effective; for example, by increasing learners' motivation, adapting to students' prior knowledge, or providing the possibility for mobile and ubiquitous learning (Greene & Larsen, 2018; Smith et al., 2017). Although digital learning environments have increased the accessibility of educational opportunities, they also require different instructional design methods than traditional face-to-face learning environments (Linder-VanBerschot & Summers, 2015; Scanlon, McAndew, & O'Shea, 2015). The lack of physical presence in digital learning environments has challenged not only adult educators, but C-Level leaders as well.

Digital learning is incredibly dynamic and constantly drivse changes in learners' demands, behaviors, and technology. Institutions need to be ready to reinvent themselves and make purposeful and directed changes in response to new technology and andragogies in order to offer high-quality learning programs. As many authors point out (Bullen & Morgan, 2011) academia has focused more on the cognitive and essential nature of the digital learner than on the consequences of their learning as digital students. For example, the constant use of the Internet, immediacy, constant socialization, and the ability to work simultaneously with different media. And yet there is no consensus on the nature of these students, which contributes to the debate about their wishes, preferences, and habits rather than on the educational implications it may have on their educational goals and work life. Recently both higher education and the professional world have been converging on a definition of education that responds closer to the needs of the labor market. The focus is no longer on capacities and qualifications, as the professional world is now regarded to be fundamentally based on competencies (Gisbert & Bullen, 2015). The specific competencies are seen as technical competencies that belong to a particular occupation and level of professional qualification. The transferable competencies are essentially social and personal in nature, and directly linked to acquired knowledge and values, such as critical thought, communication, conflict solving and negotiation, planning, and teamwork; so fundamental that they are often referred to as competencies for employability. Transferable competencies can also be broadly defined as a set of wide-ranging skills that affect various task types in different situations, which means that they are highly generalizable and transferable, and lead to efficient professional practice (Sharif, 2015). Students need to understand the wide range of creative and innovative options that being digitally competent will give them, and the need to be critical and reflexive toward the information available. Competencies have a direct effect on the use of virtual worlds: language competence, learning to learn, and the competence associated with the subject matter (Gisbert & Bullen, 2015).

One strategy to use in the 3D virtual world could be to enable students to submerge themselves in a computer-generated 3D graphic simulation, navigate through it and interact with it in real-time from a user-centered perspective. Users of a virtual world communicate synchronously, which makes it possible for them to take part in collaborative processes and provide solutions in real-time. The only drawback to being aware of this is that synchronous communication requires users to be online at the same time and this may present problems because they may be in different time zones or their timetables may be incompatible. This challenge can be overcome by using an LMS that is compatible with the virtual world to equip the system with an asynchronous connection. One such example is to use the LMS Moodle. The inclusion of the Moodle forum and a Moodle-linked blog allows the information to be processed asynchronously. Also, users who take part in the virtual world are known as avatars, human graphic representations that are associated with particular users and identify them in the virtual world (Gisbert & Bullen, 2015). Another strategy is to employ a set of tools that enables the content and activities to be displayed in a 3D context. However, not all the tools available in Moodle have a corresponding tool in the 3D environment. Thus, there needs to be an interaction between the 3D objects and 2D tools available by Sloodle. With the passage of time, more and better tools are becoming available for promoting and supporting learning using new information technologies. Virtual worlds are one of the many alternatives and, if they are used appropriately, can increase teaching possibilities enormously and democratize and improve the quality of education processes.

COLLABORATIVE DIGITAL LITERACY PRACTICES

Technology has transformed learning at the postsecondary level and significantly increased the prevalence of digital learning environments. As adult educators approach instruction design, they must consider how to apply research-based practices that preserve the quality of instruction and provide adult learners with technology-based instruction that is relevant (Sharp, 2018). One approach is the use of collaborative digital literacy practices which has been highlighted as an effective method. Sharp (2018) explored the perceived levels of confidence and importance of collaborative digital literacy practices among adult learners in digital learning environments. Before and after participating in five different collaborative digital literacy practices (i.e., blogs, asynchronous discussions, wikis, microblog messages, and paired peer feedback), participants indicated their perceived levels of confidence and importance with each practice. It was concluded that increased use of collaborative digital literacy practices leads to increased levels of confidence among adult learners. Other studies have suggested that increased levels of confidence lead to enhanced self-efficacy, which has a positive effect on learner behaviors (Chamberlain, Hillier, & Signoretta, 2015). Supplanting learner-centered approaches into digital learning environments is vital for learner success (Blakely & Sheffield, 2015; Conaway & Zorn-Arnold, 2015). Much literature has demonstrated the need for adult educators to consider perceived levels of importance among their learners as they design curricula and related learning experiences (Almeida & Vasconcelos, 2015). We live in a global world where technology, especially information and communication technology is changing the manner in which we capture educational value, and how we interact with each other. Learners are familiar with the technologies that are transforming the foundations of the global workplace and the organizations that drive it: cloud and mobile computing, big data and machine learning, sensors and intelligent manufacturing, advanced robotics and drones and clean energy technologies. These innovations are in our world, and savvy educational leader knows they have to either figure out how these technologies will transform their institutions or face disruption by others who figure it out first.

CONCLUSION

Higher education may be the context in which students consolidate their use of technology from an intentional and technological perspective. However, the coexistence of education and technology does not automatically mean that technology is correctly implemented in the educational process. Whatever the learning environment, the space designed must contain a set of activities that activate the cognitive processes that are required by most professional tasks nowadays. Emerging technologies enable specific situations to be simulated that are similar to the reality of the world of work. These activities need to follow the tenets of instructional models based on the social construction and transformation of knowledge by means of activities such as modeling, action, interaction, collaboration, experience, discovery and discussion. Educational experiences in these environments must give a sense of naturalness to the activity. The sense of autonomy and independence then depends on the fact that learners interpret and carry out the activities using an approach that is consistent with their way of performing in their real-world work.

As far back as 2008, Sternberg proposed a model that focuses on the three Rs: reasoning (a comprehensive set of thinking skills that a person needs to be an engaged, active citizen of the world), resilience (persistence in achieving goals despite the obstacles life places in our way), and responsibility (cover the ethical and moral dimension of development). However, to cultivate digital learning environments that provide rigorous academic experiences and promote student success, recent literature has described several examples of effective methods used among adult learners in digital learning environments, such as:

- Adapting cognitive apprenticeship methods (CAM) with technology tools to provide opportunities for modeling, scaffolding, coaching, articulation, and reflection.
- Creating virtual orientations and tutorials that familiarize learners with course expectations, content, and tools.
- Embedding experts as resources for learners, such as a virtual librarian.
- Employing techniques that foster the development of self-regulation.
- Facilitating collaboration and interaction through social networking and digital discourse.

The ability to digitally reimagine the educational institution is determined in large part by a clear digital strategy supported by leaders who foster a culture able to change and invent the new. What is unique to digital transformation is that risk-taking is becoming a cultural norm as more digitally advanced organizations seek new levels of competitive advantage, and higher education is no stranger to competition. The digital agenda is led from the top. Digital fluency does not demand mastery of technologies but does require the ability to articulate the value of digital technologies to the organization's future (Kane, et al., 2015).

What separates digital leaders from the rest is a digital strategy combined with a culture and leadership poised to drive the transformation. The ability to adapt quickly to change also stands out as an important capability. Perry Hewitt, a chief digital officer at Harvard University, says agility is more important than technical skills. Emory professor Konsynski concurs that the 21st century is about agility, adjustment, adaptation and creating new opportunities (Kane et al., 2015, p.). Leading by example is part of leading a digital university. You have to be an influential leader in the physical, virtual, and augmented world," says Disney senior vice president Milovich. "We have to engage people through Twitter and other social platforms, but then also stand in front of 100 people and be authentic. We toggle back and forth between virtual and physical platforms all day long" (Kane et al., 2015, p.8). To tie this all together one needs to reflect that whether culture drives technology adoption or whether technology changes the culture. Culture leads to the adoption of technology. The leader's ability to innovate depends on the impatience of the institution's culture. Leaders are aware that digital models will reach their sell-by dates quickly. Other disruptions are waiting in the wings. By the time it is obvious you need to change, it is usually too late. You have to be bold. You need a culture that lets you figure out how to win even without a blueprint (Kane et al., 2015). Research stated that digital transformation is simply not about technology.

There are two main reasons why researchers and practitioners recommend the use of digital learning environments in higher education. First, in an increasingly digitalized world, education needs to be digital as well (Fraillon et al., 2014; Kummel et al., 2020). Students should be encouraged and empowered to use digital media for communication and collaboration as well as for learning and knowledge exchange in an appropriate way to become competent and proficient members of a knowledge society. Second, digital learning environments promise to make learning and teaching more effective, for example, by increasing learner's motivation, adapting to students' prior knowledge, or providing the possibility for mobile and ubiquitous learning (Sun et al., 2009; Muenks & Miele, 2017; Hatlevik & Christophersen, 2013; Scherer et al., 2017).

By way of conclusion, the new circumstances experienced during the global pandemic represent a more significant and drastic change in the digital landscape than leaders could have previously imagined, from which it may be difficult to emerge as if nothing has happened. It has indeed become an inevitable dimension in leading in an andragogical space during the digital work intensification, especially learning and experimenting with leadership practices and discovering which approaches make it most effective to support the learners the world now needs. The adoption of e-tools, seem now to be the main path to leading the present situation and the aftermath it will inevitably have.

As a consequence of the descriptions of leaders in the digital andragogical world they need to be individuals who are competent in virtual environments, aware of current ICT tools, and capable of choosing those tools in an appropriate manner. They, as well as their own basic communication skills, social skills, team skills, change management skills, and trust-building skills, are fundamental to leading people via digital channels. Translating leadership visions into day-to-day operations, is what a leader does. Fulfilling the potential for digital transformation in their institutions, higher education leaders need to recreate the way their organizations operate in a world of digital ubiquity. This means doing things that machines cannot.

As Millennials, and next in line the iGen, begin to take over the workforce, they are bringing with them leadership trends that will shape the future of organizational leadership. Higher education must respond to an increased pace and the nature of leader tasks is increasingly complex. Despite the evidence suggesting that focusing on growth versus performance will lead to better long-term performance, business schools continue to emphasize managing performance instead of leading growth. Trends such as real-time feedback, agile networks of teams, advanced people analytics, micro-learning, personalized learning, and artificial intelligence enable the digitally minded leader to shape the future of leadership. We know that the foundational pillars of leadership, such as shared values and vision, talent development, change management, and reward and recognition, will likely continue to drive results. However, how we lead in these areas is changing (Petrucci & Rivera, 2018; Schwarzmuller, et al., 2018). Education should liberate students from passive, mindless, and uncritical acceptance of experience and how experience shapes knowledge. Learning to re-evaluate and re-story prior learning experiences can lead adults to make sense of their experience and find a new sense of identity (Foote, 2015).

REFERENCES

Agarwal, S. D., Barthel, M. L., Rost, C., Borning, A., Bennett, W. L., & Johnson, C. N. (2014). Grassroots organizing in the digital age: Considering values and technology in Tea Party and Occupy Wall Street. *Information Communication and Society*, *17*(3), 326–341. doi:10.1080/1369118X.2013.873068

Almeida, A., & Vasconcelos, C. (2015). Geothics: Maser's students' knowledge and perception of its importance. *Research in Science Education*, *45*(6), 889–906. doi:10.100711165-014-9449-3

Arghode, V., Brieger, E., & McLean, G. N. (2017). Adult learning theories: Implications for online instruction. *European Journal of Training and Development*, *41*(7), 593–609. doi:10.1108/EJTD-02-2017-0014

Avolio, B. J., Sosik, J. J., Kahai, S. S., & Baker, B. (2014). E-leadership: Re-examining transformations in leadership source and transmission. *The Leadership Quarterly*, *25*(1), 105–131. doi:10.1016/j.leaqua.2013.11.003

Barley, S. R. (2015). Why the Internet makes buying a car less loathsome: How technologies change role relations. *Academy of Management Discoveries*, *1*(1), 31–60. doi:10.5465/amd.2013.0016

Bekkhus, R., & Hallikainen, R. (2017). A new dualistic CIO toolbox: Towards ambidexterity in the digital business transformation. *Proceedings of the Conference of the Portuguese Chapter of the Association of Information Systems, 17*, 23-41.

Bell, K., & Gilliam, A. (2011). *Index-Region 10 Website*. Region10.org. https://www.region10.org/digital-fluency/index

Berman, S., & Korsten, P. (2014). Leading in the connected era. Strategy & Leadership. *Emerald Group Publishing*, *42*(1), 37–46.

Blackley, S., & Sheffield, R. (2015). Digital andragogy: A richer blend of initial teacher education in the 21st century. *Issues in Educational Research*, *25*(4), 397–414.

Bolden, R., & O'Regan, N. (2016). Digital disruption and the future of leadership: An interview with rick haythornthwaite, chairman of centrica, and masterCard. *Journal of Management Inquiry*, *25*(4), 438–446. doi:10.1177/1056492616638173

Briggs, C., & Malice, K. (2011). *Digital fluency: Building success in the digital age*. Social Lens.

Bullen, M., Morgan, T., & Qayyum, A. (2011). Digital learners in higher education: Generation is not the issue. *Canadian Journal of Learning and Technology*, *37*(1), 2. doi:10.21432/T2NC7B

Cascio, W. F., & Montealegre, R. (2016). How technology is changing work and organizations. *Annual Review of Organizational Psychology and Organizational Behavior*, *3*(1), 349–375. doi:10.1146/annurev-orgpsych-041015-062352

Chamberlain, J. M., Hillier, J., & Signoretta, P. (2015). Counting better? An examination of the impact of quantitative method teaching on statistical anxiety and confidence. *Active Learning in Higher Education*, *16*(1), 51–66. doi:10.1177/1469787414558983

Chigona, A. (2018). Digital fluency: Necessary competence for teaching and learning in connected classrooms. *The African Journal of Information Systems*, *10*(4), 7. https://digitalcommons.kennesaw.edu/ajis/vol10/iss4/7

Cifuentes, L. (2013). *A guide to administering distance learning*. Brill.

Conaway, W., & Zorn-Arnold, B. (2015). The keys to online learning for adults: The six principles of andragogy. *Distance Learning*, *12*(4), 37–42.

Cortellazzo, L., Bruni, E., & Zampieri, R. (2019). The role of leadership in a digitalized world: A review. *Frontiers in Psychology*, *10*, 1938. doi:10.3389/fpsyg.2019.01938 PMID:31507494

Draper, J. (1998). The metamorphoses of andragogy. *Canadian Journal for the Study of Adult Education*, *3*(26).

Duke, S. T., & Ward, J. D. (2009). Preparing information literate teachers: A metasynthesis. *Library & Information Science Research*, *31*(4), 247–256. doi:10.1016/j.lisr.2009.04.003

Foote, L. S. (2015). Transformational learning: Reflections of an adult learning story. *Adult Learning*, *26*(2), 84–86. doi:10.1177/1045159515573017

Forehand, M. (2017). Bloom's taxonomy: Original and revised. In M. Orey (Ed.), *Emerging perspective on learning, teaching, and technology* (pp. 41–44). CreateSpace Independent Publisher.

Fraillon, J., Ainley, J., Schulz, W., Friedman, W., & Gebhardt, E. (n.d.). *Preparing for life in a digital age*. The IDEA International Computer and Information Literacy Study International Report.

Gerth, A. B., & Peppard, J. (2016). The dynamics of CIO derailment: How CIOs come undone and how to avoid it. *Business Horizons*, *59*(1), 61–70. doi:10.1016/j.bushor.2015.09.001

Giannoukos, G., Besas, G. M., Galiropoulos, C., & Hioctour, V. (2015). The andragogy, the social change, and the transformative learning educational approaches in adult education. *Journal of Education and Practice*, *46*(50).

Gisbert, M., & Bullen, M. (Eds.). (2015). *Teaching and learning in digital sorlds: Strategies and issues in higher education*. Publicacions Universitat Rovira I Virgili. doi:10.17345/9788484243762

Grafstrom, M., & Falkman, L. L. (2017). Everyday narratives: CEO rhetoric on Twitter. *Journal of Organizational Change Management*, *30*(3), 312–322. doi:10.1108/JOCM-10-2016-0197

Greene, K., & Larsen, L. (2018). Virtual Andragogy: A new paradigm for serving adult online learners. *International Journal of Digital Society*, *9*(2), 1376–1381. doi:10.20533/ijds.2040.2570.2018.0169

Gurbaxani, V., & Dunkle, D. (2019). Gearing up for successful digital transformation. *MIS Quarterly Executive*, *18*(3), 209–220. doi:10.17705/2msqe.00017

Hatlevik, O. E., & Christophersen, K. A. (2013). Digital competence at the beginning of upper secondary school: Identifying factors explaining digital inclusion. *Computer Education*, *63*, 240–247. doi:10.1016/j.compedu.2012.11.015

Huang, H. M. (2002). Toward constructivism for adult learners in online learning environments. *British Journal of Educational Technology*, *33*(1), 27–37. doi:10.1111/1467-8535.00236

Kane, G.C., Palmer, D., Phillips, A.N., Kiron, D., & Buckley, N. (2015, July). Strategy, not technology, drives digital transformation. *MIT Sloan Management Review*.

Knowles, M. S., Holton, E. F., & Swanson, R. A. (2015). *The adult learner. The definitive classic in adult education and human resource development* (8th ed.). Routledge.

Kummel, E., Moskaliuk, J., Cress, U., & Kimmerle, J. (2020). Digital learning environments in higher education: A literature review of the role of individual vs. social seetings for measuring learning outcomes. *Education Sciences*, *10*(3), 78. doi:10.3390/educsci10030078

Laal, M., & Ghodsi, S. M. (2011). Benefits of collaborative learning. *Procedia: Social and Behavioral Sciences*, *31*, 486–490. doi:10.1016/j.sbspro.2011.12.091

Latchem, C. (2012). *Quality assurance toolkit for open and distance non-formal education*. Commonwealth of Learning.

Linder-VanBerschot, J. S., & Summers, L. L. (2015). Designing instruction in the face of technology transience. *Quarterly Review of Distance Education*, *16*(2), 107–118.

McDonough, D. (2014). Providing deep learning through active engagement of adult learners in blended courses. *Journal of Learning in Higher Education*, 9-16.

Mihelj, S., Leguina, A., & Downey, J. (2019). Culture is digital: Cultural participation, diversity, and the digital divide. *New Media & Society*, *21*(7), 1465–1485. doi:10.1177/1461444818822816

Muenks, K., & Miele, D. B. (2017). Students' thinking about effort and ability: The role of development, contextual, and individual difference factors. *Review of Educational Research*, *87*(4), 707–735. doi:10.3102/0034654316689328

NCTE. (2013). *National Council of Teachers of English 21st Century literacies*. www.ncte.org/positions/statements/21stcentdefiniton

Niessen, S. (2013). *What is digital fluency?* http://www.shuananiessen.ca/what-is-digital-fluency/

Obschonka, M., Fisch, C., & Boyd, R. (2017). Using digital footprints in entrepreneurship research: A Twitter-based personality analysis of superstart entrepreneurs and managers. *Business Venturing Insights*, *8*, 13–23. doi:10.1016/j.jbvi.2017.05.005

Office of Inspector General. (n.d.). *Semiannual Report to Congress, 73*. U.S. Department of Education.

Petrucci, T., & Rivera, M. (2018). Leading growth through the digital leader. *The Journal of Leadership Studies, 12*(3), 53–56. doi:10.1002/jls.21595

Reischmann, J. (2004). *Andragogy. History, meaning, context, function*. Andragogy.net.

Scanlon, E., McAndrew, P., & O'Shea, T. (2015). Designing for educational technology to enhance the experience of learners in distance education: How open educational resources, learning design and MOOCs are influencing learning. *Journal of Interactive Media in Education, 6*(1), 1–9. doi:10.5334/jime.al

Scherer, R., Rohatgi, A., & Hatlevik, O. E. (2017). Students' profiles of ICT use: Identification, determinants, and relations to achievement in a computer and information literacy test. *Computers in Human Behavior, 70*, 486–499. doi:10.1016/j.chb.2017.01.034

Schwarzmuller, T., Brosi, P., Duman, D., & Weloe, I. M. (2018). How does the digital transformation affect organizations? Key themes of change in work design and leadership. *Management Review, 29*(20), 114–138.

Sharif, A. (2015). Quality assurance in e-learning programs. *Teaching and learning in digital worlds: Strategies and issues in higher education*, 63-70.

Sharp, L. A. (2018). Collaborative digital literacy practices among adult learners: Levels of confidence and perceptions of importance. *International Journal of Instruction, 11*(1), 153–166. doi:10.12973/iji.2018.11111a

Siemens, G., GaSevil, D., & Dawson, S. (2015). *Preparing for the digital university: A review of the history and current state of distance, blended, and online learning*. Link Research Lab.

Simonson, M., Schlosser, C., & Orellana, A. (2011). A distance education research: A review of the literature. *Journal of Computing in Higher Education, 23*(2-3), 124–142. doi:10.100712528-011-9045-8

Smith, S.U., Hayes, S., & Shea, P. (n.d.). A critical review of the use of Wenger's Community of Practice (CoP) theoretical framework in online and blended learning research. *Online Learning*, 209-237.

Sun, P. C., Tsai, R. J., Finger, G., Chen, Y., & Yeh, D. (2008). What drives successful e-learning? An empirical investigation of the critical factors influencing learner satisfaction. *Computer Education, 50*(4), 1183–1202. doi:10.1016/j.compedu.2006.11.007

Torre, T., & Sarti, D. (2020). The "Way" toward e-leadership: Some evidence from the field. *Frontier Psychology*. https://www.ncbi.nlm.nih.gov/pmc/articles/PMC7685990/

Tsai, W. H. S., & Men, L. R. (2017). Social CEOs: The effects of CEOs' communication styles and parasocial interaction on social networking sites. *New Media & Society, 19*(11), 1848–1867. doi:10.1177/1461444816643922

Van Outvorst, F., Vistker, C., & de Waal, B. (2017). Digital leadership: The consequences or organizing and working in a digital society. *Proceedings of the 5th International Conference on Management, Leadership and Governance*, 463-471.

Veletsianos, G. (Ed.). (2016). *The emergence and innovation in digital learning*. AU Press.

Wilson, J. H. (2019). *Andragogy and the learning-tech culture revolution*. https://www.igi-global.com/chapter/andrafogy-and-the-learning-tech-culture-revoution/207508

Wooldridge, A. (2015). The Icarus syndrome meets the wearable revolution. *Korn/Ferry Briefings Talent Leaders, 6*, 27-33.

KEY TERMS AND DEFINITIONS

Academic C-Level Leaders: The most common c-level titles are: Chief Executive Officer (CEO), President, Chief Operations Officer (COO).

Digital Andragogy: Adult education develops ways of working higher education based upon skills and dispositions of 21st-century learners through the lens of adult education and the affordances of readily accessible digital technology.

Digital Learning: An incredibly dynamic and constant drive by changes in learners' demands, behaviors, and technology.

Digital Learning Environment: The impact of digitized learning on traditional education that changes the way teachers and learners learn.

Digital Ubiquity: The pervasive proliferation of technology influences leadership.

E-Leadership: A social influence process, mediated by technology, to produce a change in attitudes, feelings, thinking, behavior and performance with individuals, groups, or organizations to direct them toward achieving a specific goal.

Chapter 16
Navigating Uncertainty and the Unknown:
Andragogical Education Leadership in Times of Change

Wendy M. Green
Cleveland State University, USA

Catherine A. Hansman
Cleveland State University, USA

Anil Lalwani
University of Vermont, USA

ABSTRACT

Within the global community, the only constant is change. In Spring 2020, as COVID-19 became a global pandemic, countries closed their borders and closed schools, businesses, and governments. Universities that already had significant online presences and available courses/programs could be reassured that at least some of their programming would continue with minimal interruption. Effective university leaders were able to identify paths forward for their institutions that encompassed their education and were much more able to support faculty, staff, and students with the abrupt change. The authors look at professors as learning leaders, the leadership in higher education institutions, as well as leadership in the private sector. They examine how andragogical leadership can be utilized to shift current educational practices and consider how we might shape learning environments in order to foster more robust learning environments. But essentially, they explain how education leaders develop and organize institutional learning as technology continues to play an essential role.

DOI: 10.4018/978-1-6684-7832-5.ch016

INTRODUCTION

Change can occur incrementally, with small steps taken over time. One example is the establishment of same-sex marriage rights (*Obergefell v. Hodges, 2015*). This is also true for the contraction of rights, such as the recent US Supreme Court decision that removed the fifty-year right to abortion (*Dobbs v. Jackson Women's Health Organization*, 2022). It can occur without warning, requiring immediate responses. Shifting societal viewpoints may fuel these changes, but at other times, shifts occur swiftly as a response to environmental factors. For example, when American men left the country to fight in the world wars, women were brought into factories and workplaces to take on the roles these men had left vacant. Another example of profound change was the global introduction of desk-top computers in the early 1980's, leading to the proliferation of many households buying this "new" technology and learning to use computers for various everyday professional and personal tasks. The Covid-19 pandemic is the most recent example of swift change. In Spring 2020, as Covid-19 became a global pandemic, countries closed their borders and closed schools, businesses, and governments. Essential services remained open, but most other non-essential employees from various industries were relegated to working indefinitely from their homes (Centers for Disease Control, n.d.). This policy had a significant impact on all facets of life. Organizations and institutions scrambled to identify ways to provide goods and services to customers, clients, students, and citizens while also supporting their employees to learn innovative technology and to continue their work from their homes.

Organizations and institutions weathered the changes wrought by Covid-19 with varying degrees of success. We believe individuals who adopted andragogical leadership approaches during this time were able to support their employees and make decisions that embraced the challenges brought forth by Covid-19. Our chapter will explore the need for adult educators to develop learners not only to be able to undertake the day-to-day responsibilities of their current or immediate employment, but also to prepare them to embrace future work environments which are, in essence, unknown. We believe ground-shifting events like Covid-19, will continue to occur and that it is essential to have leaders and learners who can flourish in uncertainty. We use Covid-19 as a case example to understand how andragogical leadership, coupled with Lewin's theory of change (Burnes, 2004) and Schein's (2010) further development of Lewin's model, can encourage leaders to embrace challenging situations and use them as a platform for positive change. We will emphasize the need to develop an orientation of adaptability and creativity as we may not be able to envision what our workplaces will be like in 10, 20, or 50 years. We link this argument to the role of educational leaders and innovative approaches taken by them when faced with swiftly changing circumstances.

Uncertain Times and Swift Changes

The pandemic affected all areas of life and similar challenges were faced by leaders across many industries. Due to government lock-down orders, employees were required to shift quickly from in-person work, meetings, and interactions with co-workers to working from home on computers, possibly having to learn unfamiliar software and computer applications to accomplish their daily tasks and projects. Industry leaders and managers had to make rapid decisions on how to best train their workers to use innovative technology for their ongoing or new work while at the same time encouraging continual employee development. Employees had to learn new skills and technologies while continuing to not only meet their job requirements, but also to contend with working from home or at a distance from their usual workplaces. Many faced competing and multiple responsibilities of child and/or elder care along with the blurring of work and home boundaries.

In higher education, in March 2020, universities closed to in-person learning and many did not reopen prior to the end of the Spring 2020 semester (Neuwirth et al., 2021). Universities were forced to re-imagine their approaches to teaching and learning, moving to virtual and online learning modalities (Gnanadass & Merriweather, 2022; Green et al., 2022). (Note: for the purposes of this chapter, virtual learning spaces refer to synchronous sessions that are mediated by a technology platform and require participants to be present at a specific time. Online, asynchronous learning refers to technology-mediated learning that occurs through learning management systems like Blackboard. Learners complete work at their own pace and meet specified deadlines). This complete shift to virtual and online learning at institutions had varying degrees of success (Piotrowski et al., 2020). Some universities were better prepared to navigate a move to virtual environments than others (Hawes, 2022; Schmidt, English, & Carr-Chellman, 2022). Universities that already had significant online presences and available courses and programs as well as faculty who were experienced in teaching online courses prior to the pandemic could be reassured that at least some of their programming would continue with minimal interruption. In addition, universities that were well-resourced were able to provide supportive services and material aid to students and faculty. University leadership played a key role in how these changes were implemented. Leadership telegraphed to students and faculty how they intended to navigate uncertainty and how they would ensure the university would continue to effectively operate and potentially flourish in a challenging situation.

There are elements of andragogical leadership that enabled leaders to approach these challenges from a position of strength. Leaders operating from this perspective seek to develop employees who are self-actualized and focused on their growth (Wang, Introduction). We argue, andragogical leaders can also nurture innovation, creativity, adaptability, and a tolerance for uncertainty and ambiguity.

What Is Andragogy?

Before we can discuss andragogical leadership, it is essential to understand the concepts of andragogy and self-directed learning. First evolved in Europe and then introduced to North America by adult education leaders during the early 20th century, a number of scholars in the field of adult education helped promote assumptions about adult learners that later evolved to the concept of andragogy. Lindeman (1926), who was a well-known, early- 20th-century American adult education scholar, posited several key assumptions about adult learners. These were:

1. Adults are motivated to learn as they experience needs and interests that learning will satisfy;
2. Adult orientation to learning is life-centered;
3. Experience is the richest source for adult learning;
4. Adults have a deep need to be self-directing;
5. Individual differences among people increase with age (Knowles et al., 2005, p. 40).

Later in the 20th century, andragogy was defined by Knowles (1984) as "the art and science of helping adults learn" (p. 6). In the context of this definition, art here refers to "style," and science here refers to "method." (Wang & Hansman, 2017). Humanistic concepts of communication and lifelong learning influenced Knowles as he developed his assumptions about adult learners (Hansman, 2017), expanding

Lindemann's list by adding one more supposition related to adult learners' motivations to learn (Knowles, 1970; 1980). The key assumptions of Knowles' andragogical model are:

1. The need to know: adults need to know why they need to learn something before undertaking to learn it;
2. The learners' self-concept: adults have a self-concept of being responsible for their own decisions, for their own lives; once they have arrived at that self-concept, they develop a deep psychological need to be seen by others and treated by others as being capable of self-direction;
3. The role of the learners' experiences: adults come into educational activity with both a greater volume and a different quality of experience from youths; by virtue of simply having lived longer, they have accumulated more experience than they had as youths;
4. Readiness to learn: adults become ready to learn those things they need to know and be able to do in order to cope effectively with their real-life situations;
5. Orientation to learning: in contrast to children's and youths' subject-centered orientation to learning, adults are life-centered in their orientation to learning;
6. Motivation: while adults are responsive to some external motivators, the most potent motivators are internal pressures (Knowles et al., 2005; 2020).

Knowles continued to build notions of andragogy for education and training of adult learners throughout the 20th century (Bash, 2003). Both Lindeman's and Knowles' andragogical assumptions about adult learners "contributed to later understandings of how adults continue to learn and develop throughout their lives and make decisions concerning what and how they learn, the contexts in which they engage for learning, and how learning is assessed and evaluated" (Hansman, 2017, p. 59). Although these assumptions about learners have been critiqued by many adult educators (i.e., Hansman & Mott, 2010; Merriam & Bierema, 2014; Merriam & Baumgartner, 2020) for promoting general assumptions about adult learners and applying them to all learners in diverse learning contexts, no matter the learners' race, gender, ability, sexuality, ethnicity, or country of origin, Knowles' andragogical principles concerning adult learners continue to frame the program planning process for educational courses, programs, and workshops for adult learners.

SELF-DIRECTED LEARNING

Emerging from Knowles' assumptions of andragogy is the concept of self-directed learning, described as "a process in which individuals take the initiative, with or without the help of others" (Knowles, 1975, p. 18) in diagnosing their learning needs, planning learning goals, activities, and materials to carry out their learning, and then self-evaluating their own learning outcomes. Literature reviews and research studies focusing on how adult learners actually are self-directed in their learning (i.e., Brockett, 1985; Caffarella & O'Donnell, 1987; Owen, 2002) have resulted in the development of models of self-directed learning (i.e., Brookfield, 1986; Candy, 1991; Garrison, 1997). A Charungkaittikul and Henschke (2018) study contends that "self-direction in learning refers to both the external characteristics of an instructional process and the internal characteristics of the learner, where the individual assumes primary responsibility for a learning experience."

More recent studies concerning self-directed learning have focused on adults using self-directed learning to learn new technology (Merriam & Baumgartner, 2020). For example, King (2017) has examined how adults may learn to use the ever-changing and updating new technologies, designing their own self-directed learning plans to remain "current" with technology for their workplaces or own personal interests. Self-directed learning can play an important role in adults learning new technologies and skills such as those required in 2020 during the Covid-19 pandemic when many jobs and schools depended on workers and students being capable of using these technologies (i.e. Zoom, Google Chat, Blackboard, Canvas, Google Docs, etc.) to accomplish work and school tasks. More recently, artificial intelligence (AI) and chatbot applications have challenged workplaces and higher education institution administrators, management, faculty members, and employees to learn new technologies that may profoundly transform the ways classes are taught or business is conducted.

Adult educators' role in self-directed learning is to support adult learners to "plan, carry-out, and evaluate their own learning" (Merriam & Baumgartner, 2020, p. 141). Educators take on facilitator roles assisting, but not directing or dictating to learners to take on these tasks regarding their own learning. Kegan (2018), however, cautioned adult educators to not assume that all adults are self-directed simply by virtue of being adults, and to be mindful that some adult learners are neither capable, nor willing to engage in self-directed learning. He further contends that educators who require self-directed learning with their adult learners "are not merely asking them to take on new skills, modify their learning style, or increase their self-confidence" (p. 44); rather encouraging adult learners to be self-directed may require learners to re-examine their understandings of themselves and the world contexts in which they live while engaging with concepts of how they learn differently from those they have used in the past. Ideally, as Merriam & Baumgartner (2020) contend, self-directed learning should be promoted from childhood through post-secondary education so that, by the time learners reach adulthood, they are comfortable as self-directed learners. Charungkaittikul and Henschke (2018) propose that "self-directed learning is the beginning and end of lifelong learning; the keystone of the learning society; a supplement to and sometimes substitute for the formal education system; a vehicle for the mastering of established knowledge; and for the transformation of personal understanding."

ANDRAGOGICAL LEADERSHIP

So, now that we have discussed andragogy and self-directed learning, what does this mean for leadership? How can an andragogical approach and knowledge about self-directed learning translate into andragogical leadership, particularly in institutions of higher education? In this section, we discuss how andragogical approaches to leadership may transform institutions of higher education through engaging faculty, staff, and students in essential decisions concerning planning, programs, program delivery, curriculum design, and instructional technology.

Workplaces and higher education institutions are constantly evolving in the face of social, cultural, and technological challenges, requiring essential and timely decisions to be made. For example, during the beginning of the Covid-19 pandemic, most higher education institutions quickly pivoted to using only online learning platforms (i.e. Blackboard, Canvas, etc.). Our experiences as university faculty members was that, because of the seriousness of the pandemic, decisions were made by high-level administrators. There was little involvement of faculty members, even though faculty members were the ones on the front lines of interacting with students and using new technology skills to engage learners.

While we understand the seriousness of the pandemic crisis which necessitated these quick decisions by administrators, we envision a reform of higher education leadership that, through embracing andragogical principles, might be more inclusive of faculty input.

Other examples of decisions made with little regard for student or faculty input include higher education administrators making key decisions concerning major curriculum changes, credit hour changes, and faculty hiring decisions. In addition, student interests should not be ignored, but considered, especially concerning curriculum, program, and financial decisions which directly affect them. When higher education administrators ignore faculty, staff, and student input, leaders show how little they value their contributions to the university, and instead prioritize hierarchical forms of control and power

Employing andragogical leadership principles means that leaders "involve their followers in every step of the planning process, assessing needs, formulating goals, designing lines of action, carrying out activities, and evaluating results" (Wang, Introduction). In higher education, this means that andragogical leaders value the input of those who will be most affected by administrative decisions—faculty and staff members, and students. Leaders can show that they have confidence in, and value their faculty members' abilities through delegating challenging responsibilities and decisions. This may encourage faculty members to engage in self-directed learning on their campuses in order to meet these challenges, solve problems, engage in reflection, and proactively prepare for future expected or unexpected changes and crises that may occur in the future. Andragogical leaders model innovative approaches and offer opportunities for growth; however, they do not force these opportunities. Instead, they offer mentoring and other forms of support so that growth, development, and self-directed learning can take place should a faculty member choose to engage in these opportunities. Diverse and pluralistic views and contributions are welcome and valued by andragogical leaders. Through andragogical leaders' actions, faculty members may understand that administrative leaders are one source of knowledge, but that they themselves are knowledge generators, both alone and with their faculty and administrative colleagues.

If concepts of andragogy are embedded in higher education leadership, then perhaps universities will be better equipped to be inclusive of others beyond administrators as they make decisions in response to sudden crises, such as the Covid-19 pandemic. Through employing thoughtful, reflective, and inclusive opportunities for discourse that leads to informed decision making, all levels in higher education can then be engaged in making the best decisions possible to build proactive and responsive universities.

CHANGE PROCESSES

Universities occupy several organizational identities. Universities are viewed as spaces for innovation through research and the exploration of ideas. Universities, however, are bound by governance processes which require multiple levels of engagement, consensus, and approvals which result in measured implementation of new operational ideas. The adoption of new technologies to augment learning environments is one example of change across a continuum. Some institutions, driven by student demand and concerns for student access, adopted online learning operations when they were first introduced. Other universities remained steadfast in their objection to online learning due to their belief that it would degrade the quality of their educational experience. The Covid-19 pandemic superseded any institutional belief systems around the efficacy of technology-mediated learning as the shut-down required full and immediate adoption.

Immediate Change in Response to COVID-19

The prior section highlighted andragogical leadership as an effective way to navigate an uncertain and highly fluid situation. In this section, we introduce two theories of change. We will connect these theories to andragogical leadership and consider how leaders can map a path forward that embraces change and supports faculty, staff, and students.

When higher education abruptly moved to remote teaching and learning, there were levels of intervention and support that needed to be addressed. University leadership had to navigate the operations of the institution as a whole. This encompassed staff operations which included human resources, technological services, and facilities. It also included student services which comprised residence halls, sports and recreation, mental health and disability offices, and the needs of international students (Lalwani et al., 2022). Faculty who operated research labs had to be supported in ways to maintain their work and to ensure they were sufficiently protected from the risk of contracting Covid-19. Finally, and arguably the most essential aspect of a university's operation, there were the processes of teaching and learning. Faculty, without any preparation, were required to pivot their coursework to a virtual format (Green et al., 2022; Hansman, 2022). Faculty experiences with technology-mediated teaching and learning was varied and therefore they required different levels of technological and material assistance. There were innumerable ways to teach online and faculty chose whatever suited their purpose and that which they were most familiar with.

From an operations standpoint, universities needed to ensure students were protected from contracting Covid-19 and also able to continue with coursework. In residentially oriented schools, students had to be transferred out of university housing. For some students, this was simply a matter of returning home; for others, particularly graduate and international students, this was more challenging (Gnanadass & Merriweather, 2022; Lalwani et al., 2022). Importantly, universities had to ensure that all students had the ability to access virtual learning. Many students had the financial support to be able to access the necessary hardware and internet so they could continue with their studies. In other cases, universities had to provide resources, including laptops and hotspots as some students experienced financial pressures as a result of the pandemic (Hawes, 2022; Neuwirth et al., 2021). In addition, students required significant technological support to understand how to navigate online spaces with sometimes limited instruction from their universities. Students also had to discover and learn about the various tools that faculty were using to teach. This included learning management systems (Blackboard, Canvas, Moodle), meeting platforms (Google Meet, Microsoft Teams, Zoom), and streaming sites (Twitch). In addition, there were a variety of apps that were introduced, (Kahoot, Padlet, Quizlet, Whiteboards) during the pandemic which were unfamiliar to both students and faculty members. Universities had to find ways to facilitate learning for those who had accommodations through Disability Offices. International students, who were required to participate in face-to-face classes, also experienced issues when all classes went virtual. Faculty and students looked to university leadership to provide insight and guidance as to how to navigate these issues. This was even more significant since there were limited opportunities for interpersonal interactions during the initial lockdown.

Lewin provides a framework to understand how organizations or institutions engage in change. His three steps explain the change process as unfreezing, moving, and refreezing (Muldoon, 2020). Field theory enables us to understand how environmental stimuli can bring about changes in individuals' behaviors (Burnes, 2004). Organizations remain in a position of stasis or equilibrium until there is opportunity to make positive changes or there is a threat or situation that destabilizes the state of equilibrium and requires a more immediate response. Lewin, however, acknowledges that organizations may appear to be in a state of equilibrium, they are always engaged in smaller changes in response to environmental factors—in essence, continually reacting and adapting to change. This unfreezing process occurs when employees or members of an institution are open to change or are responding to environmental stimuli. Behavioral interventions are used to encourage employees to adopt new ideas and ways of operating. When the organization refreezes, providing support to employees will aid in the longer-term adoption of new ideas. Without this support, the newly introduced attitudes, skills, and behaviors will not translate to long-term change.

Lewin's framework implies change may be slow and take time; however, if "there are certain circumstances, such as a personal, organizational or societal crisis, the various forces in the field can shift quickly and radically" (Burnes, 2004, p. 982). The Covid-19 pandemic was an example of a societal crisis that required immediate change. Organizations became unfrozen with minimal preparation or a clear path forward. Closing down the business and educational operations in an entire country and across the world was unprecedented. University operations were completely overhauled in a short period of time and this included the instructional milieu. Steadfast opponents to technology-mediated teaching and learning found themselves in a position where they had to immediately adopt new practices. Although universities had been integrating the use of technology in teaching and learning, there was disparate implementation both across the industry and within the institutions as faculty and staff were at different levels of acceptance and skill sets (Hawes, 2022). The pandemic and closure of schools created an environment where change was inevitable and required immediate implementation. The continued health of the universities depended on their ability to identify and implement best practices.

Schein (2010) expanded Lewin's framework and situated it in the area of organizational change. If we apply Schein's (2010) expanded framework, it provides a pathway for leaders when they are trying to manage change through learning. Stage One, unfreezing, is a response to an incongruity, a disorienting dilemma, or a potential threat that induces disequilibrium. This disconfirming information or data is linked to organizational goals. Leaders who are attempting to foster transformative change recognize that it is important for employees to learn new ways of being and operating and further, to unlearn ways that are no longer effective for the institution. The disequilibrium is manifested when employees recognize shortcomings within their systems and feel compelled to make change. The third element is the institutional environment. The environment should support problem solving, innovation, and creativity while encouraging employees to engage with thorny issues without fear of retribution or humiliation. If we consider this stage in relation to Covid-19, there was an immediate need for change. Uniquely, the university shut-down was non-negotiable. Even if there were dissenting voices, they were powerless in the face of government mandates. By necessity, universities adopted a stance that was open to immediate change. However, it appeared that some university leadership welcomed ideas from a select few and favored administrators and faculty.

Stage Two of Schein's (2010) framework involves employees learning new skills, attitudes, and behaviors. Schein argues the learning process is best conducted with the use of role models or trial-and-error learning. As learning and development specialists, we might view these processes through the lens of andragogy, self-directed, and experiential learning. Employees have the opportunity to engage with experts in their arena and to adopt their actions and orientations. Employees are encouraged to be problem-solvers and to test out ideas in real work contexts. In the example of Covid-19, faculty and staff were engaged in a variety of learning activities and their backgrounds, knowledge, and capacity varied greatly. Faculty were given the directive to shift their courses online and, in many cases, this directive came with minimal guidance or support. This trial-by-fire learning may have resulted in positive experiences when the faculty were successful with the transition. Other faculty may have struggled because they did not have the technological knowledge or skills to teach remotely, or perhaps they were trying to balance working from home with childcare or eldercare. This resulted in mixed outcomes in respect to how the institution addressed the pandemic-related issues. In some cases, administrators expected faculty to solve these issues while failing to provide the necessary resources and support. In other cases, university administration provided directives without collaborating with those who would have to implement these new policies, faculty and staff.

Lastly, Stage Three refers to the refreezing process. This stage entails learners' incorporation of new skills, attitudes, and behaviors. This can include developing new mental models of one's role in the organization or one's skill sets. In our Covid-19 example, professors who perceived technology-mediated learning as inferior may begin to recognize benefits to the modality. They may understand that it promotes better student access. This may include students who do not reside near educational institutions or working adults who experience time conflicts when attempting to attend in-person classes. They may come to see themselves as adept online instructors. Andragogical leaders who take an active role in promoting a vision and a clear direction forward can facilitate a longer-term internalization of new attitudes moving forward.

ROLE OF LEADERSHIP

There were distinct areas leaders needed to address during the Covid-19 crisis. First, university stakeholders had to be provided guidance on the university's approach to the lockdowns. Leaders were required to be decisive in their decision-making process and to communicate these decisions effectively to faculty, staff, students, and for colleges and universities with significant undergraduate enrollment, to parents (Marshall et. al., 2020). Leaders had to encourage faculty to innovate and be creative in delivery of coursework while providing logistical and technical support for them to do so. Additionally, leaders had to navigate the financial pressures experienced by universities due to loss of revenues and additional costs accumulated due to the transition to remote learning. Leaders could approach the Covid-19 crisis as solely an issue to be navigated or as an opportunity to strengthen educational systems while integrating new approaches to instruction and general operating structures of the university.

In much of the research that occurred during and after Covid-19 pandemic lockdown, leaders who could envision opportunities in challenging situations were held up as exemplars of effective stewards. Importantly, these leaders partnered with faculty in reimagining new ways of doing their jobs and identifying ways to support their endeavors (Marshall et al, 2020). We argue that this type of support is also needed in times of stability or times of slower change. Institutions should be alert to ways of strengthening the ways in which they accomplish their main goals, whether that is teaching, research, or a combination of both. Faculty can be supported and encouraged to navigate ambiguities and uncertainties while identifying ways to strengthen university responsibilities to faculty, staff, students, and the communities in which they reside. We argue andragogical leadership is an optimum approach in higher education so creativity, adaptability, and innovation are valued in the institution rather than an adherence to the status quo. Sudden changes, like the Covid-19 pandemic, may force the institution out of its current state and into embracing new opportunities for change. A key question for andragogical leaders is:

how can university leaders maintain this forward momentum without creating an undue burden of tasks and responsibilities on faculty members, which would lead to the opposite of the intended outcomes?

LEADING CHANGE THROUGH UNCERTAINTY

When characterizing system leadership, Senge and colleagues (2015) indicated that change often occurs in response to undesirable outcomes and unexpected environmental demands. They believe that effective system leaders have a proactive rather than a reactive approach when handling change. These leaders help mobilize their stakeholders beyond just reacting to problems that demand change. Leading change can be challenging, especially when problems demanding change accompany uncertainty and ambiguity. Some scholars believe that uncertainty is simply the reaction to ambiguity and that "ambiguity is inherent in leadership" (White & Shullman, 2010, p. 95). Leading change arguably involves accepting ambiguity and developing tolerance for uncertainty.

Some degree of uncertainty in response to ambiguity is expected and even desirable. In White et. al.'s (1996) research, they interviewed top executives in North America and the United Kingdom asking for ideas on making choices in situations involving uncertainty. They then tested those ideas with participants in Asia and Latin America. The results showed that indeed for many executives, making choices under ambiguous conditions was immensely challenging, but it was critical for them to own their ambiguity and use it in seeking learning opportunities while advancing the decision-making process. This recognition of leading with due acknowledgement of ambiguity was termed by the authors as a 'learning style' and defined as the ability "to identify productive areas of uncertainty and confusion and to lead the organization into those areas to gain competitive (and other kinds of) advantage" (White et al., 1996, p. 89).

Despite the learning potential that accompanies uncertainty and ambiguity, many leaders would likely engage in avoidance and use their familiar style of leadership. Reacting from a place of utmost certainty in the face of ambiguity is likely to present as a veneer, where the decision makers are likely too afraid to step outside their comfort zones and make decisions that accurately address the problems. Reacting from the standpoint of an awareness of uncertainty allows for leaders to step outside their familiar modus operandi and engage in choices that might be intimidating, but approach the problems in ways that are novel and targeted. Often, the leader is not the only person making the decisions. They might be leading a team of individuals where decisions are made based on group discussion and mutual consultation. Demonstrating an authentic reaction to uncertainty can likely inspire confidence and a healthy regard for ambiguity among the various stakeholders involved.

There are different ways in which tolerance for ambiguity and resulting uncertainty can be harnessed and used to advance change in organizational settings. Creativity, innovation, flexibility, and adaptability have been associated with positive outcomes involving change (White & Shullman, 2010). In organizational contexts, creativity and innovation refer to a generative stance aimed at leading the organizations toward the path of continuous improvement and developing solutions that are original and novel. Their usage tends to overlap in matters of organizational decision-making. Flexibility and adaptability are two related concepts that are also used interchangeably in organizations, especially in the context of conflict that can be interpersonal or systemic. Flexibility refers to the stance of openness where actions leading up to a desired outcome are open for modifications. Likewise, adaptability is the openness for amends in planning and execution to meet a desired outcome. One can reasonably draw parallels between all these terms that refer to an overall forward-looking approach of actively seeking and accepting change.

At an individual level, there are certain qualities that can be honed and used to advance change amid ambiguity. Hodgson and White (2001) conducted interviews, analyzed results from a scale assessing ambiguity, and drew references from their training work with executive coaching and leadership development to identify the following list of "eight clusters" that infer actions and behaviors which exemplify leaders that make good use of ambiguity and the resulting uncertainty:

1. Leaders present with a high degree of curiosity;
2. Leaders are able and willing to make choices that accompany a reasonable amount of risk;
3. Leaders are invested in finding out what the future entails;
4. Leaders present with tenacity to solve problems;
5. Leaders are energized by the work they do;
6. Leaders show flexibility to changes in their own decisions as well as to others;
7. Leaders break down complex ideas into simple terms;
8. Leaders prioritize actions.

To adopt these principles of learning amid uncertainty (like Covid-19) in higher education would be to expect administrators to demonstrate a healthy regard and handling of uncertainty when planning Covid-19 protocols. These leaders must validate the uncertainty embedded in circumstances and provide reassurance that flexibility is the need of the hour. Furthermore, higher education administrators must leverage Covid-19 to discover innovative and sustainable forms of learning and development. For instance, pay attention to course offerings that are better suited to remote, synchronous modes of learning. While doing so, they must continually engage with the different stakeholders—faculty, staff and students. By doing so, the administrators can show a healthy regard for collaboration that is consistent with the tenets of both andragogical and leadership in times of uncertainty.

Andragogical Leadership in Institutions During Times of Change

In this paper we have introduced andragogical leadership, organizational change theories, and leading through ambiguity. The pandemic was an impetus to make significant changes to the educational operations of universities through the adoption and utilization of new technologies (Piotrowski et al. 2021). Universities have transitioned to *their* new normal. For some, this might be a return to full face-to-face instruction with students on campus. For others, it may reflect a more hybrid model. If university leadership (chairs, deans, provosts, presidents) takes the opportunity to review and debrief their operation and engagement through the pandemic, they may be able to identify the best practices that emerged in technology-mediated teaching and learning and to ensure these practices continue beyond the pandemic response. This debriefing would not be complete without the engagement of faculty, staff, and students. Including the major stakeholders would allow all parties to present their concerns and identify solutions while minimizing miscommunication. In reconceptualizing healthy working environments and positive relationships across administration, faculty, and staff, university leadership could facilitate the development of a self-actualized employee base who are focused on providing meaningful experiences to students while engaging with surrounding communities. The principles of andragogical leadership could be utilized to consider how we shape our vision of higher education. Our visions create a path forward as we foster robust learning environments and encourage students to develop the orientation to learning we, as faculty, believe is essential. If faculty are viewed as learning leaders and partners with the university administration, this would be a step to solidifying any potential gains that were made during Covid-19.

Challenges to Faculty

We have presented potential outcomes and paths forward that emphasize the benefits of change. However, we would be remiss if we failed to be explicit about challenges faculty are facing in a post Covid-19 environment. These challenges may limit faculty's ability to fully engage in large-scale change processes. We argue administrators are in a position where they can support faculty during turbulent political times.

Universities and programs are experiencing financial challenges as a result of decreases in student enrollment and associated costs related to moving to an online format (Hill & Isaac-Savage, 2022). One response has been universities' decisions to re-organize and realign colleges, schools, and departments. Administrators' goals with reorganization are to streamline processes, provide faculty with better opportunities to connect and collaborate, and importantly, to cut costs (Cherry et al 2022). Other universities have declared financial exigency as a result of the pandemic and have taken this opportunity to limit faculty numbers by leaving positions unfilled or closing programs. In response to lower enrollment, faculty are expected to market and recruit for their programs while building networks and identifying new trends in their areas of specialization (Collins & Zackarakis 2022). These changes and expectations add responsibilities to faculty who struggle to maintain work-life balance and are under increased stress (Schmidt-Crawford, 2021).

In addition to structural issues, faculty were navigating student issues through and post Covid19. Faculty have been asked not only to adapt their courses and delivery to online platforms, but they have also been asked to provide needed support to learners (Green et al., 2022). Students have experienced upheaval and for many adult students, shouldered additional responsibilities that included working from home while managing child and adult care. Faculty too, have had similar experiences (Schmidt-Crawford et al., 2021). While faculty have been asked to be flexible with students, they may not have had the support to manage the increases in their workloads in addition to navigating work-like balance issues (Schmidt-Crawford et al, 2021).

Finally, faculty find themselves under increased pressure as the climate of higher education, under the guise of neoliberalism, becomes more inhospitable. This framework has been uncritically adopted by much of our national and state leadership (Giroux, 2018). Universities, in parallel with US society, have incorporated neoliberal lenses to frame how we understand the role of government and education, among other entities (Giroux, 2014). Neoliberalism and marketization now influence much of our understanding regarding government and university operations with the assumptions that efficiencies are of paramount importance. This is coupled with the idea that private industry is best positioned to provide goods and services to the populace (Dougherty & Natow, 2019). University performance has been more tightly linked to the concept of students as clients and their value "is to improve the country's economic performance, and that considers that the most valuable outcomes of higher education are exploitable knowledge and credentialled graduates" (Brown, 2018, p. 9). Faculty find themselves in positions where student engagement is more contractual in nature and benefits of a university education are viewed through individualistic versus societal lenses. What is removed from this framework is "the notion of the university as a center of critique and a vital democratic public sphere that cultivates the knowledge, skills, and values necessary for the production of a democratic polity" and is replaced by the "view of the university as a marketing machine essential to the production of neoliberal subjects" (Brown, 2018, p. 1).

The adoption of neoliberal frameworks in conjunction with the rise of legislation to interfere with or mandate topics for university teaching creates challenging environments for faculty. Florida is one example where legislators have taken steps to regulate what content is taught in classrooms as well as how content is taught (Flaherty, 2022). Florida passed the Stop WOKE act which prohibits the use of critical theory in classrooms. In addition, Florida passed CS/CS/HB 233 (2021) - Postsecondary Education, a bill that requires universities conduct annual assessments on diversity of viewpoints and allows students to covertly record class sessions. This bill seeks to regulate what can be taught and how faculty can be monitored in classroom settings. In turn, faculty are conscious of the possibility that their words might be taken out of context and posted on social media platforms or they may be listed on conservative watch lists resulting in targeted harassment (AAUP, n.d.). Similar bills have been introduced in state legislatures in Iowa (Cox & Tachau, 2023) and Ohio (Vukovich, 2022).

Faculty engagement in any change process is limited by their capacity to complete job responsibilities as well as having a sense of security in their position. It is also dependent on the relationship with the administration and the belief that any changes will be of benefit to the educational mission of the university rather than a focus on cost-cutting. Andragogical leaders can begin to address these systemic issues while engaging faculty in ways that encourage their development and self-actualization while strengthening the university.

CONCLUSION

As we try to address society's significant challenges (health care disparities, educational disparities, climate change), we need to foster innovative thinking and creativity within our student bodies and support faculty. We need to build on prior instructional norms and consider how we might amplify learning with new technologies. Leadership can support these ideas by viewing faculty and staff as partners and creating spaces for wider engagement. Leadership can advocate for faculty and address political issues that diminish faculty security. Leadership can role model attributes that reflect these values by being curious, open to ideas, and having flexibility in how they engage with staff and faculty. University leaders who operate from an andragogical leadership perspective are grounded in humanistic perspectives which value the educational mission and vision and also the people who are responsible for the implementation of both. Lastly, we argue, andragogical leaders should be in the forefront as we try to re-envision and re-center our universities as spheres for criticality and democratic spaces.

REFERENCES

AAUP. (2017). *Statement on targeting harassment of faculty*. AAUP Updates.

Bash, L. (2003). *Adult learners in the academy*. Anker Publishing Company, Inc.

Brockett, R. G. (1985). Methodological and substantive issues in the measurement of self-directed learning readiness. *Adult Education Quarterly*, *36*(1), 15–24. doi:10.1177/0001848185036001002

Brookfield, S. (1986). *Understanding and facilitating adult learning*. Jossey-Bass.

Brown, R. (2018). Neoliberalism, marketisation and higher education: How neoliberalism is reshaping the provision of English higher education. Professorial lecture at the University of West London, London, UK.

Burnes, B. (2004). Kurt Lewin and the planned approach to change: A re-appraisal. *Journal of Management Studies*, *41*(6), 977–1002. doi:10.1111/j.1467-6486.2004.00463.x

Caffarella, R. S., & O'Donnell, J. M. (1987). Self-directed learning: A critical paradigm revisited. *Adult Education Quarterly*, *37*(4), 199–211. doi:10.1177/0001848187037004002

Candy, P. C. (1991). *Self-direction for lifelong learning*. Jossey-Bass.

Centers for Disease Control. (n.d.). *CDC Museum COVID-19 Timeline.* https://www.cdc.gov/museum/timeline/covid19.html

Charungkaittikul, S., & Henschke, J. (2018). Applying an andragogical approach to foster lifelong self-directed learning in the 21st century higher education classroom settings. *Journal of Educational Leadership in Action*, *5*(2), 5. https://digitalcommons.lindenwood.edu/ela/vol5/iss2/5

Cherry, B., Graves, B.M., & Grasse, N. (2022). Causes, processes, and effects of academic reorganization at public master's universities in the United States. *Perspectives: Policy and Practice in Higher Education*, 1-9. doi:10.1080/13603108.2022.2121779

Collins, R. A., & Zacharakis, J. (2022). Continuous Challenges: Case Study of a Sustainable University Adult Education Graduate Program. *Adult Learning*, *0*(0). Advance online publication. doi:10.1177/10451595221106171

Cox, L. K., & Tachau, K. H. (2023). *The erosion of support for education and tenure in Iowa*. American Association of University Professors (AAUP). https://www.aaup.org/article/erosion-support-education-and-tenure-iowa#.ZBW07XbMKUk

Dobbs, State Health Officer of the Mississippi Department of Health, et al., *v.* Jackson Women's Health Organization et al. (date of publication) 945 F. 3d 265. https://www.supremecourt.gov/opinions/21pdf/19-1392_6j37.pdf

Dougherty, K. J., & Natow, R. S. (2019). *Analyzing neoliberalism in theory and practice: The case of performance-based funding for higher education.* Center for Global Higher Education. Working Paper #44. https://apo.org.au/node/232706

Flaherty, C. (2022). Mixed Message. *Inside Higher Education.* https://www.insidehighered.com/news/2022/05/09/uf-seems-endorse-new-state-anti-crt-law

Garrison, D. R. (1997). Self-directed learning: Toward a comprehensive model. *Adult Education Quarterly*, *48*(1), 61–72. doi:10.1177/074171369704800103

Gill, R. (2016). Breaking the silence: The hidden injuries of neo-liberal academia. *Feministische Studien*, *34*(1), 39–55. doi:10.1515/fs-2016-0105

Giroux, H. A. (2014). *Neoliberalism's war on higher education.* Haymarket Books.

Giroux, H. A. (2018). *Terror of neoliberalism: Authoritarianism and the eclipse of democracy.* Routledge. doi:10.4324/9781315631653

Gnanadass, E., & Merriweather, L. R. (2022). Restorying COVID-19: Faculty and graduate students teaching and learning in crisis. In R. Wlodarsky & C. Hansman (Eds.), *The Covid-19 pandemic and other ongoing crises: Reflections, innovations, and adaptations. New Directions for Adult and Continuing Education*. Wiley. doi:10.1002/ace.20449

Green, W. M., Hoffman, A., Manlapaz, M., & Foshee, C. (2022). Health professions education cohort programs: A case study of pivoting from in-person to online learning in cohort-based programs. In R. Wlodarsky & C. Hansman (Eds.), *The Covid-19 pandemic and other ongoing crises: Reflections, innovations, and adaptations.* Wiley. doi:10.1002/ace.20452

Hansman, C. A. (2017). Assessing schools and organizations. In V. Wang, B. Bain, J. Hope, & C. A. Hansman (Eds.), *Educational leadership and organizational management: Linking theories to practice* (pp. 53–80). Information Age Publishing.

Hansman, C. A. (2022). "We're all in this together"? Reflections on inequity during tumultuous times. In R. Wlodarsky & C. Hansman (Eds.), *The Covid-19 pandemic and other ongoing crises: Reflections, innovations, and adaptations.* Wiley. doi:10.1002/ace.20448

Hansman, C. A., & Mott, V. (2010). Adult learners. In C. Kasworm, A. Rose, & J. Ross-Gordon (Eds.), *Handbook of Adult and Continuing Education* (pp. 13–23). Sage.

Hawes, J. (2022). The challenges of emergency online instruction at a Small HBCU. *Journal of Higher Education Theory & Practice*, *22*(18), 1–6.

Hill, L. H., & Isaac-Savage, E. P. (2022). Doors slam shut: Adult education program closures. In R. Wlodarsky & C. Hansman (Eds.), *The Covid-19 pandemic and other ongoing crises: Reflections, innovations, and adaptations.* doi:10.1002/ace.20453

Hodgson, P., & White, R. P. (2001). *Relax, it's only uncertainty.* Financial Times Prentice Hall.

James Obergefell v. Richard Hodges, no. 14-3057 (6th cir. 2014). Justia Law. https://law.justia.com/cases/federal/appellate-courts/ca6/14-3057/14-3057-2014-11-06.html

Kegan, R. (2018). What "form" transforms?: A constructive-developmental approach to transformative learning. In K. Illeris (Ed.), *Contemporary theories of learning* (pp. 29–45). Routledge. doi:10.4324/9781315147277-3

King, K. (2017). *Technology and innovation in adult learning.* Jossey-Bass.

Knowles, M. S. (1970). *The modern practice of adult education: Andragogy versus pedagogy.* Association Press.

Knowles, M. S. (1975). *Self-directed learning.* Association Press.

Knowles, M. S. (1980). *The modern practice of adult education: From pedagogy to andragogy.* Follett.

Knowles, M. S. (1984). *Andragogy in action.* Jossey-Bass.

Knowles, M. S., Holton, E., & Swanson, A. (2005). *The adult learner* (6th ed.). Elsevier Butterworth Heinemann. doi:10.4324/9780080481913

Knowles, M. S., Holton, E., Swanson, A., & Robinson, P. (2020). *The adult learner: The definitive classic in adult education and human resource development* (9th ed.). Routledge. doi:10.4324/9780429299612

Lalwani, A., & Green, W. M., & Hamlen Mansour, K. (2022). How campus alienation exacerbated international students' difficulties in accessing campus services remotely during COVID-19: Notes on policy and programming. *The American Behavioral Scientist, 0*(0), 1–17. doi:10.1177/00027642221118257

Lindeman, E. (1926). *The meaning of adult education*. New Republic, Incorporated.

Marshall, J., Roache, D., & Moody-Marshall, R. (2020). Crisis leadership: A critical examination of educational leadership in higher education in the midst of the COVID-19 pandemic. *International Studies in Educational Administration, 48*(3), 30–37.

Merriam, S. B., & Baumgartner, L. M. (2020). *Learning in adulthood: A comprehensive guide*. Jossey-Bass.

Merriam, S. B., & Bierema, L. L. (2014). *Adult learning: Linking theory and practice*. Jossey-Bass.

Muldoon, J. (2020). Kurt Lewin: Organizational change. In B. Bowden, J. Muldoon, A. Gould, & A. M. McMurray (Eds.), *The Palgrave Handbook of Management History* (pp. 615–632). Palgrave MacMillan. doi:10.1007/978-3-319-62114-2_32

Neuwirth, L. S., Jović, S., & Mukherji, B. R. (2021). Reimagining higher education during and post-COVID-19: Challenges and opportunities. *Journal of Adult and Continuing Education, 27*(2), 141–156. doi:10.1177/1477971420947738

Owen, T. R. (2002). Self-directed learning in adulthood: A literature review. ERIC Clearinghouse on Adult, Career, and Vocational Education. (ERIC Document Reproduction Service No. ED 461 050)

Piotrowski, C., & King, C. (2020). COVID-19 pandemic: Challenges and implications for higher education. *Education, 141*(2), 61–66.

Schein, E. H. (2010). *Organizational culture and leadership* (Vol. 2). John Wiley & Sons.

Schmidt, S. W., English, L. M., & Carr-Chellman, A. (2022). Conversations with leaders: Sharing perspectives on the impact of and response to COVID-19 and other crises. In *The Covid-19 Pandemic and Other Ongoing Crises: Reflections, Innovations, and Adaptations*. Wiley. doi:10.1002/ace.20455

Schmidt-Crawford, D. A., Thompson, A. D., & Lindstrom, D. L. (2021). Condolences and congratulations: COVID-19 pressures on higher education faculty. *Journal of Digital Learning in Teacher Education, 37*(2), 84–85.

Senge, P., Hamilton, H., & Kania, J. (2015). The dawn of system leadership. *Stanford Social Innovation Review, 13*(1), 27–33.

Vukovich, C. (2022). *As if Covid wasn't enough: A continuous assault on student's rights and education*. ACLU. https://www.acluohio.org/en/news/if-covid-wasnt-enough-continuous-assault-students-rights-and-education

Wang, V. (2023). *Introduction (from call for chapters). In Handbook of research on andragogical leadership and technology in a changing society*. Information Science Reference.

Wang, V., & Bain, B. (2017). Andragogical leadership and management in the workplace. In V. Wang (Ed.), *Theory and practice of adult and higher education* (pp. 249–273). Information Age Publishing.

Wang, V., & Hansman, C. A. (2017). Pedagogy & andragogy in higher education. In V. Wang (Ed.), *Theory and practice of adult and higher education* (pp. 87–111). Information Age Publishing.

White, R. P., Hodgson, P., & Crainer, S. (1996). *The future of leadership: Riding the corporate rapids into the 21st century*. Pitman Publishing.

White, R. P., & Shullman, S. L. (2010). Acceptance of uncertainty as an indicator of effective leadership. *Consulting Psychology Journal, 62*(2), 94–104. doi:10.1037/a0019991

KEY TERMS AND DEFINITIONS

Andragogical Leadership Principles: Leaders "involve their followers in every step of the planning process, assessing needs, formulating goals, designing lines of action, carrying out activities, and evaluating results" (Wang, Introduction).

Andragogy: Defined by Knowles (1984) as "the art and science of helping adults learn" (p. 6).

COVID-19 Global Pandemic: Began in late 2019 to early 2020 as a result of the spread of an infectious disease caused by the SARS-CoV-2 virus, causing severe illness and deaths. Countries closed their borders, schools, businesses, and governments. Essential services remained minimally open, but non-essential employees, teachers, and students were quickly relegated to working indefinitely remotely from their homes or other "safe" spaces.

Field Theory: Concepts that enable us to understand how environmental stimuli can bring about changes in individuals' behaviors. If there are "certain circumstances, such as a personal, organizational or societal crisis, the various forces in the field can shift quickly and radically" (Burnes, 2004, p. 982).

Lewin's Theory of Change: A framework to understand how organizations or institutions engage in change. His three steps explain the change process as unfreezing, moving, and refreezing (Muldoon, 2020).

Online, Asynchronous Learning: Refers to technology-mediated learning that occurs through learning management systems like Blackboard. Learners complete work at their own pace and meet specified deadlines.

Schein's Model: An expansion of Lewin's framework and situated it in the area of organizational change. It provides a pathway for leaders when they are trying to manage change by learning concepts (Schein, 2010).

Self-Directed Learning: "A process in which individuals take the initiative, with or without the help of others" (Knowles, 1975, p. 18) in diagnosing their learning needs, planning learning goals, activities, and materials to carry out their learning, and then self-evaluating their own learning outcomes. Charungkaittikul and Henschke (2018) contend that "self-direction in learning refers to both the external characteristics of an instructional process and the internal characteristics of the learner, where the individual assumes primary responsibility for a learning experience."

System Leadership: Peter Senge et al. (2015) indicated that change often occurs in response to undesirable outcomes and unexpected environmental demands. They believe that effective system leaders have a proactive rather than reactive approach when handling change. These leaders help mobilize their stakeholders beyond just reacting to problems that demand change.

Virtual Learning Spaces: Refer to synchronous sessions that are mediated by a technology platform and require participants to be present at a specific time.

Chapter 17
Andragogical Leaders Immersing Into the Metaverse

Linda Marie Ellington
Southern New Hampshire University, USA

ABSTRACT

An effective and impactful andragogical leader must have many key characteristics and competencies. They must be open to immersing into the Metaverse with the spirit of optimism, have the ability to adapt to change, and know that innovation and creativity in organizations are pivotal to survival as human activities fuse with technological domains. These leaders will use wisdom, experiences, intellectual insights, and curiosity to drive actions to achieve the future of their institutions. They need to be capable of leading their organizations through times of uncertainty, being self-confident during the Metaverse disruption, and doing so to be courageous action drivers as well as strategic thinkers. Be the action strategist, prehistorically known as the Allosaurus.

INTRODUCTION

This chapter will not go into the principles of andragogy, as this author assumes the readers of this book collectively know more about andragogy than one single author could include in a chapter. What this chapter will do, however, is to interpret progress, direction, and current research on the immersion of Metaverse into organizations and the impact it will have, and is having, on andragogical leaders.

The Metaverse is the post-reality universe, a perpetual and persistent multiuser environment merging physical reality with digital virtuality. It is based on the convergence of technologies that enable multisensory interactions with virtual environments, digital objects, and people such as virtual reality (VR) and augmented reality (AR). Hence, the Metaverse is an interconnected web of social, networked immersive environments in persistent multiuser platforms. It enables seamless embodied user communication in real-time and dynamic interactions with digital artifacts (Mystakidis, 2022).

Significant investments in the Metaverse by leading technology companies indicate that humanity's next emerging global trend is here (Pritchard, 2022). It is intended to be a quantum leap toward virtual and augmented reality (VR and AR), supported by all existing technologies such as blockchain, 5G,

DOI: 10.4018/978-1-6684-7832-5.ch017

and AI. The evolutionary progress of both hardware and software will offer a new experience that will be called a progression of virtual worlds as a replacement or substitution for direct reality, which is an ongoing trend (Feezell, 2018; Richter, 2020).

Current and emerging technologies such as artificial intelligence and blockchain, represented by NFT (digital objects), sensory-rich environments, virtual reality headsets, and brain-computer interfaces, are envisioned to be embedded in the foundations of the Metaverse (Musk, 2019; Zuckerberg, 2021). Ongoing global wars, dreadful pandemics, ecological challenges, and economic regrouping may also be in line with the development of the Metaverse, which could provide answers to some of these issues (Saxon et al., 2004; Smith, 2021). For example, decreasing public transportation and the use of personal vehicles can help deal with climate change. The same can be said for the spread of viruses, which are seen as a threat in the future as well (Smith, 2021).

Tools in the Metaverse are "rapidly becoming as infrastructural as electricity," (Cascio & Montealegre, 2016, p. 350). The essential AI technologies that will serve as the basis for the Metaverse include visual perception, speech recognition, decision-making, and language translation. Through learning from humans and their digital footprints for the last 30 years, AI algorithms have acquired the knowledge needed for the Metaverse to be set up. Examples of these are Google's question-and-answer algorithm based on its search engine, as most human beings are unconscious volunteers through their trillions of searches (Soricut & Brill, 2006). Some platforms currently offer Metaverse-like experiences, mainly in the realm of video games, which are combined with virtual economies (cryptocurrencies), enabling some of the users to make a living by taking part in these virtual realities including Axie Infinity, a play-to-earn games, Second Life, where users socialize, learn, and do business and Decentraland, a digital world that merges social life with cryptocurrencies, NFTs, and virtual real estate.

The potential popularity of the Metaverse can be envisioned by the number of people using VR games and apps that can be considered predecessors to the Metaverse. For example, there are 150 million Roblox monthly active users. A large portion of this is children, 2/3 of them aged 9-12 in the USA, and 1/3 of them under 16 (Meier et al., 2020).

Central claims made about the Metaverse are (1) richer ways of self-expression, (2) better immersion, (3) better socializing, (4) symmetric relation of physical and virtual spaces, (5) independent markets (via NFTs), (6) better user interfaces, and (7) high demands for regulation and governance (Bojic, 2022). The next platform is seen as even more immersive (Pritchard, 2022). Screens cannot convey realistic experiences, as much as the Metaverse technology will be capable of doing. Metaverse will generate embodied technology in which the user is inside the experience, not just looking at it. The feeling of presence is a major defining quality of the Metaverse, according to Zuckerberg, because one will be able to really feel being there with other people through facial expressions, gestures, and photo-realistic visual experiences (Zuckerberg, 2021). The Metaverse will provide better user interfaces when compared to existing communication technologies, Interacting with devices and people in the Metaverse will also be more natural, so instead of tapping or typing, one will be able to interact by gestures with hand movements, through speech, and even thoughts. This means the new technology will have features and subtle ways of communicating that cannot be delivered by existing solutions, for example, smartphones.

Numerous emerging technologies play important roles in the development and leading and learning processes. In the context of these developments, learning becomes the main axis in the organization of the educational process. Virtual worlds, Metaverses, and 3D virtual environments are now demonstrating huge potential for educational purposes because they enable analogical environments and processes to be recreated with a high degree of realism and where physical and communicative interactions closely approximate the effects of interactions that occur in the real world, For example, avatars bearing the identities of the users in these worlds take on important roles for the development of training strategies in environments where simulation is fundamental where all agents of education must assume the roles to where they are assigned, and where leaders are obliged to redefine their professional role while also becoming a learner like andragogical students in the same learning environment.

BACKGROUND

Although the Metaverse is a relatively recent addition to the everyday lexicon of technology commentators and academics alike, it has been exploited as a concept throughout the 1980s in technoscience and in literary works, the term 'Metaverse' was coined in the science fiction book Snow Crash (Stephenson, 1992). The story depicted the interaction of humans through avatars on the street, which was space owned by a single corporation. After that point, there have been many stories, films, and other pieces of art and entertainment related to virtual worlds and the struggle between the dystopian control of machines and human anarchist-survivalist (Stephenson, 1992).

Ball (2022) tells the story of Vannevar Bush in the 1930s, then president of the Carnegie Institution of Washington, who began work on a hypothetical electromechanical device that would store all books, records, and communications, and mechanically link them together by keyword association, rather than traditional mostly hierarchical storage models. Despite the enormity of its archive, Bush stressed that this *memex* (short for memory extender) could be consulted with exceeding speed and flexibility. Fast forward to 2021, the term Metaverse almost became a punchline as every company and its executives seemed to trip over themselves to mention it as something that would make their company more profitable, their customers happier, and their competitors less threatening. Prior to 2020, the term Metaverse had appeared only five times in US Securities and Exchange Commission filings. In 2021 the term was mentioned more than 260 times. That same year, Bloomberg, a software company that provides financial data and information to investors, cataloged more than a thousand stories containing the word Metaverse. The prior decade had only seven (Ball, 2022).

Interest in the Metaverse was not limited to Western nations and corporations. In May 2021, China's largest company, the internet gaming giant Tencent, publicly described its vision of the Metaverse, calling it "Hyper Digital Reality." The following day, Southern Korea's Ministry of Science and ICT (Information and Communications Technology) announced The (South Korean) Metaverse Alliance" spanning over 450 companies including SK Telecom, Woori Bank, and Hyundai Motor. In early August, South Korean gaming giant Krafton, maker of Player Unknown's Battlegrounds (also known as PUBG) completed its IPO, the second largest in the country's history. Krafton's investment bankers made sure to tell would-be investors that the company would also be a global leader in the Metaverse.

Auletta (2018) stated, trying to lasso a definition of the Metaverse is as baffling as a swollen campaign speech. Will the Metaverse replace the internet? Will it fizzle like a bright firecracker? What do the hype and the potential seismic change ahead look like?

Whatever the differences between researchers' and scholars' visions, there is yet no reason to assume that the synthetic world, presented as a dystopia, is an inevitable outcome for the actual Metaverse. As a point of contrast, French philosopher Jean Baudrillad, who coined the term *hyperreality* in 1981 described hyperreality as a state in which reality and simulations were so seamlessly integrated that they were indistinguishable. Though many scholars find this idea frightening, there are those who argue that what matters is where individuals derive more meaning and people such as Musk, Zuckerberg, and Branson conceived the meaning to be infinitely interconnected worlds (Ball, 2022).

The Metaverse is emerging at a fast pace without being properly analyzed in terms of its social impact and adequately regulated by societies. It has been found that the Metaverse does contain elements of optimism, epochalism, and inventibility, which corresponds to the concept of future essentialism (Bojic, 2022). On the other hand, power relations in society indicate that current trends in the concentration of power among Big Tech could expand even more if the Metaverse becomes mainstream. Technology deployed by the Metaverse may create an attractive environment that would directly mimic reality and further stimulate media attention in society (Bojic, 2022).

Indeed, the Metaverse can seem daunting and scary, but it also offers an opportunity to transform industries that have long resisted disruption and that must evolve in order to build a more equal global economy. It will be a place of existence for millions if not billions, of people who sit atop the digital and physical economies. The very idea of the Metaverse means an ever-growing share of our lives, labor, education, leisure, time, wealth, happiness, and relationships will be spent inside virtual worlds, rather than just extended, or aided through digital devices and software (Ball, 2022). With the importance of building, maintaining, and growing meaningful human social connections and interactions, the creation of digital, virtual, and immersive environments, such as the Omniverse (Nvidia, 2021) or the Multiverse (Carter, 2022), has recently flourished.

The launch of Horizon Worlds in 2021 by Meta Platforms and the vision of how the Metaverse could potentially shape many aspects of how we work and learn, has engendered an increasing level of questioning and debate from academics and practitioners on the numerous societal implications for many people worldwide (Fernandez & Hui, 2022). The Metaverse concept as outlined by Mark Zuckerberg, describes an integrated immersive ecosystem where the barriers between the virtual and real worlds are seamless to users when sharing experiences, a space that can provide new sensorial experiences by adding new haptic, sensory, and cognitive feedback.

Considering the behavioral and social lens of inclusion, diversity, equity, accessibility, and safety and their impact in other fields, such as the physical built environment (Zallio & Clarkson, 2021), it appears that little effort has been made to advance knowledge and practices to design safe, accessible, and inclusive digital, virtual, and immersive environments within institutions in higher education (Mahlich, 2021). Gilbert posits that universities that fail to understand, adapt, and adopt the Metaverse technologies will ultimately struggle to teach well and to graduate students with the skills needed in society (2023). The Metaverse will decide many of the hard problems of digital existence today, such as data rights, data security, misinformation, radicalization, platform, regulations, and abuse. Being aware of these decisions is key to surviving in and through artificial intelligence disruption.

The dramatic response to the Metaverse reflects the growing belief that it is the next great computing and networking platform, similar in scope to the transition from the personal computer and fixed-line internet of the 1990s to the era of mobile and cloud computing we live in today. That shift popularized a once-obscure business school term = 'disruption' – and transformed almost every industry while reshaping modern society and politics. Yet, there is a critical difference between that shift and the impending shift to the Metaverse, timing. Most industries and individuals did not foresee the significance of mobile and cloud, and consequently were stuck reacting to changes and fighting off disruption from those who better understood them.

IMPLICATIONS OF THE METAVERSE

Organizations are starting to assess the potential of the Metaverse and how it can be integrated into their existing business model. At the current point in time which could be called the renaissance of immersive technology, a fervent period of technological 'rebirth' will not be just a single digital immersive environment owned by one company. There will be a variety of digital worlds, just as there are currently several online social media platforms, productivity software, e-commerce shops, and e-higher educational institutions. The Metaverse is rapidly advancing, and its definition will evolve over time, particularly as gaming designers, engineers, researchers, policymakers, educators, leaders, and many other members of the community will gather greater knowledge as to what it can offer and what challenges are staring at them.

In the study conducted by Zallio and Carlson (2021), they found that the perception of the Metaverse can be imagined as a layer on top of the real world that does not have to entirely replace the real world and could empower users to better understand the physical world by revealing accentuated emotions, feelings, and develop ideas which are difficult to capture with human senses. People just might spend time differently than they might do in the physical world, where new heuristics will be developed and new interaction paradigms will have to be designed. A key factor in enabling the growth of learning experiences lies in the creation of a constructive narrative around the Metaverse scholars who need to spark conversations that will create awareness and develop digital educational opportunities in an andragogical environment.

The Metaverse, etymologically a combination of meta (i.e., beyond) and the universe, refers to a computer-generated world that flexibly blends physical reality and digital virtuality in order to provide immersive, interactive, realistic, and augmented digital experiences, with a wide spectrum of consumer, industrial and educational application. Cai et al. (2022) describe four applications:

Social: The Metaverse which aggregates and transcends traditional media (text, audio, image, video), will shape the future of online social and educational networks, initially built to connect individuals for communication and content sharing. For example, Meta, formerly known as Facebook, activity developing social VR platforms where users can have digital representations (avatars), interact with each other, and participate in shared activities like shopping, tourism, watching movies, and attending events.

As the vital foundation for building the Metaverse, social virtual reality (VR) has been extensively examined by the computer graphics and Human-Computer Interaction (HCI) communities. HCI researchers have been investigating a wide range of topics in social VR, including privacy and security concerns. Moreover, the HCI community studies various aspects of organizing different events on social VR platforms, such as remote learning (Maloney et al., 2020). Although the concept of the Metaverse has been around for about half a century, the outbreak of the Covid-19 pandemic accelerated the digitization of daily lives, leading to the renaissance of the Metaverse and the prosperity of social VR platforms (Chen et al., 2022).

Gaming: To implement ideal Metaverse applications, the features, i.e., shared, persistent, and decentralized, need to be taken into account. For example, a Metaverse gaming environment should allow multiple players to participate in the game to fulfill the shared feature; moreover, individual players can continuously play the game to follow the persistent feature. More importantly, the gaming environment needs to guarantee that the players can work for and own their own treasures, and their logs must be safely kept using some security technology, such as blockchains, to fulfill the decentralized feature (Vergne, 2021).

Collaboration: In the Metaverse, people can create a personalized workspace or virtual office and even an academic chat room to facilitate collaboration regardless of geographical restrictions. Digital assets representing people, such as avatars, and objects can be incorporated into the 3D space, giving a new dimension to today's academics and work life.

Education: Owing to the features of the Metaverse, it is expected that the Metaverse in education is different from traditional VR or AR-based education. For example, for a language learning course for EFL (English as Foreign Language), the purpose of the course is to enable EFL learners to have another life, a living environment using English for working, learning, social events, and entertainment, just as if they were native English speakers (Chien et al., 2020). Another example is using AR in skills training, such as the study reported by Webe, et al., (2013). In such an AR-based training mode, the learners are merely supported by AR in a short-term (i.e., 60 min) activity to complete a specified skill practice. In comparison, from the perspective of the Metaverse, this could be done better; for example, if the learners were pre-service trainees in a professional training institute, they could take a formal training program with instruction from a subject matter expert (SME), say for 3 months, to experience the authentic training process rather than a short-term activity that is far from a practical training situation. Therefore, from the perspective of AI, there are several roles of the Metaverse in education, namely, the provision of an authentic world that enables learners to "work" and "learn" with non-player characters (NPC) metaverse tutors, peers, as well as other human learners.

The biggest difference between the Metaverse and the current application of VR or AR in education lies in experiencing time and the inclusion of AI technology. First, in the Metaverse, students tend to have a life that is separate from the real world and runs for a period of time. The current applications of VR and AR in education tend to engage students in experiencing specific situations over a short time period where they can pause or start over. On the contrary, the Metaverse aims to provide an authentic life, implying that the experience inside generally cannot be paused or restarted. In addition, in the Metaverse, AI technology is needed to help maintain the authentic world. One can imagine that the NPC (non-player characters) in the Metaverse can "learn (be trained)" after interacting with users and "grow" along with the timeline. Moreover, the NPC characters in the Metaverse could remember individual users' dialogue and behaviors after encountering them in the Metaverse. On the other hand, in VR, virtual characters usually only interact with users according to the original settings and do not "learn" from users' interactions.

Hwang and Chien (2022) tell us that there are potential applications of the Metaverse in various fields, such as nursing, and healthcare education, science education, military training, and manufacturing training as well as language learning. In fact, the Metaverse enables learners to have more opportunities to experience, explore, learn, and teach in a new world, as well as work and interact with people. They can even learn or practice in those contexts that they are unable to experience in the real world. For example, most people might not have opportunities to serve at the managing level or practice flying a plane. However, this could happen in the Metaverse if the creator aims to provide the experience or learning opportunities to the users. Therefore, there are many potential applications of the Metaverse in education. According to Hwang and Chien (2022), without the loss of generality, there are several reasons for adopting the Metaverse for educational purposes:

1. To constantly situate learners in a cognitive or skill-practicing environment that could be risky or dangerous in the real world.

2. To constantly situate learners in the contexts to experience and learn what they generally do not have the opportunity to be involved in the real world.
3. To enable learners to perceive or learn something that requires long-term involvement and practice.
4. To encourage learners to try to create or explore something that they cannot afford to do in the real world owing to some practical reasons, such as the cost or the lack of real materials.
5. To enable learners to have alternative thoughts and attempts regarding their careers or personal lives.
6. To enable learners to perceive, experience, or observe things from different perspectives or roles.
7. To enable learners to learn to interact and even collaborate with people that they might not have opportunities to work with in the real world.
8. To explore the potential of higher-order thinking by engaging them in complex, diverse, and authentic tasks.

Andragogical leaders need not be ostriches, who instinctively bury their heads in the sand in the hopes that trouble will pass them by, but to welcome the presence of the Metaverse as it is going to provide a brand-new perspective on educational technology (Díaz et al., 2020; Rospigliosi, 2022). It will provide new training opportunities and contexts for learners. Many educational objectives that cannot be achieved in the real world could be conducted in the world of the Metaverse (Siyaev & Jo, 2021).

ARTIFICIAL INTELLIGENCE PERSPECTIVE

In this new world, some barriers that prevent people from learning some materials or skills, such as the limitations of time, space, or even dangers during the learning process, could be overcome (Jeong et al., 2022; Wang et al., 2022). It can be foreseen that, in the coming decade, the amount of research and the number of applications related to the use of the Metaverse in education will increase at a rapid pace. On the other hand, there are challenges in creating a Metaverse, no matter whether for educational or for other purposes (Davis et al., 2009; Falchuk et al., 2018).

In addition to ethical issues, the lack of relevant technical support is a problem at the present (Dionisio et al., 2013). From the perspective of a Metaverse creator, a good authoring system as well as powerful computer hardware and high-speed networks are required (Wiederhold, 2022). From the perspective of users, low-cost and lightweight equipment that enables people to continuously immerse themselves in and interact with the Metaverse with high-resolution digital content is required (Park & Kim, 2022). Although the cost of relevant equipment (e.g., the head mount) is still expensive to most people; several companies are trying to develop low-cost solutions. In terms of education, it is important to start considering the potential learning designs and curriculum designs in the world of the Metaverse.

So far there is no particular model or framework for developing a Metaverse andragogical world. This implies diverse possibilities for developing Metaverse-based environments for different educational purposes and applications. On the other hand, Metaverse could serve different educational purposes since it is a "world" that could contain several educational institutes (e.g., schools, universities, and professional training institutes) or practicing spaces (e.g., science parks, hospitals, and museums). One could envision it could be an alternative learning space.

As the Metaverse is such an innovative technological environment for researchers from the fields of both computer science and educational technology, it is important to compare the performances of learners using the Metaverse-based educational approach and those using the conventional technology-enhanced learning approach. From the perspective of AI technology, it is also expected that, by adopting the Metaverse-based educational approach, learners can have continuous learning opportunities without being limited by space and time, as well as receive full support or guidance for making reflections based on the analysis results of their learning logs. Employing the Metaverse allows for the learners' higher-order thinking competencies, such as problem-solving and critical thinking, which is generally a challenging task.

AI-powered applications for grading have a growing body of research that indicates their efficacy. However, there is another growing subfield in this area of research in which AI-power applications are used to support the acquisition of writing skills (Rudolph et al., 2023). For example, ChatGPT has the potential to profoundly transform educational assessment practices. The use of ChatGPT offers exceptional capacity for improving intelligent student support systems, and scaffolding student learning in adaptive and personalized ways (Zawacki-Richter et al., 2019). While ChatGPT has shown impressive capabilities in helping students in writing, AI-powered writing applications have already been widely used for quite some time. It is ironic that ChatGPT has caused so many anxieties in the academic community and yet it may prove most useful to teachers when it comes to facilitating more innovative teaching and learning. It is evident that prior to the introduction of ChatGPT, a number of AI-based writing tools had already been developed to facilitate writing practices and to promote self-directed learning by users, particularly in higher education (Nazari et al., 2021; Zhai, 2022). Such as automated writing evaluation (AWE), automated essay scoring (AES), and automated written corrective feedback (AWCF) have been increasingly adopted as alternatives to facilitate the process of writing by facilitating automated feedback and assessing items. In terms of AI-powered digital writing assistants, Grammarly is one of the most popular and well-researched ones, offering a wide range of applications, including AWE, AES, and AWCF all-in-one digital tools for writers (Taguma et al., 2018). As Grammarly has more than 20 million users worldwide, the corpus amount of data generated by its users provides the base for it to continually improve the application. A team of computer linguists and deep learning engineers at Grammarly analyze millions of sentences from academic journals in order to build cutting-edge algorithms that analyze the rules and latent habits of effective writing (Fitria, 2021). Some academics anticipate the demise of conventional educational assessment methods; thus, educators should take a holistic approach when developing strategies for leveraging AI innovation in education. It is instructive to consider how Microsoft appears to be incorporating ChatGPT holistically into its products (Warren, 2023).

The Metaverse is an imagined world with immersive digital spaces that increase, allowing a more interactive environment in educational settings. It is an expansion of synchronous communication that embraces an effective number of users to share different experiences and one that is already populated by students (Akour et al., 2022). In the form of avatars, students' unique personas happily collaborate in spaces such as Roblox and Minecraft (Mistretta, 2022). It is to the advantage of institutions to consider the adoption of customizable Metaverse platforms in which there is the support of multiple entry points for engagement, representation, action, and expression.

Various universities and educational institutions conducted several studies with Metaverse as their core (Akour et al., 2022). Researchers used Metaverse in an educational environment, adopting a problem-based approach, where teachers and students can put forward the problem and find possible solutions in the imaginary world using 3D classes and avatars (Farjami et al., 2011; Kanematsu et al., 2012, 2013).

Virtual environments and immersive games (such as Second Life, Fortnite, Roblox, and VRChat) have been described as antecedents of the Metaverse and offer some insight into the potential socio-economic impact of a fully functional persistent cross-platform Metaverse. Separating the hype and "meta..." rebranding from current reality is difficult, as "big tech" paints a picture of the transformative nature of the Metaverse and how it will positively impact people in their work, leisure, and social interaction. The potential impact on the way we conduct business, interact with brands and others, and develop shared experiences is likely to be transformational as the distinct lines between physical and digital are likely to be somewhat blurred from current perceptions. However, although the technology and infrastructure do not yet exist to allow the development of new immersive virtual worlds at scale - one that our avatars could transcend across platforms, researchers are increasingly examining the transformative impact of the Metaverse. Impacted sectors include marketing, education, and healthcare as well as societal effects relating to social interaction factors from widespread adoption, and issues relating to trust, privacy, bias, disinformation, application of law as well as psychological aspects linked to addiction and impact on vulnerable people. Andragogical leaders may want to combine the informed narrative and multi-perspective approach from experts with varied disciplinary backgrounds on many aspects of the Metaverse and its transformational impact.

GOVERNANCE

Currently, there is an intense race to construct the infrastructure, protocols, and standards that will govern the Metaverse. Large corporations are striving to construct their closed, proprietary hardware and software ecosystems so as to attract users and become the de facto Metaverse destination (Mystakidis, 2022). The debate is that others outside the large corporations note the existence of the Metaverse requires its own regulations and governance. This introduces the need to have firm rules of digital ownership, ecosystem building, norm-setting, and new digital government (Zwart & Lindsay, 2010). Also, privacy and safety need to be built into the Metaverse to prevent misuse, illegal activities, and unwanted communication, which means, that similar to current social networking sites, options such as blocking or connecting with someone will be available (Su et al., 2022). Basically, the characteristic of the Metaverse will be a feeling of presence, avatars, home space, teleporting, interoperability, privacy, safety, virtual goods, and natural interfaces (Zuckerberg, 2021).

There is a strong need for establishing governance for a variety of reasons. There is no prior experience to arrange governance in the physical and virtual worlds that are interrelated. There is tension between centralized and decentralized governance in the Metaverse. There seems to be no insight into what needs to be governed. Risk, regulations, compliance accountability, decision-making, processes and procedures, and incentives mechanisms are among the typical elements of governance. Rules of behavior, code of conduct, options to enforce rules and to put sanctions in place are needed. The Metaverse might have its own rules of law by the big tech companies, who do have their own empire and are ruling the Metaverse space. Yet, sound governance requires checks and balances and countervailing forces, critical to leading in an andragogical space.

Each of the andragogical spaces, whether academia, corporate, or country, can encounter and reproduce various kinds of social inequities from the physical world and may create new problems that are hard to foresee. The Metaverse might be viewed as a public space in which persons can move freely and securely. Like humans have rights, avatars might have rights in the Metaverse too.

At least some basic governance should be in place before the Metaverse can be launched in the andragogical environment. Some basic issues need to be addressed in advance to avoid problems at a later stage. The connection between the real and virtual world results in major governance and the questions leaders need to research:

- Will the rules and regulations of the online world be the same in the digital world?
- How will smooth operating and adhering to requirements like privacy, security, interoperability, and scalability be arranged?
- What are the risks, legal, and ethical consequences of the avatar connected to a real person?
- What governance models and principles are suitable for the Metaverse?
- Who will take the lead in setting standards?
- How to deal with the diversity of worlds having their own distinctive characteristics and rules?
- How will the governance of the physical and Metaverse be connected?
- How will we deal with conflicting values and regulations?

There are many open research questions. It is unclear what kind of governance, regulations, and policies are needed. The Metaverse does not stop at the geographical borders of a jurisdiction. It does, however, need to be only one Metaverse, not many Metaverses or Multiverses. As such, the framework of a governance model should be for everyone, open, hardware-agnostic, networked, and collectively controlled (Parisi, 2021).

CONCLUSION

Leaders need to make sure their organizations do not forget about the human aspects. We live in a turbulent time with many potential life-changing events occurring within a very short time. The development of the Metaverse can be seen as a natural extension of the platform economy with the potential for even more life-changing consequences. The Metaverse has been defined as the moment at which our digital lives – our online identities, experiences, relationships, and assets – become more meaningful to us than our physical lives (Shaan Puri cited in Hall & Moritz Baier-Lentz, 2022).

The focus of the activity on online platforms up until now has mostly been on the utilitarian exchange of value, e.g., creating networks that facilitate purchasing or advertising products and/or services from peer-to-peer and/or exposing firms to an audience around the globe. This is commonly done through large platform companies such as Airbnb, Amazon, or Facebook. As we transfer more of the activity towards a Metaverse we are likely to see less focus on utilaritian exchange and hopefully upgrade the hedonic aspects that help us as humans (rather than consumers) and the Metaverse as an extension of ourselves and the experiences of being a human. This in turn implies a larger focus on sociological aspects rather than technological ones as the leaders immerse their organizations into the Metaverse.

The potential for organizations to adapt their business models and operational capacity to function on the Metaverse is significant, with transformational impacts. For individuals to choose to interact with the Metaverse in the future, the seamless nature of the transition between physical and virtual and the multimodal enhancement of experiences and interactions opens an endless scope of possibilities, many of which perhaps beyond our current comprehension (Dick, 2021; Mystakidis, 2022). Sustainability is an essential factor for the Metaverse (Papagiannidis & Bourlakis, 2010). The Metaverse serves as a tool to complement the real world. The Metaverse enables exchanges of various experiences and new knowledge among users, this way, users build financial wealth, create new things, and have the opportunity to show a different side of themselves. However, there are platform restrictions in these social activities, and consistency is needed (e.g., worldview). Interdisciplinary research is a medium that allows the Metaverse to develop as a society beyond the level of simply being a 3D environment and physical applications (Rehm et al., 2015). Beyond simple games and social media, the Metaverse requires a variety of values and novel concepts. The philosophy, psychology, sociology, culture, economics, and politics of the Metaverse require a new perspective. It is necessary to consider an advanced perspective rather than simply substituting formulates from the real world.

The promise of the Metaverse is limitless, and basically, anything that we now do in a physical realm can be done in a Metaverse, at least in theory. At present, new worlds and new countries are re(created) in an online setting. The promise is that we can meet anyone from anywhere in the world for work or play in a 3D realm. The next potential future is that one part of a Metaverse reality can enable us to move out to the physical realm in the form of holograms, which makes it possible to be present in meetings and at workspaces and long-term even more freely anywhere around the globe without moving from our home (Chayka, 2021). The Metaverse is likely to be a reality as its development is very difficult to stop. As humans and researchers, we need to focus on how we can use the Metaverse to improve our lives and create a better society. We should focus on how the Metaverse can be used to strengthen human aspects that can only be seen as commendable such as inclusion, happiness, empowerment, or creativity. The Metaverse has the potential to improve the access and experience of several services in sectors such as education, healthcare, and culture. It is also one potential solution that enables remote working in a better and more immersive context, making it possible to work from anywhere, but still be present at the office or in meetings.

We tend to view all new technologies as a mechanism to create new, better, and brighter futures and we generally neglect the downsides and potential pitfalls. Let us not forget that the organizations that are the most important stakeholders in this transformation are able to embrace the concept of the Metaverse to create new business opportunities and be financially profitable (Chayka, 2021). The underlying thought of the Metaverse is to make a positive impact on people and to do good, but we should not be naïve in considering the underlying risk and potential harm to users. We have seen that social media has had negative consequences for many users – effects that were difficult to predict at the onset of the social media age (D'Arienzo et al., 2019).

The emergence of the Metaverse has stimulated increasing levels of academic discussion and debate on the benefits and potential transformational impact at a societal level. Whilst the emergence of the Metaverse poses many challenges related to governance, ethics, safety and security, acceptable behaviors, privacy, and potential disenfranchisement of sections of the population unable to access the necessary infrastructure to access the Metaverse (Fernandez & Hui, 2022; Haimson & Hoffman, 2016). On a broader scale, Metaverse-related technologies are expected to cross-pollinate, expand, and be further amplified by exponential technologies that will mark a transition to a fourth industrial revolution (Salmon, 2019).

The Metaverse's main dimensions are illustrated below. Most importantly it can become a democratizing factor in andragogical leadership, enabling worldwide participation on equal footing, unbound by geographical restrictions.

Figure 1. Metaverse main dimensions
Source: Mystakis (2022).

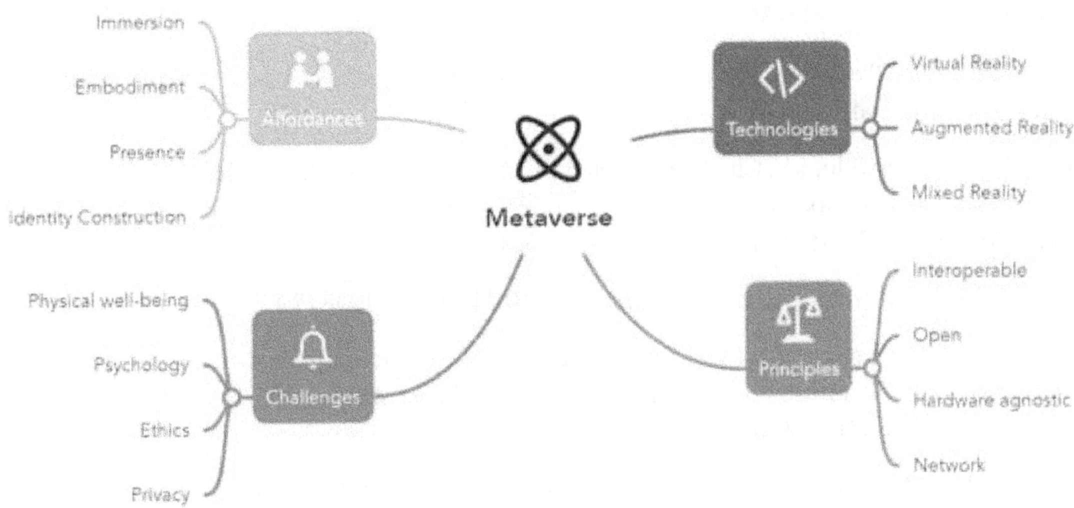

REFERENCES

Akour, I.A., Al-Maroof, R.S., Alfaisal, R., & Salloum, S.A. (2022). A conceptual framework for determining metaverse adoption in higher institutions of gulf area: An empirical study using hybrid SEM-ANN approach. *Computers and Education: Artificial Intelligence, 3.*

Auletta, K. (2018). *Frenemies: The epic disruption of the Ad business.* Penguin.

Ball, M. (2022). *The Metaverse and How It Will Revolutionize Everything.* Liveright Publishing.

Basu, T. (2021). The metaverse has a groping problem already. *MIT Technology Review, 16*(3).

Beckert, J. (2016). *Imagined futures.* Harvard University Press. https://www.hup.harvard.edu/catalog.php?isbn=9780674088825

Benbast, I. (2010). HCI Research: Future challenges and directions. *AIS Transactions on Human-Computer Interaction, 2*(2), 16–21. doi:10.17705/1thci.00011

Bojic, L. (2022). Metaverse through the prism of power and addition: What will happen when the virtual world becomes more attractive than reality? *European Journal of Future Research, 10*(22).

Bojic, L., & Marie, J. L. (2017). Addiction to old versus new media. *Serbian Political Thought, 56*(2), 33–48.

Cai, Y., Llorca, J., Tulino, A. M., & Molisch, A. F. (2022). Compute-and-data-intensive networks: The key to the Metaverse. *1st International Conference on 6G Networking (6GNet).* 10.1109/6GNet54646.2022.9830429

Carter, J. (2022). *Metaverse Vs Multiverse: How to Tell Apart The 'Parallel Universe' Concepts We Can't Stop Talking About.* https://www.forbes.com/sites/ jamiecartereurope/2022/02/07

Cascio, W. F., & Montealegre, R. (2016). How technology is changing work and organizations. *Annual Review of Organizational Psychology and Organizational Behavior*, *3*(1), 349–375. doi:10.1146/annurev-orgpsych-041015-062352

Chayka, K. (2021). *We already live in Facebook's Metaverse*. https://www.newyorker.com/culture/infinite-scroll/we-already

Chen, R., Wu, N., Varvello, M., Chens, S., & Han, B. (2022, October). Are we ready for the Metaverse? A measurement study of social virtual reality platforms. *Proceedings of the 22nd ACM Internet Measurement Conference,* 504-518. 10.1145/3517745.3561417

ChohanU. W. (2022). *Metaverse or Metacurse*? doi:10.2139/ssrn.4038770

D'Arienzo, M. C., Boursier, V., & Griffiths, M. D. (2019). Addiction to social media and attachment styles: A systematic literature review. *International Journal of Mental Health and Addiction, 17*(4), 1094–1118. doi:10.100711469-019-00082-5

deZwart, M. J., & Lindsay, D. F. (2009). Governance and the global metaverse. In D. Riha & A. Maj (Eds.), *The real and the virtual critical issues in cybercultures* (pp. 173–182). Inter-Disciplinary Press.

Diaz, J., Saldana, C., & Avila, C. (2022). Virtual world as a resource for hybrid education. *International Journal of Emerging Technology Learn*, *15*(15), 94–109. doi:10.3991/ijet.v15i15.13025

Dick, E. (2021). *Public Policy for the Metaverse: Key Takeaways. AR/VR Policy Conference.* Information Technology and Innovation Foundation.

Farjami, S., Taguchi, R., Nakahira, K. T., Nunez, R., Fukumura, Y. M., & Kanematsu, H. (2011). Multilingual problem-based learning in Metaverse. *Technology and Society Magazine*, *37*(2), 52–61.

Feezell, J. T. (2018). Agenda Setting through social media: The Importance of Incidental News Exposure and Social Filtering in the Digital Era. *Political Research Quarterly*, *71*(2), 482–494. doi:10.1177/1065912917744895

Fernandez, C.B. & Hui. (2022). Life, the Metaverse and Everything: An Overview of Privacy, Ethics, and Governance in Metaverse. *2022 IEEE 42nd International Conference on Distributed Computing Systems Workshops (ICDCSW)*, 272-277.

Ferrara E., & Yang, Z. (2015) Measuring emotional contagion in social media. *PLOS One, 10*(11). 1.pone.0142390 29. doi:0.1371/journal

Fitria, T. N. (2021). Grammarly as AI-powered English writing assistant: Students' alternative for writing English. *Metathesis: Journal of English Language, Literature, and Teaching*, *5*(1), 65–78. doi:10.31002/metathesis.v5i1.3519

Franks, M. A. (2011). Unwilling avatars: Idealism and discrimination in cyberspace. *Columbia Journal of Gender and Law*, *20*, 224.

GilbertN. (2023). *Blog*. linkedin.com/in/nikkgilbert.

Haimson, O. L., & Hoffmann, A. L. (2016). Constructing and enforcing" authentic" identity online: Facebook, real names, and non-normative identities. *First Monday, 21*(6). Advance online publication. https://journals.uic.edu/ojs/index.php/fm/article/download/6791/5521. doi:10.5210/fm.v21i6.6791

Hall, S.B., & Baier-Lentz, M. (2022). *3 technologies that will shape the future of the Metaverse – and the human experience.* World Economic Forum. https://www.weforum.org/agenda/2022/02/future-of-the-Metaverse-vr-ar-and-brain-computer/

Hwang, G.-J., & Chien, S.-Y. (2022). Definition, roles, and potential research issues of the Metaverse in education: An artificial intelligence perspective. *Science Direct, 4.* https://www.sciencedirect.com/journal/computers-and-education-artificial-intelligence

Jasanof, S. (2015). Future imperfect: science, technology, and the imagination of modernity. In *Dreamscapes of modernity: Sociotechnical imaginaries and the fabrication of power.* University of Chicago Press. doi:10.7208/chicago/9780226276663.003.0001

Jaynes, C., Seals, W. B., Calvert, K., Fei, Z., & Griffloen, J. (2003). *The Metaverse: a networked collection of inexpensive, self-configuring, immersive environments.* Proceedings on Virtual Environments. doi:10.1145/769953.769967

Jeon, H. J., Youn, H. C., Ko, S. M., & Kim, T. H. (2022). Blockchain and AI meet in the metaverse. In T. Fernández-Caramés & P. Fraga-Lamas (Eds.), *Blockchain Potential in AI.* doi:10.5772/intechopen.91580

Jeong, H., Yi, Y., & Kim, D. (2022). An innovative e-commerce platform incorporating metaverse to live commerce. *International Journal Computer Information Control, 18*(1), 221–229.

Jones, D., Snider, C., Nassehi, A., Yon, J., & Hicks, B. (2020). Characterizing the Digital Twin: A systematic literature review. *CIRP Journal of Manufacturing Science and Technology, 29*, 36–52. doi:10.1016/j.cirpj.2020.02.002

Kaur, M., & Gupta, B. (2021). *Metaverse technology and the current market.* National Institute of Technology Kurukshetra.

Kelly, K. (2016). *The inevitable: understanding the 12 technological forces that will shape our future.* Penguin Publishing Group.

Kelly, K. (2019) *AR will spark the next big tech platform—call it mirror world.* https://www.wired.com/story/mirrorworld-ar-next-big-tech-platform/

Mahlich, H. (2021). *How to ensure that the Metaverse is an inclusive space.* https://www.showstudio.com/news/how-to-ensure-the-metaverse-is-an-inclusive-space

Maloney, D., Zamanifard, S., & Freeman, G. (2020). Anonymity vs. Familiarity: Self-Disclosure and Privacy in Social Virtual Reality. *Proceedings of ACM Symposium on Virtual Reality Software and Technology (VRST).*

Meier, C., Saorín, J., de León, A. B., & Cobos, A. G. (2020). Using the Roblox video game engine for creating virtual tours and learning about the sculptural heritage. *International Journal of Emerging Technology, 5*(20), 268–280. doi:10.3991/ijet.v15i20.16535

Mistretta, S. (2022). The Metaverse – An alternative education space, AI. *Computer Science and Robotics Technology*, *0*, 1–23.

Musk E (2019) an integrated brain-machine interface platform with thousands of channels. *Journal Medical Internet Research, 21*(10).

Mystakidis, S. (2022). Metaverse. *Metaverse Encyclopedia*, *2*(1), 486–497. doi:10.3390/encyclopedia2010031

Nazari, N., Shabbir, M. S., & Setiawan, R. (2021). Application of artificial intelligence powered digital writing assistant in higher education: Randomized controlled trial. *Heliyon*, *7*(5), e07014. doi:10.1016/j.heliyon.2021.e07014 PMID:34027198

Nica, E., Poliak, M., Popescu, G. H., & Pârvu, I.-A. (2022). Decision Intelligence and Modeling, Multisensory Customer Experiences, and Socially Interconnected Virtual Services across the Metaverse Ecosystem. *Linguistic and Philosophical Investigations*, *21*(0), 137–153. doi:10.22381/lpi2120229

Nvidia. (2021). *Develop with NVIDIA Omniverse*. https://www.nvidia.com/en-gb/omniverse/.u

Papagiannidis, S., & Bourlakis, M. A. (2010). Stating the new retail drama: At a Metaverse near you. *Journal of Virtual Worlds Research*, *2*(5), 425–446. doi:10.4101/jvwr.v2i5.808

Parisi, T. (2021). *The Seven Rules of the Metaverse*. https://medium.com/meta-verses/the-seven-rules-of-the-metaverse-7d4e06fa864c

Prabhakaran, A., Mahamadu, A.M., & Mahdjoubi, L. (2022). Understanding the challenges of immersive technology use in the architecture and construction industry: A systematic review. *Automation in Construction Journal, 137.*

Pritchard, C. (2022) *Top 10 companies investing in the metaverse*. https://www.parcl.co/blog/top-10-companies-investing-in-the-metaverse

Rehm, S. V., Goel, L., & Crespi, M. (2015). The metaverse as mediator between technology, trends, and the digital transformation of society and business. *Journal of Virtual Worlds Research*, *8*(2), 1–6. doi:10.4101/jvwr.v8i2.7149

Richter, F. (2020) *The end of the TV era?* https://www.statista.com/chart/ 9761/daily-tv-and-internet-consumption-worldwide/

Rospigliosi, P. A. (2022). Metaverse or simulacra? Roblox, Minecraft, Meta, and the turn to virtual reality for education, socialization, and work. *Interactive Learning Environments*, *30*(1), 1–3. doi:10.1080/10494820.2022.2022899

Rudolph, J., Tan, S., & Tan, S. (2023). ChatGPT: Bullshit spewer or the end of traditional assessments in higher education? *Journal of Applied Learning & Teaching*, *6*(1).

Salmon, G. (2019). May the fourth be with you: Creating education 4.0. *Journal of Learning Development*, *6*(2), 95–115. doi:10.56059/jl4d.v6i2.352

Saxon, E., Baker, B., Hargrove, W., Hofman, F., & Zganjar, C. (2004). Mapping environments at risk under different global climate change scenarios. *Ecology Letters*, *8*(1), 53–60. doi:10.1111/j.1461-0248.2004.00694.x

Shriram, K., & Schwartz, R. (2017, March). All are welcome: Using VR ethnography to explore harassment behavior in immersive social virtual reality. *IEEE Virtual Reality*, 225-226.

Siyav, A., & Jo, G. S. (2021). Neuro-symbolic speech understanding in aircraft maintenance metaverse. *IEEE Access : Practical Innovations, Open Solutions*, *9*, 154484–154499. doi:10.1109/ACCESS.2021.3128616

Smith, J. (2021). *Q&A: future pandemics are inevitable, but we can reduce the risk.* https://ec.europa.eu/research-and-innovation/en/horiz on-magazine/qa-future-pandemics-are-inevitable-we-can-reduce-risk

Soricut, R., & Brill, E. (2006). Automatic question answering using the web: Beyond the factoid. *Information Retrieval*, *9*(2), 191–206. doi:10.100710791-006-7149-y

Stephenson, N. (1992). *Snow crash.* Bantam Books.

Stern, S. E. (1999). Addiction to technologies: A social psychological perspective of Internet addiction. *Cyberpsychology & Behavior*, *2*(5), 419–424. doi:10.1089/cpb.1999.2.419 PMID:19178214

Su, Z., Zhang, N., Liu, D., Luan, T. H., & Shen, X. (2022). A survey on metaverse: fundamentals, security, and privacy. arXiv:2203.02662. doi:10.48550/arXiv.2203.02662

Taguma, M., Feron, E., & Lim, M. H. (2018). *Future of education and skills 2030: Conceptual framework.* Organization of Economic Co-operation and Development.

Vergne, J. P. (2021). *The future of trust will be dystopian or decentralized: Escaping the metaverse.* Revista de Occidente on the Future of Trust.

Warren, T. (2023). Microsoft is looking at OpenAI's GPT for Word, Outlook, and PowerPoint. *The Verge.* https;//www.theverge.com/2023/1/9/23546144/micorsoft-openai-word-powrpoint-outlook-gpt-integration-rumor

Webel, S., Bockholt, U., Engelke, T., Gavish, N., Olbrich, M., & Preusche, C. (2013). An augmented reality training platform for assembly and maintenance skills. *Robotics and Autonomous Systems*, *61*(4), 398–403. doi:10.1016/j.robot.2012.09.013

Whittaker, J., Looney, S., Reed, A., & Votta. F. (2021). Recommender systems and the amplification of extremist content. *Internet Policy Review, 10*(2), 1–29.

Zallio, M., & Clarkson, P. J. (2021). Inclusion, diversity, equity, and accessibility in the built environment: A study of architectural design practice. *Building and Environment*, *206*, 206. doi:10.1016/j.buildenv.2021.108352

Zawacki-Richter, O., Marin, V. I., Bond, M., & Gouverneur, F. (2019). A systematic review of research on artificial intelligence applications in higher education-where are the educators? *International Journal of Educational Technology in Higher Education*, *16*(1), 1–27. doi:10.118641239-019-0171-0

Zhai, X. (2022). *ChatGPT user experience: Implications for education.* https://papers.ssrn.com/sol3/papers.cfm?abstract_id=4312418

Zuckerberg, M. (2021) The metaverse and how we will build it together. *Connect 2021*. https://www.youtube.com/watch?v=Uvufun6xer8

KEY TERMS AND DEFINITIONS

Andragogy: The method and practice of teaching adult learners.

Augmented Reality: A technology that superimposes a computer-generated image on a user's view of the real world, thus providing a composite view.

Avatars: An icon or figure representing a particular person in internet forums.

Extended Reality: Covers all forms of computer-generated environments or objectives that combine or create virtual and physical realities.

Metaverse: A virtual reality space in which users can interact with a computer-generated environment and other users.

Virtual Reality: A computer-generated simulation of a three-dimensional image or environment that can be interacted with in a seemingly real or physical way by a person using special electronic equipment, such as a helmet with a screen inside or gloves fitted with sensors.

Chapter 18
Facilitating Conditions for Hybrid Teaching Among Academics

Geraldine Torrisi-Steele
Griffith University, Australia

ABSTRACT

Hybrid teaching has come to the fore post pandemic disruption. Although universities are making an effort to 'return to normal', the legacy of the pandemic, alongside the ever-present need to remain competitive, is the demand for flexibility in learning and access to online learning. Many academics are now facing the challenge of hybrid teaching. It is much more than simply using technology to transmit the lecturer's voice and classroom images between physical and virtual locations. Educators need to consider how to develop effective learning strategies that provide quality and equal opportunity. This requires creativity alongside knowledge of pedagogy and technology. It must be considered that educators need to be committed to hybrid teaching and fully engage. There are existing studies in the related domain of blended learning which can help create insights into what factors will support effective hybrid teaching. The author reports on a study of factors academics to blended learning strategies and highlights how the results of the study are useful to hybrid teaching.

INTRODUCTION

After the upheaval of the pandemic, many universities have found themselves in a new teaching context. Prior to the pandemic, universities were slowly embedding technologies into learning and teaching. The integration of technology was being fueled mostly by the need for flexibility, to meet the expectations of students, and the need to improve the quality and efficiency of teaching. During the pandemic, universities had to plunge into online learning, and the many universities that were predominantly in-person campuses struggled with going online (but did it nonetheless). Now, post-pandemic disruption, universities are faced with a complex problem – students have raised expectations about flexibility and the ability to

DOI: 10.4018/978-1-6684-7832-5.ch018

study online if they want/need to and there are also students who expect to be on-campus now that the 'pandemic is over' but still want some flexibility. University leadership is also keen to return students to campus but at the same time, they do not want to lose the online cohorts and provide flexibility with online options to bolster the student experience. It seems that once students got a "taste for it [online learning], it seems the vast majority of students are keen to continue to embrace the benefits of online teaching" (Morrison, 2022).

Many universities are therefore looking towards keeping the hybrid teaching models used during the pandemic in the hope of striking a balance among flexibility, on-campus student expectations, online student expectations, and academic teacher workloads. The term hybrid teaching is being used here to mean:

An educational model in which some students attend a lecture or seminar in-person as normal, while others join virtually from home. Lecturers, therefore, teach remote and in-person students at the same time using tools like Teams or Zoom, alongside the full range of digital tools...As a variation, lecturers may also work remotely to deliver sessions to students in a classroom and online...In some cases, hybrid events might include asynchronous learning elements, including online exercises and pre-recorded video materials, to support face-to-face classroom sessions. (Centre for teaching alearning, n.d.)

The term hybrid teaching is sometimes used interchangeably with blended learning, especially in pre-Covid times (Ulla & Perales, 2022). Post-pandemic the two terms have become more distinct. Hybrid teaching and blended learning differ in that hybrid teaching refers to synchronous teaching of students in face-to-face and online modes, but blended learning refers to a mix of technology-rich strategies with face-face teaching strategies, some of which are synchronous and others asynchronous. Hybrid teaching can be considered to include the use of blended learning strategies.

A report by the BBC estimates that as a follow on from the pandemic, nearly a third of universities continue to offer hybrid teaching as of January 2023 (compared with only ~4% prior to the pandemic) (Standley, 2023). The pressure from students to keep hybrid models is only one of the factors contributing to the increased uptake of hybrid teaching among universities. Some previously face to face only universities are keeping purely online options as well as delivering on campus, which can translate into a greater workload with more course offerings. For example, universities such as the author's home institution are multi-campus, running the same degree program plus a fully online option – effectively three offerings of each course. As universities seek to recover from the pandemic, competitiveness is more aggressive, and academic teachers are now under greater pressure than ever to evolve their teaching practice, embrace technology to provide excellent learning experiences to students online and on-campus, and maintain reasonable workloads.

Considering that prior to the pandemic, only a minority of academics were successfully combining technology with their face-to-face teaching to provide effective, high-quality learning experiences (Collis & Van Der Wende, 2002; Graham & Robison, 2007; Driscoll, 2002; Hoffman, 2006; Jeffrey, Milne, & Suddaby, 2014). While it sounds simple to teach both online and face-face students at the same time, hybrid teaching is challenging. Some of the challenges of hybrid teaching include: "unequal opportunity among students to be engaged in class activities, inefficient support for online students, technical issues with the platform, lack of devices for online learning, an additional time during the weekend, unresponsive online students, inability to monitor online students, low level of readiness and hesitation among lecturers, unstable internet connection, and sound-related issues" (Nebrida & Bangud, 2022, p.

51), inadequately equipped teaching rooms or teaching rooms laid out in ways not conducive to hybrid teaching.

Educational leaders are therefore now urgently tasked with providing the facilitating conditions that will help academic teachers fully engage with hybrid teaching and develop the necessary capabilities for handling technology-rich environments along the challenges of hybrid teaching. Unfortunately, hybrid teaching, despite its uptake, has been 'undertheorized' both before, during, and after the pandemic (Ulla & Perales, 2022). However, the challenges (lack of training, experience, lack of knowledge, lack of technical confidence) faced by educators during the pandemic are similar to those observed whenever educators are seeking to integrate technology effectively into their practice. Hence, although hybrid teaching is considered under-theorized, we can draw on empirical research-related domains, such as blended learning, to lay a foundation for understanding the requirements of professional development. In the present chapter, the author reports on a study conducted to understand the factors influencing the use of technologies as part of blended strategies.

The research took place prior to the pandemic and pivots on the question of "Why do some academics use technology to achieve learning outcomes while others use technology as an add-on at a superficial level?

A discussion of how the study findings can inform professional development for hybrid teaching follows.

FACTORS INFLUENCING ACADEMICS' INTEGRATION OF TECHNOLOGY TO ACHIEVE LEARNING (BLENDED STRATEGIES): CONCEPTUAL FRAMEWORK

The conceptual framework for the research and the initial research model constructs it precipitates is shown in Figure 1.

Emanating predominantly from Davis (1986)'s technology acceptance model, but also drawing on other frameworks of constructivism, Diffusion of innovations, the evolution of teaching practice, self-efficacy, teacher competence, and teaching style, the factors that may significantly influence academics' use of blended strategies are hypothesized to be:

Normative Influences

Normative influences refer to an individual's belief about whether people important to them think that the behavior should or should not be enacted (Fishbein & Ajzen, 1975). Prior research on the acceptance of technology has investigated the role that normative influences play in academics' acceptance of technologies (e.g., Chang Lieu, Liang, Liu & Wong (2011); Tarcan, Varok, & Toker (2010)). Important to the concerns of the present study is identifying those groups whose opinions one might expect to be important to academics. Literature related to professional support of academics shows that supervisors and peers may be two such important groups. For example, Chew, Jones, and Blackey (2010) report on the implementation of online assessment as an institutional strategy at the University of Glamorgan and found that having senior university staff members using blended strategies had a generally positive influence on the uptake of blended learning by more junior faculty members.

Figure 1. Conceptual framework elements and the precipitating predictor constructs of the theoretical model

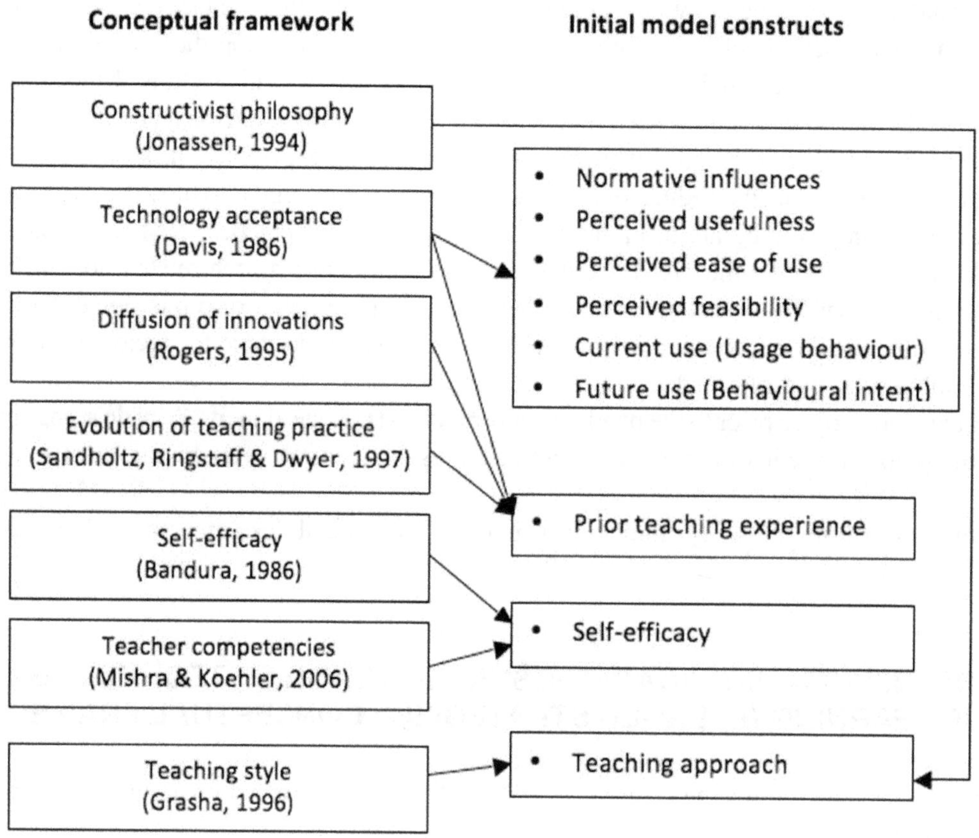

Perceived Ease of Use

The amount of effort an academic believes he or she needs to expend to implement teaching strategies that require the use of both technology and face-to-face teaching is captured by the perceived ease of use construct. Existing literature on technology acceptance points to perceived ease of use as being a significant influence on the acceptance of technology (e.g., Halawi & McCarthy, 2000; Kripanont & Tatnall, 2009). The present study explores the role perceived ease of use may play in how technology is used.

Perceived Usefulness

Perceived usefulness is a core construct of the technology acceptance models. In technology acceptance models perceived usefulness positively affects behavioral intention to use technology. For the present study, perceived usefulness is the degree to which academics believe that using technology together with face-to-face teaching strategies will enhance teaching effectiveness. Researchers such as Halawi and McCarthy (2010), Chang et al. (2011), and Tarcan, Varol and Toker (2010) have found that perceived usefulness positively influences the intention to use technology. Perhaps then, perceived usefulness may also significantly influence how the technology is used.

Perceived Feasibility

Present in some technology acceptance models (Venkatesh & Bala, 2008), feasibility is considered to influence usage behavior. Feasibility conditions such as time, infrastructure, professional support, technical support, and funding have been found to be significant in influencing academics' acceptance of technology (e.g., Bagher, Marek, & Sibbald, 2007; Davis & Fill, 2007; Kistow, 2009; Ocak, 2010; Stewart, Bachman, & Johnson, 2010; Wang, 2009). Given that implementing blended approaches generally requires a significant investment of time and resources, it is likely that academics' perceptions of feasibility conditions such as time, funding, support, and infrastructure play a role in shaping the extent to which technology is used by academics in their teaching.

Prior Teaching Experience

Sandholtz, Ringstaff, and Dwyer (1997) observe that teachers using technology tend to progress through a number of stages of technology usage with increased experience in teaching with technology. Experience as an influence on usage behavior has also been identified in the technology acceptance models described in the previous chapter.

Self-Efficacy

Self-efficacy is the belief in one's capacity to succeed in an undertaking (Bandura, 1986). Within the present study, self-efficacy refers to academics' belief in their own ability to implement teaching strategies, in a specific content area, using both technology and face-to-face teaching. The construct of self-efficacy in this study has been framed in terms of the technology, pedagogy, and content knowledge framework (Mishra & Koehler, 2006) which suggests that for teachers to use technology effectively they must possess technological, pedagogical, and content knowledge.

Self-efficacy in this study is conceptualized in terms of academics' beliefs about their ability to:

- Use technology (technical knowledge);
- Design effective teaching strategies (pedagogical knowledge);
- Understand course content (content knowledge);
- Make good decisions about teaching approaches appropriate to teaching the content (pedagogical-content knowledge);
- Identify which technologies can be used in the content area (technological-content knowledge);
- Understand the attributes of the technology and how they can be used to form new teaching strategies (pedagogical-technological knowledge);
- Identify the attributes of technology as well as design new strategies using those attributes to best teach the required content knowledge (technological-pedagogical- content knowledge).

Teaching Approach

Aligned with Grasha's (1996) conception of teaching style, the 'teaching approach' construct is used in the present study to refer to the characteristic manner in which individual teachers design the instructional process. The decision to use the teaching approach as a predictive construct is made on the observation that, in some existing studies, constructivist teaching approaches were found to be conducive to using technology as an intrinsic part of learning activities (Judson, 2006; Grasha, 1996; Weitkamp, 2006).

METHODOLOGY

A two-phase approach was used in the study. The objective of the first phase was data gathering using an online survey instrument designed for the study. The survey distributed as a survey link was distributed to Griffith university academics via email. A door-to-door paper survey was also used in an attempt to boost response rates. A total of 53 academics responded to the survey. The survey included demographic information such as gender, faculty, class size, years teaching in/out of higher education, a matrix to measure the actual current use of blended strategies which captured the extent and purpose of usage of various technologies as well as face-to-face, and 7 point Likert Scale items to measure intended future use and each of the hypothesized influencing factors.

Instrument reliability needed to be addressed. Instrument reliability was demonstrated by undertaking item analysis of each of the construct item groups, that is, item analysis (Cronbach's alpha and corrected item-total correlation) was conducted to check that each item, in each set, was a good measure of construct it intended to measure (Colten & Covert, 2007; De Vaus, 2002; Hair, Black, Babin & Anderson, 2010).

In the second phase, semi-structured interviews were conducted. A subsample (N = 8) of survey respondents was interviewed. The selection of interview participants was based on the quantitative results. Outlier sampling was used to purposefully select interview participants from the pool of survey respondents willing to participate in interviews. The qualitative data yielded from the interviews was used to add depth to the understanding of the quantitative findings.

Study Limitations

Whilst the study expands understanding of factors that influence how academics use technology after adoption and led to some recommendations for the provision of more effective professional support for academics, the study's limitations must be acknowledged before concluding. The limitations potentially restrict the generalizability of results but also direct attention to further research. In the described study, limitations related to sample size, sample acquisition, institutional culture and context, and the limited number of constructs investigated.

QUANTITATIVE FINDINGS

The means for construct scores across all 53 respondents are given in Table 1.

Regression was undertaken for constructs as predictors of 1) the *current use* of blended strategies, and 2) the *intended future use* of blended strategies. The resulting model parameters for the *current use* of blended strategies are given in Table 2.

Table 1. Construct means across all study participants (N = 53)

Construct	Mean
Perceived usefulness	5.58
Perceived feasibility	4.81
Self- efficacy	5.40
Current use score	7.20
Future use score	5.42
Perceived current use score	4.61
Number of years of teaching experience in higher education	13.31 years

Table 2. Model parameters for the current use of blended strategies

Fixed Effects				
	Estimate coefficient	Std. Error	t Value	Pr(>|t|)
(Intercept)	-5.50264	7.97975	-0.690	0.4940
Gender = female	-5.40926	10.30914	-0.525	0.6024
Perceived usefulness	2.57841	1.04254	2.473	0.0172 *
Teaching experience in higher education	-0.15763	0.07296	-2.160	0.0361 *
Self-efficacy	1.22298	1.43084	0.855	0.3972
Perceived feasibility	-1.35281	0.97083	-1.393	0.1703
Self-efficacy: Gender = female	4.82572	1.80222	2.678	0.0103 *
Perceived usefulness: Gender = female	-3.36219	1.55028	-2.169	0.0354 *

Significance codes: 0 *** 0.001 ** 0.01 * 0.05. 0.1 ' ' 1

The positive estimate coefficients in Table 2 indicate a favoring of the use of blended strategies. Those academics with high perceived usefulness scores are predisposed towards using blended strategies in their current practice (note the positive coefficient). On the other hand, the negative coefficient for teaching experience in higher education indicates academics that have been teaching in higher education longer are pre-disposed to lesser use of blended strategies.

A particularly interesting outcome is the presence of interactions with gender. In particular, there is *self-efficacy: gender* interaction and *perceived usefulness: gender* interaction. The positive coefficient for the *self-efficacy: gender* interaction indicates females with higher self-efficacy with regard to the use of blended strategies are more inclined to use blended strategies in current practice. The negative coefficient of the *perceived usefulness: gender* interaction indicates that for female academics perceived usefulness is not an as important influence on the use of blended strategies as it is for male academics.

Table 3 gives the resulting model parameters for the *intended future use* of blended strategies.

Table 3. Model parameters for intended future use of blended strategies

Fixed Effects				
	Estimate Coefficient	**Std. Error**	**t Value**	**Pr(>|t|)**
(Intercept)	-0.71953	1.99170	-0.361	0.7198
Gender = female	2.55681	2.83239	0.903	0.3721
Perceived usefulness	0.81384	0.17064	4.769	0.0000 ***
Current use score	0.09387	0.02508	3.742	0.0006 ***
Perceived feasibility	-0.86361	0.27757	-3.111	0.0034 **
Normative influences	0.42743	0.23285	1.836	0.0739.
Teaching approach	0.04828	0.19796	0.244	0.8086
Perceived ease of use	0.14647	0.14591	1.004	0.3215
Number course offerings	0.23738	0.20866	1.138	0.2620
Self-efficacy	0.25206	0.24958	1.010	0.3186
Teaching approach: gender = female	-0.72585	0.48000	-1.512	0.1383
Perceived ease of use: gender = female	-0.37108	0.25728	-1.442	0.1570
Perceived Feasibility: gender= female	0.67729	0.31988	2.117	0.0405*

Significance codes: 0 *** 0.001 ** 0.01 * 0.05. 0.1 ' ' 1

An examination of Table 3 shows that perceived usefulness is a highly significant predictor of the *intended future use of blended strategies*. The coefficients of both perceived usefulness and current use are positive indicating academics who perceive blended strategies as users are more likely to use blended strategies in the future, and those academics who are currently using blended strategies are also more likely to use blended strategies in the future.

Interestingly, perceived feasibility has a negative coefficient. This is seemingly counter-intuitive in that it suggests that those academics that perceive blended strategies to be more feasible are less likely to use them. One possible explanation can be arrived at by firstly, noting that perceived feasibility interacts with gender and the *perceived feasibility: gender = female* interaction has a positive coefficient, thus female academics who perceive the task as highly feasible are more likely to intend to use blended strategies in the future. Secondly, observe that perceived usefulness is of greater importance for male academics for the *current use* of blended strategies than it is for female academics. Possibly, for male academics, perceived usefulness is the primary concern, and they will tend not to use blended strategies, even if feasible unless they are perceived as useful.

Finally, normative influences are not significant but trending towards having a positive influence on the *intended future use* of blended strategies.

The resultant models are visually represented in Figure 2.

Figure 2. The model of factors influencing academics' inclination to integrate technologies for learning in current and future practice. Coefficients and p values are shown.

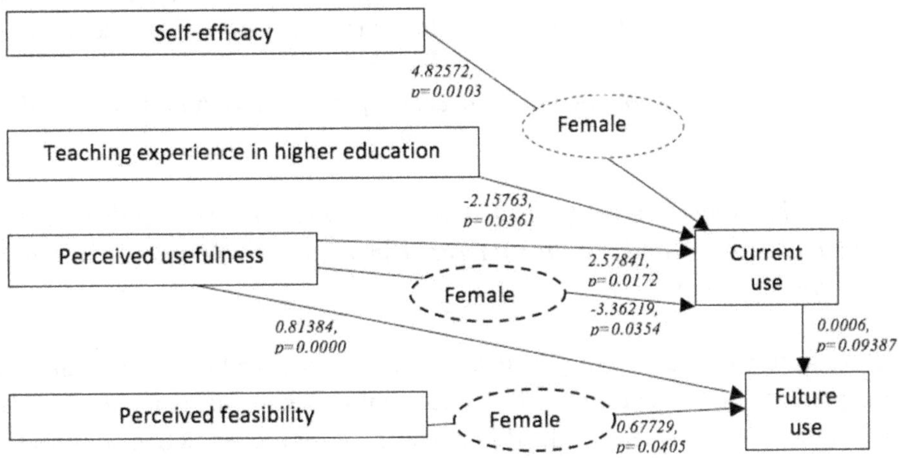

DISCUSSION

Quantitative analysis revealed perceived usefulness as a significant predictor for the use of blended strategies. For male academics especially perceived usefulness is a significant predictor of the *current use* of blended strategies. Perceived usefulness was found to be a highly significant predictor of the *intended future use* of blended strategies for both male and female academics. Given its significance, the concept of perceived usefulness warrants some discussion. During interviews, a distinction emerged between perceived usefulness for the purpose of teaching and learning versus perceived usefulness in terms of administrative efficiency or ease of access. It became abundantly obvious that when academics view the use of technology with face-to-face teaching as useful for enhancing learning, it is a strong driver for the use of blended strategies. Those academics with high perceived usefulness scores acknowledged the usefulness of technology for purposes of efficiency and flexibility but most importantly placed emphasis on use targeted towards enhancing learning experiences. For example, one participant (perceived usefulness score= 7, current use score =10) noted that blended learning is "a marriage of tech with face-to-face making it the best learning outcome". Another participant (perceived usefulness score = 7, current use score = 20) explained her integrated use of technology with face-to-face teaching in terms of learning activities and teaching:

I upload previous student essays and get them to mark them and explain why so they engage in discussion about this [discussion boards] ... I also use online case studies linked to book chapters and question banks. It's very integrated...I found it [technology] very useful ...as we move through the course, I see them [students] challenging each other...The online discussion feeds into the face-to-face teaching and learning.

In contrast, those academics with lower perceived usefulness scores focused on efficiency and access and neglected the learning experience in their comments. One participant (perceived usefulness score 4.8, current use score=0) felt that technology was "nothing more than an administrative device". Other participants with low perceived usefulness scores and low current use scores expressed a similar sentiment:

Recording lecture audio is useful for students to access" (perceived usefulness score =4.6, current use score =2).

My main motivation for using [technology] is that we have a large cohort of students ...that way I don't need to use the board as much and I use the internal communication thing on Blackboard (perceived usefulness =5, current use score =0).

An interesting feature of the resulting models for both *current* and *intended future use* of blended strategies is the presence of gender differences in the significant constructs. In the predictive model for the *current use* of blended strategies, gender differences exist for the predictive constructs of perceived usefulness and for self-efficacy. Perceived usefulness is a significant predictor of current blended practice adoption for male academics but not so important for female academics. Self-efficacy, on the other hand, is more important as a predictor of the *current use* of blended strategies for female academics. These quantitative outcomes manifested in interviews. Female academics placed emphasis on feeling confident and having the required skills. For example, female academics made comments such as:

I've got the skills and if don't have the skills then I take the time to learn to do them ..., if it's too difficult or complex then I don't bother;

I'm pretty new here, it's a constant battle to find out what I can do and what's available and how to use it... I feel quite comfortable doing stuff once I find out;

I'm still learning...things go wrong. If things go wrong, I use the tech person, he's very good. If I can't get him I use my colleagues who also have those skills so it's no problem.

An important observation is that although female academics elaborated in some detail on self-efficacy and technical support, the three male participants did not raise the issue of self-efficacy and support but rather focused almost exclusively on usefulness:

You shouldn't use it just for those reasons anyway [efficiency]. I mean it is something you can do better with technology. That's really the question. (male academic)

In the predictive model for the *intended future use* of blended strategies, gender differences are observed with respect to perceived feasibility. Perceived feasibility is a significant predictor of the intention to use blended strategies in the future for female academics but not so for male academics. The emphasis that female academics put on feasibility during interviews supports the quantitative result. In addition, when female academics discussed feasibility, the issue of time was often raised:

My focus is student-centric, make it fun, engaging, keep my students, I'm prepared to do what it takes... I make the time. (Female; perceived usefulness = 7, perceived feasibility = 6.5)

Any plans I have for the future involve more work for me – that will require a lot of extra effort and time. There are lots of things I'd like to do, but I need to draw the line somewhere. (Female, perceived usefulness=6, perceived feasibility= 3.7)

In relation to time, female academics frequently identified professional support as a mechanism to reduce time investment:

Again it's time, if someone would come to me and say well here are all your options that'd be great. (Female, perceived usefulness =6.4, perceived feasibility 2.3)

In contrast, male academics made brief mention of time, and when they did it was only in reference to technical failures leading to loss of time:

Technology has 'gotten in the way' of student engagement to a certain extent, fiddling and messing around can slow you down, the equipment can fail and nothing happens, so it has risks; in terms of a lecture does it really make you do more, I don't think so. (Male, perceived usefulness=4.6, perceived feasibility=4.0)

The remaining significant predictor constructs present in the models are teaching experience as a predictor of *current blended practice,* and current practice as a significant predictor of the *future* intention to use blended strategies. Understanding of both these predictors is also enhanced by discussion in light of interview data.

With respect to the teaching experience predictor, it was found that the greater number of years of teaching experience in higher education were less likely to be currently using blended strategies. Interview data added an interesting perspective to this finding. The participant with the greatest number of years (38 years) of teaching experience viewed the use of technology in teaching as being in competition with traditional face-to-face teaching – a question of which is better:

Is it [technology] better than carefully planned face-to-face teaching ... Can it be shown that bringing in technology is going to improve the learning experience? (Perceived usefulness score =4.8, current use score=0)

In contrast, those with the least teaching experience considered if the use of technology was appropriate to certain situations making comments such as

You've got more options [using technology with teaching]"; "this subject can't really be taught any other way...if it was a different subject then there would be more that could be done with the technology.

The importance of current practice as a significant predictor in the predictive model for the *intended future use* of blended strategies was clearly demonstrated during interviews. As the model predicted, those with positive experiences in their current practice are more likely to use blended strategies in the future, whilst those negative experiences 'put academics off future use':

I upload case studies and ask them to respond to them.[in another course] I upload previous student essays and get them to mark and explain why so they engage in discussion about this...I've found it [discussion boards] quite useful...I use them in another course [with case studies]. (Female, perceived usefulness=7, current use score=20, future use =6.2);

I don't use discussion boards [anymore]...they [students] just used them to complain or ask when marks were coming out or things like that... I have used online quizzes. It worked but it wasn't that useful because they'd [students] sit in the library [with] a set of computers. One was Googling and the other was doing the test. (Female, perceived usefulness=6, current use score=4, future use = 5.3);

I once borrowed a roving microphone to use with Lectopia. It was so complicated to use I never used it...Once [when the technology went wrong] I had to apologize to a large class. A lock-up like that rips out lots of time; I've set up discussion groups...but students remained silent and didn't use them. (Male, Perceived Usefulness=4.8, current use score=0, future use score=5.2)

A discussion of the quantitative results in light of interview data both supported findings and enriched understanding of factors shaping how academics use technology in their teaching. The most important insights from interview data are added to the theoretical model to give the enhanced model shown in Figure 3.

APPLYING STUDY RESULTS TO HYBRID TEACHING

Drive the teaching with learning objectives (rather than technology) and then explore the possibilities.

Hybrid teaching refers to the synchronous teaching of face-to-face classes with online classes. Providing both the face-to-face cohort and the online cohort with a quality learning experience is challenging because the physical environment and virtual environments are inherently different and therefore it is necessary to transform rather than 'translate' teaching strategies from one environment to the other. Hybrid teaching involves more than simply using technology to allow the image and voice of the teacher to be transmitted from the physical classroom to the virtual classroom. Seemingly, technology is the enabler to help create a learning space at the intersection of physical and virtual environments. However, the emphasis must be on the pedagogy not solely on the technology. In the study perceived usefulness was found to be a significant predictor of current and *intended future use* of blended strategies. Perceptions of the usefulness of the use of technology with face-to-face teaching that related to improving student learning (and not just efficiency and flexibility of access) were most conducive to the use of blended strategies. Similarly, to support academics to engage meaningfully with hybrid teaching can begin with first clearly defining the learning outcomes, then opening a discussion about how these learning outcomes might be achievable in both physical and virtual classrooms. The discussion needs to be informed by knowledge of the capabilities of various tools and technologies. For example, if we wish students to form a conceptual map of their prior knowledge, then tools such as Miro (https://miro.com/) can be used for collaborative concept mapping. Naturally, this requires access to technology by both students in the physical classroom and those in the virtual classroom. There are implications here for the provision of the required infrastructure within the university to enable this.

Figure 3. The model enhanced by qualitative insights

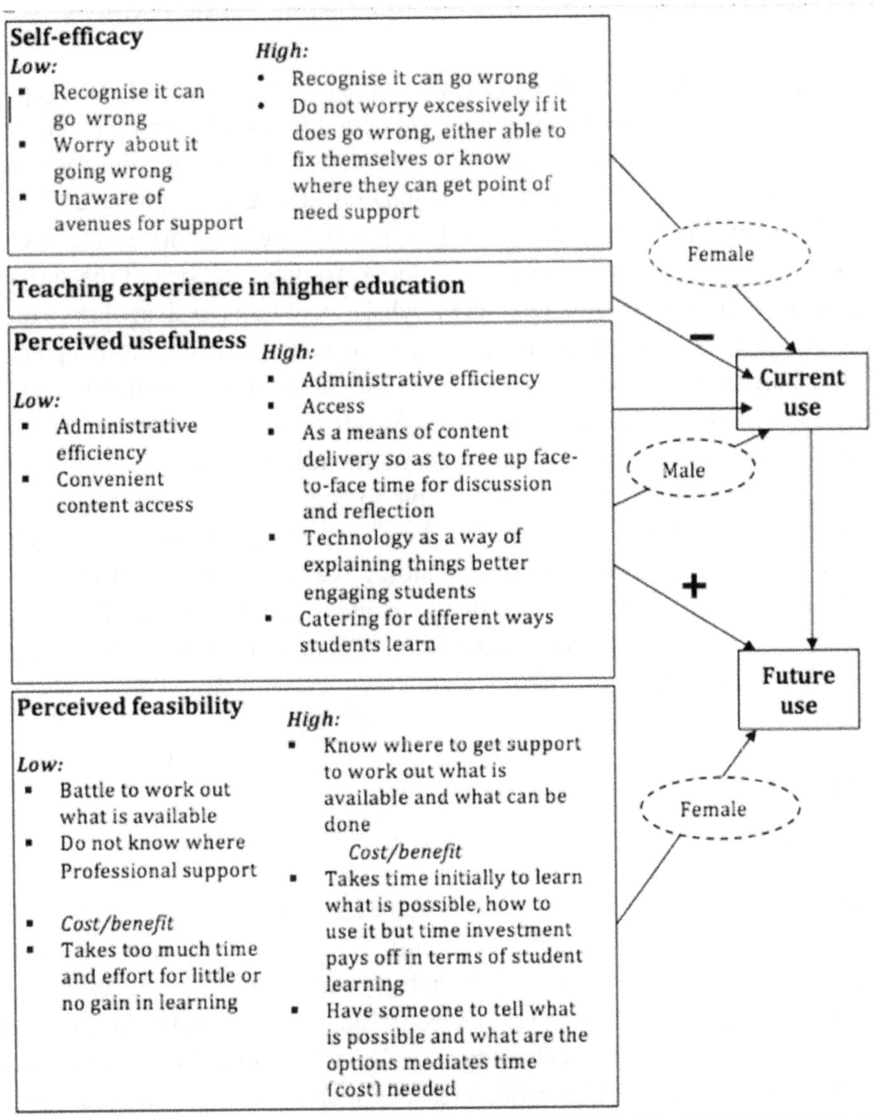

Understand Current Practice and Engage in Reflective Practice

The importance of academics' current practices with the use of technology with face-to-face teaching was made evident by the results of the present study. Current usage was found to be a highly significant predictor of the intent to use blended strategies in the future. It is recommended that there is a focus on understanding academics' current practices, and it is considered important to gain insight into prior experiences of using technology. Information about current practice may be collected during workshops and serve as a launching point for more individualized support later. As academics grapple with hybrid teaching models and blended strategies to use as part of their hybrid teaching, the importance of reflective practice cannot be understated.

Point of Delivery Support Is Important: Consider Formal Technical Support but Don't Underestimate the Value of Peer-Peer Support

Professional development is often conceptualized as a workshop or other activity taking place separately from teaching practice. However, the results of the study point to self-efficacy and perceived feasibility as important factors. Support mechanisms for point of delivery or time just-in-time situations should be integrated into professional development efforts and considered, not separate to, but as part of the professional development exercise. Academics value 'just-in-time' support, especially when there is often anxiety about 'technology going wrong' (as it does). To illustrate: one of the author's colleagues recently told the story of how in the first attempt at hybrid involving physical classes on two different campuses and an online class (in the trimester starting at the time of writing this chapter), the technical support person was only able to be present for a few minutes at the start of lecture to help with basic set up. Unfortunately, a few minutes after the support person left, technical issues occurred, and IT support was not available for some time. Obviously, it was very stressful with ~100 students in total waiting for the class to restart. Fortunately, the academic was able to contact another colleague who was able to help fix the problem and enable the class to go on. This story brings two important points to the fore: (i) the necessity of accessible, in-time support – technology can and does 'go wrong' and (ii) the value of peer-peer support as a strategy to develop hybrid teaching capabilities within faculties. With respect to the latter, it is an opportune time to mention the value of communities of practice as a mechanism of professional development (Jocius et al., 2022).

CONCLUSION

Amid the disruption of the pandemic, in education, the value of technology to enable continuity of physical connection between educator and student became apparent. At the same time, students became more aware of the flexibility offered by online learning. The legacy of the pandemic disruption is the increased uptake of online learning and especially hybrid teaching models. Historically, the uptake of technology in education is challenging because 1) effective integration of technology requires knowledge of teaching alongside knowledge of the technology and 2) educators need to be motivated to meaningfully engage with it and use it beyond superficial reasons such as ease of access. In the present paper, the author has argued that prior research related to blended learning can be used as a foundation to understand the factors which facilitate educator engagement with hybrid teaching. The author reported on a blended learning study undertaken prior to the pandemic and relates the outcomes of the study to hybrid teaching, drawing insights into factors that may support engagement with hybrid teaching. These factors should be considered in professional development and other strategies to support academics' use of hybrid teaching. Insights from previous studies in blended learning form a starting point for further empirical research to inform hybrid learning strategies and the development of hybrid learning capabilities. Alongside empirical research, it is important to not discount the value of the student's voice. As academic educators continue to explore hybrid teaching and other technology-rich teaching models, they are encouraged to be alert to and invite, conversations with students about their experiences as a basis for the continual evolution of quality teaching practices.

REFERENCES

Bagher, M., Marek, S. A., & Sibbald, A. (2007). Implementing web-assisted learning and engaging academic staff in the change process. *Journal of Organisational Transformation & Social Change*, *3*(3), 269–284. doi:10.1386/jots.3.3.269_1

Bandura, A. (1986). *Social foundations of thought and action: A social cognitive theory*. Prentice Hall.

Carmen, J. M. (2002). Blended learning design: Five key ingredients. *Journal of Educational Media*, 165–178.

Centre for Teaching and Learning. (n.d.). *What is hybrid teaching?* University of Oxford. https://www.ctl.ox.ac.uk/what-is-hybrid-teaching

Chang, J.-L., Lieu, P.-T., Liang, J.-H., Liu, H.-T., & Wong, S.-L. (2011). Factors influencing technology acceptance decision. *African Journal of Business Management*, *5*(7), 2901-2909. http://www.academic-journals.org/ajbm/abstracts/abstracts/abstracts2011/4Apr/Chang%20et%20al%202.htm

Collis, B., & Moonen, J. (2001). *Flexible learning in a digital world: Experiences and Expectations*. Kogan Page.

Davis, H. C., & Fill, K. (2007). Embedding blended learning in a university's teaching culture: Experiences and reflections. *British Journal of Educational Technology*, *38*(5), 817–828. doi:10.1111/j.1467-8535.2007.00756.x

De Vaus, D. A. (2002). *Surveys in social research*. Allen & Unwin. doi:10.4135/9781446263495

Driscoll, M. (2002, March 1). *Blended learning: let's get beyond the hype*. http://elearningmag.com/ltimagazine

Dzubian, C., Hartman, J. L., & Moskal, P. D. (2004). Blended Learning. *Educause*, (7), 1–12.

Fishbein, M., & Ajzen, I. (1975). *Belief, Attitude, Intention and Belavior: An Introduictionto Theory and Research*. Addison-Wesley.

Garrison, D. R., & Kanuka, H. (2004). Blended learning: Uncovering its transformative potential in higher education. *The Internet and Higher Education*, *7*(2), 95–105. doi:10.1016/j.iheduc.2004.02.001

Garrison, R., & Vaughan, H. (2008). *Blended learning in higher education: Framework, principles and guidelines*. Jossey-Bass.

Grasha, A. F. (1996). *Teaching with style*. Alliance Publishers.

Hair, J. F., Black, W. C., Babin, B. J., & Anderson, R. E. (2010). *Multivariate Data Analysis* (7th ed.). Prentice Hall.

Judson, E. (2006). How teachers integrate technology and their beliefs about learning: Is there a connection? *Journal of Technology and Teacher Education*, *14*(3), 581–597.

Kistow, B. (2009). E-learning at the Arthur Lok Jack Graduate School of Business: A survey of faculty members. *International Journal of Education and Development using ICT, 5*(4). http://ijedict.dec.uwi.edu/viewarticle.php?id=845

Mishra, P., & Koehler, M. J. (2006). Technological pedagogical content knowledge: A framework for teacher knowledge. *Teachers College Record, 108*(6), 1017–1054. doi:10.1111/j.1467-9620.2006.00684.x

Morrison, N. (2022, May 22). *If Students Had Their Way, Hybrid Learning Would Be Here To Stay*. https://www.forbes.com/sites/nickmorrison/2022/05/27/if-students-had-their-way-hybrid-learning-would-be-here-to-stay/?sh=604f22435876

Nebrida, J., & Bangud, A. (2022). Challenges in hybrid teaching amidst pandemic: The proposed mode. *Journal of Engineering Research and Reports, 22*(9), 51–64. doi:10.9734/jerr/2022/v22i917564

Ocak, M. (2010). Why are faculty members not teaching blended courses? Insights from faculty members. *Computers & Education, 56*(3), 689–699. doi:10.1016/j.compedu.2010.10.011

Sandholtz, J., Ringstaff, C., & Dwyer, D. (1997). *Teaching with technology*. Teachers College Press.

Standley, N. (2023, January 6). *Nearly a third of university courses still have hybrid teaching*. https://www.bbc.com/news/education-64130367

Stewart, C., Bachman, C., & Johnson, R. (2010). Predictors of faculty acceptance of online education. *Journal of Online Learning and Teaching, 6*(3), 597–616.

Tarcan, E., Varol, E. S., & Toker, B. (2010). A study on the acceptance of information technologies from the perspective of academicians in Turkey. *Ege Academic Review, 10*(3), 791–812. doi:10.21121/eab.2010319615

Uğur, B., Akkoyunlu, B., & Kurbanoğlu, S. (2011). Students' opinions on blended learning and its implementation in terms of their learning styles. *Education and Information Technologies, 16*(1), 5–23. doi:10.100710639-009-9109-9

Ulla, M., & Perales, W. (2022). Hybrid teaching: Conceptualization through practice for the post COVID19 pandemic education. *Frontiers in Education, 7*.

Venkatesh, V., & Bala, H. (2008). Technology acceptance model 3 and a research agenda on interventions. *Decision Sciences, 39*(2), 273–315. doi:10.1111/j.1540-5915.2008.00192.x

Verkroost, M., Meijerink, L., Linsten, H., & Veen, W. (2008). Finding a balance in dimensions of blended learning. *International Journal on E-Learning Academic Research Library, 7*(3), 499–522.

Wang, S.-C. (2009). University instructor perceptions of the benefits of technology use in e-learning. In K. Jusoff, S. S. Mahmoud, & R. Sivakumar (Ed.), *Proceedings of the Second International Conference on Computer and Electrical Engineering* (pp. 580-585). IEEE Computer Society Washington. 10.1109/ICCEE.2009.275

Weitkamp, E. W. (2006). *Teaching style and technology: The role of educational values in the adoption of information technology innovations in the classroom.* http://gateway.proquest.com/openurl%3furl_ver=Z39.88-2004%26res_dat=xri:pqdiss%26rft_val_fmt=info:ofi/fmt:kev:mtx:dissertation%26rft_dat=xri:pqdiss:3183581

KEY TERMS AND DEFINITIONS

Blended Learning: Learning is achieved through enriched, student-centered, experiences made possible by the harmonious integration of various strategies combining face-to-face interaction with information and communication technology.

Flipped Classroom: A commonly adopted, relatively new approach, in blended learning. Content is delivered digitally thus freeing up more time for face-to-face interaction. Flipped classrooms typically do not have traditional lectures.

Hybrid Teaching: Teaching to students in a physical classroom at the same time as students in a virtual classroom to provide engaging and quality learning experiences to both classrooms. It entails the use of technology to facilitate communication between the two classrooms and the innovative use of blended strategies to engage students effectively in both classrooms.

Technology Acceptance Model (Davis, 1986): Originally an information systems model describing how people come to accept the use of technology. Has since been extensively modified and used by many researchers in a great variety of contexts to examine the factors influencing various technology-related activities.

Chapter 19
Technology Use and Its Changing Role in Community Education

Lesley Farmer
California State University, Long Beach, USA

ABSTRACT

Technology, particularly as it supports interactive learning communities, now plays a central role in community education. While technology affords access to information more efficiently than ever before, it also requires additional skills. For technology to play out its role effectively, community educators need to design instruction, curriculum, and programs that offer customizable learning and socially rich learning environments. Every effort needs to be made to make sure that learners have opportunities to access and learn with technology. Successful community efforts to incorporate technology tend to include these factors: local needs, local resources, local broad-based planning, partnerships among organizations, accessible and appropriate technologies, opportunities for community building and socializing, and locally meaningful content and outcomes. Two community education foci are examined in detail: women's empowerment and health literacy.

INTRODUCTION

As the sense of community has expanded to encompass both physical and virtual mutual spaces and agreed-upon identities, the need for community education has also increased. Communities need to access, evaluate, and use increasing amounts of information for individual and group sustenance and improvement. Technology has served as a catalyst for such information transformation and can also be used to educate communities. For example, government agencies increasingly include e-government initiatives that use information technology tools and systems to provide better public services that can transform relations with citizens and businesses. For technology to play its role in community, several factors need to be in place, which this chapter discusses: changes in access issues, technology tools, instructional design, and topical applications of community education.

DOI: 10.4018/978-1-6684-7832-5.ch019

COMMUNITY EDUCATION DEFINITIONS

Most simply, community education is education within the community, for the community, by the community. Community education is usually associated with community-based non-formal adult education, which forms the basis of this chapter's discussion. However, the scope of community education can vary from providing all the educational needs of the community to providing strictly informal adult education for self-development. Sometimes community education has a governmental or political agenda, be it citizenship education, "liberation" connotation, or a sense of social issues-based community development. In other circles, community education reflects moral, social, or personal development. Issues of health, environment, economics, and career development can also fall under the umbrella of community education.

The Scottish Community Education defined community education as "a process designed to enrich the lives of individuals and groups by engaging with people living within a geographical area, or sharing a common interest, to develop voluntarily a range of learning, action and reflection opportunities, determined by their personal, social, economic and political needs" (Scottish Community Education Council, 1990, p. 2). United States adult educators defined community education as: "the process of identification of community needs and the marshalling of resources to meet those needs so that the community and all its members can grow through social and educational programs" (Fellenz & Coker, 1979, p. 319).

In the United States, community education may relate to public education, as defined by Minzey (1972):

Community Education is a philosophical concept which serves the entire community by providing for all of the educational needs of all its community members. It uses the local school to serve as the catalyst for bringing community resources to bear on community problems in an effort to develop a positive sense of community, improve community living and develop the community process toward the end of self-actualization. (p. 150)

A leader in the field, Wisconsin Department of Public Instruction (2008) listed five components of community education, which serve as their philosophical foundation: citizen involvement, needs assessment and planning, extended use of public education facilities, interagency coordination and cooperation, and leadership and accountability. They also supported Decker's 1992 community education principles:

- **Self-Determination**. An assessment should identify local needs and wants.
- **Self-Help**. People should be encouraged to assume more responsibility for their own well-being.
- **Leadership Development**. Community development should incorporate local citizen leadership.
- **Localization**. Education should be offered as close as possible to the intended community.
- **Integrated Delivery of Service**. Local organizations should collaborate, pooling their resources and meeting their own goals.
- **Maximum Use of Resources**. Local material and human resource should be integrated and maximized.
- **Inclusiveness**. Community education should involve all community constituents.
- **Responsiveness**. Public institutions should develop programs that are responsive to community needs.
- **Lifelong Learning**. All ages of residents in various community settings should have access to learning.

GOVERNMENTAL CONTEXTS OF COMMUNITY EDUCATION

Community education has ebbed and flowed over the last fifty years, often impacted by governmental contexts. Governments have encouraged community education to address high illiteracy rates and low employment rates. As countries have changed government reins or gained independents, community education has helped to redefine government ideologies and practices. More recently, as governments have embraced e-government procedures to gather, process and act on date, community education has been seen as a way to help citizens access those technology-based government services. Crises such as natural disasters, endemics, and human-made problems (e.g., polluted water, gangs, terrorism) also activate community education to mitigate and salvage the results at the local level.

Sometimes community education is activated to complement or counter government action or inaction. As an example, women in some countries do not have equal access to education. In some cases, governments might ignore local drug abuse. Health issues might not be adequately addressed at the government level. Especially when governments are unstable or in conflict, non-formal education is particularly needed for sharing knowledge (Maber, 2016). Recent activism issues such as #MeToo, #Black Lives Matter, and climate change occur at several levels, including the community level, reflecting the adage: Think Global, Act Local." This attitude reflects a change from education with the goal of promoting obedient self-regulating citizens to a goal of improving the community through social change (O'Grady, 2018).

Relative to government influence, the attitude towards community education has ranged from collective to individualistic approaches. The latter attitude puts the responsibility on the individual, usually or work-related skills. As a result, governments do not feel that they are responsible for continuing education, which then raises the issues of the digital divide and other inequities (Wamuyu, 2017). An informative example is Portugal's rise and fall of adult community education (Fragoso & Valadas, 2018). Coming out of a dictatorship in 1974, the country's civic organizations worked with community groups to advance education as part of direct democracy. However, in rebuilding a capitalist state, centralization cut down local reform. In 1979, the nation's government created a National Plan for Literacy and Basic Education of Adults, for which it was accountable by law; however, it failed because of lack of financial resources; again, regional efforts took up the slack and were supported by the European Council. More recently, the government has applied neoliberalism philosophy that put the educational impetus and responsibility on individuals, usually to advance their own careers; the reoccurrence of centralized education also marginalized community education.

One international significant initiative that has impacted community education is United Nation (UN)'s 2030 agenda for sustainable development goals (SDG), adopted in 2015. Its seventeen goals serve as pathways to make the planet sustainable economically, socially, and environmentally so that all people can reach their potential in a peaceful world. Goal 4 specifically focuses on inclusive and equitable lifelong education, but the overall vision asserts that universal literacy is needed to achieve the other goals, such as ending poverty, promoting well-being and healthy lives, and sustaining ecosystems. Moreover, the agenda also states that interconnected information and communications technology accelerates human progress and knowledge. Several community education projects use SDG as their framework (e.g., Holst et al, 2022; Kanwar, Balasubramanian & Carr, 2019; Wals, Mochizuki & Leicht, 2017).

More specifically, SDG goal #11 aims to make cities safe, inclusive, resilient and sustainable. One technology-centric strategy to meet that goal is the role of smart cities. A smart city is defined as one that

Balances economic, social, and environmental development, and if it links up to democratic processes through a participatory government. A smart city involves the implementation and deployment of information and communication technology (ICT) infrastructures to support social and urban growth through improving the economy, citizens' involvement and government efficiency. (Yeh, 2017, p. 556)

Some of the indicators of a smart city include water and air quality, renewable and efficient energy, green technology. In identifying factors that advance smart cities, Ismagiloiva et al. (2019) conducted a literature review. They found the IoT (Internet of Things) and cloud computing played important roles, and saw that monitoring the environment required expert knowledge. The researchers recommended more user-friendly systems to help inform residents and enable them to participate in decision-making and implementation. To make those recommendations a reality, community education, including technology aspects, are vital.

IMPACT OF TECHNOLOGY ON LEARNING

The realm of digital technology has substantially changed the face of learning. First, technology significantly expands and speeds up access to the world of information. Telecommunications has collapsed time and space. Moreover, people can respond to each other and share group information much more easily than in the past. The convergence of communication industries such as telephony and television further expanded one-way dissemination of information, and social media has greatly facilitated multi-way sharing of information – and misinformation. In addition, the production of information has been simplified and enhanced by numerous technology tools (e.g., photo editing, podcasting, screen casting) and mobile devices.

A useful way to examine the impact of technology on learning is to employ Schramm's 1954 communication cycle framework. In this model, information is first internally conceptualized and expressed externally. A communications channel such as radio or the Internet disseminates the information, which the audience then accesses and receives. At that point, the information is processed and perhaps acted upon, which is when learning can occur.

The nature of the information itself has been affected by digital technology. How does one express an idea? Depending on the available tools and the idea creator's technical skill, the idea is shaped by the medium's properties. Besides the obvious combination of text, image, and sound, digital technology facilitates the repurposing and transformation of information to address different objectives or different audiences. For that reason, the creator should consider of the interface between the information product and the user, which often supports That same interface also gives rise to linked, interactive and dynamic information.

In some cases, the creator of an idea needs to collaborate with another person in order to express the idea to the satisfaction of the creator. Perhaps the creator does not have access to the desired technology or the skill to use it. Book publishing is an obvious example where the author creator submits a manuscript for the publisher to polish and produce the book in desired formats. The creation expression process becomes much more sophisticated for films. Indeed, these examples constitute a subset information cycle.

The next stage in the information cycle is the dissemination of that information via some kind of communications channel. Older technological channels were largely mass media outlets such as radio and television stations along with film studio outlets; telecommunications systems tended to be one-to-one channels via telephones – with the form of expression being the voice. Electro-mechanic storage devices such as videotape, audiotape and records became ways to disseminate information more conveniently, although they required equipment to access the information itself. Digital technology greatly expanded options for existing communications channels and opened the world to the Internet. Social media was another breakthrough channel as it facilitated two-way and multi-way communication and sharing of information.

Access to these communication channels has also been impacted by digital technologies. On one hand, people have much greater potential access to information around the world. On the other hand, the continuing digital divide demonstrates the preconditions for access: appropriate equipment and Internet access. In that respect, the near ubiquity of mobile "smart" phones has lowered the barrier to digital access and the production of digital information.

The last stage of the information cycle involves human processing of the accessed information. Especially with the deluge of information and fewer intermediaries to pre-select and curate that information, more than ever, the information receiver needs to interpret the format of information as well as its content, and their interdependence. How those individuals deal with the information completes the cycle, informing others.

Instruction, then, constitutes one example of dealing with information and informing others to "feed" the cycle. As with the other stages, digital technology impacts the nature of instruction, highlighting the issue of standardization versus customization (Martinez, 2005). On one hand, technology enables instruction to be mass-produced and standardized. Course design and resources can be easily duplicated and disseminated. To a degree, grading can be done automatically once a quiz is set up. An online video conference can be scheduled for one time, with the understanding that learners anywhere in the world can meet simultaneously. On the other hand, with so many more resources available because of telecommunications and digitization, learners are more likely to find information that fits their particular needs, and educators can provide learners with more choices in what resource they use or how they can demonstrate knowledge. Technology-based instruction can be structured so that learners can self-pace their work. Instructors can develop automated needs assessment tools that direct the learner to the most appropriate resources and activities based on their responses. Likewise, instructors can store digital learning objects that can be accessed for a variety of applications across curricula, which can lead to a one-size-fits-all mentality or can support a learner's specific needs, depending on how those objects are used.

In the process of educating, technology can change the relationship of teacher and learner – and the nature of learning processes themselves (Wang, Fong, & Kwan, 2010). Because more resources can be accessed independently, and learners can access and manipulate those resources at any time, learners have the opportunity to be more independent. Because access and communication can be asynchronous but still timely (as opposed to traditional correspondence courses), learners may be expected to accomplish more because they have a more flexible window of opportunity. Essentially, learners are not constrained by class time. Indeed, the wall between class time and daily life becomes more porous conversations can continue non-stop if one so wishes. Moreover, the interactivity of the Internet, particularly social media tools, enables learners to relate to one another more fully. Group projects become more feasible because students do not have to meet at one place at one time. On the other hand, when trying to assess learning progress, instructors have to take additional measures to decipher how each person is contributing to the end result. The instructor's status is likely to shift from sage on the stage to guide on the side, providing an educational environment in which learners can explore.

In short, the incorporation of digital technology fosters more independent, learner-centered learning. Maier and Warren (2000) listed several technology-enhanced instructional design strategies that support such a model:

- Flexible learning through ubiquitous access to resources, including the instructor
- Resource-centered learning provided within a virtual environment, which enables learners to use tools to manipulate resources according to individual needs and interests
- Self-paced learning, unhampered by class time or one-shot lectures
- Collaborative learning using web-based tools to compare learning and to generate knowledge together
- Distance learning because course materials and structure can be accessed remotely.

THEORIES AND MODELS THAT IMPACT TECHNOLOGY IN COMMUNITY EDUCATION

Several social theories and models apply to technology-enhanced community education.

Bandura's 1996 social cognitive theory investigated causal relationships between personal behavior and events with environmental factors as they influence the adoption of new competencies and attitudes. The social aspect of the theory refers to the social original of thought and action, and the cognitive aspect of the theory reflects how the mind impacts affect and action. Bandura stated that self-efficacy beliefs helped individuals plan and optimize personal change. Collectively, people self-organize and self-regulate; they are producers as well as products of their environment, which can extend to virtual environments.

Social capital theory addresses the collective benefits of "social capital" assets such as family background, norms of reciprocity, community ties, shared vision, social networks, and status in a social space. Individuals with high social capital tend to contribute more to collective intelligence (Chiu, Hsu, & Wang, 2006).

Freire's theory of critical pedagogy (2017) advocated transformative change through individually constructed meaning, whereby people become more aware of their socio-cultural reality and their ability to transform it. More generally, the growing recognition of contextualized or situated learning points to the socio-cultural aspects of planning that are needed in order to affect change such as community improvement. Where communities can exist online, context takes on an addition dimension.

Generative learning theory asserts that individuals are active participants in the learning process, connecting presented information with their prior knowledge. In the process, they generate new understanding of information (Wittrock, 1991). Generative learning theory asserts three key processes are involved in learning: motivation processes, learning processes, and knowledge creation processes. The learner's interest and self-attribution of effort leads to motivation, which activates the learning process. The learner gives sustained attention to receive, analyze and store information. Next, the generative process coordinates and generates information, integrating, elaborating and reconceptualizing it. The activities that involve creating relations and meaning are considered generative learning strategies. Community educators should design experience that encourages such generative processes.

A Community of Practice (CoP) consists of a group of people with common values and goals. Unlike in formal education courses, most CoPs are voluntarily self-established and self-defined. However, other elements of CoP can apply to community educational settings: interdependent relationships as they learn from each other, a spectrum of knowledge and experience, and a social dimension (Lave & Wenger, 1991). When community educators design instruction with a CoP mind set, they set up an expectation that learners are active contributors to the community knowledge base. Learner expertise can serve as a key element in this process. Learners can locate, evaluate and selecting relevant community resources to add to the instructional material collection. Learners can create information resources to demonstrate their own content knowledge and skills as well as to add to contribute to the community body of knowledge, including local resource repositories.

One framework that can serve as a bridge between learning communities and technology-enhanced education is the Community of Inquiry framework (Thompson, Vogler & Xiu, 2017). This model frames online learning as a collaborative process supported by social presence, teaching presence, and cognitive presence. These presences work together to facilitate critical thinking and learning. The affordance of technology helps educators match community goals, promote critical thinking, and build a sense of community.

A similar model that captures the idea of community education expertise needed by educators is TPACK: Technological Pedagogical Content Knowledge (Koehler & Mishra, 2009). This model shows the kinds of knowledge needed for instruction in the Digital Age (see Figure 1). An example of intersecting CK and PK is teaching how to read nutrition labels, and an example of intersecting CK and TK is maintaining a database of daycare providers. An example of PK and TK is teaching how to download an ereader, and an example of intersecting three components is instruction on evaluating educational apps. These components together comprise a model for technology-enhanced instruction, which can include information, digital and media literacies. Ideally, community educators need to be content experts, know how to teach, and able to integrate technology. However, one person might know be competent in all three areas; If TPACK is applied to instruction within a course, theoretically several people could contribute in their areas of knowledge to the course. A good exercise is for educators to map their knowledge onto TPACK.

The transtheoretical stages of change model developed by Prochaska and DiClemente (1986) includes six stages of change, each having unique properties and strategies for transitioning to the next stage, which applies directly to community education, especially when designing learning experiences for people at different levels of awareness and implementation.

1. **Pre-Contemplation**. The individual is not considering change. He or she should be encouraged to self-explore and re-evaluate their current behavior. The person might need to confirm a lack of readiness to change. The educator introduces the new concept or technology to awake consciousness.
2. **Contemplation**. The individual is ambivalent about change, such as learning a new skill. The educator helps the individual to identify the pros and cons of behavior, and consider new expectations about the possible outcomes of change.
3. **Preparation**. The educator provides the individual with experiences with the new knowledge or tool, thus enabling the individual to test behaviors or attitudes. The educator can help such individuals locate social support, address obstacles, and take small steps for immediate success.
4. **Action**. The individual practices new behavior for 3-6 months. One needs to deal with feelings of loss, seek support from the educator or community, and concentrate on long-term benefits.
5. **Maintenance**. The individual is committed to sustaining the new behavior. He or she should feel comfortable discussing with the educator about ways to deal with possible regression, and internally reinforce sustained effort.

Figure 1. Intersections of technical, pedagogical, and content knowledge

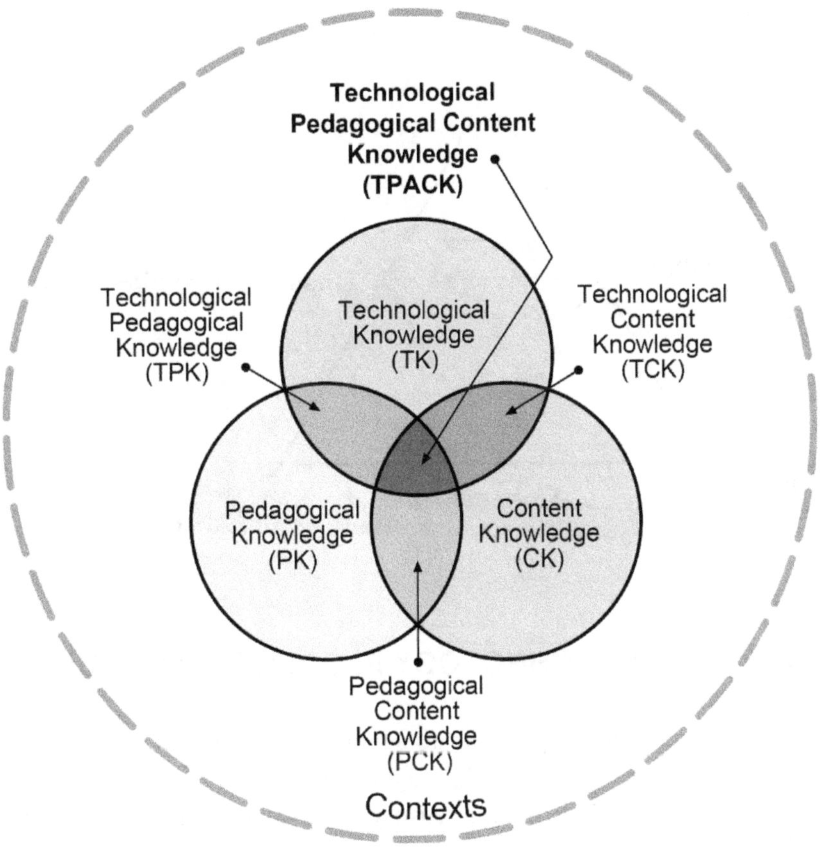

6. **Relapse**. The individual resumes old behavior. With the help of the educator, he or she should identify triggers for regression, evaluate incentives and obstacles, and plan stronger coping strategies.

Engeström's (1999) activity theory offers an effective instructional design structure for community-based problem-solving. Engeström's posited that persons actively engage with the environmental context using resources, such as technology tools, in order to reach a goal, which can include learning. The person acts according to rules, within a community or organization that determines the division of labor. For example, health literacy educators establish the conditions such that the person needs to work with the community effectively in order for the outcome to be achieved. As such, the educators need to address each component of the activity system. The goal should be engaging, such as a community-based problem or project to solve such as child malnutrition. The activity space needs to include the project context, the project object, and the space within which to manipulate objects and produce. The organizational context, including the rules and social interaction, help define the project and parameters; this aspect needs to consider local norms and expectations. The subjects (i.e., learners) also need to have the tools to be able to act upon the object to reach the goal; this part of the system benefits the most from technology, which can facilitate access to resources, optimize collaboration, and routinize some operations. One important aspect of activity theory is the interdependence of each component, so community educators need to apply systems thinking to support and make adaptations to ensure optimum outcomes (Aguayo, 2016).

Figure 2. Engeström's activity theory

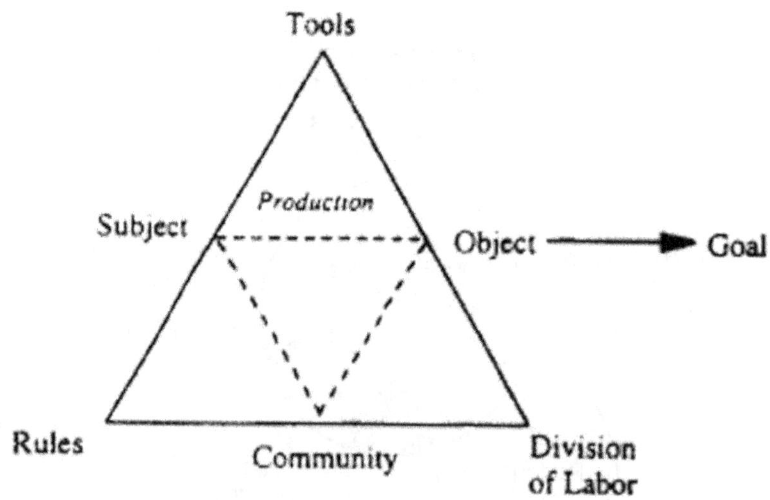

HISTORICAL TRENDS OF TECHNOLOGY IN COMMUNITY EDUCATION

When discussing instructional technology, educators and researchers tend to focus on the advances of the last hundred years, starting with mass media such as film, radio, and television. These nonprint or audio-visual technologies extended information sensory experience to encompass sound and movement. They also enabled mass delivery of instruction such as civic education and emergency preparedness training, which could be disseminated to communities across the nation. However, the last three decades have witnessed the greatest impact on communication, largely because of seminal developments: desktop computers (personal sized, easy-to-use hardware), productivity software (such as the spreadsheet program VisiCalc), and Ethernet for communication (including agreed upon transfer protocols), Several other recent technologies also may be considered disruptive; that is, they significantly change how people operate: broadband, which facilitates content streaming; global positioning systems (GPS), which facilitate logistics; virtual and augmented reality (XR), which simulate the real world; smart phones, which provide portable Internet access as well as other support digital-based functions including digital cameras; and social media, which facilitate user-generated content and multi-way communication.

Technology also exists within the larger framework of society at large. In his book *The Third Wave,* Toffler (1980) asserted that the U. S. was transitioning from an industrial society to a super- or post-industrial society. He noted the accelerated change of pace and people's sense of information overload. He thought that people saw themselves as consumers and producers, helping themselves to improve. The SCANS (Secretary's Commission on Achieving Necessary Skills) Report of the U. S. Department of Labor (1991) recommended five overarching competencies for the entire workforce: resource management, interpersonal skills, information use, systems understanding, and technology use. Specifically, workers should be able to select equipment and tools, apply technology appropriately to specific tasks, and maintain and troubleshoot equipment. By 2005, Friedman (2005) asserted that technology served as a catalyst for globalization, facilitating outsourcing and supply-chaining to streamline operations. Online collaboration and sharing have become ubiquitous at all levels, from convergent corporations to individuals using personal digital devices such as cell phones and iPods.

Instructional Changes

In the last thirty years, technology-enhanced community education has shifted from a top-down to broad-based basis, from one-directional to two- and multi-directional communication, from static to dynamic content, from lecturer-dependent to content-dependent resources, and from administrative- to learning-oriented focus. With the advent of social media, the conditions for a true learning community involving educators, learners, and practitioners can emerge, depending on each stakeholder's expertise and support.

As early as 1960, it was predicted that teachers and even the school system could be eliminated and replaced by educational technology developed alternative institutions. Such predictions have largely fallen by the wayside, but in 1979 the Association of Educational Communications and Technology contended that educational technology affected the organizational structure of education at the curricular level such that an array of facilitated learning could exist in a variety of institutions. Instructional design, delivery and assessment could change dramatically because of educational technology. In terms of community education, technology applications were more likely to occur in governmental initiatives that used mass media to educate their population (Jonassen, 2004).

In the 1980s video entered community education. Videotapes were attractive for distance learning because presentations could be recorded, duplicated, and disseminated at relatively low cost. The size of an LP record (an even older technology), videodiscs could hold 30-60 minutes of full-motion video or other information, and could support random access. Other low-tech instructional tools included audiocassettes, overhead projectors and lap-dissolve projection (i.e., two slide projectors). Instructors were expected to produce content that could be displayed with the appropriate equipment. Similarly, instructors developed content and used authoring languages to communicate with the student. (Chambers, 1982). Community educators used technology unsystematically, largely because they usually had to learn the technology themselves; their practice largely reflected financial and administrative support.

In the 1990s the term "learning organization" emerged, reflecting the need for organizations as a whole to develop skills and manage knowledge systematically. Instructional technology became more integrated in the 1990s, with desktop computers being able to connect peripherals such as CDs and link to the Internet. More flexibility in instructional delivery facilitated greater instructor-learner interaction, through digital forms of synchronous conferencing and asynchronous communication. Course management systems appeared, which provided for a coordinated approach to course content, communication, and management (Rossett, 2002). These systems were usually Web-based, which enabled users to communicate across computer platforms.

In the 21st century the Internet constituted the main technology content vehicle for community education. As speed and bandwidth improved, the learning community could access information in a wide variety of formats online from around the world. Developing instructional content also became easier with "natural" programming language and templates, robust productivity applications, and lower-cost equipment such as digital cameras and camcorders. Uploading and sharing content was also facilitated by the Web's graphical interface. As a result, more applicable learning aids such as tutorials and interactive simulations were available, feeding the production and training cycle.

Probably the most recent significant change in teaching with technology within community education has been the incorporation of social media. The term web 2.0 is an earlier term given to express the idea of interactivity that now marks web-based collaborative tools such as Facebook and Google Education. Rather than one-way communication, social media enables learners to participate in two-way (or multi-way) active communication to share and create information. In studying information management systems, JISC and SCONUL (2008) noted the impact of web 2.0 in concentrating the aggregation of information; and diffusing the dissemination and reuse of content. The comparatively easy access of the Internet further broadens the base of participation. Another key feature of social media and networking is collective intelligence, so that it can resemble mass media channels by creating knowledge at the same time that it represents interpersonal channels that influence attitudes of peers. Especially in today's digital world, community education should consider incorporating interactivity and social networking features, such as messaging and sharing options, incorporation of learning community repositories, enterprise mash-ups, and user-customizable subject tags.

At this point in history, instructional technology has become a mature tool. Instructional design aligns outcomes with student needs, resources, learning activities, and assessment. Course and training delivery ranges from one-time workshops anywhere/anytime to permanent learning communities. Courses may incorporate technology in class or online, with hybrid delivery systems constituting the preferable stance. COVID-19 served as a catalyst for pivoting to online education, which continues to influence decisions about educational delivery modalities. The underlying instructional question is: how does technology enhance community education? Community learners also make more choices as to their form and source of learning, often creating a collage of educational venues provided by a variety of content vendors.

CONDITIONS FOR TECHNOLOGY INCORPORATION INTO COMMUNITY EDUCATION

For technology to play a significant role in community education, several conditions have to be met (Kamssu, Siekpe, & Ellzy, 2004):

- Available and affordable IT (information technology) equipment and telecommunications service
- Convenient access to Internet-connected computers (or equivalent)
- Technology competence of education providers and learners
- Basic literacy skills
- Ideally, political and economic stability on several levels: nation, region, and community
- Technology management policies and procedures.

At the organizational level, technology integration depends on several factors:

- Sufficient material and human resources to provide technology enhanced education
- Organizational technology readiness
- Top management support
- Organizational culture that supports technology.

Space constitutes another factor that organizations need to consider when providing community education; technology adds another dimension to that factor in that it requires equipment and usually Internet connectivity. A typical arrangement is a community technology-enhanced learning center. In some cases, such centers are located within libraries that are considered the local information access center. In other cases, the emphasis is on formal educational settings or technology training. In the latter case, technology skills are taught for a variety of reasons: workplace skills, personal digital literacy, media literacy, basic literacy, family education or consumer activity (Nurhayati et al., 2021).

Learning About Technology, and Learning With Technology

In the digital age, communities need to gain technological competency; the vast majority of jobs involve technology, and information largely exists and is transmitted in digital form. Even face-to-face community education involves technology at some point, most probably in accessing information. As information and communication technology (ICT) literacy pertains to the ability to use digital technology and communication tools to deal with information and create knowledge, it is increasingly a prerequisite skill for access to, and participation in, much of community education. In those cases, where people are not ICT literate, then community education should focus on teaching such ICT skills.

Learning *about* technology differs from learning *with* technology. Particularly if the concepts to be learned involve the use of technology, that added skill needs to be addressed explicitly. For instance, if learners do not know how to create a wiki, then they will need to be opportunity -- and the time -- to learn and practice that skill so they can participate in collective intelligence work. Educators need to figure out how that learning will occur. Do participants have access to the hardware and software? Do they have the time to access and learn the application beyond the work day or in addition to daily life demands? Does instructional design need to include resources and explicit activities to address that technical learning?

It should be noted that even before community members engage in technology learning experiences, they must be able to physically access the relevant technology. For that reason, community educators should survey learner access to technologies (hardware, software, connectivity), and frontload the conditions for learning. Additionally, ways need to be found to ensure equitable access to these technologies; for instance, educators should be sensitive to populations who may special needs such as English language learners and the physically challenged.

- **Physical Access**. The educator should find out what public technology access is available: such as libraries, recreation centers, Internet cafes, universities. Instructional resources should be accessed by low-performance equipment and dial-up connectivity. Text needs to be saved in formats that can be read by open source software; .rtf files are usually an acceptable format. Unfortunately, while image-rich and multimedia documents take advantage of multiple learning styles, they also require broader bandwidth and may need plug-ins in order to play correctly. Furthermore, instructional materials should be accessible for individuals with disabilities.
- **Intellectual Access**. Educators may need to provide directions on the use of technology tools such as spreadsheets or authoring programs. In some cases, even simple operations such as attaching files need to be explained. Educators can help learners be self-sufficient learners by showing them how to get technical assistance through help screens, manuals, web FAQs, and online tutorials. While some learners are motivated to use technology, and will make the extra effort to

process information, others have a negative attitude toward technology that will impede information processing.

- **Building Community**. One of the main drawbacks of e-learning (i.e., digital-based learning) can be the feeling of isolation. With the advent of social media, learners can interact with their online peers to build a community of learners. Educators need to set up the conditions for interactive learning by designing opportunities for group discussion and collaborative learning activities. Community building also supports social-emotional learning and well-being. Threaded discussion forums, file sharing tools, cloud storage, and online conferencing are a few of the tools available. These community-based structures have to be supported through the online training of these tools.

Even with these general provisions, learning might not succeed because community members have differing pre-existing attitudes and experiences with technology, particularly if they are adult learners. Some people lack technology competence, or do not want to learn these skills. Some fear technology, thinking it might supplant valuable social or labor-intensive job skills. For some adults, technology itself poses a barrier to learning; if they cannot accept the online environment, learners will not be able to have physical and intellectual access to the information itself. Yocco at al. (2011) noted several factors that led to the public's use of technology in environmental education:

- **Personal Variables:** age, learning style, prior technology use
- **Perception of Technology Variables:** perceived usefulness, perceived ease of use, its attractiveness
- **Public Behavior:** intent to use, initial use, attrition or continued use.

At the very least, educators need to provide learners with a variety of instructional materials and learning activities to choose from. Relating to technology is affected by emotional states, so educators need to make sure that learners can feel safe in taking intellectual risks in trying technology, and are rewarded for their efforts as well as their final products (Rogers, 1962). To that end, learners should have opportunities to:

- Operate computer and other technological systems (i.e., open and close an application, save a document, print, use a mouse, use a menu and navigation bar, etc.)
- Use productivity tools
- Navigate the Internet (i.e., use browsers and search engines, use email, etc.)
- Evaluate information critically
- Communicate digitally.

INSTRUCTIONAL DESIGNER INCORPORATION OF TECHNOLOGY

How does technology-enhanced community education reflect the community? Technology can lead to standardized, one-size-fits-all instruction that ignore local interests and resources. Fear rightly exists that technology-based community education can homogenize cultures and downplay local cultural heritages. As noted before, instructional strategies should involve both cognitive and social aspects of the local community, reflecting effective learning principles.

Instructional Design Technology Decisions

Throughout the community education instructional design process, educators need to determine which technologies will be used – and to what extent – in light of ICT competency and community interests. At the minimum, instructors need to determine whether technology will be used as a tool to deliver instruction, as a learning aid, or as the outcome itself. Even the instructional focus, whether to emphasize knowledge and skills or problem-solving, requires careful consideration. Technology-based choices apply at each step in designing instruction, as shown here.

- **Assess Needs:** Online survey, video capture of behavior, analysis of website "hits"
- **Identify Learners:** Online solicitations, social media subscriptions, organizational lists
- **Identify Outcomes:** Technology standards, online content standards, online education syllabi
- **Identify Indicators:** Technology rubrics, technology products, electronic Delphi method (i.e., getting experts to come to consensus on key points)
- **Identify Prerequisite Skills:** Web-based assessment, performance observation, content analysis of sample work
- **Identify Content:** Online training documents, electronic journal articles, virtual chat
- **Identify the Instructional Format:**
    - Resources (Internet, DVDs, e-books)
    - Instructor (technician, instruction designer, remote-site expert)
    - Methods (web tutorial, hands-on manipulation, video presentation)
    - Timeframe (self-paced, real-time, just-in-time)
    - Location (computer lab, classroom, home)
    - Grouping (individual, online group, work unit)
    - Individualization (programmed instruction, choice of technology, coaching)
    - Affective domain (wiki, threaded discussion, virtual chat)
- **Contextualize Instruction:** Web page within a community website, community services database, digital images of applications
- **Implement the Plan:** Project planning software program, web page development, video conferencing
- **Assess the Plan:** Learning management system statistics, videotape, online survey.

Throughout the process, instructional design should build in learner choice in order to respect different paths to knowledge and contextual-sensitive learning.

Special attention needs to be given to the learning system interface. Major online learning systems usually provide standardized templates to structure content efficiently. Homegrown websites are less likely to have such organizational tools; however, they offer more opportunities to design interfaces and include information that is community-centered. In producing a self-directed educational website to address the local lake's environmental issues, Aguayo and Eames (2017) included several features: a broad range of content from basic to complex, local indigenous knowledge, sustainability issues to provide the big picture, guidelines and solutions about the lake issue, a question/answer section, use of multimedia to provide choice in ways to understand content, links to local organizations to facilitate partnerships, interactive elements to enhance engagement, and supporting literacies.

Representative Technology Tools

Thousands of technology tools may be leveraged for community education. Each type of technology tool lends itself to specific learning approaches. Whenever possible, the unique properties of each tool should be exploited to optimize these experiences. Educators also need to determine participants' level of digital literacy and Internet access capability to in order to incorporate tools that align with their situation. UNESCO (2022) identified tasks that indicate digital skill levels as follows: basic (e.g., keyboarding, email, word processing, managing privacy settings), intermediate (e.g., digital publishing, digital graphic design), and advanced (cybersecurity, big data, Internet of things, digital entrepreneurship, artificial intelligence).

Each of the following approaches fosters deep engagement with curriculum because the format lends itself well to packaging and communicating content. Each approach also fosters interactivity between learner and content, and facilitates collaboration. Learners can work with each other both in real time and asynchronously, anytime and anywhere. Here are several interactive tools that work well for community education.

- **Threaded Discussions.** Archive individual contributions and responses chronologically by subject.
- **Online Chats.** Such as LiveChat and Zendesk enable individuals to discuss issues in real time. Some chat programs include features that allow documents and web pages to be viewed in common.
- **Blogs.** (Web logs) such as BlogSpot and Blogger enable individuals to write about their experiences chronologically, and others can comment on the entries.
- **Social Networking and Microblog Platforms.** Such as Twitter and Mastodon allow users to post, receive and follow short messages; they can serve as useful "back channels" during and after group meetings.
- **Online Collaborative Boards.** Such as whiteboards, Jamboard and Padlet enable learners to post brainstorm and other ideas simultaneously.
- **Social Bookmarking.** (Public lists of recommended websites) such as Reddit and StumbleUpon help build sources of peer-reviewed information.
- **Group Writing Tools.** Such as with Google Docs and LibreOffice allows learners to share and edit documents using a suite of productivity tools.
- **Collaborate Website Builders.** Such as Weebly and WordPress offer a platform for group-generated websites to share in-depth information for the community. These platforms used to be called wiki (as in the Hawaiian term "wiki-wiki," which means fast) because they do not require contributors to know markup language. Such community-built websites serve many purposes: group projects for which each person can focus on one aspect for one web page; point-of-view studies where an issue can be investigated from different stakeholders' perspectives; literature circles in which each learner can lead a discussion on some aspect of a book; document commentary for which learns can post comments about a text of image; case studies that learners can develop collaboratively or can each share their perspectives on an existing case study; and knowledge management websites to organize collective intelligence.

- **Audio Files.** Archive lectures and shorter tutorials, and learners can listen to the audio files at their convenience for as many times as they wish; Audio file software such as Audacity and Wavosaur can capture and edit audio files. These days, the most common audio files, which are broadcast, are called podcasts.
- **Video.** Captures movement and processes, and records events as they happen. Communities can document local history, and demonstrate important procedures, such as emergency measures. Platforms such as YouTube, Vimeo and TikTok enable the community to share their videos. Screencasting is a subset of video capturing, which creates videos of computer screen information; representative tools are Camtasia, Loom and Screen-o-Matic.
- **Multimedia.** Programs such as Voicethread enable users to develop, share and comment on online presentations.
- **Image Sharing.** Programs such as Instagram and Flickr enable users to upload and share photos and other images; these items can be combined into albums that reflect content matter learned.
- **Polling Tools.** Such as Poll Everywhere and Poll Maker enable educators and learners to get instant feedback from their target group; this tool is particularly useful for baseline data and checks for understanding.
- **Survey Tools.** Such as Survey Monkey and Google Forms enable learners to gather and analyze information about the community.
- **Serious (Educational) Games.** Such as iCivics and Extreme Event actively engage learners; they provide both textual and visual cues, often require collaboration in order to accomplish a task, usually demand clear communication, provide immediate feedback, foster attention to detail, and facilitate problem-solving skills, (Gros, 2003). Gamification in general has become an engaging way to learn problem-solving skills, probably as a result of more adults playing electronic games.
- **Web Conferencing.** Tools such as Zoom and GoToMeeting incorporate text, image and sound to more closely approximate physical interaction.
- **Virtual Environments.** (Object-oriented online environments) enable individuals to interact virtually using avatars. Virtual environments usually have ways for applications to be embedded and documents to be stored. Second Life was an early entry into this field; IMVU and Active Worlds are more current alternative platforms.
- **Online Team Management Tools.** Such as Trello, Slack and Bascamp enable groups to plan and implement projects; these tools usually include communication, timeline, and shared document features.
- **Cloud Storage Services.** Such as Good Drive, DropBox and OneDrive enable individuals to store and share all kinds of documents securely with designated people.

CURRENT PRACTICES IN TECHNOLOGY'S ROLE IN COMMUNITY EDUCATION

Communities around the world are incorporating technology into non-formal education. Successful efforts tend to include these factors: local needs, local resources, local broad-based planning, partnerships among organizations, accessible and appropriate technologies, opportunities for community building and socializing, and locally meaningful content and outcomes. More specifically, effective technology use in community education speeds up learning, increases educators' productivity, provides more choice in content and assessment modalities, increases the potential for more differentiated and personalized learning, and facilitates learning analytics to optimize teaching and learning (Mursid, Saragih & Sitompul, 2022). To illustrate how technology-enhanced community education can impact the community as a whole and its residents, two examples are detailed here: health literacy and women's empowerment.

Health Literacy Education

UNESCO has a long-term interest in community health literacy, which is also reflected in the UN's SDG goal #3 to ensure well-being and healthy lives for all. To that end, the World Health Organization defined health literacy as entailing "people's knowledge, motivation and competences to access, understand, appraise and apply health information to make judgements and take decisions in everyday life concerning health care, disease prevention and health promotions to maintain or improve quality of life" (Kickbusch et al., 2013, p. 4).

Providing community health education requires several components: a community-based health literacy framework and action plan, committed community partners and leaders, personnel who have health literacy and training skills, curated resources that align with the target audience, organizational infrastructure, and community-based communication channels such as social media and other media outlets (including schools and social groups) (Bensley & Brookins-Fisher, 2018). For example, in 2018 the American Library Association partnered with the National Network of Libraries of Medicine to produce a Libraries Transform Health Literacy Toolkit to help libraries provide trusted health information and facilitate community health education.

Fischer et al. (2014) conducted a literature review to identify several barriers to accepting and using health information, focusing on older individuals. The researchers found that elders may be unfamiliar with some technology, unwilling to ask for technical help, distrustful of technology, and be concerned about privacy issues. To address these barriers, the researchers recommended learner-sensitive instructional design, digital literacy education, research on effective practices, and inclusive policy.

In their research, Bensley and Brookins-Fisher (2018) also focused on community health educator education. They noted the benefits of having local community members serve as health educators, but stated that these workers needed the following competencies in order to be effective: multicultural literacy, ability to create an inclusive learning environment, presentation and facilitation skills, instructional design ability, product creation skills, knowledge of disabilities, and group management skills. Burhansstipanov et al. (2014) detailed their successful training efforts of community health educators, which included meeting quarterly and sharing lessons about training, content curation, workshop logistics, problem-solving, and building on community needs. Wilson et al. (2020) also focused on community educator education, and noted the barriers of transformation needs, lack of patient care supplies, and noncompliance by care givers. Their training included health information, technology us, and communication with hospitals.

In guiding learning cities, the World Health Organization (WHO) (2022) prescribed a three-step process to develop a policy for lifelong learning for health: developing awareness of a city's role in health literacy, developing supporting policies, and putting those policies into action. Several data points about the city are needed to proceed: resident demographics and epidemiology, the city's historic and current situation, the city's physical environment, the city's living conditions (e.g., utilities, housing, food, pets), the city's infrastructure (e.g., zoning, transposition, telecommunication), health care, budget, and legislation. WHO also identified factors for successful health literacy policies and their implementation: local political control; a coordinated formal, informal and non-formal education system; incorporation of several types of sites and modalities for learning; effective communication in terms of health information in relevant languages; capacity building of non-formal educators and other community members; broad participation and collaboration; attention to cultural and contextual factors; commitment to inclusion and empowerment; a health equity agenda for social justice; and research on impactful interventions. Cities need to support such efforts by facilitating resource allocations, mobilizing stakeholders, and creating platforms for communication and collaborations. WHO also encouraged citizenship for health, where cities provide its community members the knowledge, skills and capabilities to participate in health-related decisions and actions.

The following examples demonstrated how the above recommendations could be successfully implemented. Higa, Davidson and Loos (2021) focused on factors for successful diabetes self-management education. Community health services developed telehealth education using mobile technology, building on community strengths. For instance, they activated community social capital: a sense of community and trust, common values, leveraging of reciprocal practices, and friend/family support. They also customized education, providing participant options. Rural communities in Tanzania gained specific knowledge of HIV/AIDS and other diseases through simple short basic animated videos that could be viewed on tablets, coupled with free access to community information hotspots with an integrated digital health education platform (Holst et al, 2022). Similarly, American rural communities learned how to self-manage their diabetes through mobile apps that were customized according to individuals' needs; local school of nursing and engineers developed the interactive modules, which included information about nutrition, medications, exercises, goal-setting, and self-monitoring for complications with accompanying interventions (Hunt, Henderson & Chapman, 2018). Mobile apps for cancer education were also developed by local health workers, based on Malay community use of m-technology (Marzuki et al, 2019). Also addressing cancer, focusing on cancer, community public health and video experts developed and broadcast eleven short social media video scenarios for people with low health literacy (Valera et al., 2021).

Women's Empowerment

Another UN Sustainable Development Goal is gender equality (#3), supported by SDG #4 quality education and SDG #8 decent work and economic growth. Women's status is impacted by both legislation and social expectations and norms. In many parts of the world, women's status is tied to community participation and power. Community education helps women increase access to information, assess their own life situation, meet people, communicate more effectively, improve social practice, get and give support, gain self-confidence and agency, manage themselves and their homes more effectively, gain marketable skills, build solidarity, and participate more in decision-making (Leahy, 2016; Mengistie, 2022). With the recent advent of women's activism, such as the #metoo movement and reproductive justice, feminist community education has increased, and has focused more on practicalities than academics (Aitken, 2017). Such feminist andragogy starts with the individuals critically reframes issues, examines and addresses power structures, considers gender contexts, shares and validates personal experiences as legitimate sources of knowledge, and challenges public-private dichotomy. Topics range from identity politics and voting, to challenging patriarchies and sexism as well as economic equity and social justice.

General education for women remains viable and valued. Functional literacy programs such as reading, writing, and practical mathematics continue to be needed. Practical knowledge such as parenting, health practices, and agriculture are also popular. Learning about technology and learning with technology is also increasing (Megistie, 2022).

When planning communication education for women, developers should be aware of women's preference in their educational experiences. In researching gender differences in self-directed learning practices, Grover and Miller (2014) discovered that women were more likely than men to consult more options before deciding which educational opportunities to attend. They also used selected resources and suggested practices than men. Men are more interested in credential-linked formal classes than women, while women were more likely than men to join an organization more for interacting than for learning. In turn, Leahy (2016) identified factors that foster women's participation in education: a safe, open, friendly and positive environment; educators who are caring, encouraging, understanding, and non-judgmental; sensitivity to home and other demands; peer support and friendship; and childcare. Likewise, in terms of educational strategies, O'Grady (2018) found that women valued making connections and doing group work, including self-help groups. They also appreciated interacting in an environment that transgresses dominant culture norms and nurtures their own social capital.

Pamphilon and Mikhailovich (2017) discussed the benefits of gender-inclusive community education, built on the experience of farm families in Papua New Guinea. Prior to the training program, males tended to make all the decisions and outside experts gave one-way presentations. Program planning was culturally appropriate, and built local capacity by using a train-the-trainer approach with local peers who gained facilitation and andragogy skills. Participants consisted of coed family teams, and the education was asset-based to build community expertise. The educational experience consisted largely of group dialogue, social learning and locally-situated analysis. For instance, participants shared their own best practices, which could then be consulted after the training. As a result of the program, women became more skilled and participated more in farming and family decisions.

FUTURE TRENDS IN TECHNOLOGY'S FUTURE ROLE IN COMMUNITY EDUCATION

UNESCO's sustainable development goals, governmental situations, and globalization will likely frame future technology-enhanced community education. The trajectory of technology may drive how community education is implemented. Several possibilities are projected: a wider range and use of non-textual matter, greater use of XR (that is, augmented and virtual reality), greater use of technology to teach soft skills, more specific training and micro-credentials. Three additional trends are detailed here.

Cross-Cultural Community Education

Globalization has led to more cross-fertilization of technology integration practice. With the rapid pace of technology advancement and consequential change, communities need to constantly learn in order to maintain their societal "edge." With today's instant telecommunications, organizations and other entities have risen in number, and can significantly impact technology integration advances. Web-based communities of practice can exist anywhere, crossing national borders almost seamlessly. In addition, technology-based translation tools facilitate comprehension of information in foreign languages.

Therefore, cultural literacy is increasingly important. Not only should community educators design curriculum that is sensitive to the community's own cultural values and norms, but with cross-cultural access to information and other communities, educators need to incorporate cultural literacy into the curriculum content itself. Multicultural education elements distilled from Lowenstein (2009) include:

- Cultural pluralism theory, commitment, and affirmation
- Knowledge and identity construction from cultural assumptions and social positions
- Critical awareness
- Social and educational issues facing minorities
- Prejudice reduction, social justice, and equity
- Reconstruction of social structures to facilitate individual and group empowerment
- Skills and strategies for addressing multiculturalism.

Collective Intelligence

Collective intelligence may be defined as the ability of a group to think, learn and create collectively. Collective intelligence serves as a model of the adage "the whole is greater than the sum of its parts." Collective intelligence is distinguished from collaboration in that a specific goal is identified, processes of interaction are aligned with that goal, and decisions are made as a unified group.

While the idea of collective intelligence is old, the incorporation of digital technology has transformed it because collective intelligence can now transcend time and space to speed up access and processing, and it can be scaled up to much larger groups. Furthermore, the expanded repertoire of digital tools, including the use of social media and cloud technologies, offers new ways to interact and generate collective intelligence, be it in the representation, conveyance processing, or management of information. For example, automated agents can classify and filter participants' input, thus facilitating expert validation and organizing of results to offer new collective intelligence. In issue-based information systems (IBIS), participants suggest contributors to identified problems, and pose arguments in favor or challenging those factors. Virtual voting leads to ideas based on several contributions, and challenges can lead to decision paths to test commitment, both of which lead to identifying promising solutions. These collective intelligence deliberation platforms provide argumentation feedback, analytics, and visualizations to improve collection reflection and decision-making (Shum, De Liddo and Klein, 2014).

To enhance community disaster resilience, Kankanamge, Yigitcanlar Goonetilleke (2021) used crowdsourcing to identify ways to manage natural disasters. Using relevant GPS data, the research developed an artificial intelligence game to collect community information. The crowdsourcing raised community awareness of the program, facilitated community sharing of information, and leveraged the responses to advance disaster planning. Interestingly, their government looked askance at crowdsourcing because they were afraid that it would result in mission information and sustainability problems.

Artificial Intelligence (AI)

Technological advances have led to many artificial intelligence (AI) applications: from virtual assistance agents or "smart speakers" such as Siri and Alexa to recommendation systems for consumerism and healthcare. AI is also known for facial recognition that may profile people unjustly, bots that swarm on social media, and deep fake videos. In community education, AI is like to increase its presence as a subject matter as well as an instructional and learning tool. For instance, AI tutoring systems can be used to give personalize instruction and assessment by branching to appropriate content based on user response. AI-enhanced robots are already used as teaching assistants or learner support. AI can also be used as a management tool to efficiently keep records of – and analyze – individuals' educational experiences and behaviors; technically, all the participant data in a community education program could be analyzed leveraging AI in order to determine group patterns and gaps, along with recommended targeted interventions (Zhang & Aslan, 2021).

CONCLUSION

Technology, particularly as it supports interactive learning communities, now plays a central role in community education. While technology affords access to information more efficiently than ever before, it also requires additional skills. Successful community efforts to incorporate technology tend to include these factors: local needs, local resources, local broad-based planning, partnerships among organizations, accessible and appropriate technologies, opportunities for community building and socializing, and locally meaningful content and outcomes. As such, technology-enhanced community education demonstrates unique characteristics that fit the community.

For technology to play out its role effectively, community educators need to design instruction, curriculum, and programs that offer customizable learning and socially-rich learning environments. Technology should be incorporated in meaningful and effective ways that match learning objectives and learner needs. Every effort needs to be made to make sure that learners have opportunities to access and learn with technology, and that they can choose the type of learning that best meets their needs. To that end, organizations should ensure that a robust and equitable technology infrastructure is in place, and that training be appropriated supported as a right of every citizen.

Some of the current trends to foster technology integration while paying attention to the affective domain include:

- Enriching the repertoire of instructional strategies, including interactive online tutorials, gamification, and community-based problem-based learning
- Developing communities of practice and peer-to-peer learning
- Creating databases and repositories to facilitate community-supported learning.

REFERENCES

Aguayo, C. (2016). Activity theory and online community education for sustainability: When systems meet reality. In D. Gedera & J. Williams (Eds.), *Activity theory in education* (pp. 139–151). Brill. doi:10.1007/978-94-6300-387-2_9

Aguayo, C., & Eames, C. (2017). Promoting community socio-ecological sustainability through technology: A case study from Chile. *International Review of Education, 63*(6), 871–895. doi:10.100711159-017-9685-7

Aitken, M. (2017). Feminism: A fourth to be reckoned with? Reviving community education feminist pedagogies in a digital age. *Concept, 8*(1), 1–18.

American Library Association & National Network of Libraries of Medicine. (2018). *Libraries transform health literacy toolkit*. American Library Association. https://ilovelibraries.org/librariestransform/health-literacy-toolkit/

Bandura, A. (1986). *Social foundations of thought and action: A social cognitive theory*. Prentice-Hall.

Bensley, R., & Brookins-Fisher, J. (2018). *Community and public health education methods a practical guide* (4th ed.). Jones & Bartlett Learning.

Burhansstipanov, L., Krebs, L. U., Harjo, L., Watanabe-Galloway, S., Pingatore, N., Isham, D., Duran, F. T., Denny, L., Lindstrom, D., & Crawford, K. (2014). Providing community education: Lessons learned from native patient navigators. *Journal of Cancer Education*, *29*(3), 596–606. doi:10.100713187-014-0690-2 PMID:25087698

Chambers, G. (1982). A model for the development of technical instructors. *Master Abstracts International*, *20*(4), 342.

Chiu, C., Hsu, M., & Wang, E. (2006). Understanding knowledge sharing in virtual communities: An integration of social capital and social cognitive theories. *Decision Support Systems*, *42*(3), 1872–1888. doi:10.1016/j.dss.2006.04.001

Decker, L. (1992). *Community education: Building learning communities* (Rev. ed.). Community Education Association.

Engeström, Y. (1999). Activity theory and individual and social transformation. *Perspectives on Activity Theory*, *19*(38), 19–30. doi:10.1017/CBO9780511812774.003

Fellenz, R., & Coker, L. (1979). Toward a definition of adult education, community education and community development. In P. Cunningham (Ed.), *Yearbook of adult and continuing education* (pp. 366–369). Marquis Academic Media.

Fischer, S., David, D., Crotty, B. H., Dierks, M., & Safran, C. (2014). Acceptance and use of health information technology by community-dwelling elders. *International Journal of Medical Informatics*, *83*(9), 624–635. doi:10.1016/j.ijmedinf.2014.06.005 PMID:24996581

Fragoso, A., & Valadas, T. (2018). The rise and fall of adult community education in Portugal. *Social Sciences*, *7*(11), 239. doi:10.3390ocsci7110239

Freire, P. (2007). Pedagogy of the oppressed. *Continuum*.

Friedman, T. (2005). *The world is flat*. Farrar, Straus and Giroux.

Gros, B. (2003). The impact of digital games in education. *First Monday*, *8*(7). https://firstmonday.org/issues/issue8_7/gros/index.htm

Grover, K., & Miller, M. (2014). Gender differences in self-directed learning practices among community members. *PAACE Journal of Lifelong Learning*, *23*, 19–31.

Higa, C., Davidson, E., & Loos, J. (2021). Integrating family and friend support, information technology, and diabetes education in community-centric diabetes self-management. *Journal of the American Medical Informatics Association : JAMIA*, *28*(2), 261–275. doi:10.1093/jamia/ocaa223 PMID:33164074

Holst, C., Stelzle, D., Diep, L., Sukums, F., Ngowi, B., Noll, J., & Winkler, A. (2022). Improving health knowledge through provision of free digital health education to rural communities in Iringa, Tanzania: Nonrandomized intervention study. *Journal of Medical Internet Research, 24*(7), e37666. doi:10.2196/37666 PMID:35900820

Hunt, C. W., Henderson, K., & Chapman, R. (2018). Using technology to provide diabetes education for rural communities. *Online Journal of Rural Nursing and Health Care : the Official Journal of the Rural Nurse Organization, 18*(2), 134–151. doi:10.14574/ojrnhc.v18i2.525

Ismagiloiva, E., Hughes, L., Rana, N., & Dwivedi, Y. (2019, June). Role of smart cities in creating sustainable cities and communities: a systematic literature review. In *International Working Conference on Transfer and Diffusion of IT* (pp. 311-324). Springer.10.1007/978-3-030-20671-0_21

JISC & SCONUL. (2008). *Library management systems study*. Sheffield, UK: Sero Consulting. 10.1007/978-3-030-20671-0_21

Jonassen, D. (Ed.). (2004). *Handbook of research on educational communications and technology* (2nd ed.). Lawrence Erlnbaum Associates.

Kamssu, A., Siekpe, J., & Ellzy, J. (2004). Shortcomings to globalization: Using Internet technology and electronic commerce in developing countries. *Journal of Developing Areas, 38*(1), 151–169. doi:10.1353/jda.2005.0010

Kankanamge, N., Yigitcanlar, T., & Goonetilleke, A. (2021). Public perceptions on artificial intelligence driven disaster management: Evidence from Sydney, Melbourne and Brisbane. *Telematics and Informatics, 65*, 101729. doi:10.1016/j.tele.2021.101729

Kanwar, A., Balasubramanian, K., & Carr, A. (2019). Changing the TVET paradigm: New models for lifelong learning. *International Journal of Training Research, 17*(sup1), 54–68. doi:10.1080/14480220.2019.1629722

Kickbusch, I., Pelikan, J., Apfel, F., & Tsouros, A. (Eds.). (2013). *Health literacy*. World Health Organization.

Koehler, M. J., & Mishra, P. (2009). What is technological pedagogical content knowledge? *Contemporary Issues in Technology & Teacher Education, 9*(1), 60–70.

Lave, J., & Wenger, E. (1991). *Situated learning: Legitimate peripheral participation*. Cambridge University Press. doi:10.1017/CBO9780511815355

Leahy, J. (2016). *The impact of adult and community education on women's lives* [Unpublished thesis]. University of Canterbury.

Lowenstein, K. (2009). The work of multicultural teacher education: Reconceptualizing white teacher candidates as learners. *Review of Educational Research, 79*(1), 163–196. doi:10.3102/0034654308326161

Maber, E. (2016). Finding feminism, finding voice? Mobilising community education to build women's participation in Myanmar's political transition. *Gender and Education, 28*(3), 416–430. doi:10.1080/09540253.2016.1167175

Maier, P., & Warren, A. (2013). *Integrating technology in learning and teaching*. Kogan Page. doi:10.4324/9781315042299

Martinez, M. (2005). Mass customization: Designing for successful learning. *International Journal of Educational Technology*, *2*(2). http://www.ed.uiuc.edu/ijet/v2n2/martinez/index.html

Marzuki, M., Yaacob, N., bin Yaacob, N., Hassan, M., & Ahmad, S. (2019). Usable mobile app for community education on colorectal cancer: development process and usability study. *JMIR Human Factors, 6*(2). https://humanfactors.jmir.org/2019/2/e12103/

Mengistie, A. (2022). Exploring women's agency through adult education program in Ethiopia. *Cogent Social Sciences*, *8*(1), 2073672. doi:10.1080/23311886.2022.2073672

Minzey, J. (1972). Community education: An amalgam of many views. *Phi Delta Kappan*, *54*(3), 150–153.

Mursid, R., Saragih, A., & Sitompul, H. (2022). *The role of educational technology in learning innovation for community education development.* Paper presented at ICONSEIR 2021, Medan, Indonesia. 10.4108/eai.21-12-2021.2317244

Nurhayati, S., Musa, S., Boriboon, G., Nuraeni, R., & Putri, S. (2021). Community learning center efforts to improve information literacy in the community for cyber-crime prevention during a pandemic. *Journal of Nonformal Education*, *7*(1), 32–38. doi:10.15294/jne.v7i1.26883

O'Grady, M. (2018). An institutional ethnography of a feminist organization: a study of community education in Ireland. *European Journal for Research on the Education and Learning of Adults, 9*(1), 29-44.

Pamphilon, B., & Mikhailovich, K. (2017). Bringing together learning from two worlds: Lessons from a gender-inclusive community education approach with smallholder farmers in Papua New Guinea. *Australian Journal of Adult Learning*, *57*(2), 172–196.

Prochaska, J., & DiClemente, C. (1986). Toward a comprehensive model of change. In W. Miller & N. Heather (Eds.), *Treating addictive behaviors: processes of change* (pp. 3–27). Plenum Press. doi:10.1007/978-1-4613-2191-0_1

Rogers, E. (1962). *Diffusion of innovation*. Free Press.

Rossett, A. (Ed.). (2002). *ASTD e-learning handbook*. McGraw-Hill.

Schramm, W. (1954). *The process and effects of mass communication*. University of Illinois Press.

Scottish Community Education Council. (1990). *CeVe Scotland: Preservice training for community education work*. Scottish Community Education Council.

Shum, S., De Liddo, A., & Klein, M. (2014). *DCLA meet CIDA: Collective intelligence deliberation analytics*. Paper presented at 2nd International Workshop on Discourse-Centric Learning Analytics, LAK14: 4th International Conference on Learning Analytics & Knowledge, Indianapolis, IN.

Thompson, P., Vogler, J. S., & Xiu, Y. (2017). Strategic tooling: Technology for constructing a community of inquiry. *Journal of Educators Online*, *14*(2), n2. doi:10.9743/jeo.2017.14.2.10

Toffler, A. (1980). *The third wave*. Bantam Books.

U. S. Department of Labor. (1991). *Secretary's Commission on Achieving Necessary Skills*. Government Printing Office.

UNESCO. (2022). *Making lifelong learning a reality: A handbook*. UNESCO.

United Nations. (2015). *Agenda for sustainable development*. United Nations.

Valera, P., Acuna, N., Alzate-Duque, L., Liang, L. E., Cupertino, P., & Merulla, J. (2021). The development and prototype feedback of digital cancer 101 videos to enhance cancer education for marginalized communities with limited health literacy. *Cancer Control*, *28*, 1–10. doi:10.1177/10732748211006055 PMID:33784858

Wals, A., Mochizuki, Y., & Leicht, A. (2017). Critical case-studies of non-formal and community learning for sustainable development. *International Review of Education*, *63*(6), 783–792. Advance online publication. doi:10.100711159-017-9691-9

Wamuyu, P. (2017). Bridging the digital divide among low income urban communities. Leveraging use of community technology centers. *Telematics and Informatics*, *34*(8), 1709–1720. doi:10.1016/j.tele.2017.08.004

Wang, F., Fong, J., & Kwan, R. R. (2010). Handbook of research on hybrid learning models: Advanced tools, technologies, and applications. Information Science Reference.

Wilson, E., Lee, L., Klas, R., & Nesbit, K. (2020). Technology and rehabilitation training for community health workers: Strengthening health systems in Malawi. *Health & Social Care in the Community*, *28*(3), 833–841. doi:10.1111/hsc.12914 PMID:31808218

Wisconsin Department of Public Instruction. (2008). *Wisconsin's components of communication education*. Wisconsin Department of Public Instruction.

Wittrock, M. (1991). Generative teaching of comprehension. *The Elementary School Journal*, *92*(2), 167–182. doi:10.1086/461686

World Health Organization. (2022). *Enactment guide for the Yeonsu declaration for learning cities*. World Health Organization.

Yeh, J. (2017). The effects of successful ICT-based smart city services: From citizens' perspectives. *Government Information Quarterly*, *34*(3), 556–565. doi:10.1016/j.giq.2017.05.001

Yocco, V., Danter, E., Heimlich, J., Dunckel, B., & Myers, C. (2011). Exploring use of new media in environmental education contexts: Introducing visitors' technology use in zoos model. *Environmental Education Research*, *17*(6), 801–814. doi:10.1080/13504622.2011.620700

Zhang, K., & Aslan, A. (2021). AI technologies for education: Recent research & future directions. *Computers and Education: Artificial Intelligence*, *2*, 100025. doi:10.1016/j.caeai.2021.100025

KEY TERMS AND DEFINITIONS

Collective Intelligence: The ability of a group to think, learn and create collectively.

Community Education: The process of identifying community needs and coordinating resources to provide social and educational programs that meet those needs.

Digital Literacy: The ability to use technology to access, evaluate, use, and generate information effectively and responsibly.

Health Literacy: The ability to access, understand, appraise, and apply health information to make decisions about health maintenance and health care.

Information Literacy: The ability to locate, access, evaluate, use, manage, and create information effectively and responsibly.

Instructional Design: The creation of learning experiences and materials resulting in the acquisition and application of knowledge and skills.

Sustainable Development Goals: A collection of 17 interlinked global goals developed by United Nations a serve as a global framework for current and future peace and prosperity for people and the planet.

Chapter 20
Examining Adult Learning Assumptions and Theories in Technology-Infused Communities and Professions

Judith E. Parker
Teachers College, Columbia University, USA

Sandra L Williamson-Leadley
University of Canterbury, New Zealand

ABSTRACT

Adult learning in today's society is primarily focused on the needs of adults as they function in their communities and professions. Adult learning principles and theories have been the foundation of adult education practice for over a century. However, a review of the historical context reveals that the technology-infused learning landscape of today is dramatically different from its predecessor. This makes a reexamination of these ideas important for their future application. This chapter will reexamine the adult learning principles of transformative learning, andragogy, and critical reflection in this technology-infused world and propose a new paradigm and corresponding practices for the new learning landscape in communities and professions.

INTRODUCTION

Adults in today's society are primarily members of two groups: their community and their profession. Community education brings images of non-profit, interest/hobby classes and social service organizations while professional development suggests more formal classroom settings and work-related. However, the effective and successful participation of adults within these various groups requires continuous learning in diverse venues.

DOI: 10.4018/978-1-6684-7832-5.ch020

Technology's influence is pervasive in adult education – with what they do, how they do it when they do it, and how they continually learn to accomplish their personal and professional goals. While technology is often treated as a piece of equipment or a methodology, in fact, it has become infused into everyone's actions and thoughts. It is not technology in the world but technology as a part of the world (Parker, 2010). As people stare at personal handheld devices (smart phones or tablets) or LED displays in public places, for example, Manhattan's Times Square or Tokyo's Shibuya Crossing, it is easy to take for granted the access to up-to-the-minute information about where friends and family are, what is happening in the global news, or when the next train will arrive. Technology has changed the approach to communication, and searching for information, and reshaped the thinking process. Adapting to this technological environment provides a new idea of space, time, community, and society. In the late 1990s, Loader (1998) noted that numerous social scientists shared the idea that "society is being transformed by a revolution in information technology which is creating an entirely new social structure" (p. 4). However, with this 'information revolution' and 'new social structure' also came the challenges created by the 'digital divide' (Zdjelar & Zadjela Hrustek, 2021). These challenges include access, affordability, and an individual's ability to use digital technology; all of which are essential aspects in the creating of an information/knowledge society (Bornman, 2015).

The objective of this chapter is to examine several traditional assumptions and theories that have shaped adult learning practice for over a century. However, they will be examined through the technology-infused lives of today's adult learners. Technology is not just a tool or a lens. As the term infusion implies, technology has permeated the very essence of the theories; they may appear different and offer new challenges and opportunities to both the theorist and the practitioner.

Background/Theoretical Framework

Historically, adult education was practiced before its theories were developed. Now that technology-enhanced learning has been practiced for several decades, it seems timely to examine the basic principles of adult learning in light of this practice. This chapter will reexamine three ideas that are key elements in the theory and practice of adult learning in community education and professional development: Mezirow's transformative learning; Knowles' self-directedness; and, Brookfield's critical reflection.

Mezirow's Transformative Learning

Dr. Jack Mezirow, an Emeritus Professor at Teachers College/Columbia University, first proposed his theory of transformative learning in the 1980s. The theory's hallmark of a disorienting dilemma as the source for reevaluating one's assumptions and progressing to learning has been a much-debated issue over the decades. Was it a singular event or a series of events that caused the dilemma? Mezirow (2000) defines learning as the 'process of using a proper interpretation to construe a new or a revised interpretation of the meaning of one's experience in order to guide future action (p.5). The part that experience plays in Adult learning has long been seen as a "fundamental role" (Merriam & Baumgartner, 2020, p.197). Whether in communities or professions, transformative learning became a desirable goal of adult learning.

Knowles's Andragogy and Self-Directedness

Many theories of adult learning are grounded in the work of Malcolm Knowles and colleagues. By the 1980s, he had delineated several assumptions about adult learners. He formulated them into six core andragogical principles: "the learner's need to know, self-directed learning, prior experience of the learner, readiness to learn, orientation to learning and problem solving, and motivation to learn" (Knowles, Holton, & Swanson, 2005, p. 183). Several of Knowles' principles of adult learning suggest a connection between experience and learning. Knowles indicated that adult learners had experience that could be a resource for their learning, they needed to feel ready to learn, and saw an immediate application for their new knowledge (Knowles et al., 2005; Knowles, Holton III, Swanson, & Robinson, 2020). Discussion around these principles continues in the field of adult learning today.

Brookfield's Critical Reflection

A major contributor to those discussions was Stephen Brookfield. Brookfield (1986) particularly questioned Knowles' ideas of self-directedness, his assumption of relating learning to particular social roles, and his focus on adults' need for immediate application of learning. Brookfield (1987) described nine key concepts of the critical thinking process but then synthesized them into two key activities: "identifying and challenging assumptions" and "exploring and imagining alternatives" (p. 15). Brookfield (2005) moved the focus of adult learning to critical reflection. Critical thinking or critical reflection is "reflecting on the assumptions underlying our own and others' ideas and actions, and contemplating alternative ways of thinking and living" (Brookfield, 1986, p. x).

A New Paradigm for Adult Learning in a Technology-Infused World

Communities and Professions

The three theories and assumptions that were examined in the previous section were selected because they have been core tenants of adult learning throughout its development and evolution. They provide the foundation for much of the practice of education and training in communities and professional settings for adult learners. As technology has become more of an integral part of those communities and professions – how have these theories about learning and practice been impacted? The original premise of the authors is that technology is not just a device that is utilized as a tool, but rather, technology has infused every aspect of society to essentially change the thought process in learning.

Consider the following example. A Victorian home, built in 1900, had a unique chandelier in the foyer. The central portion illuminated at the flip of the switch on the wall, but the arms that extended on the edges appeared to be missing bulbs. Further examination revealed sleeves over the gas jets. Typical of light fixtures of the early 1900s, the chandelier could function on gas or electricity, just in case the newly invented electric lights failed. Today, the stability of electricity is assumed and the backup gas jets are redundant. Just as it is inconceivable to consider building a new house without electricity, it is just as inconceivable to consider functioning in a world without technology and the connectedness of the Internet.

Compare the above analogy to today's information and digital technology. Many can remember the days of balancing a weighty laptop in one hand and a folder full of backup sheets of paper in the other–just in case the technology failed. Connectivity through handheld devices is assumed to be the norm. Today's communities include wifi, available everywhere from coffee shops to parks. Smartphones blur the line between phones and computers and make connectivity even more accessible through 'hot-spotting'. And this connectivity serves to even broaden the very sense of community. Online social networks and messaging apps connect the world.

With the ubiquitous presence of digital technology, the line between one's personal life and their 'work' hours seems to be less defined. For instance, adult learners working on assignments or projects and educators being available to answer questions out of class hours. Or employees working for offshore companies where meetings are not always able to be held at a time during 'office' hours for the employee due to the time difference. The lockdown periods during the COVID-19 global pandemic meant that these lines were less definitive due to the number of people who were working or studying from home.

There is also a blurring between community and professional lives. One's physical presence in one venue or community does not preclude the availability to the other. The community and profession are no longer two silos of one's life. Yermish (2000) labeled this a "passing of remoteness" and described it as a "phenomenon of the rise of the Internet and other distance-shrinking technologies" (p. 208). Although written over 20 years ago, the same could still be said of today's use of technology and the increase in connectivity.

The history of adult teaching and learning has numerous examples of shifts and drifts in its focus (Chen, 2014). However, adult teaching and learning has not just 'shifted' or 'drifted' in response to the introduction of technology. Because of technology, the essence of how theorists examine their basic foundational assumptions and how practitioners put them to use has been inextricably altered.

Historical Context for the New Paradigm of Adult Learning

Examining the history of adult education reveals only a few other events that have had as dramatic an effect as modern technology. Knowles (1989) traces the roots of adult education to the great teachers of ancient China, Rome, and Greece. He notes that it was because of their experience with adults that these educators "perceived learning to be a process of active inquiry, not passive reception of transmitted content" and therefore "invented techniques for engaging learners in active inquiry" (p. 61). He specifically credits the Chinese and Hebrews with the "case method or critical incident", the Greeks with the "Socratic dialogue" in which a dilemma was posed to the group who would pool their knowledge and experience to develop a solution, and the Romans with the methodology that involved forcing the students to state positions and to defend them.

Adult learning as an academic field has its roots in the early 1900s. The Association of Adult and Continuing Education began to sponsor studies in the field and in 1934 and later published the first *Handbook of Adult and Continuing Education*. The 1960 edition of this Handbook was the first to include individual essays by adult educators in order to provide an overview of the current nature and trends in the field of adult education. Cyril Houle's (1961), *The Inquiring Mind* and (1972) *Design of Education* laid the foundation for Tough's (1967) *Learning Without a Teacher* and (1979) *The Adult's Learning Projects,* which introduced the concept of self-directed learning and later influenced Knowles. Yermish (2000) reminds us that "ever since the advent of television, educators have wrestled with the viability of using this technology to reach wider audiences. Educational television facilitated the distribution of high-quality program content in a one-directional fashion" (p. 208). In this way, new technology continues to be developed, replacing older technology for teaching, continuing professional development, and business. Such technologies came to the fore during the lockdowns due to the global COVID-19 pandemic. Zoom™ and Microsoft's Teams™ are just two examples of technology (platforms) that had been in use, mostly in business, before the pandemic. However, once the lockdowns happened and companies and schools/educational institutions closed their doors and workers and students were at home, these platforms became a way of keeping staff working/communicating with colleagues and clients as well as making online classes available.

Merriam (2008) identified two changes in the focus of adult learning. One is from the individual learner to the learner within the various contexts in which learning takes place. This new perspective considers "learning as part of the system's cultural and historical norms…(and) how physical space and spatiality encourages or inhibits learning" (p. 94). The second shift is from learning as a purely cognitive activity to one that is multidimensional in nature. In this more holistic approach, "learning is construed as a much broader activity involving the body, the emotions, and the spirit as well as the mind" (p. 95). Illeris' (2004) three dimensions of learning (cognitive, emotional, social) offer a similar approach to the discussion of learning. Mazarr's (1999) earlier work echoes this picture by explaining that "knowledge-era learning" must become more "holistic" and "high-tech"; it must "emphasize creativity and participation" and be "characterized by more choice and competition" (p. 285).

But between the ancient civilizations and the development of the field of adult learning, a major event impacted teaching and learning; Guttenberg invented the printing press. Lesgold (2000) notes that before the printing press, "direct discussion with a wise person was the primary way of gaining knowledge" (p. 399). Knowles (1989) credits the invention of the printing press with having a great "impact on the advancement of adult education" (p. 62). This changed every aspect of people's community and professional lives. But even with the advent of the early printing presses, books were expensive and rare, and only certain key books were reproduced widely. The industrial revolution enhanced this process by making it faster and more efficient but essentially it was the same process.

The second major event does not surface until the late 1900s. Between the printing press of the mid-1400s and today, history is filled with technological innovations. Parker (2009) notes that "Whether it was the printing press, telephone, copying machine, or computer, educators have always been challenged to use new technology for effective teaching and learning" (p. 223). However, none created the quantum leap effect of information and communication technologies (ICT). The new "information society" had arrived. Loader (1998) provides a sense of just how dramatic this impact was by announcing that "the emergence of the new information and communications technologies, such as the Internet, are said to herald the coming of the 'information society': a new social and economic paradigm restructuring the traditional dimensions of time and space within which we live, work, and interact" (p. 3). Mazarr (1999) notes that "the knowledge era is an interdisciplinary time" (p. 11) and that the "new sciences of complexity remind us that boundaries between problems and disciplines are less important than the threads that connect them" (p. 12). All these suggest that Loader's earlier statement should be expanded to include a new educational paradigm that restructures the traditional dimensions of time and space within which we live, work, interact and learn.

However, as early as 1985, Jarvis (1985) suggested that this "information society" was being replaced by the "information technology society". Alheit (2009) suggested that the "communication and interaction networks of the IT age" will "create the "future form of knowledge" which is "*doing knowledge*, a kind of lifestyle that determines the structures of society far beyond the purely occupational domain" (p. 119).

Jarvis (1985) noted that as society moves toward this new information technology society, "it is quite essential that the designs of the adult learning be such that they encourage critical awareness about all forms of information they receive, whether their teacher is in the classroom, the purveyor of information through the media, or the billions of pages of circulars, or even the programmer of a computer learning package" (p. 14). Fenwick (2008) lists four emerging trends in adult learning:

- Areas of definitions;
- An increased emphasis on practice-based learning;
- The importance of identity and literacy; and,
- Power and politics.

He defines learning not as an outcome but as a process that is "understood to involve not just human change, but interconnections of humans and their actions with rules, tools and texts, cultural, and material environments" (p. 19). These authors have suggested new views of knowledge, society, and learning in the communities in which we live.

Examining this new paradigm makes it obvious that inquiry, problem-solving, and communication skills are not just catchwords in today's K-12 curriculum guides. They must be essential elements of adult education as well. They are essential to everyone's survival in communities and professions. The three issues outlined earlier in this chapter have changed. The frames of reference are expanded, self-directedness feels very different, and reflection seems time-consuming.

Transformative Learning

Does transformative learning look different today than it did as originally proposed by Mezirow? Is there a Transformative Learning 2.0? From its beginning, Transformational Learning has had a longstanding impact on adult education. Sharan Merriam (2008) notes that, as she reviewed three successive updates of *New Directions for Adult and Continuing Education,* she observed that a chapter on transformational learning was the only constant across these updates. However, she did note differences in emphasis in how transformative learning was discussed. The 1993 edition was mostly descriptive. The 2001 chapter focused on empirical research. While empirical research continued to be an emphasis in the 2008 update, that latest edition included diverse theoretical perspectives about transformative learning. The continued global interest in transformative learning is evident in numerous publications on the topic (for example, Cranton, 2016; Taylor & Cranton, 2012; Wang, Torrisi-Steele & Reinsfeld, 2021). Various aspects continue to be discussed and argued.

The disorienting dilemma has been a much-discussed component of transformative learning theory. The theory originally suggested a dilemma limited to personal interactions and experiences. In today's connected world, the source of that dilemma could be the local community, the online community, or the global community. Or the source could be the physical workplace, the virtual workplace, or colleagues/competitors around the globe. All of these sources of disorienting dilemmas have the potential for severely impacting lives and possibly be the source of transformative learning.

The framework of globalization is important not just as it reflects economic and environmental issues, but as it frames the individual knowledge base which could be the source of a dilemma. With the plethora of information available, it might even be difficult to isolate 'the dilemma'. The almost immediate influx of data about one's community and profession on a local to a global level could be overwhelming.

The disorienting dilemma at a distance could produce its own set of dilemmas. It could be difficult for the learner to get enough information to assess its accuracy and to have little, if any, control over the circumstances. Earlier learners might have had a sense of control over the transformation process, but this whirlwind of mostly opinionated information could in itself pose a dilemma. Adult learners need to develop information literacy skills to be discerning in what they view and how it is perceived.

Another view of the influence of technology could, however, put a positive spin on the similar issues. The following example illustrates some of these ideas. In teaching adult learning courses, in vocational or professional continuing education, there is usually an online assignment or task to provide students with the experience of learning online (authors' own experiences). Using technology in an education course offers a unique advantage. In most courses, technology is utilized as a course delivery system (e.g. Blackboard™, Moodle™, etc.) with course materials, readings, resource materials as well as the online submission of assignments all in one place. This use of technology for accessing course materials is distinct from "the use of technology as a learning tool, integrated into learning strategy" (Wang, Torrisi-Steele & Reinsfeld, 2021, p. 325). In an adult education course, the course objectives and tasks usually facilitate the inclusion of an analysis of the effectiveness of the use of technology. The presence of technology in a learning mode is often new to the students, which in itself poses a dilemma. But students are continually required to reflect on and discuss the learning advantages and disadvantages of technology as they experience it in their own learning. Their preconceived assumptions about online assignment often change as a result of the experience and the course assignments that utilize technology enhance this learning (Parker, 2010).

Since the process itself assumes a reevaluation of one's assumptions and changing one's taken-for-granted frames of reference, the continued connectedness of today's adult learner cannot help but impact this process. With a constant influx of information, one's assumptions might be continually changing, making it nearly impossible to reevaluate. How one moves from dilemma to transformative learning becomes a very different process. While the transformative nature of learning is especially important, it might be more difficult to achieve in today's communities and professions (Merriam & Baumgartner, 2020).

Andragogy and Self-Directedness

Knowles et al. (2005, 2020) saw technology as providing learning opportunities in the "andragogical tradition" (p. 237) and as consistent with the adult learning idea of self-directedness. The ideas of andragogy, and particularly the self-directedness of adult learners, created a power shift in the classroom. Technology has changed the very paradigm of thinking about power and brought the idea of self-directedness to a new level. If knowledge is truly power, then the speed at which information can be obtained and the bulk of the information available online today certainly alters the power structure of learning (Knowles et al., 2015).

The first power shift came with the printing press when, as Lesgold noted earlier in this chapter, it was now no longer necessary to find a knowledgeable person to answer a question or solve a problem. The self-directed learner now only needed a book and could control how and when it was read. The availability of today's ubiquitous handheld devices sign also the expectation that knowledge is always instantly available. Whether in the communities in which they live or in the professions in which they work, everyone is expected to be more self-directed in accessing information and gaining knowledge (Knowles et al., 2015).

The assumption is that this self-directed adult will be motivated to use technology to gain status in the community or as a competitive tool in a profession. But motivation and learning are more complex. Several authors support the idea that learners are most successful when other learners are involved whether they are neighbors in a community or colleagues at work. As Palloff and Pratt (1999) suggest, "key to the learning process are the interactions among students themselves, the interactions between faculty and students, and the collaboration in learning that results from these interactions" (p. 5). Nilson (2016) confirms the idea that students learn best when they are actively engaged, when learning evokes emotional and not just intellectual involvement. This is consistent with Illeris's three dimensions of learning (referred to earlier in the chapter). Cranton (2016) emphasizes the importance of empowering the student through interactions in the learning environment and being aware of power relationships. "The creation of a learning community supports and encourages knowledge acquisition. It creates a sense of excitement about learning together and renews the passion involved with exploring new realms in education" (Palloff & Pratt, 1999, p. 163). But introducing technology includes an added dimension (Palloff & Pratt, 2007).

One example that technology introduces is the possibility of asynchronous (as well as synchronous) interactions among students. A task requires students to post their reflections on the week's readings on a discussion board. They must then select another person's posting and respond to it. In a classroom, the student is required to listen to every student's comments. The online discussion board requires the students to be self-directed in selecting an entry of interest to respond to and fosters critical thinking by allowing the students time to develop their thoughts. It fosters more engagement by each student (authors' own experiences). A quote from a student (used with permission) in the lead author's online class provides some details. The student comments on a discussion held in an asynchronous discussion board as part of a class assignment.

The discussion conducted here is very involving; everybody could get a chance to express his own ideas. Moreover, the discussion board online gives us a further opportunity to share ideas with all of the class. It has been developed into a real learning forum. Everybody chose their favorite articles about learning and training in their fields, and then shared their own ideas on the "Blackboard", thus evokes a real open discussion. This learning style makes me feel that I can learn anytime anywhere from so many people of diverse fields. By posting, reading, and replying online, our learning location has burst out of the limited classroom and lecture time boundary, thus it has given us an authentic flexibility and motivation to learn.

It seems especially interesting to see words like 'real open discussion,' 'learning forum', 'motivation', and 'authentic flexibility' as part of the student reflections on the online experience. In the chapter "The Passion of Pluralism," Greene (1995) comments about community: "We are in search of what John Dewey called 'the Great Community," but at the same time, we are challenged as never before to confront plurality and multiplicity" (p. 155). "To open up our experience to existential possibilities of multiple kinds is to extend and deepen what each of us thinks of when he or she speaks of a community" (p. 161). Elements of adult learning theory in action are evidenced in the comments and actions of participants in the technology-enhanced learning cited above.

At the same time that the positive aspects of technology are applauded, there are serious questions to be addressed. Is there a dichotomy developing of not just have and have not access to technology, but a dichotomy in thinking patterns with ethical and social justice implications? Communities provide tools such as computers and primarily instruction on the mechanics of their use for simple tasks. But little attention is paid to the thinking process or decision-making process involved in their effective use.

Critical Reflection

What does critical reflection look like in a 2.0 world? Is it qualitatively different than originally proposed? Is reflection-in-action rather than reflection-on-action becoming the norm? What is the role of critical reflection? Previously, researchers meticulously plotted data points on an x-y axis and, as time allowed, analyzed the resulting trends. Now the ubiquitous graphing calculator and Excel software effortlessly plot the data points and offer the options for plotting trend lines or cycles, releasing time for more in-depth analysis and reflection on the results. Lesgold (2000) suggests that just as the book "removed some of the need for memorization as a force for knowledge distribution, so the computer removes some of the need for overlearning of routine information processing procedures, since these can be accomplished by computers" (p. 401). He cites the example of whether or not students need to know how to perform complex algebraic calculations if they can be done by computer or calculator. However, real-world problems do not have simple answers that can be searched on Google. Jarvis (1985) noted that "learners must be aware of the processes of the manufacture of information and be confident that they can process it adequately to make it their own personal knowledge in which they may have the certainty that the phenomena are real and have the characteristics attributed to them. To achieve this, adult educators must design learning experiences that encourage the creation of critical awareness necessary to ensure the continuation of democracy in an information society" (p. 14).

An example of an activity that facilitates critical reflection is a task that the authors have both used in their online courses. An asynchronous Discussion Board is used where students post their own ideas and comments about reading and then reply to other student's ideas. The task requires students to read a journal article of their choosing related to staff development. They are then to summarize their article and post their reflection on it. Following are a few representative student comments on this assignment.

One student in the lead author's class (used with permission) commented on the advantages of computer technology in eliminating communication barriers and fostering reflection.

It is a medium that does promote student engagement in discourse without the normal bias of face-to-face communication (because our appearance is reduced to letters in a computer screen). And although we have the opportunity to influence and suggest tone, etc. by the use of color, sizes, etc. the initial barriers to traditional communication are somehow diminished. The use of discussion boards allows for a lot of reflection prior to committing to opinions. The student has the time and the resources to build a message that will convey every idea that s/he wants to communicate.

Other students noted the importance of the well-constructed questions and the depth of the reflection and conversations. "The process of responding to skilled questions posed by co-group members allowed me to consider and deeply reflect on my actions with respect to learning and how it is applied in the workplace." "The conversations were not superficial interactions but purposeful, focused, and useful. The instructions allowed all group members an equal opportunity to have their 'voices' heard." While today's technological advancement is characterized by electronics and digitization, implying working faster, it is interesting to see that it can also be used to slow down and reflect before constructing a thoughtful response in a discussion. Kelly (2010) notes that "book reading strengthened our analytical skills, encouraging us to pursue an observation down to the footnote. Screen reading encourages rapid pattern-making, associating this idea with another, equipping us to deal with the thousands of new thoughts expressed every day. The screen rewards and nurtures thinking in real time" (p. 125). However, it is now possible to slow down and reflect on the thought before posting the response for others to read.

This new social paradigm changes our entire sense of space and time. Sitting with an individual or group of colleagues in a room has been replaced by typing on a computer keyboard or any of the numerous hand-held internet-accessible devices, reading text on a screen, or listening to voices on the phone. Instead of feeling the presence of other collaborators, participants are connected by voice or text and might be easily distracted by daily tasks. While the efficiency of such a round-the-clock network is dramatically increased, the issues of ownership and credit can become more complicated in this collaborative global world. It seems that digital connectedness, with its podcasts, blogs, twitters, YouTube, virtual realities, and global, non-stop researching has become and will remain a part of the future of adult education. The greatest challenge, as professionals in the field of adult education, may be to ensure that the essential component of critical reflection continues to have a place in this world.

Possible Solutions and Recommendations

One view of this new learning landscape places the student behind the technology in importance. A closer look at the three ideas of adult learning that are the focus of this chapter, namely transformative learning, self-directedness, and critical reflection, reveals that technology can empower the student. Below are four issues that might be addressed with this idea in mind.

Issue 1: The learner must develop a filter to discern important issues and focus efforts. This is important both in focusing on the dilemma that is the source of the transformative learning or being able to make the required and informed decisions needed by the self-directed student. Likewise, technology has made tasks and decision-making in our communities and professions easier and faster. But as more time becomes available, it must be channeled into reflection. This personal filter can only be developed by critical reflection and will be a useful information literacy tool when working with information about and for the individual's community and profession.

Issue 2: The old educational goal: get the right answer. Achieving this goal usually took time and effort to get the answer. It often required a trip to the library or pages of numerical calculations. The focus on the process of finding the answer was isolated to a few specialized areas. Now answers are everywhere. New educational goal: learn the process, evaluate the source, and question the information (development of information literacy skills). Critical thinking again seems at the heart of this issue. While digital technology can undoubtedly save much of the time that might have been used to explore issues and search for answers, it is important that the conscious decision be made to use the released time for analysis and reflection and not fill it with additional tasks.

Issue 3: Information cannot be taken at face value. This is not a new problem, e.g., information on Wikipedia can be can be edited by anyone. Questioning information on any screen must be part of the learning process. Who is the author? Is it a reliable source? (hint - look at the URL address).

Issue 4: Current technology may promote breadth more than depth. It is easy to search for a topic and find links to related topics but more challenging to find the interrelationship between topics and ideas.

This list of issues can certainly be supplemented by additional thoughts of every reader. The overarching issue, though relates to the ideas of transformative learning, self-directedness, and critical reflection within the communities and professions of the adult learner.

FUTURE TRENDS

In the rapidly changing pace of new technology, the term future has a new meaning. It does not refer to years, months, or days from now. Time really is more of a continuum. The future starts from a second from now and moves forward. Therefore, starting from today might be the best way to expand the view into the future. But what does this really mean in communities and professions?

Since books have been printed or lettered by a scribe, people have been 'lost in a good book.' Today, it is just as easy to get 'lost' in an online game. Some authors caution that emerging new technologies will not replace the old. Valmont (2003) reminds us that "oral storytelling did not die when Gutenberg created the printing press" and "novels did not go away when films became popular. Literacies simply evolve" (p. 298). While this is true on a universal scale, as evidenced by the examples above, in isolated venues, technology does, in fact, replace more traditional forms of communication.

So, to succeed in the communities and professions of tomorrow, some authors offer valuable advice. Bray (2009) noted that "career and technical education is always about the future" because its focus is on preparing "lifelong learners who can transition today's knowledge to the needs of tomorrow" (p. 6).

CONCLUSION

Theories are certainly essential in any field, and examining them is essential to moving the field forward. Elias and Merriam (2005) articulate the need for theory as a practice foundation. "Theory without practice leads to empty idealism, and action without philosophical reflection leads to a mindless activism" (p. 4). Parker (2009) noted that "much of the work of the theorists has been developed through the lens of the practitioner. As theories developed, practice changed. As practice changed, new theories developed" (pp. 224-225). Today's technology-infused environment in communities and professions provide opportunities for the development of even more sophisticated and insightful theories as the foundation for more effective practice. It is a new normal with new paradigms.

The three adult learning ideas that are addressed in this chapter focus on three critical elements of community and professional education:

- The disorienting dilemma of transformative learning;
- The power issues of the self-directed learner; and,
- The decision-making skill of critical reflection.

While this chapter has focused on three theorists and their contributions to adult learning, Heaney (2000) reminds us that:

Individual practitioners do not define the field of adult education, nor do experts. A definition of a field of practice is the social product of many individuals who negotiate the values and meaning of work they come to see as serving a common purpose over time. (p. 561)

The interactions between these individuals that result in that social product have been and will continue to be impacted by digital technology. This chapter's authors' predictions might be a future with even more connections, and more blurring of boundaries, a future that values personal learning but shared experiences and goals.

REFERENCES

Alheit, P. (2009). Biographical learning – within the new lifelong learning. In K. Illeris (Ed.), *Contemporary theories of learning: Learning theorists...in their own words* (pp. 116–128). Routledge.

Bornman, E. (2015). Information society and digital divide in South Africa: Results of longitudinal surveys. *Information Communication and Society*, *19*(2), 264–278. doi:10.1080/1369118X.2015.1065285

Bray, J. (2009). Embrace the future now! *Techniques*, *84*(5), 6.

Brookfield, S. (1986). *Understanding and facilitating adult learning*. Jossey-Bass.

Brookfield, S. (1987). *Developing critical thinkers*. Jossey-Bass.

Brookfield, S. (2005). *The power of critical theory: Liberating adult learning and teaching*. Jossey-Bass.

Chen, J. (2014). Teaching nontraditional adult students: Adult learning theories in practice. *Teaching in Higher Education*, *4*(19), 406–118. doi:10.1080/13562517.2013.860101

Cranton, P. (2016). *Understanding and promoting transformative learning* (3rd ed.). Stylus Publishing.

Elias, J. L., & Merriam, S. B. (2005). *Philosophical foundations of adult education*. Krieger Publishing Company.

Fenwick, T. (Fall 2008). Workplace learning: emerging trends and new perspectives. In *New directions for adult and continuing education No. 119* (pp. 17–26). Jossey-Bass. doi:10.1002/ace.302

Greene, M. (1995). *Releasing the imagination*. Jossey-Bass.

Heaney, T. W. (2000). Adult education and society. In A. L. Wilson & A. R. Hayes (Eds.), *Handbook of adult and continuing education* (pp. 559–572). Jossey-Bass.

Houle, C. O. (1961). *The inquiring mind*. University of Wisconsin Press.

Houle, C. O. (1972). *The design of education*. Jossey-Bass.

Illeris, K. (2004). *The three dimensions of learning*. Krieger.

Jarvis, P. (1985). Thinking critically in an information society: A sociological Analysis. *Lifelong Learning*, 11-14.

Kelly, K. (2010). From print to pixel. *Smithsonian, 41*(4), 122–128.

Knowles, M. (1989). *The making of an adult educator*. Jossey-Bass.

Knowles, M., Holton, E., & Swanson, R. (2015). *The adult learner* (8th ed.). Elsevier.

Knowles, M. S., Holton, III, Swanson, & Robinson, P. A. (2020). *The adult learner the definitive classic in adult education and human resource development* (9th ed.). Routledge.

Lesgold, A. (2000). What are the tools for? Revolutionary change does not follow the usual norms. In S. Lajoie (Ed.), *Computers as cognitive tools* (pp. 399–408). Lawrence Erlbaum Associates.

Loader, B. (1998). *Cyberspace divide: Equality, agency, and policy in the information society*. Routledge. doi:10.4324/9780203169537_chapter_1

Mazarr, M. (1999). *Global trends 2005: An owner's manual for the next decade*. St. Martin's Press.

Merriam, S., & Baumgartner, L. (2020). *Learning in adulthood: A Comprehensive guide* (4th ed.). John Wiley & Sons.

Merriam, S. B. (2008). Adult learning theory for the twenty-first century. In *New Directions for adult and continuing education* (pp. 93-98). Jossey-Bass.

Mezirow, J., & ... (2000). *Learning as transformation: Critical perspectives on a theory in progress*. Jossey-Bass.

Nilson, L. B. (2016). *Teaching at its best* (4th ed.). Jossey-Bass.

Palloff, R., & Pratt, K. (1999). *Building learning communities in cyberspace*. Jossey-Bass.

Palloff, R., & Pratt, K. (2007). *Building online learning communities: effective strategies for the virtual classroom*. Jossey-Bass.

Parker, J. (2009). Adult learning and CTE: A shared history influenced by technology. In V. Wang (Ed.), *Definitive Readings in the History, Philosophy, Practice and Theories of Career and Technical Education* (pp. 217–236). Information Science Reference.

Parker, J. (2010). Technology as integral to a new paradigm of adult education. *International Journal of Adult Vocational Education and Technology, 1*(2), 10–18. doi:10.4018/javet.2010040102

Taylor, E. W., & Cranton, P. (2012). *The handbook of transformative learning: theory, research and practice*. Jossey-Bass.

Tough, A. (1967). *Learning without a teacher*. Institute for Studies in Education.

Tough, A. (1979). *The adult's learning project*. Institute for Studies in Education.

Valmont, W. (2003). *Technology for literacy teaching and learning*. Houghton Mifflin.

Wang, V., Torrisi-Steele, G., & Reinsfeld, E. (2021). Transformative learning, epistemology and technology in adult learning. *Journal of Adult and Continuing Education, 27*(2), 324–340. doi:10.1177/1477971420918602

Yermish, I. (2000). A case for case studies via video-conferencing. In L. Lau (Ed.), *Distance learning technologies: Issues, trends and opportunities* (pp. 208–217). Idea Group Publishing. doi:10.4018/978-1-878289-80-3.ch015

Zdjelar, R., & Zajdela Hrustek, N. (2021). Digital divide and e-inclusion as challenges of the information society – research review. *Journal of Information and Organizational Sciences*, *45*(2), 601–638. doi:10.31341/jios.45.2.14

KEY TERMS AND DEFINITIONS

Critical Reflection: The Victorian Government Department of Education and Training in Australia, defines critical reflection as an extension of critical thinking. Critical reflection asks us to think about our own practice and ideas and then it challenges us to step back and examine our thinking by asking probing questions.

Digital Divide: According to the Close that Gap organization's website, the digital divide is defined as the gap that exists between those who have reliable internet access and those with very limited access of no access at all.

Disorienting Dilemma: The IGI Global website states that a disorienting dilemma is experience of crisis/new information that causes an individual to question their values, beliefs or assumptions and leads to transformation.

Information Literacy: According to the American Library Association website, information literacy is a set of abilities requiring individuals to recognize when information is needed and have the ability to locate, evaluate, and use effectively the needed information.

Self-Directed Learning: The BetterUp Professional Development blog describes self-directed learning as a process in which an individual takes the initiative, with or without the help of others, in diagnosing their learning needs, formulating learning goals, identifying resources for learning, choosing and implementing appropriate learning strategies, and evaluating learning outcomes.

Technology-Infused Teaching: The extent to which technology permeates into teaching.

Transformative Learning: The Western Governors University blog states that transformative learning is one theory of learning that particularly focuses on adult education. Transformative learning focuses on the idea that learners can adjust their thinking based on new information.

Compilation of References

AAUP. (2017). *Statement on targeting harassment of faculty*. AAUP Updates.

Abet, T. (2021). Heart, vessel diseases leading cause of deaths, say experts. *Daily Monitor*. https://www.monitor.co.ug/uganda/special-reports/heart-vessel-diseases-leading-cause-of-deaths-say-experts-3366312

Ablin, F. (1963). *Education in the USSR: A collection of readings from soviet journals*. International Arts & Sciences.

Aboramadan, M., & Dahleez, K. A. (2020). Leadership styles and employees' work outcomes in nonprofit organizations: The role of work engagement. *Journal of Management Development*, *39*(7/8), 869–893. doi:10.1108/JMD-12-2019-0499

Acquaro, D. (2019). Preparing the next generation of educational leaders: Initiating a leadership discourse in initial teacher education. *International Studies in Educational Administration*, *47*, 107–124.

Adaileh, M. J., Alrawashdeh, M., Elrehail, H., & Aladayleh, K. J. (2020). Assessing the nexus between knowledge management and firm performance: A data article. *Data in Brief*, *32*(10), 1–8. doi:10.1016/j.dib.2020.106283 PMID:32984475

Aepli, M., Angst, V., Iten, R., Kaiser, H., Lüthi, I., & Schweri, J. (2017). *Die entwicklung der kompetenzanforderungen auf dem arbeitsmarkt im zuge der digitalisierung. Schlussbericht. Arbeitsmarktpolitik No. 47*. SECO Publikation.

Agarwal, S. D., Barthel, M. L., Rost, C., Borning, A., Bennett, W. L., & Johnson, C. N. (2014). Grassroots organizing in the digital age: Considering values and technology in Tea Party and Occupy Wall Street. *Information Communication and Society*, *17*(3), 326–341. doi:10.1080/1369118X.2013.873068

Aguayo, C. (2016). Activity theory and online community education for sustainability: When systems meet reality. In D. Gedera & J. Williams (Eds.), *Activity theory in education* (pp. 139–151). Brill. doi:10.1007/978-94-6300-387-2_9

Aguayo, C., & Eames, C. (2017). Promoting community socio-ecological sustainability through technology: A case study from Chile. *International Review of Education*, *63*(6), 871–895. doi:10.100711159-017-9685-7

Ahmad, S., Meher, S., Siddique, N., & Khan, M. A. (2021). *Global research on andragogy: A bibliometric analysis*. Library Philosophy and Practice.

Ainalem, M. (2020). *Programmers are not designers, or are they?* https://uxdesign.cc/programmers-are-not-designers-or-are-they-7c0cbe5fca7

Aitken, M. (2017). Feminism: A fourth to be reckoned with? Reviving community education feminist pedagogies in a digital age. *Concept*, *8*(1), 1–18.

Ajzen, I. (1991). The theory of planned behavior. *Organizational Behavior and Human Decision Processes*, *50*(2), 179–211. doi:10.1016/0749-5978(91)90020-T

Akgün, A. E., & Keskin, H. (2021). Team intuition and creativity in new product development projects: A multi-faceted perspective. *Journal of Engineering and Technology Management*, *62*, 101660. doi:10.1016/j.jengtecman.2021.101660

Akour, I.A., Al-Maroof, R.S., Alfaisal, R., & Salloum, S.A. (2022). A conceptual framework for determining metaverse adoption in higher institutions of gulf area: An empirical study using hybrid SEM-ANN approach. *Computers and Education: Artificial Intelligence, 3.*

Akyol, Z., Garrison, D. R., & Ozden, M. Y. (2009). Online and blended communities of inquiry: Exploring the developmental and perceptional differences. *International Review of Research in Open and Distance Learning, 10*(6), 65–83. doi:10.19173/irrodl.v10i6.765

Alavi, M., & Leidner, D. E. (2001). Knowledge management and knowledge management systems: Conceptual foundations and research issues. *Management Information Systems Quarterly, 25*(1), 107. doi:10.2307/3250961

Albort-Morant, G., Ariza-Montes, A., Leal-Rodríguez, A., & Giorgi, G. (2020). How does positive work-related stress affect the degree of innovation development? *International Journal of Environmental Research and Public Health, 17*(2), 520. doi:10.3390/ijerph17020520 PMID:31947589

Albrecht, T. L., & Adelman, M. B. (1987). Communicating social support: A theoretical perspective. In T. L. Albrecht & M. B. Adelman (Eds.), *Communicating social support* (pp. 18–38). Sage.

Alger, C., & Kopcha, T. J. (2011). Technology supported cognitive apprenticeship transforms the student teaching field experience: Improving the student teaching field experience for all triad members. *Teacher Educator, 46*(1), 71–88. doi:10.1080/08878730.2010.529986

Alheit, P. (2009). Biographical learning – within the new lifelong learning. In K. Illeris (Ed.), *Contemporary theories of learning: Learning theorists…in their own words* (pp. 116–128). Routledge.

Allen, I., & Seaman, J. (2013). *Changing course: Ten years of tracking education in the United States.* https://files.eric.ed.gov/fulltext/ED541571.pdf

Allen, S. (2016). Applying adult learning principles to online course design. *Distance Learning., 13*(3), 25–31.

Allen, T. D., Herst, D. E., Bruck, C. S., & Sutton, M. (2000). Consequences associated with work-to-family conflict: A review and agenda for future research. *Journal of Occupational Health Psychology, 5*(2), 278–308. doi:10.1037/1076-8998.5.2.278 PMID:10784291

Almeida, A., & Vasconcelos, C. (2015). Geothics: Maser's students' knowledge and perception of its importance. *Research in Science Education, 45*(6), 889–906. doi:10.100711165-014-9449-3

Alzayed, M. A., Miller, S. R., & McComb, C. (2021). Empathic creativity: Can trait empathy predict creative concept generation and selection? *Artificial Intelligence for Engineering Design, Analysis and Manufacturing, 35*(4), 369–383. doi:10.1017/S0890060421000196

American Library Association & National Network of Libraries of Medicine. (2018). *Libraries transform health literacy toolkit.* American Library Association. https://ilovelibraries.org/librariestransform/health-literacy-toolkit/

American Psychological Association. (2014). *The road to resilience. American Psychological Association.* http://www.apa.org/helpcenter/road-resilience.aspx

American Trade Union Delegation to the Soviet Union. (1928). *Soviet Russia in the second decade.* John Day Publishing.

Anderson, E., Winn, K. M., Young, M. D., Groth, C., Korach, S., Pounder, D., & Rorrer, A. K. (2018). Examining university leadership preparation: An analysis of program attributes and practices. *Journal of Research on Leadership Education, 13*(4), 375–397. doi:10.1177/1942775117735873

Anderson, S. E., Mascall, B., Stiegelbauer, S., & Park, J. (2012). No one way: Differentiating school district leadership and support for school improvement. *Journal of Educational Change, 13*(4), 403–430. doi:10.100710833-012-9189-y

Anderson, T. D., & Garrison, D. R. (1995). Critical thinking in distance education: Developing critical communities in an audio teleconference context. *Higher Education, 29*(2), 183–199. doi:10.1007/BF01383838

Anderson, T., Rourke, L., Garrison, D. R., & Archer, W. (2001). Assessing teaching presence in a computer conferencing context. *Journal of Asynchronous Learning Networks, 5*(2), 1–17. doi:10.24059/olj.v5i2.1875

Andrew, G. J. (1980). *Continuing education accreditation. Power and conflict in continuing education.* Wadsworth Publishing Co.

Antonakis, J., Cianciolo, A. T., & Sternberg, R. J. (2004). *The nature of leadership*. Sage Publications.

Anyan, F., Hjemdal, O., Bizumic, B., & Friborg, O. (2020). Measuring resilience across Australia and Norway. *European Journal of Psychological Assessment, 36*(2), 280–288. doi:10.1027/1015-5759/a000509

Appleyard, M., Enders, A., & Velazquez, H. (2020). Regaining R&D leadership: The role of design thinking and creative forebearance. *California Management Review, 62*(2), 12–29. doi:10.1177/0008125619897395

Arain, G. A., Bhatti, Z. A., Ashraf, N., & Fang, Y. H. (2020). Topdown knowledge hiding in organizations: An empirical study of the consequences of supervisor knowledge hiding among local and foreign workers in the Middle East. *Journal of Business Ethics, 164*(11), 611–625. doi:10.100710551-018-4056-2

Archbald, D. (2016). System-Level Instructional Leadership—A District-Level Leadership Case: Implementing PLCs in Schools. *The International Journal of Educational Leadership Preparation, 11*(2), 116–151.

Archer, D. E., & Zhang, Y. (2021). The critical role of global and culturally responsive leadership for higher education in the 21st century. *Journal of Interdisciplinary Studies in Education, 10*(1), i–v.

Arghode, V., Brieger, E., & McLean, G. N. (2017). Adult learning theories: Implications for online instruction. *European Journal of Training and Development, 41*(7), 593–609. doi:10.1108/EJTD-02-2017-0014

Armstrong, M. (2012). *Armstrong's handbook of management and leadership: Developing effective people skills for better leadership and management*. Kogan Page.

Artino, A. R. (2007). Online military training: Using a social cognitive view of motivation and self-regulation to understand students' satisfaction, perceived learning, and choice. *Quarterly Review of Distance Education, 8*(3), 191–202.

Ashforth, B. E., & Mael, F. (1989). Social identity theory and the organization. *Academy of Management Review, 14*(1), 20–39. doi:10.2307/258189

Athanassoulis, N., & Wilson, J. (2009). When is deception in research ethical? *Clinical Ethics, 4*(1), 44–49. doi:10.1258/ce.2008.008047

Aubrey, K., & Riley, A. (2015). *Understanding and using educational theories*. SAGE Publications.

Auletta, K. (2018). *Frenemies: The epic disruption of the Ad business*. Penguin.

Avery, G. C., & Bergsteiner, H. (2011). Sustainable leadership practices for enhancing business resilience and performance. *Strat Leader, 39*(3), 5–15. doi:10.1108/10878571111128766

Avissar, I., Alkaher, I., & Gan, D. (2018). The role of distributed leadership in mainstreaming environmental sustainability into campus life in an Israeli teaching college: A case study. *International Journal of Sustainability in Higher Education, 19*(3), 518–546. doi:10.1108/IJSHE-07-2017-0105

Avolio, B. J., & Gardner, W. L. (2005). Authentic leadership development: Getting to the root of positive forms of leadership. *The Leadership Quarterly*, *16*(3), 315–338. doi:10.1016/j.leaqua.2005.03.001

Avolio, B. J., & Hannah, S. T. (2008). Developmental readiness: Accelerating leader development. *Consulting Psychology Journal*, *60*(4), 331–347. doi:10.1037/1065-9293.60.4.331

Avolio, B. J., Sosik, J. J., Kahai, S. S., & Baker, B. (2014). E-leadership: Re-examining transformations in leadership source and transmission. *The Leadership Quarterly*, *25*(1), 105–131. doi:10.1016/j.leaqua.2013.11.003

Avolio, B., Waldman, D., & Yammarino, F. (1991). Leading in the 1990s: The four Is of transformational leadership. *Journal of European Industrial Training*, *15*(4), 9–16. doi:10.1108/03090599110143366

Ayman, R., & Lauritsen, M. (2018). Contingencies, context, situation, and leadership. In *The nature of leadership* (3rd ed., pp. 138–166). Sage Publications. doi:10.4135/9781506395029.n6

Baas, T., Dewhurst, S., & Peyre, S. (2020). Leadership and management for scientists. In L. Infante Lara, L. Daniel, & R. Chalkley (Eds.), *BEST* (pp. 87–101). Academic Press. doi:10.1016/B978-0-12-820759-8.00006-1

Backus, C., Keegan, K., Gluck, C., & Gulick, L. (2010). Accelerating leadership development via immersive learning and cognitive apprenticeship. *International Journal of Training and Development*, *14*(2), 144–148. doi:10.1111/j.1468-2419.2010.00347.x

Bagher, M., Marek, S. A., & Sibbald, A. (2007). Implementing web-assisted learning and engaging academic staff in the change process. *Journal of Organisational Transformation & Social Change*, *3*(3), 269–284. doi:10.1386/jots.3.3.269_1

Baglione, S., & Nastanski, M. (2007). Superiority of online discussion: Faculty perceptions. *Quarterly Review of Distance Education*, *8*(2), 139–150. https://www.infoagepub.com/index.php?id=89&i=4

Bain, R., Packer, J., & Mills, C. (1999). Using journal writing to enhance student teachers' reflectivity during field experience placement. *Teachers and Teaching*, *5*(1), 51–73. doi:10.1080/1354060990050104

Ball, M. (2022). *The Metaverse and How It Will Revolutionize Everything*. Liveright Publishing.

Bandura, A. (1977). *Social learning theory*. Prentice Hall.

Bandura, A. (1986). *Social foundations of thought and action: A social cognitive theory*. Prentice-Hall, Inc.

Banfield, R. (2015). *Design leadership*. O'Reilly Media, Inc.

Barley, S. R. (2015). Why the Internet makes buying a car less loathsome: How technologies change role relations. *Academy of Management Discoveries*, *1*(1), 31–60. doi:10.5465/amd.2013.0016

Barth, R. S. (2013). The time is ripe (again). *Educational Leadership*, *71*(2), 10–16.

Bash, L. (2003). *Adult learners in the academy*. Anker Publishing Company, Inc.

Bass, B. M. (1998). Two decades of research and development in transformational leadership. *European Journal of Work and Organizational Psychology*, *8*(1), 9–32. doi:10.1080/135943299398410

Bass, B. M. (2008). *The Bass Handbook of Leadership: Theory, Research, and Managerial Applications* (4th ed.). Free Press.

Basu, T. (2021). The metaverse has a groping problem already. *MIT Technology Review, 16*(3).

Batagiannis, S. C. (2011). Promise and possibility for aspiring principals: An emerging leadership identity through learning to do action research. *Qualitative Report*, *16*(5), 1304–1329.

Battersby, D. (1987). From androgogy to gerogogy. *Journal of Educational Gerontology, 2*(1), 4–10.

Beckert, J. (2016). *Imagined futures*. Harvard University Press. https://www.hup.harvard.edu/catalog.php?isbn=9780674088825

Bekkhus, R., & Hallikainen, R. (2017). A new dualistic CIO toolbox: Towards ambidexterity in the digital business transformation. *Proceedings of the Conference of the Portuguese Chapter of the Association of Information Systems, 17*, 23-41.

Bell, K., & Gilliam, A. (2011). *Index-Region 10 Website*. Region10.org. https://www.region10.org/digital-fluency/index

Bell, D. (1960). *The end of ideology: On the exhaustion of political ieas in the fifties*. Free Press.

Bell, D. (1973). *The coming of post-industrial society*. Basic Books.

Bell, L. A. (2016). Theoretical foundations for social justice education. In *Teaching for diversity and social justice* (pp. 3–26). Routledge.

Bem, D. J. (1967). Self-perception: An alternative interpretation of cognitive dissonance phenomena. *Psychological Review, 74*(3), 183–200. doi:10.1037/h0024835 PMID:5342882

Benbast, I. (2010). HCI Research: Future challenges and directions. *AIS Transactions on Human-Computer Interaction, 2*(2), 16–21. doi:10.17705/1thci.00011

Benjamin, H. (1935). Revolutionary education in Mexico. *The Annals of the American Academy of Political and Social Science, 182*(1), 181–189. doi:10.1177/000271623518200121

Bennet, N., & Lemoine, G. J. (2014). What VUCA really means for you. *Harvard Business Review*. https://hbr.org/2014/01/what-vuca-really-means-for-you

Bennett, E., & McWhorter, R. (2021). Virtual HRD's role in crisis and the post Covid-19 professional lifeworld: Accelerating skills for digital transformation. *Advances in Developing Human Resources, 23*(1), 5–25. doi:10.1177/1523422320973288

Bennett, N., Woods, P., Wise, C., & Newton, W. (2007). Understandings of middle leadership in secondary schools: A review of empirical research. *School Leadership & Management, 27*(5), 453–470. doi:10.1080/13632430701606137

Bensley, R., & Brookins-Fisher, J. (2018). *Community and public health education methods a practical guide* (4th ed.). Jones & Bartlett Learning.

Benson, A., Li, D., & Shue, K. (2018). Research: Do people really get promoted to their level of incompetence? *Harvard Business Review, 8*. https://hbr.org/2018/03/research-do-people-really-get-promoted-to-their-level-of-incompetence

Benson, P. (2001). *Teaching and researching autonomy in language learning*. Longman.

Berard, J. (2013). *Accelerating leadership development: Practical solutions for building your organization's potential*. Jossey-Bass.

Berdahl, R. O., Altbach, P. G., & Gumport, P. J. (2011). Introduction: The contexts of American higher education. In P. G. Altbach, P. J. Gumport, & R. O. Berdahl (Eds.), *American higher education in the twenty-first century: Social, political, and economic challenges*. Johns Hopkins University Press.

Bereday, G. Z. F., Brickman, W., & Read, G. H. (1960). *The changing Soviet school*. Houghton Mifflin.

Berestova, A., Gayfullina, N., & Tikhomirov, S. (2020). Leadership and Functional Competence Development in Teachers: World Experience. *International Journal of Instruction, 13*(1), 607–622. doi:10.29333/iji.2020.13139a

Berkowsky, R. W., Cotton, S. R., Yost, E. A., & Winstead, V. P. (2013). Attitudes towards and limitations to ICT use in assisted and independent living communities: Findings from a specially designed technological intervention. *Educational Gerontology, 39*(11), 797–811. doi:10.1080/03601277.2012.734162 PMID:24244065

Berman, S., & Korsten, P. (2014). Leading in the connected era. Strategy & Leadership. *Emerald Group Publishing, 42*(1), 37–46.

Bertrand, M., & Rodela, K. C. (2018). A framework for rethinking educational leadership in the margins: Implications for social justice leadership preparation. *Journal of Research on Leadership Education, 13*(1), 10–37. doi:10.1177/1942775117739414

Bettez, S. C. (2008). Social justice activist teaching in the university classroom. *Unsettling beliefs: Teaching theory to teachers*, 279-296.

Billingsley, B. S. (2007). Recognizing and supporting the critical roles of teachers in special education leadership. *Exceptionality, 15*(3), 163–176. doi:10.1080/09362830701503503

Bishop, J. H., & Mane, F. (2004). The impacts of career-technical education on high school labor market success. *Economics of Education Review, 23*(4), 381–402. doi:10.1016/j.econedurev.2004.04.001

Blackley, S., & Sheffield, R. (2015). Digital andragogy: A richer blend of initial teacher education in the 21st century. *Issues in Educational Research, 25*(4), 397–414.

Blanchard, K., Zigarmi, P., & Zigarmi, D. (1985). *Leadership and the one-minute manager: Increasing effectiveness through situational leadership.* William Morrow.

Blaschke, L. (2012). Heutagogy and lifelong learning: A review of Heutagogical practice and self-determined learning. *IRRODL*, 57-71. https://files.eric.ed.gov/fulltext/EJ979639.pdf

Blaschke, L., & Hase, S. (2016). Heutagogy: A holistic framework for creating twenty-first-century self-determined learners. In B. Gros, Kinshuk, & M. Maina (Eds.), *The future of ubiquitous learning*. Springer doi:10.1007/978-3-662-47724-3_2

Bliss & Lawrence. (2009). Is the whole greater than the sum of its parts? A comparison of small group and whole class discussion board activity in online courses. *Journal of Asynchronous Learning Networks, 13*(4), 25–39.

Bloom, B. S. (Ed.). (1956). *Taxonomy of educational objectives*. Longman.

Bloom, P. (2017). *Against empathy: The case for rational compassion.* Random House.

Bloomquist, C. D., & Georges, L. (2022). Interdisciplinary leadership: A leadership development model for scholar-practitioners. *Journal of Leadership Education, 21*(4), 58–75. doi:10.12806/V21/I4/A4

Blume, B., Ford, J., Baldwin, T., & Huang, J. (2010). Transfer of training: A meta-analytic review. *Journal of Management, 36*(4), 1065–1105. doi:10.1177/0149206309352880

Blumenthal, D. M., Bernard, K., Bohnen, J., & Bohmer, R. (2012). Addressing the leadership gap in medicine: Residents need for systematic leadership development training. *Academic Medicine, 87*(4), 513–522. doi:10.1097/ACM.0b013e31824a0c47 PMID:22361800

Boateng, C. (2012). *Restructuring vocational and technical education in Ghana: The role of leadership development.* Academic Press.

Boboc, M. (2015). Online course dynamic design informed by student response and formative assessment. In S. Koc, X. Liu, & P. Wachira (Eds.), *Assessment in online and blended learning environments* (pp. 103–124). Information Age Publishing.

Bogotch, I. E. (2002). Educational leadership and social justice: Practice into theory. *Journal of School Leadership, 12*(2), 138–156. doi:10.1177/105268460201200203

Bohinski, C. A., & Mulé, N. (2016). Telecollaboration: Participation and negotiation of meaning in synchronous and asynchronous activities. *MEXTESOL Journal, 40*(3).

Boisot, M., & Cox, B. (1999). The I-Space: A framework for analyzing the evolution of social computing. *Technovation, 19*(9), 525–536. doi:10.1016/S0166-4972(99)00049-8

Bojadziev, M., Stefanovska-Petkovska, M., Handziski, V., & Barlakoska, G. (2016). Age related preferences of leadership style: Testing McGregor theory X and Y. *Journal of Management Research, 8*(4), 187. doi:10.5296/jmr.v8i4.10088

Bojic, L. (2022). Metaverse through the prism of power and addition: What will happen when the virtual world becomes more attractive than reality? *European Journal of Future Research, 10*(22).

Bojic, L., & Marie, J. L. (2017). Addiction to old versus new media. *Serbian Political Thought, 56*(2), 33–48.

Bolden, R. (2011). Distributed leadership in organizations: A review of theory and research. *International Journal of Management Reviews, 13*(3), 251–269. doi:10.1111/j.1468-2370.2011.00306.x

Bolden, R., & O'Regan, N. (2016). Digital disruption and the future of leadership: An interview with rick haythornthwaite, chairman of centrica, and masterCard. *Journal of Management Inquiry, 25*(4), 438–446. doi:10.1177/1056492616638173

Bolliger, D. U., & Martindale, T. (2004). Key factors for determining student satisfaction in online courses. *International Journal on E-Learning, 3*(1), 61–67.

Bonfield, C. A., Salter, M., Longmuir, A., Benson, M., & Adachi, C. (2020). Transformation or evolution?: Education 4.0, teaching and learning in the digital age. *Higher Education Pedagogies, 5*(1), 223–246. doi:10.1080/23752696.2020.1816847

Bong, C. L., Fraser, K., & Oriot, D. (2016). Cognitive Load and Stress in Simulation. In V. J. Grant & A. Cheng (Eds.), *Comprehensive healthcare simulation: Paediatrics* (pp. 3–17). Springer International Publishing. doi:10.1007/978-3-319-24187-6_1

Bonilla y Segura, G. (1945). Report on the cultural missions of Mexico. U. S. Gov't. Printing Office.

Booth, G. C. (1941). *Mexico's school-made society.* Stanford University Press.

Bornman, E. (2015). Information society and digital divide in South Africa: Results of longitudinal surveys. *Information Communication and Society, 19*(2), 264–278. doi:10.1080/1369118X.2015.1065285

Borradaile, K., Martinez, A., & Schochet, P. (2021). *Adult education strategies: Identifying and building evidence of effectiveness. Study Snapshot. NCEE 2021-007.* Distributed by ERIC Clearinghouse.

Borrego, M., Douglas, E. P., & Amelink, C. T. (2009). Quantitative, qualitative, and mixed research methods in engineering education. *Journal of Engineering Education, 98*(1), 53–66. doi:10.1002/j.2168-9830.2009.tb01005.x

Boston, W., Diaz, S. R., Gibson, A. M., Ice, P., Richardson, J., & Swan, K. (2009). An exploration of the relationship between indicators of the community of inquiry framework and retention in online programs. *Journal of Asynchronous Learning Networks, 13*(3), 67–83. doi:10.24059/olj.v13i3.1657

Bouchrika, I. (2022). *The andragogy approach: Knowles' adult learning theory principles.* https://research.com/education/the-andragogy-approach

Boyce, J., & Bowers, A. J. (2018). Different levels of leadership for learning: Investigating differences between teachers individually and collectively using multilevel factor analysis of the 2011-2012 Schools and Staffing Survey. *International Journal of Leadership in Education, 21*(2), 197–225. doi:10.1080/13603124.2016.1139187

Branson, C. M., Marra, M., Franken, M., & Penney, D. (2018). *Leadership in higher education from a transrelational perspective*. Bloomsbury Academic. doi:10.5040/9781350042414

Braun, S., Peus, C., Frey, D., & Knipfer, K. (2016). Leadership in academia: Individual and collective approaches to the quest for creativity and innovation. *Leadership Lessons from Compelling Contexts, 8*, 349–365. doi:10.1108/S1479-357120160000008013

Bray, E., Aoki, K., & Dlugosh, L. (2008). Predictors of learning satisfaction in Japanese online distance learners. *International Review of Research in Open and Distance Learning, 9*(3). Advance online publication. doi:10.19173/irrodl.v9i3.525

Bray, J. (2009). Embrace the future now! *Techniques, 84*(5), 6.

Brazier, D. K. (2005). Influence of contextual factors on health-care leadership. *Leadership and Organization Development Journal, 26*(2), 128–141. doi:10.1108/01437730510582563

Breen, J. M., & Gleason, M. C. (2022). A call for the leadership education field. *New Directions for Student Leadership, 2022*(176), 111–122. doi:10.1002/yd.20535 PMID:36565148

Briggs, C., & Malice, K. (2011). *Digital fluency: Building success in the digital age*. Social Lens.

Brockett, R. G. (1985). Methodological and substantive issues in the measurement of self-directed learning readiness. *Adult Education Quarterly, 36*(1), 15–24. doi:10.1177/0001848185036001002

Brookfield, S. (1987). *Developing critical thinkers*. Jossey-Bass.

Brookfield, S. (2005). *The power of critical theory: Liberating adult learning and teaching*. Jossey-Bass.

Brookfield, S. (2015). *The skillful teacher*. Jossey-Bass.

Brookfield, S. D. (1984). *Understanding and facilitating advanced learning*. Jossey-Bass.

Brookfield, S. D. (1986). *Understanding and facilitating adult learning*. Jossey-Bass.

Brookfield, S. D. (1995). *Becoming a critically reflective teacher*. Jossey-Bass.

Brown, R. (2018). Neoliberalism, marketisation and higher education: How neoliberalism is reshaping the provision of English higher education. Professorial lecture at the University of West London, London, UK.

Brown, J., Gosling, T., Sethi, B., Sheppard, B., Stubbings, C., Sviolka, J., Williams, J., Zarubina, D., & Fisher, L. (2018). *Workforce of the future. The competing forces shaping 2030*. PricewaterhouseCoopers.

Brown, K. M. (2006). Leadership for social justice and equity: Evaluating a transformative framework and andragogy. *Educational Administration Quarterly, 42*(5), 700–745. doi:10.1177/0013161X06290650

Bruner, J. (1960). *The process of education*. Harvard College Press.

Bühring, J., & Liedtka, J. (2018). Embracing systematic futures thinking at the intersection of Strategic Planning, Foresight and Design. *Journal of Innovation Management, 6*(3), 134–152. doi:10.24840/2183-0606_006-003_0006

Bühring, J., & Moore, P. (2018). Emotional and Social Intelligence as 'Magic Key' in Innovation: A Designer's call toward inclusivity for all - Letter from Academia. *Journal of Innovation Management, 6*(2), 6–12. doi:10.24840/2183-0606_006.002_0002

Bukharin, N. I., & Preobrazhensky, E. A. (1969). The ABCs of communism. Penguin. (Original publication 1920)

Bullen, M., Morgan, T., & Qayyum, A. (2011). Digital learners in higher education: Generation is not the issue. *Canadian Journal of Learning and Technology, 37*(1), 2. doi:10.21432/T2NC7B

Buller, J. L. (2015). The only thing we have to change Is—change itself. *Change Leadership in Higher Education: A Practical Guide to Academic Transformation*, 1-27.

Burawat, P. (2019). The relationships among transformational leadership, sustainable leadership, lean manufacturing, and sustainability performance in Thai SMEs manufacturing industry. *International Journal of Quality & Reliability Management, 36*(6), 1014–1036. doi:10.1108/IJQRM-09-2017-0178

Burhansstipanov, L., Krebs, L. U., Harjo, L., Watanabe-Galloway, S., Pingatore, N., Isham, D., Duran, F. T., Denny, L., Lindstrom, D., & Crawford, K. (2014). Providing community education: Lessons learned from native patient navigators. *Journal of Cancer Education, 29*(3), 596–606. doi:10.100713187-014-0690-2 PMID:25087698

Burnes, B. (2004). Kurt Lewin and the planned approach to change: A re-appraisal. *Journal of Management Studies, 41*(6), 977–1002. doi:10.1111/j.1467-6486.2004.00463.x

Burton, S., Bessette, D., Brown-Jackson, K., & Grimm, F. (2013). Transforming the mental paradigm on digital literacy through understanding andragogy and maximizing continuous education. *ICERI2013 6th International Conference of Education, Research, and Innovation.*

Bush, T., & Glover, D. (2003). School leadership: Concepts and evidence. *National College for School Leadership*, 1-42.

Bush, T. (2003). *Theories of educational leadership and management* (3rd ed.). SAGE Publications, Inc.

Bush, T. (2007). Educational leadership and management: Theory, policy and practice. *South African Journal of Education, 27*(3), 391–406.

Bush, T. (2008). *Leadership and management development in education.* SAGE Publications, Inc. doi:10.4135/9781446213605

Bush, T., & Glover, D. (2014). School leadership models: What do we know? *School Leadership & Management, 34*(5), 553–571. doi:10.1080/13632434.2014.928680

Cabrera, N. L., Jeffrey, M. F., Jaquette, O., & Marx, R. M. (2014). Missing the (student achievement) forest for all the (political) trees: Empiricism and the Mexican American studies controversy in Tucson. *American Educational Research Journal, 51*(6), 1084–1118. doi:10.3102/0002831214553705

Cadorin, L., Suter, N., Dante, A., Williamson, S. N., Devetti, A., & Palese, A. (2012). Self-directed learning competence assessment within different healthcare professionals and amongst students in Italy. *Nurse Education in Practice, 12*(3), 153–158. doi:10.1016/j.nepr.2011.10.013 PMID:22112861

Caena, F., & Redecker, C. (2019). Aligning teacher competence frameworks to 21st century challenges: The case for the European digital competence framework for educators (DIGCOMPEDU). *European Journal of Education, 54*(3), 356–369. doi:10.1111/ejed.12345

Caffarella, R. S., & O'Donnell, J. M. (1987). Self-directed learning: A critical paradigm revisited. *Adult Education Quarterly, 37*(4), 199–211. doi:10.1177/0001848187037004002

Cai, Y., Llorca, J., Tulino, A. M., & Molisch, A. F. (2022). Compute-and-data-intensive networks: The key to the Metaverse. *1st International Conference on 6G Networking (6GNet).* 10.1109/6GNet54646.2022.9830429

Callahan, C., & Grunberg, N. E. (2018). *Military medical leadership.* Academic Press.

Cambron-McCabe, N., & McCarthy, M. M. (2005). Educating school leaders for social justice. *Educational Policy, 19*(1), 201–222. doi:10.1177/0895904804271609

Campanotta, L., Simpson, P., & Newton, J. (2018). Program quality in leadership preparation programs: An assessment tool. *Education, 138*(3), 219–228.

Candy, P. C. (1991). *Self-direction for lifelong learning*. Jossey-Bass.

Cannon-Bowers, J. A., Salas, E., & Converse, S. A. (1993). Shared mental models in expert team decision making. In N. J. Castellan (Ed.), *Individual and group decision making* (p. 221246). Lawrence Erlbaum Associates.

Carbery, R., & Garavan, T. N. (2011). Conceptualizing participation in formal training and development activities: A planned behavior approach. In R. F. Poell & M. van Woerkom (Eds.), *Supporting workplace learning* (pp. 27–45). Springer Netherlands. doi:10.1007/978-90-481-9109-3_3

Carbone, A., Evans, J., Ross, B., Drew, S., Phelan, L., Lindsay, K., & Ye, J. (2017). Assessing distributed leadership for learning and teaching quality: A multi-institutional study. *Journal of Higher Education Policy and Management, 39*(2), 183–196. doi:10.1080/1360080X.2017.1276629

Carless, D., & Boud, D. (2018). The Development of Student Feedback Literacy: Enabling Uptake of Feedback. *Assessment & Evaluation in Higher Education, 43*(8), 1315–1325. doi:10.1080/02602938.2018.1463354

Carmeli, A., Gelbard, R., & Reiter-Palmon, R. (2013). Leadership, creative problem-solving capacity, and creative performance: The importance of knowledge sharing. *Human Resource Management, 52*(1), 95–121. doi:10.1002/hrm.21514

Carmen, J. M. (2002). Blended learning design: Five key ingredients. *Journal of Educational Media*, 165–178.

Carroll, J. M. (2001). *Human-computer Interaction in the new millennium*. Addison-Wesley.

Carter, J. (2022). *Metaverse Vs Multiverse: How to Tell Apart The 'Parallel Universe' Concepts We Can't Stop Talking About*. https://www.forbes.com/sites/ jamiecartereurope/2022/02/07

Caruth, G. D. (2014). *Meeting the needs of older students in higher education*. Participatory Educational Research (PER). http://www.partedres.com/archieve/issue_1_2/3-per_14-09_volume_1_issue_2 _page_21_35.pdf

Carver, C. L. (2016). Transforming identities: The transition from teacher to leader during teacher leader preparation. *Journal of Research on Leadership Education, 11*(2), 158–180. doi:10.1177/1942775116658635

Carver, R., King, R., Hannum, W., & Fowler, B. (2007). Toward a model of experiential e-learning. *Journal of Online Learning and Teaching, 3*(3), 247–256.

Cascio, W. F., & Montealegre, R. (2016). How technology is changing work and organizations. *Annual Review of Organizational Psychology and Organizational Behavior, 3*(1), 349–375. doi:10.1146/annurev-orgpsych-041015-062352

Caso, A., & Toledano, V. L. (2017). The university debate between Antonio Caso and Vicente Lombardo Toledano [1933]. In C. Sanchez & R. E. Sanchez Jr., (Eds.), *Mexican philosophy of the 20th Century: Essential readings* (pp. 73–92). Oxford University Press.

Cassar, V., Bezzina, F., & Buttigieg, S. C. (2017). The relationship between transformational leadership and work attitudes: Comparing mediating influences of social identity and the psychological contract. *Leadership and Organization Development Journal, 38*(5), 646–661. doi:10.1108/LODJ-11-2015-0248

Ceballos, M., Buckridge, H., & Taylor, R. T. (2020). Educational Leadership Students and Mixed Reality Experiences: Building Student Confidence to Communicate with Parents and Teachers. *The International Journal of Educational Leadership Preparation, 15*(1), 58–71.

Celi, M., & Rudkin, J. (2015). Towards user involvement in envisioning practices: Opening the use of the future. In D. Bihanic (Ed.), *Empowering users through design: Interdisciplinary studies and combined approaches for technological products and services* (pp. 59–78). Springer., doi:10.1007/978-3-319-13018-7_4

Centers for Disease Control. (n.d.). *CDC Museum COVID-19 Timeline*. https://www.cdc.gov/museum/timeline/covid19.html

Centre for Teaching and Learning. (n.d.). *What is hybrid teaching?* University of Oxford. https://www.ctl.ox.ac.uk/what-is-hybrid-teaching

Cercone, K. (2008). Characteristics of adult learners with implications for online learning design. *AACE Journal, 16*(2), 137–159.

Cervera-Torres, S., Ruiz-Fernández, S., Godbersen, H., Massó, L., Martínez-Rubio, D., Pintado-Cucarella, S., & Baños, R. M. (2021). Influence of resilience and optimism on distress and intention to self-isolate: Contrasting lower and higher COVID-19 illness risk samples from an extended health belief model. *Frontiers in Psychology, 12*, 662395. doi:10.3389/fpsyg.2021.662395 PMID:34108919

Challen, C. (2020). Rethinking higher education policy and leadership in the 21st century: Enhancing strategies for global citizenship and justice. *Journal of Higher Education Policy and Leadership Studies, 1*(1), 77–81. doi:10.29252/johepal.1.1.77

Chamberlain, J. M., Hillier, J., & Signoretta, P. (2015). Counting better? An examination of the impact of quantitative method teaching on statistical anxiety and confidence. *Active Learning in Higher Education, 16*(1), 51–66. doi:10.1177/1469787414558983

Chambers, G. (1982). A model for the development of technical instructors. *Master Abstracts International, 20*(4), 342.

Chance, N. L. (2022). Resilient leadership: A phenomenological exploration into how black women in higher education leadership navigate cultural adversity. *Journal of Humanistic Psychology, 62*(1), 44–78. doi:10.1177/00221678211003000

Chang, J.-L., Lieu, P.-T., Liang, J.-H., Liu, H.-T., & Wong, S.-L. (2011). Factors influencing technology acceptance decision. *African Journal of Business Management, 5*(7), 2901-2909. http://www.academicjournals.org/ajbm/abstracts/abstracts/abstracts2011/4Apr/Chang%20et%20al%202.htm

Chan, S. (2010). Applications of andragogy in multi-disciplined teaching and learning. *Journal of Adult Education, 39*(2), 25–35.

Chantrill, C. (2015). *US Education Spending History from 1900: A Century of Education Spending*. http://usgovernmentspending.com

Charques, R. (1932). *Soviet education: Some aspects of cultural revolution*. L. & V. Woolf.

Chartered Institute of Personnel and Development. (2020). *Health and wellbeing at work 2022*. CIPD.

Charungkaittikul, S., & Henschke, J. (2018). Applying an andragogical approach to foster lifelong self-directed learning in the 21st century higher education classroom settings. *Journal of Educational Leadership in Action, 5*(2), 5. https://digitalcommons.lindenwood.edu/ela/vol5/iss2/5

Chatterjee, S., Chaudhuri, R., Thrassou, A., & Vrontis, D. (2021). Antecedents and consequences of knowledge hiding: The moderating role of knowledge hiders and knowledge seekers in organizations. *Journal of Business Research, 128*(5), 303–313. doi:10.1016/j.jbusres.2021.02.033

Chawdhry, A. A., Paullet, K., Baugh, J., & Nakama, D. (2022). How the COVID-19 shutdown impacted grades at the collegiate level? *Information Systems Education Journal, 20*(5), 34–41.

Chayka, K. (2021). *We already live in Facebook's Metaverse*. https://www.newyorker.com/culture/infinite-scroll/we-already

Chejlyk, S. (2006). *The effects of online course format and three components of student perceived interactions on overall course satisfaction*. https://www.editlib.org/p/117746/

Cheng, A. Y., & Szeto, E. (2016). Teacher leadership development and principal facilitation: Novice teachers' perspectives. *Teaching and Teacher Education*, *58*, 140–148. doi:10.1016/j.tate.2016.05.003

Cheng, E. W. L., & Hampson, I. (2008). Transfer of training: A review and new insights. *International Journal of Management Reviews*, *10*(4), 327–341. doi:10.1111/j.1468-2370.2007.00230.x

Cheng, J. W., Mitomo, H., Kamplean, A., & Seo, Y. (2021). Lesser evil? Public opinion on regulating fake news in Japan, South Korea, and Thailand–A three-country comparison. *Telecommunications Policy*, *45*(9), 102185. doi:10.1016/j.telpol.2021.102185

Chen, J. (2014). Teaching nontraditional adult students: Adult learning theories in practice. *Teaching in Higher Education*, *4*(19), 406–118. doi:10.1080/13562517.2013.860101

Chen, R., Wu, N., Varvello, M., Chens, S., & Han, B. (2022, October). Are we ready for the Metaverse? A measurement study of social virtual reality platforms. *Proceedings of the 22nd ACM Internet Measurement Conference,* 504-518. 10.1145/3517745.3561417

Cherry, B., Graves, B.M., & Grasse, N. (2022). Causes, processes, and effects of academic reorganization at public master's universities in the United States. *Perspectives: Policy and Practice in Higher Education*, 1-9. doi:10.1080/13603108.2022.2121779

Cherubini, L. (2009). New teachers' perceptions of induction: Insights into principles practices. *The Alberta Journal of Educational Research*, *55*(2), 185–198. http://www.ajer.ca/

Chigona, A. (2018). Digital fluency: Necessary competence for teaching and learning in connected classrooms. *The African Journal of Information Systems*, *10*(4), 7. https://digitalcommons.kennesaw.edu/ajis/vol10/iss4/7

Chirca, R. C., & Pânişoară, I.-O. (2020). Study on Romanian students' digital level and learning styles. *Journal of Educational Sciences and Psychology*, *10*(1), 42–52.

Chiu, C., Hsu, M., & Wang, E. (2006). Understanding knowledge sharing in virtual communities: An integration of social capital and social cognitive theories. *Decision Support Systems*, *42*(3), 1872–1888. doi:10.1016/j.dss.2006.04.001

ChohanU. W. (2022). *Metaverse or Metacurse*? doi:10.2139/ssrn.4038770

Choi, E. H., Kim, E. K., & Kim, P. B. (2018). Effects of the educational leadership of nursing unit managers on team effectiveness: Mediating effects of organizational communication. *Asian Nursing Research*, *12*(2), 99–105. doi:10.1016/j.anr.2018.03.001 PMID:29614322

Chomsky, N. (2014). What is the common good? *Toward Freedom*, 9. https://towardfreedom.org/story/archives/special-reports-archives/noam-chomsky-what-is-the-common-good/

Chomsky, N. (2015, May 26*). On being truly educated* [Interview]. The Brainwaves Video Anthology. https://youtu.be/eYHQcXVp4F4

Chomsky, N., Roberts, I., & Watumull, J. (2023, March 8). Opinion | Noam Chomsky: The False Promise of ChatGPT. *The New York Times*. https://www.nytimes.com/2023/03/08/opinion/noam-chomsky-chatgpt-ai.html?smid=tw-share

Choo, A. S., Linderman, K., & Schroeder, R. G. (2004). Social and method effects on learning behaviors and knowledge creation in six sigma projects. *Academy of Management Proceedings*, *2004*(1), C1–C6. doi:10.5465/ambpp.2004.13857620

Chopra, D. (2010). *The soul of leadership: Unlocking your potential for greatness.* Harmony Books.

Christian, P., & Lolas, F. (1985). The stress concept as a problem for a theoretical pathology. *Social Science & Medicine, 2*(12), 363–365. doi:10.1016/0277-9536(85)90444-7 PMID:4095592

Chung, H. (2022). Best practices teaching adult English as a second language in the virtual classroom. *New Directions for Adult and Continuing Education, 2022*(175/176), 45–57. doi:10.1002/ace.20467

Cifuentes, L. (2013). *A guide to administering distance learning.* Brill.

Clardy, A. (2000). Learning on their own: Vocationally oriented self-directed learning projects. *Human Resource Development Quarterly, 11*(2), 105–125. doi:10.1002/1532-1096(200022)11:2<105::AID-HRDQ2>3.0.CO;2-5

Clark, R. E. (1994). Media will never influence learning. *Educational Technology Research and Development, 42*(2), 21–29. doi:10.1007/BF02299088

Clayton, J. K., Sanzo, K. L., & Myran, S. (2013). Understanding mentoring in leadership development: Perspectives of district administrators and aspiring leaders. *Journal of Research on Leadership Education, 8*(1), 77–96. doi:10.1177/1942775112464959

Cleveland-Innes, M. (2019). The community of inquiry theoretical framework. Designing collaborative online and blended learning. In H. Beetham & R. Sharpe (Eds.), *Rethinking pedagogy for a digital age* (pp. 85–101). Routledge. doi:10.4324/9781351252805-6

Cohen, L., Manion, L., & Morrison, K. (2003). *Research methods in education.* Routledge Farmer.

Colcleugh, D. (2013). *Everyone a leader: A guide to leading high-performance organizations for engineers and scientists.* University of Toronto Press.

Coleman, M., & Glover, D. (2010). *Educational leadership and management: Developing insights and skills.* Open University Press.

Collins, A., & Halverson, R. (2018). *Rethinking education in the age of technology: The digital revolution and schooling in America.* Teachers College Press.

Collins, D. B., & Holton, E. F. (2004). The effectiveness of managerial leadership development programs: A meta-analysis of studies from 1982 to 2001. *Human Resource Development Quarterly, 15*(2), 217–248. doi:10.1002/hrdq.1099

Collins, J. (2001). *Good to great.* Random House Business Books.

Collins, M. (1995). Critical commentaries on the role of the adult educator: From self-directed learning to postmodernist sensibilities. In M. R. Welton (Ed.), *In defense of the lifeworld* (pp. 71–98). SUNY Press.

Collins, R. A., & Zacharakis, J. (2022). Continuous Challenges: Case Study of a Sustainable University Adult Education Graduate Program. *Adult Learning, 0*(0). Advance online publication. doi:10.1177/10451595221106171

Collis, B., & Moonen, J. (2001). *Flexible learning in a digital world: Experiences and Expectations.* Kogan Page.

Coltart, C. E., Cheung, R., Ardolino, A., Bray, B., Rocos, B., Bailey, A., Bethune, R., Butler, J., Docherty, M., Drysdale, K., Fayaz, A., Greaves, F., Hafferty, J., Malik, A. N., Moolla, A., Morganstein, L., Pathiraja, F., Shah, A., Sleat, G., ... Donaldson, L. (2012). Leadership development for early career doctors. *Lancet, 379*(9828), 1847–1849. doi:10.1016/S0140-6736(12)60271-2 PMID:22531157

Comstock, M., Supovitz, J., & Kaul, M. (2021). Exchange quality in teacher leadership ties: Examining relational quality using social network and leader-member exchange theories. *Journal of Professional Capital and Community, 6*(4), 395–409. doi:10.1108/JPCC-01-2021-0002

Conaway, W., & Zorn-Arnold, B. (2015). The keys to online learning for adults: The six principles of andragogy. *Distance Learning, 12*(4), 37–42.

Confessore, S. J., & Kops, W. J. (1998). Self-directed learning and the learning organization: Examining the connection between the individual and the learning environment. *Human Resource Development Quarterly, 9*(4), 365–375. doi:10.1002/hrdq.3920090407

Conger, J. A. (1993). The brave new world of leadership training. *Organizational Dynamics, 21*(3), 46–58. doi:10.1016/0090-2616(93)90070-H

Connelly, C. E., Zweig, D., Webster, J., & Trougakos, J. P. (2012). Knowledge hiding in organizations. *Journal of Organizational Behavior, 33*(1), 64–88. doi:10.1002/job.737

Connolly, M., James, C., & Fertig, M. (2019). The difference between educational management and educational leadership and the importance of educational responsibility. *Educational Management Administration & Leadership, 47*(4), 504–519. doi:10.1177/1741143217745880

Cook, K. (1945). *The house of the people: An account of Mexico's new schools of action.* U.S. Gov't. Printing Office.

Cooper, C. L., Dewe, P. J., & O'Driscoll, M. P. (2001). What is stress? In *Organizational stress: A review and critique of theory, research, and applications.* SAGE Publications, Inc., https://www.doi.org/10.4135/9781452231235.n1

Cooper, M. K., & Henschke, J. A. (2001, Sept. 26-28). Andragogy: Its research value for practice. In W.C. Hine (Ed.), *Annual Midwest Research-to-Practice Conference in Adult, Continuing & Community Education.* Academic Press.

Cooper, C. D., Scandura, T. A., & Schriesheim, C. A. (2005). Looking forward but learning from our past: Potential challenges to developing authentic leadership theory and authentic leaders. *The Leadership Quarterly, 16*(3), 475–493. doi:10.1016/j.leaqua.2005.03.008

Cope, R. F., Cope, R. F. III, Bass, A. N., & Syrdal, H. A. (2011). Innovative knowledge management at Disney: Human capital and queuing solutions for services. *Journal of Service Science, 4*(1). Advance online publication. doi:10.19030/jss.v4i1.4268

Corbett, F., & Spinello, E. (2020). Connectivism and leadership: Harnessing a learning theory for the digital age to redefine leadership in the twenty-first century. *Heliyon, 6*(1), e03250. doi:10.1016/j.heliyon.2020.e03250 PMID:31993523

Correia, M. S. (2019). Sustainability: An overview of the triple bottom line and sustainability implementation. *International Journal of Strategic Engineering, 2*(1), 29–38. doi:10.4018/IJoSE.2019010103

Cortellazzo, L., Bruni, E., & Zampieri, R. (2019). The role of leadership in a digitalized world: A review. *Frontiers in Psychology, 10*, 1938. doi:10.3389/fpsyg.2019.01938 PMID:31507494

Cosner, S., Tozer, S., Zavitkovsky, P., & Whalen, S. P. (2015). Cultivating exemplary school leadership preparation at a research intensive university. *Journal of Research on Leadership Education, 10*(1), 11–38. doi:10.1177/1942775115569575

Costa, A., & Garmston, R. (2002). *Cognitive coaching: A foundation for Renaissance schools.* Academic Press.

Counts, G. (1931). *The Soviet challenge to America.* John Day Publishing.

Counts, G. (1959). *Khrushchev and the central committee speak on education.* Pittsburgh University Press.

Counts, G. (1975). *The challenge of Soviet education* (Vol. 1957). Greenwood.

Covey, S. R. (1989). *The seven habits of highly effective people: Restoring the character ethic.* Simon and Schuster.

Cowling, M., & Birt, J. (2018). Pedagogy before technology: A design-based research approach to enhancing skills development in paramedic science using mixed reality. *Information (Basel), 29*(9), 29. doi:10.3390/info9020029

Cox, L. K., & Tachau, K. H. (2023). *The erosion of support for education and tenure in Iowa*. American Association of University Professors (AAUP). https://www.aaup.org/article/erosion-support-education-and-tenure-iowa#.ZBW07XbMKUk

Crane, L. (2016). *Knowledge and discourse matters: Relocating knowledge management's sphere of interest onto language* (1st ed.). Wiley.

Cranton, P. (2016). *Understanding and promoting transformative learning* (3rd ed.). Stylus Publishing.

Creswell, J. W., & Plano Clark, V. L. (2018). *Designing and conducting mixed methods research* (3rd ed.). SAGE.

Cristancho, S. (2016). Lessons on resilience: Learning to manage complexity. *Perspectives on Medical Education, 5*(3), 133–135. doi:10.1007/S40037-016-0277-1 PMID:27256287

Cristaudo, W., Fiering, N., & Leutzsch, A. (2015). *Introduction: Eugen Rosenstock-Huessy (1888-*1973*). Culture, Theory & Critique*, *56*(1), 1–12. doi:10.1080/14735784.2014.979086

Cross, K. P. (1999). *Learning is about making connections. The Cross Papers, 3.*

Cross, P. (1981). *Adults as learners*. Jossey-Bass.

Cruz-González, C., Rodríguez, C. L., & Segovia, J. D. (2021). A systematic review of principals' leadership identity from 1993 to 2019. *Educational Management Administration & Leadership, 49*(1), 31–53. doi:10.1177/1741143219896053

Cullen-Lester, K. L., Maupin, C. K., & Carter, D. R. (2017). Incorporating social networks into leadership development: A conceptual model and evaluation of research and practice. *The Leadership Quarterly, 28*(1), 130–152. doi:10.1016/j.leaqua.2016.10.005

Cunningham, K. M., & Lochmiller, C. R. (2020). Content-specific leadership: Identifying literature-based implications for principal preparation. *Journal of Research on Leadership Education, 15*(4), 261–282. doi:10.1177/1942775119845004

Cupit, T., Stout-Aguilar, J., Cannon, L., & Norton, J. (2019). Assessing the nurse managers' span of control: A partnership between executive leadership, nurse scientists and clinicians. *Nurse Leader, 17*(2), 103–108. doi:10.1016/j.mnl.2018.12.004

D'Arienzo, M. C., Boursier, V., & Griffiths, M. D. (2019). Addiction to social media and attachment styles: A systematic literature review. *International Journal of Mental Health and Addiction, 17*(4), 1094–1118. doi:10.100711469-019-00082-5

Dachner, A. M., & Polin, B. (2016). A systematic approach to educating the emerging adult learner in undergraduate management courses. *Journal of Management Education, 40*(2), 121–151. doi:10.1177/1052562915613589

Dahlman, C., Zeng, Z. D., & Wang, S. (2007). *Enhancing China's competitiveness through lifelong learning*. The World Bank.

Daniel, C. O. (2019). Effects of job stress on employee's performance. *International Journal of Business. Management and Social Research, 6*(2), 375–382.

Danişman, S., Tosuntas, S. B., & Karadag, E. (2015). The effect of leadership on organizational performance. In E. Karadag (Ed.), *Leadership and organizational outcomes* (pp. 143–168). Springer. doi:10.1007/978-3-319-14908-0_9

Darling-Hammond, L. (2000). *Teacher quality and student achievement: A review of state policy evidence*. doi:10.14507/epaa.v8n1.2000

Davidson-Shivers, G. V. (2009). Frequency and types of instructor interactions in online instruction. *Journal of Interactive Online Learning, 8*(1), 23–40.

Davidson-Shivers, G. V., & Rasmussen, K. L. (2006). *Web-based learning: Design, implementation, & evaluation.* Merrill Prentice Hall.

Davis McCauley, K., Hammer, E., & Hinojosa, A. S. (2017). An andragogical approach to teaching leadership. *Management Teaching Review*, *2*(4), 312–324. doi:10.1177/2379298117736885

Davis, H. C., & Fill, K. (2007). Embedding blended learning in a university's teaching culture: Experiences and reflections. *British Journal of Educational Technology*, *38*(5), 817–828. doi:10.1111/j.1467-8535.2007.00756.x

Davis, H. S. (2012). Discussion as a bridge: Strategies that engage adolescent and adult learning styles in the postsecondary classroom. *The Journal of Scholarship of Teaching and Learning*, *13*(1), 68–76.

Dawson, V. R. (2018). Fans, friends, advocates, ambassadors, and haters: Social media communities and the communicative constitution of organizational identity. *Social Media + Society*, *4*(1). Advance online publication. doi:10.1177/2056305117746356

Day, D. V. (2010). The difficulties of learning from experience and the need for deliberate practice. *Industrial and Organizational Psychology*, *3*(1), 41–44. . doi:10.1111/j.1754-9434.2009.01195.x

Day, C., & Baskett, H. K. (1982). Discrepancies between intentions and practice: Reexamining some basic assumptions about adult and continuing professional education. *International Journal of Lifelong Education*, *1*(2), 143–155. doi:10.1080/0260137820010205

Day, C., Gu, Q., & Sammons, P. (2016). The impact of leadership on student outcomes: How successful school leaders use transformational and instructional strategies to make a difference. *Educational Administration Quarterly*, *52*(2), 221–258. doi:10.1177/0013161X15616863

Day, C., Harris, A., & Hadfield, M. (2001). Challenging the orthodoxy of effective school leadership. *International Journal of Leadership in Education*, *4*(1), 39–56. doi:10.1080/13603120117505

Day, D. V. (2000). Leadership development: A review in context. *The Leadership Quarterly*, *11*(4), 581–613. doi:10.1016/S1048-9843(00)00061-8

Day, D. V., Harrison, M. M., & Halpin, S. M. (2009). *An integrative approach to leader development: Connecting adult development, identity, and expertise*. Routledge.

Day, D. V., & Liu, Z. (2019). What is wrong with leadership development and what might be done about it? In R. E. Riggio (Ed.), *What's wrong with leadership: Improving leadership research and practice* (pp. 226–240). Routledge. doi:10.4324/9781315163604-13

De Nobile, J. (2018). Towards a theoretical model of middle leadership in schools. *School Leadership & Management*, *38*(4), 395–416. doi:10.1080/13632434.2017.1411902

De Vaus, D. A. (2002). *Surveys in social research.* Allen & Unwin. doi:10.4135/9781446263495

Deci, E. L., & Ryan, R. M. (1993). Die selbstbestimmungstheorie der motivation und ihre bedeutung für die pädagogik. *Zeitschrift für Pädagogik*, *39*(2), 223–238.

Decker, L. (1992). *Community education: Building learning communities* (Rev. ed.). Community Education Association.

del Carmen Salazar, M. (2013). A humanizing pedagogy: Reinventing the principles and practice of education as a journey toward liberation. *Review of Research in Education*, *37*(1), 121–148. doi:10.3102/0091732X12464032

Delaney, Y. A. (2012). Research on mentoring language teachers: Its role in language education. *Foreign Language Annals*, *45*(1), 184–202. doi:10.1111/j.1944-9720.2011.01185.x

Denisova-Schmidt, E., & Leontyeva, E. (2014). The unified state exam in Russia: Problems and perspectives. *Industry and Higher Education, 76*(76), 22–23. doi:10.6017/ihe.2014.76.5530

Denson, N., Loveday, T., & Dalton, H. M. (2010). Student evaluation of courses: What predicts satisfaction? *Higher Education Research & Development, 29*(4), 339–356. doi:10.1080/07294360903394466

DeRue, D. S., & Myers, C. (2014). Leadership development: a review and agenda for future research. In D. V. Day (Ed.), *The Oxford handbook of leadership and organizations* (pp. 829–852). Open University Press.

Deterding, S., Khaled, R., Nacke, L. E., & Dixon, D. (2011, May). Gamification: Toward a definition. In *CHI 2011 gamification workshop proceedings* (Vol. 12, p. 15). ACM.

Dewald, N. H. (2003). Pedagogy and andragogy. In E. A. Dupuis (Ed.), *Developing web-based instruction: Planning, designing, managing, and evaluating for results* (pp. 47–68). Neal-Schuman.

Dewey, J. (1897, January 16). My pedagogic creed. *The School Journal, 54*(3), 77–80.

Dewey, J. (1916). *Democracy and Education.* Free Press.

Dewey, J. (1927). *The public and its problems.* Holt and Company.

Dewey, J. (1929). *Impressions of Soviet Russia and the revolutionary world.* New Republic.

Dewey, J. (1938). Experience and education. *Touchstone (Nashville, Tenn.).*

Dewey, J. (1959). My pedagogic creed. In J. Dewey (Ed.), *Dewey on education* (pp. 19–32). Teachers College, Columbia University. (Original work published 1897)

Dewey, J. (1972). *Not guilty: Report of the commission of inquiry into the charges made against Leon Trotsky in the Moscow trials.* Anchor Foundation.

deZwart, M. J., & Lindsay, D. F. (2009). Governance and the global metaverse. In D. Riha & A. Maj (Eds.), *The real and the virtual critical issues in cybercultures* (pp. 173–182). Inter-Disciplinary Press.

Di Vaio, A., Palladino, R., Pezzi, A., & Kalisz, D. E. (2021). The role of digital innovation in knowledge management systems: A systematic literature review. *Journal of Business Research, 123*, 220–231. doi:10.1016/j.jbusres.2020.09.042

Diamond, A. (2014). Pre-service early childhood educators' leadership development through reflective engagement with experiential service learning and leadership literature. *Australasian Journal of Early Childhood, 39*(4), 12–20. doi:10.1177/183693911403900403

Diaz, J., Saldana, C., & Avila, C. (2022). Virtual world as a resource for hybrid education. *International Journal of Emerging Technology Learn, 15*(15), 94–109. doi:10.3991/ijet.v15i15.13025

Dick, E. (2021). *Public Policy for the Metaverse: Key Takeaways. AR/VR Policy Conference.* Information Technology and Innovation Foundation.

Dickinson, L. (1993). Talking shop: Aspects of autonomous learning. *ELT Journal, 47*(4), 330–336. doi:10.1093/elt/47.4.330

Dickson, M., Den Hartog, D., & Deuling, J. (2003). Research on leadership in a cross-cultural context: Making progress, and raising new questions. *The Leadership Quarterly, 14*(6), 729–768. doi:10.1016/j.leaqua.2003.09.002

Dimmock, C. (2003). Leadership in learning-centred schools: cultural context, functions and qualities. In M. Brundrett, N. Burton, & R. Smith (Eds.), *Leadership in education* (pp. 3–22). SAGE Publications, Inc. doi:10.4135/9781446215036.n1

Dinh, J. E., Lord, R. G., Gardner, W. L., Meuser, J. D., Liden, R. C., & Hu, J. (2014). Leadership theory and research in the new millennium: Current theoretical trends and changing perspectives. *The Leadership Quarterly*, *25*(1), 36–62. doi:10.1016/j.leaqua.2013.11.005

DiPaola, M. F. (2003). Conflict and change: Daily challenges for school leaders. In N. Bennett, M. Crawford, & M. Cartwright (Eds.), *Effective educational leadership* (pp. 143–158). Paul Chapman.

Dirkx, J. M., Smith, R., & Amey, M. (2001, Sept. 26-28). Being a "college student": The community college experience of under-prepared adult learners. In W. C. Hine (Ed.), *Annual Midwest Research-to-Practice Conference Adult Continuing Community Education.* Academic Press.

Dirkx, J. M. (1998). Transformative learning theory in the practice of adult education: An overview. *PAACE Journal of Lifelong Learning*, *7*, 1–14.

Dixson, M. (2015). Measuring student engagement in the online course: The online student engagement scale. *Online Learning*, *19*(4), 1–14. http://www.online learningconsortium.org

Dixson, M. D. (2010). Creating effective student engagement in online courses: What do students find engaging? *The Journal of Scholarship of Teaching and Learning*, *10*(2), 1–13.

Dobbs, State Health Officer of the Mississippi Department of Health, et al., *v.* Jackson Women's Health Organization et al. (date of publication) 945 F. 3d 265. https://www.supremecourt.gov/opinions/21pdf/19-1392_6j37.pdf

Dolph, D. (2017). Challenges and opportunities for school improvement: Recommendations for urban school principals. *Education and Urban Society*, *49*(4), 363–387. doi:10.1177/0013124516659110

Domyshcva, S. A., & Kopylova, N. V. (2021). Overcoming non-linguistic barriers to effective speaking in Russian adult ESP classroom. *History, Pedagogics, Philology*, *27*(1), 87–94. doi:10.18287/2542-0445-2021-27-1-87-94

Donavant, B. W. (2009). The new, modern practice of adult education online instruction in a continuing professional education setting. *Adult Education Quarterly*, *59*(3), 227–245. doi:10.1177/0741713609331546

Dong, Y. T., Bartol, K. M., Zhang, Z. X., & Li, C. W. (2017). Enhancing employee creativity via individual skill development and team knowledge sharing: Influences of dual-focused transformational leadership. *Journal of Organizational Behavior*, *38*(3), 439–458. doi:10.1002/job.2134

Donovan, J. (2015). *The importance of building online learning communities.* https://blog.online.colostate.edu/blog/uncategorized/the-importance-of-building-online-learning-communities/

Dopson, S., Ferlie, E., McGivern, G., Fischer, M. D., Mitra, M., Ledger, J., & Behrens, S. (2019). Leadership development in higher education: A literature review and implications for programme redesign. *Higher Education Quarterly*, *73*(2), 218–234. doi:10.1111/hequ.12194

Dougherty, K. J., & Natow, R. S. (2019). *Analyzing neoliberalism in theory and practice: The case of performance-based funding for higher education.* Center for Global Higher Education. Working Paper #44. https://apo.org.au/node/232706

Douglass, A. (2018). Redefining leadership: Lessons from an early education leadership development initiative. *Early Childhood Education Journal*, *46*(4), 387–396. doi:10.100710643-017-0871-9

Drago-Seversen, E., Blum-DeStefano, J., & Brooks Lawrence, D. (2023). *Growing for justice: A developmental continuum of leadership capacities and practices.* Corwin.

Draper, J. (1998). The metamorphoses of andragogy. *Canadian Journal for the Study of Adult Education*, *3*(26).

Drew, G. (2010). Issues and challenges in higher education leadership: Engaging for change. *Australian Educational Researcher*, *37*(3), 57–76. doi:10.1007/BF03216930

Driscoll, M. (2002, March 1). *Blended learning: let's get beyond the hype.* http://elearningmag.com/ltimagazine

Duch, B. J., Groh, S. E., & Allen, D. E. (2001). Why problem-based learning? A case study of institutional change in undergraduate education. In B. Duch, S. Groh, & D. Allen (Eds.), *The power of problem based learning* (pp. 3–11). Stylus.

Duckworth, A. (2016). *Grit: The power of passion and perseverance*. Scribner.

Duke, D. L. (2019). Judgment and the preparation of educational leaders. *Journal of Research on Leadership Education*, *14*(3), 191–211. doi:10.1177/1942775117752455

Duke, S. T., & Ward, J. D. (2009). Preparing information literate teachers: A metasynthesis. *Library & Information Science Research*, *31*(4), 247–256. doi:10.1016/j.lisr.2009.04.003

Duncan-Howell, J. (2010). Teachers making connections: Online communities as a source of professional learning. *British Journal of Educational Technology*, *41*(2), 324–340. doi:10.1111/j.1467-8535.2009.00953.x

Du, S., Swaen, V., Lindgreen, A., & Sen, S. (2013). The roles of leadership styles in corporate social responsibility. *Journal of Business Ethics*, *114*(1), 155–169. doi:10.100710551-012-1333-3

Dweck, C. (2006). *Mindset: The new psychology of success: How we can learn to fulfill us potential*. Ballantine Books.

Dweck, C. (2016). *Mindset: The new psychology of success*. Penguin Random House LLC.

Dweck, C. S. (1999). *Self-theories: Their role in motivation, personality, and development*. Psychology Press.

Dyson, F. J., Hansma, P. K., Roof, W. C., & Proctor, J. D. (2001). *Technology and social justice*. UCTV, Instructional Resources, University of California.

Dzubian, C., Hartman, J. L., & Moskal, P. D. (2004). Blended Learning. *Educause*, (7), 1–12.

Eadens, D. W., & Ceballos, M. (2022). Educational Leadership Preparation and Professional Roles: Are We Serving the Needs of Leadership Roles Along the Leadership Continuum? *Journal of Research on Leadership Education*, *19427751221076416*. Advance online publication. doi:10.1177/19427751221076416

Edmondson, A. C., & Lei, Z. (2014). Psychological Safety: The History, Renaissance, and Future of an Interpersonal Construct. *Annual Review of Organizational Psychology and Organizational Behavior*, *1*(1), 23–43. doi:10.1146/annurev-orgpsych-031413-091305

Ehrich, L., Kimber, M., & Ehrich, J. (2016). The micropolitics of management in universities: Challenges and opportunities. In J. Ryan & D. Armstrong (Eds.), *Working without the system* (pp. 187–204). Information Age Publishing.

Eisenhardt, K. M. (1999c). Strategy as Strategic Decision Making. *Sloan Management Review*, *40*(3), 65–72. https://dialnet.unirioja.es/servlet/articulo?codigo=2485499

Ekoto, E., & Galkwad, P. (2015). The impact of andragogy on learning satisfaction of graduate students. *American Journal of Educational Research*, *3*(11), 1378–1386. doi:10.12691/education-3-11-6

Elbayadi, M. E. (2014). *Relational leadership, DevOps, and the Post-PC era: Toward a practical theory for 21st century technology leaders* [Unpublished doctoral thesis]. Antioch University.

Elias, J. L., & Merriam, S. B. (1995). *Philosophical foundations of adult education*. Krieger.

Ellemers, N., & Haslam, S. A. (2012). Social identity theory. In P. A. M. Van Lange, A. W. Kruglanski, & E. T. Higgins (Eds.), *Handbook of theories of social psychology* (Vol. 2, pp. 379–398). Sage Publications Ltd. doi:10.4135/9781446249222.n45

Ellinger, A. D. (2004). The concept of self-directed learning and its implications for human resource development. *Advances in Developing Human Resources, 6*(2), 158–177. doi:10.1177/1523422304263327

Ely, R. J., Ibarra, H., & Kolb, D. M. (2011). Taking gender into account: Theory and design for women's leadership development programs. *Academy of Management Learning & Education, 10*(3), 474–493. doi:10.5465/amle.2010.0046

Engeström, Y. (1999). Activity theory and individual and social transformation. *Perspectives on Activity Theory, 19*(38), 19–30. doi:10.1017/CBO9780511812774.003

Erpenbeck, J., & Sauter, W. (2013). *So werden wir lernen! Kompetenzentwicklung in einer welt fühlender Computer, kluger wolken und sinnsuchender Netze*. Springer Gabler. doi:10.1007/978-3-642-37181-3

Escobar, M., Fernandez, A., Guevara Niebla, G., & Freire, P. (1994). *Paolo Freire on higher education: A dialogue at the National University of Mexico*. SUNY Press.

Esipov, B. P., & Goncharov, N. K. (1947). *I want to be like Stalin*. Introduction by G. Counts. John Day.

EU-OSHA. (2014). *Calculating the cost of work-related stress and psychosocial risks*. Publications Office of the European Union., doi:10.2802/20493

European Parliament & Council of the European Union. (2006). *Recommendation 2006/962 – Key competences for lifelong learning*. Author.

Evans, M., Dalkir, K., & Bidian, C. (2014). A holistic view of the knowledge life cycle: The knowledge management cycle (KMC) model. *Electronic Journal of Knowledge Management, 12*(2), 85–97. https://academic-publishing.org/index.php/ejkm/article/view/1015

Fairhurst, G. T. (2009). Considering the context in discursive leadership research. *Human Relations, 62*(11), 1607–1633. doi:10.1177/0018726709346379

Fairman, J. C., & Mackenzie, S. V. (2015). How teacher leaders influence others and understand their leadership. *International Journal of Leadership in Education, 18*(1), 61–87. doi:10.1080/13603124.2014.904002

Falasca, M. (2011). Barriers to adult learning: Bridging the gap. *Australian Journal of Adult Learning, 51*(3), 583–588. http://find.galegroup.com/gtx/start.do?userGroupName=gale15691&prodId=AONE

Farjami, S., Taguchi, R., Nakahira, K. T., Nunez, R., Fukumura, Y. M., & Kanematsu, H. (2011). Multilingual problem-based learning in Metaverse. *Technology and Society Magazine, 37*(2), 52–61.

Faulconer, E. K., Griffith, J., Wood, B., Acharyya, S., & Roberts, D. (2018). A comparison of online, video synchronous, and traditional learning modes for an introductory undergraduate physics course. *Journal of Science Education and Technology, 27*(5), 404–411. doi:10.100710956-018-9732-6

Fediaevskaia, V. (1936). *Nursery school and parent education in Soviet Russia*. Dutton.

Feezell, J. T. (2018). Agenda Setting through social media: The Importance of Incidental News Exposure and Social Filtering in the Digital Era. *Political Research Quarterly, 71*(2), 482–494. doi:10.1177/1065912917744895

Feiman-Nemser, S. (2001). Helping novices learn to teach. *Journal of Teacher Education, 52*(1), 17–26. doi:10.1177/0022487101052001003

Fellenz, R., & Coker, L. (1979). Toward a definition of adult education, community education and community development. In P. Cunningham (Ed.), *Yearbook of adult and continuing education* (pp. 366–369). Marquis Academic Media.

Fenwick, T. (2008). Workplace learning: Emerging trends and new perspectives. *New Directions for Adult and Continuing Education*, 119, 17–26. doi:10.1002/ace.302

Fernandez, C.B. & Hui. (2022). Life, the Metaverse and Everything: An Overview of Privacy, Ethics, and Governance in Metaverse. *2022 IEEE 42nd International Conference on Distributed Computing Systems Workshops (ICDCSW)*, 272-277.

Ferrara E., & Yang, Z. (2015) Measuring emotional contagion in social media. *PLOS One, 10*(11). 1.pone.0142390 29. doi:0.1371/journal

Ferreira, D., & MacLean, G. (2017). Andragogy in the 21st century. *Language Research Bulletin ICU Tokyo*, *32*, 10–18.

Ferri, F., Grifoni, P., & Guzzo, T. (2020). Online learning and emergency remote teaching: Opportunities and Challenges in Emergency Situations. *Societies (Basel, Switzerland)*, *10*(4), 1–18. doi:10.3390oc10040086

Field, A. (2013). *Discovering statistics using IBM SPSS statistics: And sex and drugs and rock "N" roll* (4th ed.). Sage.

Fields, J., Kenny, N. A., & Mueller, R. A. (2019). Conceptualizing educational leadership in an academic development program. *The International Journal for Academic Development*, *24*(3), 218–231. doi:10.1080/1360144X.2019.1570211

Finkel, S. (2008). *On the ideological front: The Russian intelligentsia and the making of the Soviet public sphere*. Yale University Press.

Fink, L. D. (2013). *Creating significant learning experiences: An integrated approach to designing college courses*. Jossey-Bass.

Fischer, S., David, D., Crotty, B. H., Dierks, M., & Safran, C. (2014). Acceptance and use of health information technology by community-dwelling elders. *International Journal of Medical Informatics*, *83*(9), 624–635. doi:10.1016/j.ijmedinf.2014.06.005 PMID:24996581

Fishbein, M., & Ajzen, I. (1975). *Belief, Attitude, Intention and Belavior: An Introduictionto Theory and Research*. Addison-Wesley.

Fishbein, M., & Ajzen, I. (2011). *Predicting and changing behavior: the reasoned action approach*. Taylor & Francis. doi:10.4324/9780203838020

Fitria, T. N. (2021). Grammarly as AI-powered English writing assistant: Students' alternative for writing English. *Metathesis: Journal of English Language, Literature, and Teaching*, *5*(1), 65–78. doi:10.31002/metathesis.v5i1.3519

Fitzpatrick, S. (1979). *Education and social mobility in the Soviet Union, 1921-1934*. Cambridge University Press. doi:10.1017/CBO9780511523595

Fitzsimmons, T. W., & Callan, V. J. (2016). Applying a capital perspective to explain continued gender inequality in the C-suite. *The Leadership Quarterly*, *27*(3), 354–370. doi:10.1016/j.leaqua.2015.11.003

Fitzsimmons, T. W., & Callan, V. J. (2019). The diversity gap in leadership: What are we missing in current theorizing? *The Leadership Quarterly*, *31*(4), 101347. doi:10.1016/j.leaqua.2019.101347

Flaherty, C. (2022). Mixed Message. *Inside Higher Education*. https://www.insidehighered.com/news/2022/05/09/uf-seems-endorse-new-state-anti-crt-law

Fleming, E. A., & Caffarella, R. S. (2000). *Leadership for adult and continuing education. In Adult Education Research Conference, Vancouver, BC, Canada, 2000*. New Prairie Press.

Fleming, J., & Ledogar, R. J. (2008). Resilience, an evolving concept: A review of literature relevant to aboriginal research. *Pimatisiwin*, *6*(2), 7–23. PMID:20963184

Fletcher, D., & Sarkar, M. (2013). Psychological resilience: A review and critique of definitions, concepts, and theory. *European Psychologist, 18*(1), 12–23. doi:10.1027/1016-9040/a000124

Follett, J. (2015). Designing for emerging technologies. In J. Follett (Ed.), *Designing for rmerging technologies: UX for genomics, robotis, and the iInternet of things* (pp. 1–26). O'Reilly Media, Inc.

Foote, L. S. (2015). Transformational learning: Reflections of an adult learning story. *Adult Learning, 26*(2), 84–86. doi:10.1177/1045159515573017

Forehand, M. (2017). Bloom's taxonomy: Original and revised. In M. Orey (Ed.), *Emerging perspective on learning, teaching, and technology* (pp. 41–44). CreateSpace Independent Publisher.

Form, S., & Kaernbach, C. (2018). More is not always better: The differentiated influence of empathy on different magnitudes of creativity. *Europe's Journal of Psychology, 14*(1), 54–65. doi:10.5964/ejop.v14i1.1432 PMID:29899798

Fornaciari, C. J., & Lund Dean, K. (2014). The 21st-century syllabus: From pedagogy to andragogy. *Journal of Management Education, 38*(5), 701–723. doi:10.1177/1052562913504763

Forrest, S. III, & Peterson, T. (2006). It's Called andragogy. *Academy of Management Learning & Education, 5*(1), 113–122. doi:10.5465/amle.2006.20388390

Fraenkel, J. R., Wallen, N. E., & Hyun, H. H. (2012). *How to design and evaluate research in education* (Vol. 7). McGraw-hill.

Fragoso, A., & Valadas, T. (2018). The rise and fall of adult community education in Portugal. *Social Sciences, 7*(11), 239. doi:10.3390ocsci7110239

Fraillon, J., Ainley, J., Schulz, W., Friedman, W., & Gebhardt, E. (n.d.). *Preparing for life in a digital age.* The IDEA International Computer and Information Literacy Study International Report.

Francoeur, J. (2013). Mindset not skill set: Evaluating in new paradigms of leadership. *Advances in Developing Human Resources, 15*(1), 10–26. doi:10.1177/1523422312466835

Franks, M. A. (2011). Unwilling avatars: Idealism and discrimination in cyberspace. *Columbia Journal of Gender and Law, 20*, 224.

Freese, A. R. (2006). Reframing one's teaching: Discovering our teacher selves through reflection and inquiry. *Teaching and Teacher Education, 22*(1), 100–119. doi:10.1016/j.tate.2005.07.003

Freire, P. (2020). Pedagogy of the oppressed. In P. Kuppers & G. Robertson (Eds.), *Toward a sociology of education* (pp. 374-386). Routledge. doi:10.4324/9781003060635-5

Freire, P. (1970). *Pedagogy of the oppressed: Cultural action for freedom* (M. Bergman Ramos, Trans.). Continuum Press.

Friborg, O., Hjemdal, O., Rosenvinge, J. H., Martinussen, M., Aslaksen, P. M., & Flaten, M. A. (2006). Resilience as a moderator of pain and stress. *Journal of Psychosomatic Research, 61*(2), 213–219. doi:10.1016/j.jpsychores.2005.12.007 PMID:16880024

Fribory, O., Barlaug, D., Martinussen, M., Rosenvinge, J. H., & Hjemdal, O. (2005). Resilience in relation to personality and intelligence. *International Journal of Methods in Psychiatric Research, 14*(1), 29–42. doi:10.1002/mpr.15 PMID:16097398

Friedman, T. (2005). *The world is flat.* Farrar, Straus and Giroux.

Friedrich, J., Becker, M., Kramer, F., Wirth, M., & Schneider, M. (2020). Incentive design and gamification for knowledge management. *Journal of Business Research, 106*, 341–352. doi:10.1016/j.jbusres.2019.02.009

Fukuyama, F. (1992). *The end of history and the last man.* Free Press.

Funari, T. S., Feider, L. L., & Schoneboom, B. A. (2015). Build high-performing military nurse leaders through coaching, sponsorship, and engagement. *Nurse Leader, 13*(3), 54–59. doi:10.1016/j.mnl.2014.07.019

Furman, G. (2012). Social justice leadership as praxis: Developing capacities through. preparation programs. *Educational Administration Quarterly, 48*(2), 191–229. doi:10.1177/0013161X11427394

Fylan, F. (2005). Semi-structured interviewing. In J. Miles & P. Gilbert (Eds.), *A Handbook of research methods for clinical and health psychology* (pp. 65–77). Oxford University Press.

Gagné, M., & Deci, E. L. (2005). Self-determination theory and work motivation. *Journal of Organizational Behavior, 26*(4), 331–362. doi:10.1002/job.322

Galbraith, J. K. (1958). *The affluent society.* Houghton Mifflin.

Galbraith, J. K. (1967). *The new industrial state.* Houghton Mifflin.

Galbraith, M. W., & Gilley, J. W. (1984). Using self-directed learning contracts to improve performance and instruction. *Performance & Instruction Journal, 23*(8), 9–10. doi:10.1002/pfi.4150230809

Galey-Horn, S. (2020). Capacity-building for district reform: The role of instructional-coach teams. *Teachers College Record, 122*(10), 1–40. doi:10.1177/016146812012201003

Galloway, A. R., & Thacker, E. (2013). *The exploit: A theory of networks.* University of Minnesota Press.

Gamede, B., Ajani, O., & Afolabi, O. (2022). Exploring the adoption and usage of learning management system as alternative for curriculum delivery in South African higher education institutions during Covid-19 lockdown. *International Journal of Higher Education, 11*(1), 71–84. doi:10.5430/ijhe.v11n1p71

Gao, F., Zhang, T., & Franklin, T. (2013). Designing asynchronous online discussion environments: Recent progress and possible future directions. *British Journal of Educational Technology, 44*(3), 469–483. doi:10.1111/j.1467-8535.2012.01330.x

Garg, N., & Gera, S. (2019). Gratitude and leadership in higher education institutions: Exploring the mediating role of social intelligence among teachers. *Journal of Applied Research in Higher Education, 12*(5), 915–926. Advance online publication. doi:10.1108/JARHE-09-2019-0241

Garrison, D. R. (1997). Self-directed learning: Toward a comprehensive model. *Adult Education Quarterly, 48*(1), 61–72. doi:10.1177/074171369704800103

Garrison, D. R. (2006). Online collaboration principles. *Journal of Asynchronous Learning Networks, 10*(1), 25–34. doi:10.24059/olj.v10i1.1768

Garrison, D. R. (2022). Shared metacognition in a community of inquiry. *Online Learning, 26*(1), 6–18. doi:10.24059/olj.v26i1.3023

Garrison, D. R., & Anderson, T. (2003). *E-learning in the 21st century: A framework for research and practice.* London: Routledge/Falmer. doi:10.4324/9780203166093

Garrison, D. R., Anderson, T., & Archer, W. (1999). Critical inquiry in a text-based environment: Computer conferencing in higher education. *The Internet and Higher Education, 2*(2-3), 87–105. doi:10.1016/S1096-7516(00)00016-6

Garrison, D. R., & Kanuka, H. (2004). Blended learning: Uncovering its transformative potential in higher education. *The Internet and Higher Education*, *7*(2), 95–105. doi:10.1016/j.iheduc.2004.02.001

Garrison, D. R., & Shale, D. (1990). *A new framework and perspective. In Education at a distance: From issues to practice*. Kreiger.

Garrison, R., Anderson, T., & Archer, W. (2009). The first decade of the community inquiry framework: A retrospective. *The Internet and Higher Education*, *13*(1), 5–19.

Garrison, R., & Vaughan, H. (2008). *Blended learning in higher education: Framework, principles and guidelines*. Jossey-Bass.

Gasparini, A. (2015). Perspective and Use of Empathy in Design Thinking. In *ACHI 2015: The Eighth International Conference on Advances in Computer-Human Interactions*. IARIA.

Geerts, J. M., Goodall, A. H., & Agius, S. (2020). Evidence-based leadership development for physicians: A systematic literature review. *Social Science & Medicine*, *246*, 12709. doi:10.1016/j.socscimed.2019.112709 PMID:31887629

Gerstenschlager, N. E., & Barlow, A. T. (2019). Transitioning from practicing teacher to teacher leader: A case study. *Teacher Development*, *23*(1), 18–35. doi:10.1080/13664530.2018.1515106

Gerth, A. B., & Peppard, J. (2016). The dynamics of CIO derailment: How CIOs come undone and how to avoid it. *Business Horizons*, *59*(1), 61–70. doi:10.1016/j.bushor.2015.09.001

Ghinea, V., Cantaragiu, R., & Ghinea, M. (2019). The Peter principle and the limits of our current understanding of organizational incompetence. *Quality - Access to Success*, *20*(172), 74–77.

Giannoukos, G., Besas, G. M., Galiropoulos, C., & Hioctour, V. (2015). The andragogy, the social change, and the transformative learning educational approaches in adult education. *Journal of Education and Practice*, *46*(50).

Giber, D., & Friedman, D. (2006). Leaders of the future. *Leadership Excellence Essentials*, *23*(6), 12–13.

GilbertN. (2023). *Blog*. linkedin.com/in/nikkgilbert.

Gill, R. (2016). Breaking the silence: The hidden injuries of neo-liberal academia. *Feministische Studien*, *34*(1), 39–55. doi:10.1515/fs-2016-0105

Giroux, H. A. (2014). *Neoliberalism's war on higher education*. Haymarket Books.

Giroux, H. A. (2018). *Terror of neoliberalism: Authoritarianism and the eclipse of democracy*. Routledge. doi:10.4324/9781315631653

Gisbert, M., & Bullen, M. (Eds.). (2015). *Teaching and learning in digital sorlds: Strategies and issues in higher education*. Publicacions Universitat Rovira I Virgili. doi:10.17345/9788484243762

Gloppen, J. (2009). Perspectives on design leadership and design thinking and how they relate to European. service industries. *Design Management Journal*, *4*(1), 33–47. doi:.00005.x doi:10.1111/j.1942-5074.2009

Gmelch, W., & Wolverton, M. (2002). *College deans: Leading from within*. American Council on Education and Oryx Press.

Gnanadass, E., & Merriweather, L. R. (2022). Restorying COVID-19: Faculty and graduate students teaching and learning in crisis. In R. Wlodarsky & C. Hansman (Eds.), *The Covid-19 pandemic and other ongoing crises: Reflections, innovations, and adaptations. New Directions for Adult and Continuing Education*. Wiley. doi:10.1002/ace.20449

Gold, H. E. (2005). Engaging the adult learner: Creating effective library instruction. *Portal (Baltimore, Md.)*, *5*(4), 467–481. doi:10.1353/pla.2005.0051

Gonon, P. (2002). Der betrieb als erzieher – Knappheit als pädagogische herausforderung. *Zeitschrift für Pädagogik, 48*(3), 317–335.

Gonzalez et al. v. Arizona State Board of Education . (2017). Case no. CV 10-623 TUC AWT. United States Arizona District Court.

Goodall, A. H., & Bäker, A. (2015). A theory exploring how expert leaders influence performance in knowledge-intensive organizations. In I. M. Welpe, J. Wollersheim, S. Ringelhan, & M. Osterloh (Eds.), *Incentives and Performance* (pp. 49–67). Springer International Publishing. doi:10.1007/978-3-319-09785-5_4

Goodall, A., & Stoller, J. K. (2017). The future of clinical leadership: Evidence for physician leadership and the educational pathway for new leaders. *BMJ Leader, 1*(2), 8–11. doi:10.1136/leader-2017-000010

Gosling, J., & Mintzberg, H. (2003). The five minds of a manager. *Harvard Business Review, 81*(11), 54–63. PMID:14619151

Grafstrom, M., & Falkman, L. L. (2017). Everyday narratives: CEO rhetoric on Twitter. *Journal of Organizational Change Management, 30*(3), 312–322. doi:10.1108/JOCM-10-2016-0197

Graham, S., & Weiner, B. (1996). Theories and principles of motivation. Handbook of Educational Psychology, 4(1), 63-84.

Gramsci, A. (1961). *Selection from the prison notebooks.* International Publishers.

Grant, C. (2009). Towards a conceptual understanding of education leadership: Place, space and practices. *Education as Change, 13*(1), 45–57. doi:10.1080/16823200902933529

Grant, C. (2017). Distributed leadership in South Africa: Yet another passing fad or a robust theoretical tool for investigating school leadership practice? *School Leadership & Management, 37*(5), 457–475. doi:10.1080/13632434.2017.1360856

Grasha, A. F. (1996). *Teaching with style.* Alliance Publishers.

Green, K. C. (2000, February 17-19). *Technology and instruction: Compelling, competing, and complementary visions for the instructional role of technology in higher education.* Paper prepared for a conference on the changing American University: The future of the City of Intellect, convened by the Center for Ideas & Society at the University of California, Riverside, CA, United States.

Greene, K., & Larsen, L. (2018). Virtual andragogy: A new paradigm for serving adult online learners. *International Journal of Digital Society, 9*(2), 1376-1381.

Greene, K., & Larsen, L. (2018). Virtual Andragogy: A new paradigm for serving adult online learners. *International Journal of Digital Society, 9*(2), 1376–1381. doi:10.20533/ijds.2040.2570.2018.0169

Greene, M. (1995). *Releasing the imagination.* Jossey-Bass.

Greenhaus, J. H., & Beutell, N. J. (1985). Sources of conflict between work and family roles. *Academy of Management Journal, 10*, 76–88.

Green, W. M., Hoffman, A., Manlapaz, M., & Foshee, C. (2022). Health professions education cohort programs: A case study of pivoting from in-person to online learning in cohort-based programs. In R. Wlodarsky & C. Hansman (Eds.), *The Covid-19 pandemic and other ongoing crises: Reflections, innovations, and adaptations.* Wiley. doi:10.1002/ace.20452

Grinder, M. (1997). *The science of nonverbal communication.* Michael Grinder and Associates.

Grissom, J. A., Mitani, H., & Woo, D. S. (2019). Principal preparation programs and principal outcomes. *Educational Administration Quarterly, 55*(1), 73–115. doi:10.1177/0013161X18785865

Gromova, N. S. (2020). Digitalization of education in Russia: From modern technologies to an innovative model of the educational environment. *Advances in Economics, Business and Management Research, 156*, 54–59. doi:10.2991/aebmr.k.201205.009

Gronn, P. (2008). The future of distributed leadership. *Journal of Educational Administration, 46*(2), 141–158. doi:10.1108/09578230810863235

Gros, B. (2003). The impact of digital games in education. *First Monday, 8*(7). https://firstmonday.org/issues/issue8_7/gros/index.htm

Gross, B. (1980). *Friendly fascism: The new face of power in America*. M. Evans.

Grossman, R., & Salas, E. (2011). The transfer of training: What really matters? *International Journal of Training and Development, 15*(2), 103–120. doi:10.1111/j.1468-2419.2011.00373.x

Grover, K., & Miller, M. (2014). Gender differences in self-directed learning practices among community members. *PAACE Journal of Lifelong Learning, 23*, 19–31.

Grow, G. O. (1991). Teaching learners to be self-directed. *Adult Education Quarterly, 41*(3), 125–149. doi:10.1177/0001848191041003001

Guglielmino, L. M. (1997). *Contributions of the self-directed learning readiness scale (SDLRS) and the learning preference assessment (LPA) to the definition and measurement of self-direction in learning*. Paper presented at the First World Conference on Self-Directed Learning, Montreal, Canada.

Gumede, V. (2017). Leadership for Africa's development. *Journal of Black Studies, 48*(1), 74–90. doi:10.1177/0021934716678392

Gurbaxani, V., & Dunkle, D. (2019). Gearing up for successful digital transformation. *MIS Quarterly Executive, 18*(3), 209–220. doi:10.17705/2msqe.00017

Gusakov, A. (2019). The role of design thinking for achieving leaders in the digital transformation of business. *Advances in Social Science, Education and Humanities Research, 441*, 356–362.

Habermas, J. (1971). *Knowledge and human interests*. Beacon Press.

Habiballah, S., Bibu, N., & Danaiata, D. (2021). Educational Leadership and ICT Implementation in Israeli Arab Sector-towards a Model of Hybrid Leadership. *Revista de Management Comparat International, 22*(1), 74–86. doi:10.24818/RMCI.2021.1.74

Haimson, O. L., & Hoffmann, A. L. (2016). Constructing and enforcing" authentic" identity online: Facebook, real names, and non-normative identities. *First Monday, 21*(6). Advance online publication. https://journals.uic.edu/ojs/index.php/fm/article/download/6791/5521. doi:10.5210/fm.v21i6.6791

Hair, J. F., Black, W. C., Babin, B. J., & Anderson, R. E. (2010). *Multivariate Data Analysis* (7th ed.). Prentice Hall.

Hall, D. T. (2004). Self-awareness, identity, and leader development. In N.J. Mahwah (Ed.), Leader development for transforming organizations: Growing leaders for tomorrow (pp. 153–176). Lawrence Erlbaum Associates Publishers.

Hall, S.B., & Baier-Lentz, M. (2022). *3 technologies that will shape the future of the Metaverse – and the human experience*. World Economic Forum. https://www.weforum.org/agenda/2022/02/future-of-the-Metaverse-vr-ar-and-brain-computer/

Hallinger, P. (2013). A conceptual framework for systematic reviews of research in educational leadership and management. *Journal of Educational Administration, 51*(2), 126–149. doi:10.1108/09578231311304670

Hallinger, P. (2018). Surfacing a hidden literature: A systematic review of research on educational leadership and management in Africa. *Educational Management Administration & Leadership*, *46*(3), 362–384. doi:10.1177/1741143217694895

Hallinger, P., & Kovačević, J. (2022). Mapping the intellectual lineage of educational management, administration and leadership, 1972–2020. *Educational Management Administration & Leadership*, *50*(2), 192–216. doi:10.1177/17411432211006093

Hall, S. (1980). Encoding/decoding. In S. Hall, D. Hobson, A. Lowe, & P. Willis (Eds.), *Culture, media, language: Working papers in cultural studies* (pp. 128–138). Hutchinson.

Han, Y. J., & Bromilow, D. (2010). Graphic designers win the leadership game. *Design Management Journal*, *5*(1), 20–31. doi:.00011.x doi:10.1111/j.1948-7177.2010

Hanafi, Y., Taufiq, A., Saefi, M., Ikhsan, M. A., Diyana, T. N., Thoriquttyas, T., & Anam, F. K. (2021). The new identity of Indonesian Islamic boarding schools in the "new normal": The education leadership response to COVID-19. *Heliyon*, *7*(3), e06549. doi:10.1016/j.heliyon.2021.e06549 PMID:33851049

Hannah, S. T., Avolio, B. J., Luthans, F., & Harms, P. D. (2008). Leadership efficacy: Review and future directions. *The Leadership Quarterly*, *19*(6), 669–692. doi:10.1016/j.leaqua.2008.09.007

Hanselmann, H. (1951). Andragogik: Wesen, Möglichkeiten, Grenzen der Erwchsenenbildung [Andragogy: Essence, Potentiality and Limits of Adult Education]. Rotapfel Verlag.

Hansen, R. J., Talmage, C. A., Thaxton, S. P., & Knopf, R. C. (2019, March 26). Barriers to age-friendly universities: Lessons from Osher lifelong learning institute demographics and perceptions. *Gerontology & Geriatrics Education*, *40*(2), 1–23. doi:10.1080/02701960.2019.1572003 PMID:30688166

Hansen, R. J., Talmage, C. A., Thaxton, S. P., & Knopf, R. C. (2020). Enhancing older adult access to lifelong learning institutes through technology-based instruction: A brief report. *Gerontology & Geriatrics Education*, *41*(3), 342–351. doi:10.1080/02701960.2019.1618852 PMID:31116688

Hansman, C. A. (2017). Assessing schools and organizations. In V. Wang, B. Bain, J. Hope, & C. A. Hansman (Eds.), *Educational leadership and organizational management: Linking theories to practice* (pp. 53–80). Information Age Publishing.

Hansman, C. A. (2022). "We're all in this together"? Reflections on inequity during tumultuous times. In R. Wlodarsky & C. Hansman (Eds.), *The Covid-19 pandemic and other ongoing crises: Reflections, innovations, and adaptations.* Wiley. doi:10.1002/ace.20448

Hansman, C. A., & Mott, V. (2010). Adult learners. In C. Kasworm, A. Rose, & J. Ross-Gordon (Eds.), *Handbook of Adult and Continuing Education* (pp. 13–23). Sage.

Hanson, A. (1996). The search for a separate theory of adult learning: Does anyone really need andragogy? In R. Edwards, A. Hanson, & P. Raggatt (Eds.), *Boundaries of adult learning* (pp. 99-108). Routledge.

Harms, P. D., Spain, S. M., & Hannah, S. T. (2011). Leader development and the dark side of personality. *The Leadership Quarterly*, *22*(3), 495–509. doi:10.1016/j.leaqua.2011.04.007

Harrell, K. B., & Wendt, J. L. (2019). The impact of blended learning on community of inquiry and perceived learning among high school learners enrolled in a Public Charter School. *Journal of Research on Technology in Education*, *51*(3), 259–272. doi:10.1080/15391523.2019.1590167

Harrington, F. H. (1977). *The future of adult education: New responsibilities of colleges and universities*. Jossey-Bass Publishers.

Harris, A., & Spillane, J. (2008). Distributed leadership through the looking glass. *Management in Education, 22*(1), 31–34. doi:10.1177/0892020607085623

Hart, A. W. (1995). Reconceiving school leadership: Emergent views. *The Elementary School Journal, 96*(1), 9–28. doi:10.1086/461812

Hartree, A. (1984). Malcolm Knowles' theory of andragogy: A critique. *International Journal of Lifelong Education, 3*(3), 203–210. doi:10.1080/0260137840030304

Hartz, L. (1991). *The liberal tradition in America*. Harper. (Original work published 1955)

Harvey, S., & Chia-yu, K. (2017). Social Processes and Team Creativity. In R. Reiter-Palmon (Ed.), *Team Creativity and Innovation* (pp. 87–128). Oxford University Press. doi:10.1093/oso/9780190222093.003.0005

Hase, S., & Ellis, A. (2001). Problems with online learning are systemic, not technical. In J. Stephenson (Ed.), *Teaching & learning online: new pedagogies for new technologies* (pp. 27–30). Routledge.

Hashim, A. K. (2020). Coaching and districtwide improvement: Exploring the systemic leadership practices of instructional coaches. *Teachers College Record, 122*(10), 1–44. doi:10.1177/016146812012201005

Haslam, S. A., & McGarty, C. (2003). Research method and statistics in psychology. *Sage (Atlanta, Ga.)*.

Hassan, A., Gallear, D., & Sivarajah, U. (2018). Critical factors affecting leadership: a higher education context. *Transforming Government: People. Process and Policy, 12*(1), 110–130. doi:10.1108/TG-12-2017-0075

Hatlevik, O. E., & Christophersen, K. A. (2013). Digital competence at the beginning of upper secondary school: Identifying factors explaining digital inclusion. *Computer Education, 63*, 240–247. doi:10.1016/j.compedu.2012.11.015

Hawes, J. (2022). The challenges of emergency online instruction at a Small HBCU. *Journal of Higher Education Theory & Practice, 22*(18), 1–6.

Hayes, S. D., & Irby, B. J. (2020). Challenges in preparing aspiring principals for instructional leadership: Voices from the field. *International Journal of Leadership in Education, 23*(2), 131–151. doi:10.1080/13603124.2018.1562102

Health and Safety Executive (HSE). (2008). *HSE Management Standards Indicator Tool*. HSE, Bootle.

Health and Safety Executive (HSE). (2022). *Working days lost in Great Britain*. https://www.hse.gov.uk/statistics/dayslost.htm

Heaney, T. W. (2000). Adult education and society. In A. L. Wilson & A. R. Hayes (Eds.), *Handbook of adult and continuing education* (pp. 559–572). Jossey-Bass.

Heck, R. H., & Hallinger, P. (2014). Modeling the longitudinal effects of school leadership on teaching and learning. *Journal of Educational Administration, 52*(5), 653–681. doi:10.1108/JEA-08-2013-0097

Henderson, L. (2017). 'Someone had to have faith in them as professionals': An evaluation of an action research project to develop educational leadership across the early years. *Educational Action Research, 25*(3), 387–401. doi:10.1080/09650792.2016.1173566

Hendriks, M., & Steen, R. (2012). Results from school leadership effectiveness studies (2005–2010). In J. Scheerens (Ed.), *School Leadership Effects Revisited* (pp. 65–129). Springer. doi:10.1007/978-94-007-2768-7_4

Henschke, J. A. (2011). Considerations regarding the future of andragogy. *Adult Learning, 22*(1), 34–37. doi:10.1177/104515951102200109

Henschke, J. A. (2011). Considerations regarding the future of andragogy. *Participatory Educational Research, 1*(2), 21–35.

Henschke, J. A. (2020). *Facilitating adult and organizational learning through andragogy: A history, philosophy, and major themes*. Information Science Reference.

He, Q., Wang, G., Luo, L., Shi, Q., Xie, J., & Meng, X. (2017). Mapping the managerial areas of building information modeling (BIM) using scientometric analysis. *International Journal of Project Management, 35*(4), 670–685. doi:10.1016/j.ijproman.2016.08.001

Heron, J. (1967). *Feeling and personhood: Psychology in another key*. Sage Publications.

Herrington, J., Reeves, T., & Oliver, R. (2010). *A guide to authentic e-learning*. Routledge.

Hersey, P., & Blanchard, K. H. (1969a). Life-cycle theory of leadership. *Training and Development Journal, 23*, 26–34.

Heslin, P. A., & Keating, L. A. (2017). In learning mode? The role of mindsets in derailing and enabling experiential leadership development. *The Leadership Quarterly, 28*(3), 367–384. doi:10.1016/j.leaqua.2016.10.010

Hickey, G. (2014). The importance of learning philosophies on technology selection in education. *Journal of Learning Design, 7*(3). Advance online publication. doi:10.5204/jld.v7i3.184

Hieker, C., & Pringle, J. (2021). *The future of leadership development: Disruption and the impact of megatrends*. Palgrave MacMillan. doi:10.1007/978-3-030-53544-5

Higa, C., Davidson, E., & Loos, J. (2021). Integrating family and friend support, information technology, and diabetes education in community-centric diabetes self-management. *Journal of the American Medical Informatics Association : JAMIA, 28*(2), 261–275. doi:10.1093/jamia/ocaa223 PMID:33164074

Hilliker, S. (2020). Virtual exchange as a study abroad alternative to foster language and culture exchange in TESOL teacher education. *TESL-EJ, 23*(4).

Hill, L. H., & Isaac-Savage, E. P. (2022). Doors slam shut: Adult education program closures. In R. Wlodarsky & C. Hansman (Eds.), *The Covid-19 pandemic and other ongoing crises: Reflections, innovations, and adaptations*. doi:10.1002/ace.20453

Himsworth, J. (2018, June 19). *Why design thinking should also serve as a leadership philosophy*. https://www.forbes.com/sites/forbesagencycouncil/2018/07/19/why-design-thinking-should-also-serve-as-a-leadership-philosophy/?sh=861c66a5a90f

Hodgson, P., & White, R. P. (2001). *Relax, it's only uncertainty*. Financial Times Prentice Hall.

Hogg, M. A. (2001). A social identity theory of leadership. *Personality and Social Psychology Review, 5*(3), 184–200. doi:10.1207/S15327957PSPR0503_1

Hohepa, M. K. (2013). Educational leadership and indigeneity: Doing things the same, differently. *American Journal of Education, 119*(4), 617–631. doi:10.1086/670964

Holland, R. W., Verplanken, B., & van Knippenberg, A. (2002). On the nature of attitude-behavior relations: The strong guide, the weak follow. *European Journal of Social Psychology, 32*(6), 869–876. doi:10.1002/ejsp.135

Hollingworth, L., Olsen, D., Asikin-Garmager, A., & Winn, K. M. (2018). Initiating conversations and opening doors: How principals establish a positive building culture to sustain school improvement efforts. *Educational Management Administration & Leadership, 46*(6), 1014–1034. doi:10.1177/1741143217720461

Holmes, G., & Abington-Cooper, M. (2000). Pedagogy vs andragogy: A false dichotomy? *The Journal of Technology Studies, 26*(2), 50–55. doi:10.21061/jots.v26i2.a.8

Holmes, K., & Prieto-Rodriguez, E. (2018). Student and staff perceptions of a learning management system for blended learning in teacher education. *The Australian Journal of Teacher Education, 43*(3), 21–34. doi:10.14221/ajte.2018v43n3.2

Holst, C., Stelzle, D., Diep, L., Sukums, F., Ngowi, B., Noll, J., & Winkler, A. (2022). Improving health knowledge through provision of free digital health education to rural communities in Iringa, Tanzania: Nonrandomized intervention study. *Journal of Medical Internet Research, 24*(7), e37666. doi:10.2196/37666 PMID:35900820

Holzweiss, P. C., Joyner, S. A., Fuller, M. B., Henderson, S., & Young, R. (2014). Online graduate students' perceptions of best learning experiences. *Distance Education, 35*(3), 311–323. doi:10.1080/01587919.2015.955262

Honig, M. I., Venkateswaran, N., & McNeil, P. (2017). Research use as learning: The case of fundamental change in school district central offices. *American Educational Research Journal, 54*(5), 938–971. doi:10.3102/0002831217712466

Horkheimer, M., & Adorno, T. W. (1972). *Dialectic of enlightenment.* Herder & Herder.

Horton, M., Kohl, J., & Kohl, H. (1998). *The long haul: An autobiography.* Teachers College Press.

Hosler, K. A., & Arend, B. D. (2012). The importance of course design, feedback, and facilitation: Student perceptions of the relationship between teaching presence and cognitive presence. *Educational Media International, 49*(3), 217–229. doi:10.1080/09523987.2012.738014

Hou, F., Su, Y., Qi, M., Wang, L., & Wang, Q. (2021). Entrepreneurial team knowledge diversity and creativity: A multilevel analysis of knowledge sharing, individual creativity, and team creativity. *Frontiers in Psychology, 12*, 717756. doi:10.3389/fpsyg.2021.717756 PMID:34539518

Houle, C. O. (1961). *The inquiring mind.* University of Wisconsin Press.

Houle, C. O. (1961). *The inquiring minds.* University of Wisconsin Press.

Houle, C. O. (1972). *The design of education.* Jossey-Bass.

House, R., Javidan, M., Hanges, P., & Dorfman, P. (2002). Understanding cultures and Implicit Leadership Theories across the Globe: An Introduction to Project GLOBE. *Journal of World Business, 37*(1), 3–10. doi:10.1016/S1090-9516(01)00069-4

Huang, H. M. (2002). Toward constructivism for adult learners in online learning environments. *British Journal of Educational Technology, 33*(1), 27–37. doi:10.1111/1467-8535.00236

Hulin, C., Netemeyer, R., & Cudeck, R. (2001). Can a reliability coefficient be too high? *Journal of Consumer Psychology, 10*(1), 55–58. doi:10.1207/S15327663JCP1001&2_05

Hülsheger, U. R., Anderson, N., & Salgado, J. F. (2009). Team-level predictors of innovation at work: A comprehensive meta-analysis spanning three decades of research. *The Journal of Applied Psychology, 94*(5), 1128–1145. doi:10.1037/a0015978 PMID:19702361

Hunt, C. W., Henderson, K., & Chapman, R. (2018). Using technology to provide diabetes education for rural communities. *Online Journal of Rural Nursing and Health Care : the Official Journal of the Rural Nurse Organization, 18*(2), 134–151. doi:10.14574/ojrnhc.v18i2.525

Hunter, S. T., Neely, B. H. Jr, & Gutworth, M. B. (2018). Selection and team creativity: Meeting unique challenges through diversity and flexibility. In R. Reiter-Palmon (Ed.), *Team creativity and innovation* (pp. 259–281). Oxford University Press.

Hussar, W., & Bailey, T. (2014). *Projections of Education Statistics to 2022* (41st ed.). National Center for Education Statistics.

Husserl, E. (1931). *Ideas: General introduction to pure phenomenology*. Allen & Unwin.

Hwang, G.-J., & Chien, S.-Y. (2022). Definition, roles, and potential research issues of the Metaverse in education: An artificial intelligence perspective. *Science Direct, 4*. https://www.sciencedirect.com/journal/computers-and-education-artificial-intelligence

Hytten, K., & Bettez, S. C. (2011). *Understanding education for social justice*. Educational.

Igaz, H., & Gulbahar, Y. (2017, July 20-22). *Why do learners choose online learning: The learners' voices*. International Association for Development of the Information Society. 11th International Conference on e-Learning, Lisbon, Portugal.

Ilin, M. (1931). New Russia's primer. Houghton Mifflin.

Illeris, K. (2004). *The three dimensions of learning*. Krieger.

Ingersoll, R. (2014). *Seven trends: The transformation of the teaching force*. Consortium for Policy Research in Education.

International Society for Technology in Education. (2023). *Standards for educational leaders*. https://www.iste.org/standards/iste-standards-for-education-leaders

Iqbal, Q., & Ahmad, N. H. (2020). Sustainable development: The colors of sustainable leadership in learning organization. *Sustainable Development, 29*(1), 108–119. doi:10.1002d.2135

Iqbal, Q., Ahmad, N. H., & Li, Z. (2021). Frugal-based innovation model for sustainable development: Technological and market turbulence. *Leadership and Organization Development Journal, 42*(3), 396–407. doi:10.1108/LODJ-06-2020-0256

Irvine, P. A., & Brundrett, M. (2019). Negotiating the next step: The part that experience plays with middle leaders development as they move into their new role. *Educational Management Administration & Leadership, 47*(1), 74–90. doi:10.1177/1741143217720457

Işık, A. H., Karakış, R., & Güler, İ. (2010). Postgraduate students' attitudes towards distance learning. *Procedia: Social and Behavioral Sciences, 9*, 218–222. doi:10.1016/j.sbspro.2010.12.139

Ismagiloiva, E., Hughes, L., Rana, N., & Dwivedi, Y. (2019, June). Role of smart cities in creating sustainable cities and communities: a systematic literature review. In *International Working Conference on Transfer and Diffusion of IT* (pp. 311-324). Springer.10.1007/978-3-030-20671-0_21

Jacobson, S., McCarthy, M., & Pounder, D. (2015). What makes a leadership preparation program exemplary? *Journal of Research on Leadership Education, 10*(1), 63–76. doi:10.1177/1942775115569420

Jain, S., & Anjuman, A. S. S. (2013). Facilitating the acquisition of soft skills through training. *The IUP Journal of Soft Skills, 2*(2), 32–39.

James Obergefell v. Richard Hodges, no. 14-3057 (6th cir. 2014). Justia Law. https://law.justia.com/cases/federal/appellate-courts/ca6/14-3057/14-3057-2014-11-06.html

Jarvis, P. (1985). Thinking critically in an information society: A sociological Analysis. *Lifelong Learning*, 11-14.

Jarvis, P. (2002). *The theory & practice of teaching*. Kogan Page.

Jarvis, P. (2006). *Towards a comprehensive theory of human learning* (Vol. 1). Routledge.

Jasanof, S. (2015). Future imperfect: science, technology, and the imagination of modernity. In *Dreamscapes of modernity: Sociotechnical imaginaries and the fabrication of power*. University of Chicago Press. doi:10.7208/chicago/9780226276663.003.0001

Jaynes, C., Seals, W. B., Calvert, K., Fei, Z., & Griffloen, J. (2003). *The Metaverse: a networked collection of inexpensive, self-configuring, immersive environments*. Proceedings on Virtual Environments. doi:10.1145/769953.769967

Jeanes, E. (2021, March). A meeting of mind(sets). Integrating the pedagogy and andragogy of mindsets for leadership development. *Thinking Skills and Creativity*, *39*, 100758. https://www.sciencedirect.com/science/article/abs/pii/S1871187120302327. doi:10.1016/j.tsc.2020.100758

Jeong, H., Yi, Y., & Kim, D. (2022). An innovative e-commerce platform incorporating metaverse to live commerce. *International Journal Computer Information Control*, *18*(1), 221–229.

Jeon, H. J., Youn, H. C., Ko, S. M., & Kim, T. H. (2022). Blockchain and AI meet in the metaverse. In T. Fernández-Caramés & P. Fraga-Lamas (Eds.), *Blockchain Potential in AI*. doi:10.5772/intechopen.91580

Jiang, W., Zhao, X., & Ni, J. (2017). The impact of transformational leadership on employee sustainable performance: The mediating role of organizational citizenship behavior. *Sustainability*, *9*(9), 1567. doi:10.3390u9091567

Johnson, D. W., Johnson, R. T., & Holubec, E. (1998). *Cooperation in the classroom*. Interaction Book Company.

Johnson, L. (2016). Geragogy. In S. L. Danver (Ed.), *The SAGE encyclopedia of online education* (pp. 504–508). SAGE Publications.

Johnson, N. N., & Fournillier, J. B. (2022). Increasing diversity in leadership: Perspectives of four Black women educational leaders in the context of the United States. *Journal of Educational Administration and History*, *54*(2), 174–192. doi:10.1080/00220620.2021.1985976

Johnson, S. K., Putter, S., Reichard, R. J., Hoffmeister, K., Cigularov, K. P., Gibbons, A. M., Chen, P. Y., & Rosecrance, J. C. (2018). Mastery goal orientation and performance affect the development of leader efficacy during leader development. *Journal of Leadership & Organizational Studies*, *25*(1), 30–46. doi:10.1177/1548051817713689

Johnson, T. C. (2016). *Diversity, knowledge management, and team creativity in virtual healthcare teams: A predictive study*. ProQuest Dissertations & Theses.

Johnson, T., Wisniewski, M. A., Kuhlemeyer, G., Isaacs, G., & Krzykowski, J. (2012). Technology adoption in higher education: Overcoming anxiety through faculty bootcamp. *Journal of Asynchronous Learning Networks*, *16*(2), 63–72. doi:10.24059/olj.v16i2.240

Jonassen, D. (Ed.). (2004). *Handbook of research on educational communications and technology* (2nd ed.). Lawrence Erlnbaum Associates.

Jonassen, D. H. (1991). Objectivism versus constructivism: Do we need a new philosophy paradigm? *Educational Technology Research and Development*, *39*(3), 5–14. doi:10.1007/BF02296434

Jones, C., Penaluna, K., & Penaluna, A. (2019). The promise of andragogy, heutagogy and academagogy to enterprise and entrepreneurship education pedagogy. *Education + Training*, *61*(9), 1170–1186. doi:10.1108/ET-10-2018-0211

Jones, D., Snider, C., Nassehi, A., Yon, J., & Hicks, B. (2020). Characterizing the Digital Twin: A systematic literature review. *CIRP Journal of Manufacturing Science and Technology*, *29*, 36–52. doi:10.1016/j.cirpj.2020.02.002

Jones, I. M. (2011). Can you see me now? Defining teaching presence in the online classroom through building a learning community. *Journal of Legal Studies Education*, *28*(1), 67–116. doi:10.1111/j.1744-1722.2010.01085.x

Jones, S., Harvey, M., Hamilton, J., Bevacqua, J., Egea, K., & McKenzie, J. (2017). Demonstrating the impact of a distributed leadership approach in higher education. *Journal of Higher Education Policy and Management*, *39*(2), 197–211. doi:10.1080/1360080X.2017.1276567

Jones, S., Lefoe, G., Harvey, M., & Ryland, K. (2012). Distributed leadership: A collaborative framework for academics, executives and professionals in higher education. *Journal of Higher Education Policy and Management*, *34*(1), 67–78. doi:10.1080/1360080X.2012.642334

Jozaisse, F. (2011b). Design leadership: Current limits and future opportunities. In R. Cooper, S. Jung-inger, & T. Lockwood (Eds.), *The handbook of design management* (pp. 398–413). Berg. doi:10.5040/9781474294126.ch-025

Joziasse, F. (2011a). The soul of design leadership. *Issue Design Management Review Design Management Review*, *22*(3), 34–42. doi:.00138.x doi:10.1111/j.1948-7169.2011

Judson, E. (2006). How teachers integrate technology and their beliefs about learning: Is there a connection? *Journal of Technology and Teacher Education*, *14*(3), 581–597.

Kamssu, A., Siekpe, J., & Ellzy, J. (2004). Shortcomings to globalization: Using Internet technology and electronic commerce in developing countries. *Journal of Developing Areas*, *38*(1), 151–169. doi:10.1353/jda.2005.0010

Kane, G.C., Palmer, D., Phillips, A.N., Kiron, D., & Buckley, N. (2015, July). Strategy, not technology, drives digital transformation. *MIT Sloan Management Review*.

Kankanamge, N., Yigitcanlar, T., & Goonetilleke, A. (2021). Public perceptions on artificial intelligence driven disaster management: Evidence from Sydney, Melbourne and Brisbane. *Telematics and Informatics*, *65*, 101729. doi:10.1016/j.tele.2021.101729

Kanwar, A., Balasubramanian, K., & Carr, A. (2019). Changing the TVET paradigm: New models for lifelong learning. *International Journal of Training Research*, *17*(sup1), 54–68. doi:10.1080/14480220.2019.1629722

Kaqinari, T., Makarova, E., Audran, J., Doring, A. K., Gobel, K., & Kern, D. (2021). The switch to online teaching during the first COVID-19 lockdown: A comparative study at four European universities. *Journal of University Teaching & Learning Practice*, *18*(5), 10–21. doi:10.53761/1.18.5.10

Karadag, E. (2020). The effect of educational leadership on students' achievement: A cross-cultural meta-analysis research on studies between 2008 and 2018. *Asia Pacific Education Review*, *21*(1), 49–64. doi:10.100712564-019-09612-1

Kariel, H. S. (1977). *Beyond liberalism: Where relations grow.* Chandler & Sharp.

Kariel, H. S. (1961). *The decline of American pluralism*. Stanford University Press.

Kariel, H. S. (1966). *The promise of politics*. Prentice-Hall.

Kariel, H. S. (1989). *The desperate politics of postmodernism.* University of Massachusetts Press.

Kaur, M., & Gupta, B. (2021). *Metaverse technology and the current market.* National Institute of Technology Kurukshetra.

Kazanjian, R. K., & Drazin, R. (2012). Organizational learning, knowledge management, and creativity. In *Handbook of organizational creativity* (pp. 547–568). Academic Press. doi:10.1016/B978-0-12-374714-3.00021-5

Keeler, L. C. (2006). Student satisfaction and types of interaction in distance education courses. *Dissertation Abstracts International, 67*(9). (UMI No. 3233345)

Ke, F. (2010). Examining online teaching, cognitive, and social presence for adult students. *Computers & Education*, *55*(2), 808–820. doi:10.1016/j.compedu.2010.03.013

Kegan, R. (2018). What "form" transforms?: A constructive-developmental approach to transformative learning. In K. Illeris (Ed.), *Contemporary theories of learning* (pp. 29–45). Routledge. doi:10.4324/9781315147277-3

Kelly, K. (2019) *AR will spark the next big tech platform—call it mirror world*. https://www.wired.com/story/mirrorworld-ar-next-big-tech-platform/

Kelly, K. (2010). From print to pixel. *Smithsonian*, *41*(4), 122–128.

Kelly, K. (2016). *The inevitable: understanding the 12 technological forces that will shape our future*. Penguin Publishing Group.

Kenyon, C., & Hase, S. (2001). Moving from andragogy to Heutagogy in vocational education. *Research to Reality: Putting VET Research To Work. Proceedings of the Australian Vocational Education and Training Research Association (AVETRA) Conference.*

Kenyon, C., & Hase, S. (2001). Moving from andragogy to heutagogy. *ERIC 18*(1). https://eric.ed.gov/?idED456279

Keohane, N. O. (2013). Higher education in the 21st century: Innovation, adaptation, preservation. *PS, Political Science & Politics*, *46*(1), 102–105. doi:10.1017/S1049096512001734

Khalifa, A., Kato, T., & Yamamoto, S. (2017). *Measuring the effect of repetitive queries and implicit learning with joining-in type robot-assisted language learning system*. Paper presented at ISCA workshop on Speech and Language Technology in Education, Stockholm, Sweden. 10.21437/SLaTE.2017-3

Khilji, S. E. (2022). An approach for humanizing leadership education: Building learning. community & stakeholder engagement. *Journal of Management Education*, *46*(3), 439–471. doi:10.1177/10525629211041355

Khrushchev, N. (1958). *Proposals to reform Soviet Education*. Soviet News.

Kickbusch, I., Pelikan, J., Apfel, F., & Tsouros, A. (Eds.). (2013). *Health literacy*. World Health Organization.

Kilburn, A., Kilburn, B., & Hammond, K. (2017). The role of quality in online higher education. *Journal of Higher Education Theory and Practice*, *17*(7), 80–86.

Kim-Cohen, J., & Turkewitz, R. (2012). Resilience and measured gene-environment interactions. *Development and Psychopathology*, *24*(4), 1297–1306. doi:10.1017/S0954579412000715 PMID:23062298

King, K. (2017). *Technology and innovation in adult learning*. Jossey-Bass.

King, K. P. (2009). Teachers of adults in the global community. In V. Wang (Ed.), *Strategic approaches towards curriculum development for adult learners in the global community* (pp. 91–119). Krieger.

Kingston, D., & Siraj, I. (2017). Supporting the implementation of the foundation phase through effective professional development. *Wales J. Educ.*, *19*(1), 39–68. doi:10.16922/wje.19.1.3

Kırkgöz, Y., Ersoy, M., & Şavran, E. (2020, November 21-22). Exploring Pre-Service Teachers' Experiences during Onlıne Education. In *Proceedings of Global Conference on Advances in Education and Research (GCAER-2020), Tsuruoka, Japan (Online)* (pp. 37-42). Academic Press.

Kırkgöz, Y. (2007). Innovation as a curriculum renewal process in a Turkish University. In A. Rice (Ed.), *Revitalizing an established program for adult learners* (pp. 135–158). TESOL Publications.

Kırkgöz, Y. (2017). Working the problem: Finding solutions to student dissatisfaction in EAP for engineering. In T. Stewart (Ed.), *Voices from the TESOL classroom: Participant inquiries in higher education classes* (pp. 17–26). TESOL Publications.

Kırkgöz, Y. (2018). Problem-based learning in a teacher education program: A study of learning outcomes. In Z. S. Genç (Ed.), *Updating perspectives on English language teaching and teacher education* (pp. 79–92). Peter Lang.

Kirtman, L. (2009). Online versus in-class courses: An examination of differences in learning outcomes. *Issues in Teacher Education, 18*(2), 103–116. https://eric.ed.gov/?id=EJ858508

Kistow, B. (2009). E-learning at the Arthur Lok Jack Graduate School of Business: A survey of faculty members. *International Journal of Education and Development using ICT, 5*(4). http://ijedict.dec.uwi.edu/viewarticle.php?id=845

Kjellström, S., Stålne, K., & Törnblom, O. (2020). Six ways of understanding leadership development: An exploration of increasing complexity. *Leadership, 16*(4), 434–460. doi:10.1177/1742715020926731

Kjellström, S., Törnblom, O., & Stålne, K. (2020). A dialogue map of leader and leadership development methods: A communication tool. *Cogent Business & Management, 7*(1), 1717051. doi:10.1080/23311975.2020.1717051

Klecka, C. L., Cheng, Y.-M., & Clift, R. T. (2004). Exploring the potential of electronic mentoring. *Action in Teacher Education, 26*(3), 2–9. doi:10.1080/01626620.2004.10463327

Klimoski, R., & Mohammed, S. (1994). Team mental model: Construct or metaphor? *Journal of Management, 20*(2), 403–437. doi:10.1177/014920639402000206

Knemeyer, D. (2015). The changing role of design. In J. Follett (Ed.), *Designing for emerging technologygies* (pp. 427–438). O'Reilly Media Inc.

Kniffin, K. M., Wansink, B., & Shimizu, M. (2015). Sports at work: Anticipated and persistent correlates of participation in high school athletics. *Journal of Leadership & Organizational Studies, 22*(2), 217–230. doi:10.1177/1548051814538099

Knowles, M. (1996). *Andragogy: An emerging technology for adult learning*. Academic Press.

Knowles, M. (1996). Andragogy: An emerging technology for adult learning. *Boundaries of adult learning*, 82-98.

Knowles, M. S. (1968). Andragogy, not pedagogy. *Academic Leadership, 16*(10), 350-352. 386.

Knowles, M. S., Holton, E., & Swanson, A. (2011). *The adult learner: The definitive classic in adult education and human resource development* (7th ed.). Taylor & Francis.

Knowles, M. S., Holton, III, Swanson, & Robinson, P. A. (2020). *The adult learner the definitive classic in adult education and human resource development* (9th ed.). Routledge.

Knowles, M.S. (1968). *Andragogy, not pedagogy*. Adult Leadership.

Knowles, M. (1989). *The making of an adult educator*. Jossey-Bass.

Knowles, M. S. (1962). *The adult education movement in the United States*. Holt, Rinehart, and Winston.

Knowles, M. S. (1970). *The modern practice of adult education: Andragogy versus pedagogy*. Association Press.

Knowles, M. S. (1975). *Self-directed learning*. Association Press.

Knowles, M. S. (1975). *Self-directed learning: A guide for learners and teachers*. Association Press.

Knowles, M. S. (1979). Andragogy revisited part II. *Adult Education Quarterly, 30*(1), 52–53. doi:10.1177/074171367903000104

Knowles, M. S. (1980). *The modem practice of adult education: From pedagogy to andragogy*. Follett.

Knowles, M. S. (1980). *The modern practice of adult education: From pedagogy to andragogy* (2nd ed.). Cambridge Books.

Knowles, M. S. (1984). *Andragogy in action*. Jossey-Bass.

Knowles, M. S. (1984). Introduction: The art and science of helping adults learn. In *Andragogy in action* (pp. 1–21). Jossey-Bass.

Knowles, M. S. (1984). *The adult learner: A neglected species* (3rd ed.). Gulf.

Knowles, M. S. (1984). *Theory of andragogy*. Jossey-Bass.

Knowles, M. S. (1986). *Using learning contracts*. Jossey-Bass.

Knowles, M. S. (1989). *The making of an adult educator: An autobiographical journey*. Jossey-Bass.

Knowles, M. S. (1995). *Designs for adult learning*. American Society for Training and Development.

Knowles, M. S., Holton, E. F. III, & Swanson, R. A. (2011). *The adult learner: The definitive classic in adult education and human resource development* (7th ed.). Elsevier.

Knowles, M. S., Holton, E. F., & Swanson, R. A. (2005). *The adult learner: the definitive classic in adult education and human resource development* (6th ed.). Elsevier, Inc., doi:10.4324/9780080481913

Knowles, M. S., Holton, E. F., & Swanson, R. A. (2015). *The adult learner. The definitive classic in adult education and human resource development* (8th ed.). Routledge.

Knowles, M. S., Holton, E., & Swanson, A. (1998). *The adult learner*. Gulf.

Knowles, M. S., Holton, E., & Swanson, A. (2005). *The adult learner the definitive classic in adult education and human resource development* (6th ed.). Elsevier Butterworth Heinemann.

Knowles, M. S., & Hulda, F. (1973). *Introduction to group dynamics*. Follett.

Knutson, S. (2003). Experiential learning in second-language classrooms. *TESL Canada Journal*, *20*(2), 52–64. doi:10.18806/tesl.v20i2.948

Koehler, M. J., & Mishra, P. (2009). What is technological pedagogical content knowledge? *Contemporary Issues in Technology & Teacher Education*, *9*(1), 60–70.

Kolb, A. Y., & Kolb, D. A. (2005). Learning styles and learning spaces: Enhancing experiential learning in higher education. *Academy of Management Learning & Education*, *4*(2), 193–212. doi:10.5465/amle.2005.17268566

Komives, S. R., Lucas, N., & McMahon, T. R. (2009). *Exploring leadership: For college students who want to make a difference*. John Wiley & Sons.

Konradt, U., & Eckardt, G. (2016). Short-term and long-term relationships between reflection and performance in teams: Evidence from a four-wave longitudinal study. *European Journal of Work and Organizational Psychology*, *25*(6), 804–818. doi:10.1080/1359432X.2016.1160058

Kooij, D. T. A. M., Lange, A. H. D., Jansen, P. G. W., Kanfer, R., & Dikkers, J. S. E. (2011). Age and work-related motives: Results of a meta-analysis. *Journal of Organizational Behavior*, *32*(2), 197–225. doi:10.1002/job.665

Kotter, J. P. (2012, November). Accelerate! *Harvard Business Review*. https://hbr.org/2012/11/accelerate

Kotter, J. P. (1999). *John Kotter on what leaders really do*. Harvard Business School Press.

Koutropoulos, A. (2011). Digital natives: Ten years after. *Journal of Online Learning*.

Kouzes, J., & Posner, B. (2012). *The leadership challenge. How to make extraordinary things happen in organizations*. The Leadership Challenge, A Wiley Brand.

Kovačević, J., Mujkić, A., & Kapo, A. (2020). Examining school leadership in a transitional context: A mixed-methods study of leadership practices and school cultures as mechanisms of educational change. *Educational Management Administration & Leadership*. Advance online publication. doi:10.1177/1741143220971286

Kovács-Ondrejkovic, O., Strack, R., Antebi, P., López Gobernado, A., & Lyle, E. (2019, November 5). *Decoding Global Trends in Upskilling and Reskilling*. https://www.bcg.com/publications/2019/decoding-global-trends-upskilling-reskilling

Kozlowski, S. W., & Ilgen, D. R. (2006). Enhancing the effectiveness of work groups and teams. *Psychological Science in the Public Interest*, *7*(3), 77–124. doi:10.1111/j.1529-1006.2006.00030.x PMID:26158912

Kraft, S. (1999). Selbstgesteuertes lernen. Problembereiche in theorie und praxis. *Zeitschrift für Pädagogik*, *45*, 833–845.

Kratzer, J., & Mrożewski, M. (2021). Factors fostering creativity in start-up teams. In *Creative Success in Teams* (pp. 123–144). Academic Press. doi:10.1016/B978-0-12-819993-0.00007-2

Kropp, B., & McRae, E. R. (2022, January 13). *11 Trends That Will Shape Work in 2022 and Beyond*. https://hbr.org/2022/01/11-trends-that-will-shape-work-in-2022-and-beyond

Krouse, S. S., & Afifi, T. D. (2007). Family-to-work spillover stress: Coping communicatively in the workplace. *Journal of Family Communication*, *7*(2), 85–122. doi:10.1080/15267430701221537

Kumashiro, K. K. (2000). Toward a theory of anti-oppressive education. *Review of Educational Research*, *70*(1), 25–53. doi:10.3102/00346543070001025

Kumi-Yeboah, A. (2018). Designing a cross-cultural collaborative online learning framework for online instructors. *Online Learning*, *22*(4), 181–201.

Kummel, E., Moskaliuk, J., Cress, U., & Kimmerle, J. (2020). Digital learning environments in higher education: A literature review of the role of individual vs. social seetings for measuring learning outcomes. *Education Sciences*, *10*(3), 78. doi:10.3390/educsci10030078

Kuo, Y. C., & Belland, B. (2016). An exploratory study of adult learners' perceptions of online learning: Minority students in continuing education. *Educational Technology Research and Development*, *64*(4), 661–680. doi:10.100711423-016-9442-9

Kuo, Y. C., Walker, A., Schroder, K. E. E., & Belland, B. R. (2014). Interaction, Internet self-efficacy, and self-regulated learning as predictors of student satisfaction in online education courses. *The Internet and Higher Education*, *20*, 35–50. doi:10.1016/j.iheduc.2013.10.001

Kyaw, B., Saxena, N., Posadzki, P., Vseteckova, J., Nikolaou, C. K., George, P. P., Divakar, U., Masiello, I., Kononowicz, A. A., Zary, N., & Tudor Car, L. (2019). Virtual reality for health professions education: Systematic review and meta-analysis by the digital health education collaboration. *Journal of Medical Internet Research*, *20*(1), e12959. doi:10.2196/12959 PMID:30668519

Kythreotis, A., Pashiardis, P., & Kyriakides, L. (2010). The influence of school leadership styles and culture on students' achievement in Cyprus primary schools. *Journal of Educational Administration*, *48*(2), 218–240. doi:10.1108/09578231011027860

La Botz, D. (10 Sep 2013). Mexican teachers' massive strikes crescendo to a showdown. *Labor Notes*. http://labornotes.org/2013/09/mexican-teachers'-massive-strikes-crescendo-showdown

Laal, M., & Ghodsi, S. M. (2011). Benefits of collaborative learning. *Procedia: Social and Behavioral Sciences*, *31*, 486–490. doi:10.1016/j.sbspro.2011.12.091

Ladson-Billings, G. (2021). I'm here for the hard re-set: Post pandemic pedagogy to preserve. Our culture. *Equity & Excellence in Education, 54*(1), 68–78. doi:10.1080/10665684.2020.1863883

Laloux, F. (2014). *Reinventing Organizations. Ein Leitfaden zur Gestaltung sinnstiftender Formen der Zusammenarbeit.* Vahlen.

Lalwani, A., & Green, W. M., & Hamlen Mansour, K. (2022). How campus alienation exacerbated international students' difficulties in accessing campus services remotely during COVID-19: Notes on policy and programming. *The American Behavioral Scientist, 0*(0), 1–17. doi:10.1177/00027642221118257

Lamnek, S. (2010). *Qualitative sozialforschung*. Beltz Verlag.

Lapierre, J., & Giroux, V. (2003). Creativity and work environment in a high-tech context. *Creativity and Innovation Management, 12*(1), 11–23. doi:10.1111/1467-8691.00263

Lara-Cabrera, M. L., Betancort, M., Muñoz-Rubilar, C. A., Rodríguez Novo, N., & De las Cuevas, C. (2021). The mediating role of resilience in the relationship between perceived stress and mental health. *International Journal of Environmental Research and Public Health, 18*(18), 9762. doi:10.3390/ijerph18189762 PMID:34574684

Latchem, C. (2012). *Quality assurance toolkit for open and distance non-formal education.* Commonwealth of Learning.

Lave, J., & Wenger, E. (1991). *Situated learning: Legitimate peripheral participation.* University Press. doi:10.1017/CBO9780511815355

Leahy, J. (2016). *The impact of adult and community education on women's lives* [Unpublished thesis]. University of Canterbury.

Ledoux, S. F. (2002). Defining natural sciences. *Behaviorology Today, 5*(1), 24–36.

Lee, J. Y., Swink, M., & Pandejpong, T. (2011). The roles of worker expertise, information sharing quality, and psychological safety in manufacturing process innovation: An intellectual capital perspective. *Production and Operations Management, 20*(4), 556–570. doi:10.1111/j.1937-5956.2010.01172.x

Lee, K. (2010). Philosopher or philistine? In V. Wang (Ed.), *Assessing and evaluating adult learning in career and technical education* (pp. 27–52). doi:10.4018/978-1-61520-745-9.ch003

Leithwood, K. (2021). A review of evidence about equitable school leadership. *Education Sciences, 11*(8), 377. doi:10.3390/educsci11080377

Leithwood, K., Harris, A., & Hopkins, D. (2020). Seven strong claims about successful school leadership revisited. *School Leadership & Management, 40*(1), 5–22. doi:10.1080/13632434.2019.1596077

Leithwood, K., Mascall, B., & Strauss, T. (2009). What we have learned and where we go from here. In K. Leithwood, B. Mascall, & T. Strauss (Eds.), *Distributed leadership according to the evidence* (pp. 269–282). Routledge. doi:10.4324/9780203868539

Leithwood, K., Sun, J., & McCullough, C. (2019). How school districts influence student achievement. *Journal of Educational Administration, 57*(5), 519–539. Advance online publication. doi:10.1108/JEA-09-2018-0175

Lenin, V. I. (1975). *On public education* (Vol. 1921). Progress Press.

Leonard, E. (1991). *Friendly rebel: A personal and social history of Eric C. Lindeman.* Adamant Press.

Leong, S. (2018). Various ways of understanding the concept of andragogy. *Cogent Education, 5*(1), 1496643. Advance online publication. doi:10.1080/2331186X.2018.1496643

Lerner, M. (1992). *America as a civilization: Life and thought in the United States today*. Simon and Schuster.

Lesgold, A. (2000). What are the tools for? Revolutionary change does not follow the usual norms. In S. Lajoie (Ed.), *Computers as cognitive tools* (pp. 399–408). Lawrence Erlbaum Associates.

Levin, D. (1963). *Soviet education today*. Monthly Review Press.

Liang, J.-C., & Tsai, C.-C. (2008). Internet self-efficacy and preferences toward constructivist internet-based learning environments: A study of pre-school teachers in Taiwan. *Journal of Educational Technology & Society*, *11*(1), 226–237.

Liao, Z., & Zhang, M. (2020). The influence of responsible leadership on environmental innovation and environmental performance: The moderating role of managerial discretion. *Corporate Social Responsibility and Environmental Management*, *27*(5), 2016–2027. doi:10.1002/csr.1942

Liaw, S., Huang, H. M., & Chen, G. D. (2007). Surveying instructor and learner attitudes toward e-learning. *Computers & Education*, *49*(4), 1066–1080. doi:10.1016/j.compedu.2006.01.001

Li, M., & Yu, Z. (2022). Teachers' Satisfaction, Role, and Digital Literacy during the COVID-19 Pandemic. *Sustainability*, *14*(3), 1121. doi:10.3390u14031121

Lindeman, E. C. (1925). *What is adult education?* [Unpublished manuscript]. Columbia University, Butler Library Lindeman Archive.

Lindeman, E. C. (1926). *The meaning of adult education*. New Republic.

Linder-VanBerschot, J. S., & Summers, L. L. (2015). Designing instruction in the face of technology transience. *Quarterly Review of Distance Education*, *16*(2), 107–118.

Lineweaver, T. T. (2010). Online discussion assignments improve students' class preparation. *Teaching of Psychology*, *37*(3), 204–209. doi:10.1080/00986283.2010.488546

Lipman, M. (1991). *Thinking in education*. Cambridge University Press.

Lipset, S. M. (1960). *Political man: The social bases of politics*. Doubleday.

Lipton, L., & Wellman, B. (2013). *Learning focused supervision: Developing professional expertise in standards-driven systems*. Mira Via.

Liu, W. C., & Fang, C. L. (2010). The Effect of Different Motivation Factors on Knowledge-Sharing Willingness and Behavior. *Social Behavior and Personality*, *38*(6), 753–758. doi:10.2224bp.2010.38.6.753

Liu, Z., Venkatesh, S., Murphy, S. E., & Riggio, R. E. (2020). Leader development across the lifespan: A dynamic experiences-grounded approach. *The Leadership Quarterly*, *32*(5), 101382. doi:10.1016/j.leaqua.2020.101382

Loader, B. (1998). *Cyberspace divide: Equality, agency, and policy in the information society*. Routledge. doi:10.4324/9780203169537_chapter_1

Loeng, S. (2013). Eugen Rosenstock-Huessy: An andragogical pioneer. *Studies in Continuing Education*, *35*(2), 241–253. doi:10.1080/0158037X.2012.749850

Loeng, S. (2017). Alexander Kapp: The first known user of the andragogy concept. *International Journal of Lifelong Education*, *36*(20), 1–15. doi:10.1080/02601370.2017.1363826

London, M. (2002). *Leadership development: Paths to self-insight and professional growth*. Lawrence Erlbaum Associates, Inc.

Lord, R. G., & Brown, D. J. (2004). *Leadership processes and follower self-identity*. Lawrence Erlbaum Associates, Inc.

Lowenstein, K. (2009). The work of multicultural teacher education: Reconceptualizing white teacher candidates as learners. *Review of Educational Research, 79*(1), 163–196. doi:10.3102/0034654308326161

Lowenthal, P. R., & Dunlap, J. C. (2018). Investigating students' perceptions of instructional strategies to establish social presence. *Distance Education, 39*(3), 281–298. doi:10.1080/01587919.2018.1476844

Lowyck, J. (2014). Bridging learning theories and technology-enhanced environments: A Critical appraisal of its history. In Handbook of research on educational communications and technology (pp. 3-20). Springer. doi:10.1007/978-1-4614-3185-5_1

Lunacharsky, A. V. (1981). *On education* (Vol. 1920). Progress Press.

Lynch, K., Grummell, B., & Devine, D. (2012). *New managerialism in education commercialization, carelessness, and gender*. Palgrave Macmillan. doi:10.1057/9781137007230

Lytovchenko, I., & Voronina, H. (2020). MOOC as remote ESP learning tool at university in quarantine: Focus on Students' Attitudes. *Revista Românească pentru Educaţie Multidimensională, 12*(2), 70–76. doi:10.18662/rrem/12.2Sup1/291

Lytovchenko, I., Lavrysh, Y., Lukianenko, V., & Ogienko, O. (2020). How to teach grammar to adult ESP learners at technical university more communicatively: Task-based approach. *Multidisciplinary Journal for Education. Social and Technological Sciences, 7*(1), 54–71. doi:10.4995/muse.2020.12419

Maber, E. (2016). Finding feminism, finding voice? Mobilising community education to build women's participation in Myanmar's political transition. *Gender and Education, 28*(3), 416–430. doi:10.1080/09540253.2016.1167175

Machida, M., & Schaubroeck, J. (2011). The role of self-efficacy beliefs in leader development. *Journal of Leadership & Organizational Studies, 18*(4), 459–468. doi:10.1177/1548051811404419

Mackeracher, D. (2005). *Making sense of adult learning*. University of Toronto Press.

Mack, T. C. (2015). Leadership in the future. In M. Sowcik, A. C. Andenoro, M. McNutt, & S. E. Murphy (Eds.), *Leadership 2050: Critical challenges, key contexts, and emerging trends* (pp. 9–22). Emerald Group Publishing Limited.

Macpherson, C. B. (1962). *The political theory of possessive individualism: Hobbes and locke*. Oxford University Press.

Mahlich, H. (2021). *How to ensure that the Metaverse is an inclusive space*. https://www.showstudio.com/news/how-to-ensure-the-metaverse-is-an-inclusive-space

Maier, P., & Warren, A. (2013). *Integrating technology in learning and teaching*. Kogan Page. doi:10.4324/9781315042299

Mainemelis, C., Boyatzis, R. E., & Kolb, D. A. (2002). Learning styles and adaptive flexibility: Testing experiential learning theory. *Management Learning, 33*(1), 5–33. doi:10.1177/1350507602331001

Makadok, R., Burton, R., & Barney, J. (2018). A practical guide for making theory contributions in strategic management. *Strategic Management Journal, 39*(6), 1530–1545. doi:10.1002mj.2789

Makarenko, A. S. (1954). A book for parents. Foreign Languages. (Original publication 1939)

Makarenko, A. S. (1973). *The road to life* (Vol. 1936). Oriole.

Maloney, D., Zamanifard, S., & Freeman, G. (2020). Anonymity vs. Familiarity: Self-Disclosure and Privacy in Social Virtual Reality. *Proceedings of ACM Symposium on Virtual Reality Software and Technology (VRST).*

Mancuso, D., Chlup, D., & McWhorter, R. (2010). A study of adult learning in a virtual world. *Advances in Developing Human Resources*, *12*(6), 681–699. doi:10.1177/1523422310395368

Manikkath, J., Matuluko, A., Leong, A., Ching, D., Dewart, C., Lim, R., Meilianti, S., & Uzman, N. (2020). Exploring young pharmacists' and pharmaceutical scientists' needs and expectations within an international pharmacy organization: Findings from FIP's needs assessment survey. *Research in Social & Administrative Pharmacy*, *16*(12), 1677–1685. Advance online publication. doi:10.1016/j.sapharm.2020.02.015 PMID:32169471

Marcatto, F., Di Blas, L., Luis, O., Festa, S., & Ferrante, D. (2021). The perceived occupational stress scale: A brief tool for measuring workers' perceptions of stress at work. *European Journal of Psychological Assessment*. Advance online publication. doi:10.1027/1015-5759/a000677

Marcos Sanchez, V. (2022). The right to a digital education. *Mexico Business News*. https://mexicobusiness.news/talent/news/right-digital-education

Marcuse, H. (1964). *One-dimensional man*. Beacon Press.

Mariani, M. (2001). Distance learning in postsecondary education: Learning whenever, wherever. *Occupational Outlook Quarterly*, *45*(2), 2–10.

Markie, P. J. (1994). *A professor's duties: Ethical issues in college teaching*. Rowman & Littlefield.

Marshall, J., Roache, D., & Moody-Marshall, R. (2020). Crisis leadership: A critical examination of educational leadership in higher education in the midst of the COVID-19 pandemic. *International Studies in Educational Administration*, *48*(3), 30–37.

Martinelli, D., & Erzikova, E. (2017). Public relations leadership development cycle: A cross-cultural perspective. *Public Relations Review*, *43*(5), 1062–1072. doi:10.1016/j.pubrev.2017.09.002

Martinez, M. (2005). Mass customization: Designing for successful learning. *International Journal of Educational Technology*, *2*(2). http://www.ed.uiuc.edu/ijet/v2n2/martinez/index.html

Martin, J. S., & Marion, R. (2005). Higher education leadership roles in knowledge processing. *The Learning Organization*, *12*(2), 140–151. Advance online publication. doi:10.1108/09696470510583520

Martin, J., Cormican, K., Samaio, S. C. B., & Wu, Q. (2018). Shared leadership and team performance: An analysis of moderating factors. *Procedia Computer Science*, *138*, 671–679. doi:10.1016/j.procs.2018.10.089

Marton, F. (1986). Phenomenography: A research approach to investigating different understandings of reality. *Journal of Thought*, *21*(3), 28-49.

Marzuki, M., Yaacob, N., bin Yaacob, N., Hassan, M., & Ahmad, S. (2019). Usable mobile app for community education on colorectal cancer: development process and usability study. *JMIR Human Factors*, *6*(2). https://humanfactors.jmir.org/2019/2/e12103/

Mathieu, J. E., Heffner, T. S., Goodwin, G. F., Salas, E., & Cannon-Bowers, J. A. (2000). The influence of shared mental models on team process and performance. *The Journal of Applied Psychology*, *85*(2), 273–283. doi:10.1037/0021-9010.85.2.273 PMID:10783543

Mauno, S., & Kinnunen, U. (1999). The effects of job stressors on marital satisfaction in Finnish dual-earner couples. *Journal of Organizational Behavior*, *20*(6), 279–895. doi:10.1002/(SICI)1099-1379(199911)20:6<879::AID-JOB982>3.0.CO;2-2

Maxwell, J. C. (2013). *The 5 levels of leadership: proven steps to maximize your potential*. Center Street.

Mayring, P. (2000). Qualitative content analysis. *Forum Qualitative Social Research*, *1*(2), 20.

Mazarr, M. (1999). *Global trends 2005: An owner's manual for the next decade*. St. Martin's Press.

Mazzoccoli, A., & Wolf, G. (2016). Mentoring through the leadership journey: From novice to expert. *Nurse Leader*, *14*(4), 253–256. doi:10.1016/j.mnl.2016.05.012

Mazzoni, E., & Benvenuti, M. (2015). A robot partner for preschool children learning English using socio-cognitive conflict. *Journal of Educational Technology & Society*, *18*(4), 474–485.

McAdamis, S. (2010). Central office and school leaders create a conversation. *The Learning Professional*, *31*(3), 24.

McCaslin, M., & Hickey, D. T. (2001). Self-regulated learning and academic achievement: A Vygotskian view. *Theoretical Perspectives*, *2*, 227–252.

McClelland, J. (1971). Bolshevik approaches to higher education, 1917-1921. *Slavic Review*, *4*(4), 818–831. doi:10.2307/2493850

McDonough, D. (2014). Providing deep learning through active engagement of adult learners in blended courses. *Journal of Learning in Higher Education*, 9-16.

McGinity, R., Heffernan, A., & Courtney, S. J. (2022). Mapping trends in educational-leadership research: A longitudinal examination of knowledge production, approaches and locations. *Educational Management Administration & Leadership*, *50*(2), 217–232. doi:10.1177/17411432211030758

McKay, A. S., Reiter-Palmon, R., & Kaufman, J. C. (2021). An interdisciplinary view on team creativity: Toward integration across fields. In *Creative Success in Teams* (pp. 231–241). Academic Press. doi:10.1016/B978-0-12-819993-0.00012-6

McMahon, B., & Portelli, J. P. (2004). Engagement for what? Beyond popular discourses of student engagement. *Leadership and Policy in Schools*, *3*(1), 59–76. doi:10.1076/lpos.3.1.59.27841

McNeil, S. G., Robin, B. R., & Miller, R. M. (2000). Facilitating interaction, communication and collaboration in online courses. *Computers & Geosciences*, *26*(6), 699–708. doi:10.1016/S0098-3004(99)00106-5

McQuay, P. (2002). *A Discussion Paper on Vocational Technical Education in the United States of America*. Academic Press.

Mdingi, S. L. (2018). *Investigating the levels of management and leadership development among medical doctors* [Masters Thesis]. Nelson Mandela Metropolitan University.

Meier, C., Saorín, J., de León, A. B., & Cobos, A. G. (2020). Using the Roblox video game engine for creating virtual tours and learning about the sculptural heritage. *International Journal of Emerging Technology*, *5*(20), 268–280. doi:10.3991/ijet.v15i20.16535

Mengistie, A. (2022). Exploring women's agency through adult education program in Ethiopia. *Cogent Social Sciences*, *8*(1), 2073672. doi:10.1080/23311886.2022.2073672

Merleau-Ponty, M. (1962). *Phenomenology of perception* (C. Smith, Trans.). Routledge & Kegan Paul.

Merriam, S. B. (2008). Adult learning theory for the twenty-first century. In *New Directions for adult and continuing education* (pp. 93-98). Jossey-Bass.

Merriam, Sh. B., & Brockett, R. G. (2007). *The profession and practice of adult education*. Jossey Bass.

Merriam, S. B. (2008). Adult learning theory for the twenty-first century. *New Directions for Adult and Continuing Education*, *119*(119), 93–98. doi:10.1002/ace.309

Merriam, S. B., & Bierema, L. L. (2013). *Adult learning: Linking theory to practice*. Jossey-Bass.

Merriam, S. B., & Bierema, L. L. (2014). *Adult learning: Linking theory and practice*. Jossey-Bass.

Merriam, S. B., & Bierema, L. L. (2014). *Adult learning: Linking theory with practice*. Jossey-Bass.

Merriam, S. B., & Caffarella, R. S. (1999). *Learning in adulthood: A comprehensive guide*. Jossey-Bass.

Merriam, S., & Baumgartner, L. (2020). *Learning in adulthood: A Comprehensive guide* (4th ed.). John Wiley & Sons.

Mertens, D. M. (2007). Transformative paradigm: Mixed methods and social justice. *Journal of Mixed Methods Research, 1*(3), 212–225. doi:10.1177/1558689807302811

Mesmer-Magnus, J. R., & Dechurch, L. A. (2009). Information sharing and team performance: A meta-analysis. *The Journal of Applied Psychology, 94*(2), 535–546. doi:10.1037/a0013773 PMID:19271807

Mesterova, J., Prochazka, J., Vaculik, M., & Smutny, P. (2015). Relationship between self-efficacy, transformational leadership and leader effectiveness. *Journal of Advanced Management Science, 3*(2), 109–122. doi:10.12720/joams.3.2.109-122

Meuser, M., & Nagel, U. (1991). ExpertInneninterviews – vielfach erprobt, wenig bedacht: Ein beitrag zur qualitativen methodendiskussion. In D. Garz & K. Kraimer (Eds.), *Qualitativ-empirische Sozialforschung: Konzepte, Methoden, Analysen* (pp. 441–471). Westdt. Verlag. doi:10.1007/978-3-322-97024-4_14

Meyer, R., & Haunschild, A. (2017). Individuelle kompetenzentwicklung und betriebliche organisationsentwicklung im kontext moderner beruflichkeit – berufspädagogische und arbeitswissenschaftliche Befunde und Herausforderungen. *bwpat, 32*, 1-20.

Mezirow, J. (2008). An overview on transformative learning. *Lifelong Learning,* 40-54.

Mezirow, J. (1978). Perspective transformation. *Adult Education Quarterly, 28*(2), 100–110. doi:10.1177/074171367802800202

Mezirow, J. (1990). *Fostering critical reflection in adulthood.* Jossey-Bass Publishers.

Mezirow, J. (1991). *Transformative dimensions of adult learning*. Jossey-Bass.

Mezirow, J. (1997). Transformative learning: Theory to practice. *New Directions for Adult and Continuing Education, 74*(74), 5–12. doi:10.1002/ace.7401

Mezirow, J. (2000). *Learning as transformation: critical perspectives on a theory in progress* (1st ed.). Jossey-Bass.

Mezirow, J., & ... (2000). *Learning as transformation: Critical perspectives on a theory in progress*. Jossey-Bass.

Michlewski, K. (2015). *Design attitude*. Gower.

Mihelj, S., Leguina, A., & Downey, J. (2019). Culture is digital: Cultural participation, diversity, and the digital divide. *New Media & Society, 21*(7), 1465–1485. doi:10.1177/1461444818822816

Miles, M. B., & Huberman, A. M. (1994). *Qualitative data analysis: An expanded sourcebook* (2nd ed.). Sage Publications, Inc.

Miller, D. (2001). Successful change leaders: What makes them? What do they do that is different? *Journal of Change Management, 2*(4), 359–368. doi:10.1080/714042515

Miller, M. T., & Grover, K. S. (2014). Characteristics and training of adult education program leaders in community colleges. *International Journal of Educational Studies, 1*(2), 83–90.

Milliron, M. D., & Miles, C. L. (1998). *Technology, learning, and community (TLC)*. League for Innovation in the Community College.

Minzey, J. (1972). Community education: An amalgam of many views. *Phi Delta Kappan*, *54*(3), 150–153.

Mishra, P., & Koehler, M. J. (2006). Technological pedagogical content knowledge: A framework for integrating technology in teachers' knowledge. *Teachers College Record*, *108*(6), 1017–1054. doi:10.1111/j.1467-9620.2006.00684.x

Mistretta, S. (2022). The Metaverse – An alternative education space, AI. *Computer Science and Robotics Technology*, *0*, 1–23.

Moher, D., Liberati, A., Tetzlaff, J., & Altman, D. G. (2009). Preferred reporting items for systematic reviews and meta-analyses: The PRISMA statement. *Annals of Internal Medicine*, *151*(4), 264–269. doi:10.7326/0003-4819-151-4-200908180-00135 PMID:19622511

Monaghan, C. H., & Boboc, M. (2017). (Re) Defining Leadership in Higher Education in the US. In V. Wang (Ed.), *Encyclopedia of Strategic Leadership and Management* (pp. 567–579). IGI Global. doi:10.4018/978-1-5225-1049-9.ch040

Montuori, A., & Donnelly, G. (2018). Transformative leadership. In J. Neal (Ed.), *Handbook of personal and organizational transformation* (pp. 319–350). Springer. doi:10.1007/978-3-319-66893-2_59

Morrison, N. (2022, May 22). *If Students Had Their Way, Hybrid Learning Would Be Here To Stay*. https://www.forbes.com/sites/nickmorrison/2022/05/27/if-students-had-their-way-hybrid-learning-would-be-here-to-stay/?sh=604f22435876

Movahedazarhouligh, S., Banerjee, R., & Luckner, J. (2022). An examination of current leadership practices in early childhood and early childhood special education: A mixed methods study. *Early Education and Development*, *33*(4), 700–722. doi:10.1080/10409289.2021.1909937

mpksbohemian [screen name]. (n.d.) *Metagogy*. http://www.verbotomy.com/verbottle.php?jargonism_id=54

Mubin, O., Shahid, S., & Bartneck, C. (2013). Robot-assisted language learning through games: A comparison of two case studies. *Australian Journal of Intelligent Information Processing Systems*, *13*(3), 9–14.

Muenks, K., & Miele, D. B. (2017). Students' thinking about effort and ability: The role of development, contextual, and individual difference factors. *Review of Educational Research*, *87*(4), 707–735. doi:10.3102/0034654316689328

Mukan, N., & Lavrysh, Y. (2020). Video conferencing integration: Challenges and opportunities at universities. *Revista Românească Pentru Educație Multidimensională*, *12*(1Sup2), 108–114. doi:10.18662/rrem/12.1sup2/253

Muldoon, J. (2020). Kurt Lewin: Organizational change. In B. Bowden, J. Muldoon, A. Gould, & A. M. McMurray (Eds.), *The Palgrave Handbook of Management History* (pp. 615–632). Palgrave MacMillan. doi:10.1007/978-3-319-62114-2_32

Mumford, M. D., Marks, M. A., Connelly, M. S., Zaccaro, S. J., & Reiter-Palmon, R. (2000). Development of leadership skills: Experience and timing. *The Leadership Quarterly*, *11*(1), 87–114. doi:10.1016/S1048-9843(99)00044-2

Mumford, T. V., Campion, M. A., & Morgeson, F. P. (2007). The leadership skills strataplex: Leadership skill requirements across organizational levels. *The Leadership Quarterly*, *18*(2), 154–166. doi:10.1016/j.leaqua.2007.01.005

Murphy, S. E. (2011). Providing a foundation for leadership development. In S.E. Murphy & R. J. Reichard (Eds.), Early development and leadership: Building the next generation of leaders (pp. 3–37). Routledge/Taylor & Francis Group.

Murphy, G. (2019). A systematic review and thematic synthesis of research on school leadership in the Republic of Ireland: 2008–2018. *Journal of Educational Administration*, *57*(6), 675–689. Advance online publication. doi:10.1108/JEA-11-2018-0211

Murphy, S. E., & Johnson, S. K. (2011). The benefits of a long-lens approach to leader development: Understanding the seeds of leadership. *The Leadership Quarterly*, *22*(3), 459–470. doi:10.1016/j.leaqua.2011.04.004

Mursid, R., Saragih, A., & Sitompul, H. (2022). *The role of educational technology in learning innovation for community education development.* Paper presented at ICONSEIR 2021, Medan, Indonesia. 10.4108/eai.21-12-2021.2317244

Murugaboopathy, V., Hebbal, M., Ankola, A., Mohandoss, S., & Sharma, R. (2020). Students' perspective of need for leadership development program in dental education. *Journal of Global Oral Health, 2*(2), 93–98. doi:10.25259/JGOH_65_2019

Musk E (2019) an integrated brain-machine interface platform with thousands of channels. *Journal Medical Internet Research, 21*(10).

Mystakidis, S. (2022). Metaverse. *Metaverse Encyclopedia, 2*(1), 486–497. doi:10.3390/encyclopedia2010031

Nahavandi, A. (2008). *The art and science of leadership.* Pearson Education.

Nakanjako, D., Namagala, E., Semeere, A., Kigozi, J., Sempa, J., Ddamulira, J. B., Katamba, A., Biraro, S., Naikoba, S., Mashalla, Y., Farquhar, C., & Sewankambo, N. (2015). Global health leadership training in resource-limited settings: A collaborative approach by academic institutions and local health care programs in Uganda. *Human Resources for Health, 13*(1), 87. doi:10.118612960-015-0087-2 PMID:26581196

Nandi, D., Hamilton, M., & Harland, J. (2012). Evaluating the quality of interaction in asynchronous discussion forums in fully online courses. *Distance Education, 33*(1), 5–30. doi:10.1080/01587919.2012.667957

Narayan, V., Herrington, J., & Cochrane, T. (2019). Design principles for heutagogical learning: Implementing student-determined learning with mobile and social media tools. *Australasian Journal of Educational Technology, 35*(3), 86–101. doi:10.14742/ajet.3941

National Research Council. (2015). *Enhancing the effectiveness of team science.* The National Academies Press., doi:10.17226/19007

Natu, S., & Aparicio, M. (2022). Analyzing knowledge sharing behaviors in virtual teams: Practical evidence from digitalized workplaces. *Journal of Innovation &Amp. Knowledge (Beverly Hills, Calif.), 7*(4), 100248. doi:10.1016/j.jik.2022.100248

Nazari, N., Shabbir, M. S., & Setiawan, R. (2021). Application of artificial intelligence powered digital writing assistant in higher education: Randomized controlled trial. *Heliyon, 7*(5), e07014. doi:10.1016/j.heliyon.2021.e07014 PMID:34027198

NCTE. (2013). *National Council of Teachers of English 21st Century literacies.* www.ncte.org/positions/statements/21stcentdefiniton

Nebrida, J., & Bangud, A. (2022). Challenges in hybrid teaching amidst pandemic: The proposed mode. *Journal of Engineering Research and Reports, 22*(9), 51–64. doi:10.9734/jerr/2022/v22i917564

Neimann, T., & Wang, V. (2017). Deep Learning and Online Education as an Informal Learning Process: Is There a Relationship between Deep Learning and Online Education as an Informal Learning Process? In V. Wang (Ed.), *Adult education and vocational training in the digital age* (pp. 37–57). IGI Global. doi:10.4018/978-1-5225-0929-5.ch003

Neuwirth, L. S., Jović, S., & Mukherji, B. R. (2021). Reimagining higher education during and post-COVID-19: Challenges and opportunities. *Journal of Adult and Continuing Education, 27*(2), 141–156. doi:10.1177/1477971420947738

Newman, A., Donohue, R., & Eva, N. (2017). Psychological safety: A systematic review of the literature. *Human Resource Management Review, 27*(3), 521–535. doi:10.1016/j.hrmr.2017.01.001

Ng'weno, B. (2020). The historical, political, and demographic context of leadership in Africa. In E. Jordans, B. Ng'weno, & H. Spencer-Oatey (Eds.), Developing Global Leaders: Insights from African Case Studies (pp. 59–87). Springer International Publishing. doi:10.1007/978-3-030-14606-1_4

Ng, W. L. (2008). Transformational leadership and the integration of information and communications technology into teaching. *The Asia-Pacific Education Researcher, 17*(1), 1–14. doi:10.3860/taper.v17i1.346

Nhlapo, M. D., & Hlalele, D. J. (2023). Successes and failures of the university-community partnership: A case study of Imbali semi-rural settlement. *Education and Urban Society, 55*(2), 222–237. doi:10.1177/00131245211027368

Nica, E., Poliak, M., Popescu, G. H., & Pârvu, I.-A. (2022). Decision Intelligence and Modeling, Multisensory Customer Experiences, and Socially Interconnected Virtual Services across the Metaverse Ecosystem. *Linguistic and Philosophical Investigations, 21*(0), 137–153. doi:10.22381/lpi2120229

Nick, G., Reinhold, G., & Valerie, I. (2005). *Managing health services.* McGraw-Hill Education.

Niebuhr, R. (1944). *The children of light and the children of darkness.* https://www.semanticscholar.org/paper/The-Children-of-Light-and-the-Children-of-Darkness%3A-Niebuhr/516e53cd752cae99195320cfeeb038b1e95ca06c

Niessen, S. (2013). *What is digital fluency?* http://www.shuananiessen.ca/what-is-digital-fluency/

Nilson, L. B. (2016). *Teaching at its best* (4th ed.). Jossey-Bass.

Ni, Y., Rorrer, A. K., Pounder, D., Young, M., & Korach, S. (2019). Leadership matters: Preparation program quality and learning outcomes. *Journal of Educational Administration, 57*(2), 185–206. Advance online publication. doi:10.1108/JEA-05-2018-0093

Noble, D. J. (2015). Leading for tomorrow in a world yearning for social justice. In M. Sowcik, A. C. Andenoro, M. McNutt, & S. E. Murphy (Eds.), *Leadership 2050: Critical challenges, key contexts, and emerging trends* (pp. 43–58). Emerald Group Publishing Limited.

Nonaka, I., & Takeuchi, H. (1995). *The Knowledge-Creating Company: How Japanese Companies Create the Dynamics of Innovation* (1st ed.). Oxford University Press.

Noroozi, O., Busstra, M. C., Mulder, M., Biemans, H. J. A., Tobi, H., Geelen, A., van't Veer, P., & Chizari, M. (2011). Online discussion compensates for suboptimal timing of supportive information presentation in a digitally supported learning environment. *Educational Technology Research and Development, 60*(2), 193–221. doi:10.100711423-011-9217-2

North, C. E. (2008). What is all this talk about "social justice"? Mapping the terrain of education's latest catchphrase. *Teachers College Record, 110*(6), 1182–1206. doi:10.1177/016146810811000607

Noval, J. (2018). A qualitative study on the usage of a learning management system by allied health faculty. *Journal of Allied Health, 47*(3), 72–182. PMID:30194823

Nuckles, C. R. (2000). Student-centered teaching: Making it work. *Adult Learning, 11*(4), 5–6. doi:10.1177/104515959901100403

Núñez-Canal, M., de Obesso, M. de las M., & Pérez-Rivero, C. A. (2022). New challenges in higher education: A study of the digital competence of educators in Covid times. *Technological Forecasting and Social Change, 174*, 121270. Advance online publication. doi:10.1016/j.techfore.2021.121270

Nurhayati, S., Musa, S., Boriboon, G., Nuraeni, R., & Putri, S. (2021). Community learning center efforts to improve information literacy in the community for cyber-crime prevention during a pandemic. *Journal of Nonformal Education, 7*(1), 32–38. doi:10.15294/jne.v7i1.26883

Nvidia. (2021). *Develop with NVIDIA Omniverse.* https://www.nvidia.com/en-gb/omniverse/.u

Nworie, J. (2012). Applying leadership theories to distance education leadership. *Online Journal of Distance Learning Administration, 15*(5), 1–16.

Nyssa, Z. (2019). Why scientists succeed yet their organizations splinter: Historical and social network analyses of policy advocacy in conservation. *Environmental Science & Policy, 98*(2020), 8–94. . doi:10.1016/j.envsci.2019.05.007

O'Grady, M. (2018). An institutional ethnography of a feminist organization: a study of community education in Ireland. *European Journal for Research on the Education and Learning of Adults, 9*(1), 29-44.

Obschonka, M., Fisch, C., & Boyd, R. (2017). Using digital footprints in entrepreneurship research: A Twitter-based personality analysis of superstart entrepreneurs and managers. J. *Business Venturing Insights, 8*, 13–23. doi:10.1016/j.jbvi.2017.05.005

Ocak, M. (2010). Why are faculty members not teaching blended courses? Insights from faculty members. *Computers & Education, 56*(3), 689–699. doi:10.1016/j.compedu.2010.10.011

Oc, B. (2018). Contextual leadership: A systematic review of how contextual factors shape leadership and its outcomes. *The Leadership Quarterly, 29*(1), 218–235. doi:10.1016/j.leaqua.2017.12.004

Ode, E., & Ayavoo, R. (2020). The mediating role of knowledge application in the relationship between knowledge management practices and firm innovation. *Journal of Innovation & Amp. Knowledge (Beverly Hills, Calif.), 5*(3), 210–218. doi:10.1016/j.jik.2019.08.002

Offermann, L. R., Thomas, K. R., Lanzo, L. A., & Smith, L. N. (2019). Achieving leadership and success: A 28-year follow-up of college women leaders. *The Leadership Quarterly, 31*(4), 101345. doi:10.1016/j.leaqua.2019.101345

Office for National Statistics (ONS). (2021). *Demography question development for Census 2021.* https://www.ons.gov.uk/census/censustransformationprogramme/questiondevelopment/demographyquestiondevelopmentforcensus2021

Office of Inspector General. (n.d.). *Semiannual Report to Congress, 73.* U.S. Department of Education.

Ohlsson, S. (2011). Deep learning: How the mind overrides experience. *Choice Reviews Online, 49*(03), 49–1770. doi:10.5860/CHOICE.49-1770

Olga, T., & Yuliya, L. (2019). Humanistic trend in education in a global context. *SHS Web of Conferences, 69.* 10.1051hsconf/20196900121

Olgren, C. H. (2000). Learning strategies for learning technologies. In E. J. Burge (Ed.), The strategic use of learning technologies (pp. 7–16). Jossey-Bass. doi:10.1002/ace.8801

Open, A. I. (2023, March 3). *Introducing ChatGPT.* https://openai.com/blog/chatgpt

Orozco, L., & Allison, B. (2008). Looking Within: Examining Positive Relationships and Healthy Organizational Cultures in Departments of Educational Leadership. *Educational Leadership and Administration: Teaching and Program Development, 20*, 64–76.

Orphanos, S., & Orr, M. T. (2014). Learning leadership matters: The influence of innovative school leadership preparation on teachers' experiences and outcomes. *Educational Management Administration & Leadership, 42*(5), 680–700. doi:10.1177/1741143213502187

Orr, M. T., Hollingworth, L., & Beaudin, B. (2019). Performance assessment for school leaders: Comparing field trial and implementation results. *Journal of Educational Administration, 58*(1), 38–59. doi:10.1108/JEA-05-2018-0087

Osika, E. R., Johnson, R. Y., & Buteau, R. (2009). Factors influencing faculty use of technology in online instruction: A case study. *Online Journal of Distance Learning Administration, 12*(1), 1–14.

Otte, K. P., Konradt, U., & Oldeweme, M. (2018). Effective team reflection: The role of quality and quantity. *Small Group Research, 49*(6), 739–766. doi:10.1177/1046496418804898

Owen, T. R. (2002). Self-directed learning in adulthood: A literature review. ERIC Clearinghouse on Adult, Career, and Vocational Education. (ERIC Document Reproduction Service No. ED 461 050)

Owens, B. P., & Hekman, D. R. (2012). Modeling how to grow: An inductive examination of humble leader behaviors, contingencies, and outcomes. *Academy of Management Journal, 55*(4), 787–818. doi:10.5465/amj.2010.0441

Özdemir, N., Gümüş, S., Kılınç, A. Ç., & Bellibaş, M. Ş. (2022). A systematic review of research on the relationship between school leadership and student achievement: An updated framework and future direction. *Educational Management Administration & Leadership*. Advance online publication. doi:10.1177/17411432221118662

Palloff, R., & Pratt, K. (1999). *Building learning communities in cyberspace*. Jossey-Bass.

Palloff, R., & Pratt, K. (2007). *Building online learning communities: effective strategies for the virtual classroom.* Jossey-Bass.

Palmer, P. (1991). The courage to teach. *The National Teaching and Learning Forum, 1*(2), 2-7.

Pamphilon, B., & Mikhailovich, K. (2017). Bringing together learning from two worlds: Lessons from a gender-inclusive community education approach with smallholder farmers in Papua New Guinea. *Australian Journal of Adult Learning, 57*(2), 172–196.

Papagiannidis, S., & Bourlakis, M. A. (2010). Stating the new retail drama: At a Metaverse near you. *Journal of Virtual Worlds Research, 2*(5), 425–446. doi:10.4101/jvwr.v2i5.808

Papanaoum, Z. (1994). Adult educators: Issues of professionalization. In R. Benn & R. Fieldhouse (Eds.), *Training and professional development in Adult and Continuing Education.* University of Exeter, Center for Research in Continuing Education.

Papa, R. (Ed.). (2010). *Technology leadership for school improvement.* Sage.

Papa, R. (Ed.). (2014). *Media rich instruction: Connecting curriculum to all learners.* Springer.

Parisi, T. (2021). *The Seven Rules of the Metaverse*. https://medium.com/meta-verses/the-seven-rules-of-the-metaverse-7d4e06fa864c

Parker, J. (2009). Adult learning and CTE: A shared history influenced by technology. In V. Wang (Ed.), *Definitive Readings in the History, Philosophy, Practice and Theories of Career and Technical Education* (pp. 217–236). Information Science Reference.

Parker, J. (2010). Technology as integral to a new paradigm of adult education. *International Journal of Adult Vocational Education and Technology, 1*(2), 10–18. doi:10.4018/javet.2010040102

Park, H.-R., & Kim, D. (2011). Reading-strategy use by English as a second language learners in online reading tasks. *Computers & Education, 57*(3), 2156–2166. doi:10.1016/j.compedu.2011.05.014

Parsons, M., & Stephenson, M. (2005). Developing reflective practice in student teachers: Collaboration and critical partnerships. *Teachers and Teaching, 11*(1), 95–116. doi:10.1080/1354060042000337110

Pavlov, I. P. (1927). *Conditioned reflexes.* Routledge.

Pekel, N. (2002). *Students' attitudes towards web-based independent learning at Bilkent university school of English language* [Master's thesis]. Bilkent University.

Peng, J., Wang, Z., & Chen, X. (2018). Does self-serving leadership hinder team creativity? A moderated dual-path model. *Journal of Business Ethics, 159*(2), 419–433. doi:10.100710551-018-3799-0

Pereira, L., Fernandes, A., Sempiterno, M., Dias, L., Lopes Da Costa, R., & António, N. (2021). Knowledge management maturity contributes to project-based companies in an open innovation era. *Journal of Open Innovation, 7*(2), 126. doi:10.3390/joitmc7020126

Perrone, F., & Tucker, P. D. (2019). Shifting profile of leadership preparation programs in the 21st century. *Educational Administration Quarterly, 55*(2), 253–295. doi:10.1177/0013161X18799473

Perry, J., Mobley, F., & Brubaker, M. (2017). Most doctors have little or no management training, and that's a problem. *Harvard Business Review, 15*(December). https://hbr.org/2017/12/most-doctors-have-little-or-no-management-training-and-thats-a-problem

Perry, S. J., Hunter, E. M., Currall, S. C., & Frauenheim, E. (2017). Developing engineering leaders: An organized innovation approach to engineering education. *Engineering Management Journal, 29*(2), 99–107. doi:10.1080/10429247.2017.1298189

Peters, J., Taylor, J., & Doi, M. (2011). Reflections on three types of teaching and learning in a model-based, blended online course. *International Journal of Learning, 18*(2), 357–366. doi:10.18848/1447-9494/CGP/v18i02/47465

Peters, J. M., & Armstrong, J. L. (1998). *Facilitating Collaborative Learning*. Jossey-Bass.

Peterson, C. M., & Ray, C. M. (2013). Andragogy and metagogy: The evolution of neologisms. *Journal of Adult Education, 42*(2), 80–85.

Petrucci, T., & Rivera, M. (2018). Leading growth through the digital leader. *The Journal of Leadership Studies, 12*(3), 53–56. doi:10.1002/jls.21595

Piaget, J. (1951). *Play, dreams and imitation in childhood.* Norton.

Pietrzak, R. H., & Southwick, S. M. (2011). Psychological resilience in OEF-OIF Veterans: Application of a novel classification approach and examination of demographic and psychosocial correlates. *Journal of Affective Disorders, 133*(3), 560–568. doi:10.1016/j.jad.2011.04.028 PMID:21624683

Piotrowski, C., & King, C. (2020). COVID-19 pandemic: Challenges and implications for higher education. *Education, 141*(2), 61–66.

Piwowar-Sulej, K., & Iqbal, Q. (2023). Leadership styles and sustainable performance: A systematic literature review. *Journal of Cleaner Production, 382*, 134600. Advance online publication. doi:10.1016/j.jclepro.2022.134600

Ployhart, R. E. (2004). Organizational staffing: A multilevel review, synthesis, and model. *Research in Personnel and Human Resources Management, 23*, 121–176. doi:10.1016/S0742-7301(04)23003-1

Pöggeler, F. (1987). *Einfürung in die Andragogik: Grundfragen der Erwachsenenbildung* [Introduction to Andragogy: Fundamental Issues in Adult Education]. Düsseldorf, Germany: A. Henn Verlag.

Polanyi, M. (1966). *The tacit dimension.* P. Smith.

Polizzi, C., Lynn, S. J., & Perry, A. (2020). Stress and coping in the time of COVID-19: Pathways to resilience and recovery. *Clinical Neuropsychiatry, 17*(2), 59.

Pollock, R. V. H., Jefferson, A. McK., & Wick, C. W. (2015). *The six disciplines of breakthrough learning: how to turn training and development into business results*. John Wiley & Sons, Inc. doi:10.1002/9781119153832

Pollok, P., Amft, A., Diener, K., Lüttgens, D., & Piller, F. T. (2021). Knowledge diversity and team creativity: How hobbyists beat professional designers in creating novel board games. *Research Policy, 50*(8), 104174. doi:10.1016/j.respol.2020.104174

Porter, L. R. (1997). *Creating the virtual classroom distance learning with the internet*. John Wiley &Sons.

Po-ying, C. (2007). How students react to the power and responsibility of being decision-makers in their own learning. *Language Teaching Research, 11*(2), 225–241. doi:10.1177/1362168807074613

Prabhakaran, A., Mahamadu, A.M., & Mahdjoubi, L. (2022). Understanding the challenges of immersive technology use in the architecture and construction industry: A systematic review. *Automation in Construction Journal, 137*.

Pracana, C., & Wang, M. (2018). *Psychological applications and trends 2018*. Distributed by ERIC Clearinghouse.

Price, D. C., & Mansfield, K. C. (2021). Leadership in the Time of COVID: Connecting Community Resources to Meet the Needs of North Carolina Students. In Frontiers in Education (p. 313). Frontiers. doi:10.3389/feduc.2020.615101

Price, P. A. (2004). *Genesis and evolution of the United States Air Force academy's officer development system*. http://www.dtic.mil/dtic/tr/fulltext/u2/a428315.pdf

Pritchard, C. (2022) *Top 10 companies investing in the metaverse*. https://www.parcl.co/blog/top-10-companies-investing-in-the-metaverse

Prochaska, J., & DiClemente, C. (1986). Toward a comprehensive model of change. In W. Miller & N. Heather (Eds.), *Treating addictive behaviors: processes of change* (pp. 3–27). Plenum Press. doi:10.1007/978-1-4613-2191-0_1

Prokopenko, E. (2009). *Metagogical analysis of pedagogical apparatuses*. https://estherspace.wordpress.com/2009/01/17/metagogical-analysis-of-pedagogical-apparatuses

Punke, H. H. (1932). Recent developments in Mexican secondary education. *The School Review, 40*(9), 693–706. doi:10.1086/439401

Purcell, W. M., & Chahine, T. (2019). Leadership and governance frameworks driving transformational change in an entrepreneurial UK university. *Leadership and Organization Development Journal, 40*(5), 612–623. doi:10.1108/LODJ-07-2018-0280

Quick, J. C., Wright, T. A., Adkins, J. A., Nelson, D. L., & Quick, J. D. (2013). *Preventive stress management in organizations* (2nd ed.). American Psychological Association. doi:10.1037/13942-000

Quin, J. (2009). Growing social justice educators: A pedagogical framework for social justice education. *Intercultural Education, 20*(2), 109–125. doi:10.1080/14675980902922192

Quintana, M. G. B., & Zambrano, E. P. (2014). E-mentoring: The effects on pedagogical training of rural teachers with complex geographical accesses. *Computers in Human Behavior, 30*, 629–636. doi:10.1016/j.chb.2013.07.042

Raby, D. L. (1978). Ideology and state-building: The political function of rural education in Mexico, 1921 - 1935. *Ibero-amerikanisches Archiv, 4*(1), 21–38.

Radermacher, F. J. (1996). Cognition in systems. *Cybernetics and Systems, 27*(1), 1–42. doi:10.1080/019697296126651

Radford, A. W. (2011). *Learning at a distance: Undergraduate enrollment in distance education courses and degree programs. Stats in Brief. NCES 2012-154*. Distributed by ERIC Clearinghouse.

Ramos, S. (2017). Twenty years of education in Mexico [1941]. In C. Sanchez & R. E. Sanchez Jr., (Eds.), *Mexican philosophy of the 20th Century: Essential readings* (pp. 53–62). Oxford University Press. doi:10.1093/oso/9780190601294.003.0005

Ramsden, P. (1988). *Improving learning: new perspectives.* Kogan Page.

Rao, M. S. (2012). Myths and Truths about Soft Skills. *T+D.* Available at: https://www.td.org/magazines/td-magazine/myths-and-truths-about-soft-skills

Rathert, S., & Kırkgöz, Y. (2017). Examining leadership perceptions and practices to deliberate on educational policy implementation - an interdisciplinary approach. *International Journal on Lifelong Education and Leadership, 3*(1), 1–10. doi:10.25233/ijlel/2017-v3i1p1

Rausch, A. (2012). Task characteristics and learning potentials—Empirical results of three diary studies on workplace learning. *Vocations and Learning, 6*(1), 55–79. doi:10.100712186-012-9086-9

Ravens, R. (1982). *The origin and growth of action learning.* Chartwell Bratt.

Recigno, T. E., & Kramer, P. (2022). The development of leaders: Suggested modules for education. *The Open Journal of Occupational Therapy, 10*(4), 1–8. doi:10.15453/2168-6408.1941

Rees, T., Mitchell, I., Evans, L., & Hardy, L. (2010). Stressors, social support and psychological responses to sport injury in high-and low-performance standard participants. *Psychology of Sport and Exercise, 11*(6), 505–512. doi:10.1016/j.psychsport.2010.07.002

Rego, A., Owens, B., Leal, S., Melo, A. I., Cunha, M. P., Gonçalves, L., & Ribeiro, P. (2017). How a leader's humility helps teams to be humbler, psychologically stronger, and more effective: A moderated mediation model. *The Leadership Quarterly, 28*(5), 639–658. doi:10.1016/j.leaqua.2017.02.002

Rehm, S. V., Goel, L., & Crespi, M. (2015). The metaverse as mediator between technology, trends, and the digital transformation of society and business. *Journal of Virtual Worlds Research, 8*(2), 1–6. doi:10.4101/jvwr.v8i2.7149

Reich, J., & Ruipérez-Valiente, J. A. (2019). The MOOC pivots. *Science, 363*(6423), aav7958. Advance online publication. doi:10.1126cience.aav7958 PMID:30630920

Reischmann, J. (2004). *Andragogy. History, meaning, context, function.* Andragogy.net.

Reiser, R. A. (2001). A history of instructional design and technology: Part I: A history of instructional media. *Educational Technology Research and Development, 49*(1), 53–64. doi:10.1007/BF02504506

Reivich, K., Gillham, J. E., Chaplin, T. M., & Seligman, M. E. (2013). From helplessness to optimism: The role of resilience in treating and preventing depression in youth. In S. Goldstein & R. B. Brooks (Eds.), *Handbook of resilience in children* (pp. 201–214). Springer. doi:10.1007/978-1-4614-3661-4_12

Resnick, L. B. (1991). Shared cognition: Thinking as social practice. In L. B. Resnick, J. M. Levine, & S. D. Teasley (Eds.), *Perspectives on socially shared cognition* (pp. 1–20). American Psychological Association. doi:10.1037/10096-018

Reyes, D. L., Dinh, J., Lacerenza, C. N., Marlow, S. L., Joseph, D. L., & Salas, E. (2019). The state of higher education leadership development program evaluation: A meta-analysis, critical review, and recommendations. *The Leadership Quarterly, 30*(5), 101311. Advance online publication. doi:10.1016/j.leaqua.2019.101311

Rice, M., Davis, S. L., Soistmann, H. C., Johnson, A. H., Gray, L., Turner-Henson, A., & Lynch, T. (2020). Challenges and strategies of early career nurse scientists when the traditional postdoctoral fellowship is not an option. *Journal of Professional Nursing, 36*(6), 462–468. doi:10.1016/j.profnurs.2020.03.006 PMID:33308541

Richard, R. J., Walker, D. O., Putter, S. E., Middleton, E., & Johnson, S. K. (2017). Believing is becoming: The role of leader developmental efficacy in leader self-development. *Journal of Leadership & Organizational Studies, 24*(2), 137–156. doi:10.1177/1548051816657981

Richter, F. (2020) *The end of the TV era?* https://www.statista.com/chart/ 9761/daily-tv-and-internet-consumption-worldwide/

Riley, K., & MacBeath. (2003). Effective leaders and effective schools. In N. Bennett, M. Crawford, & M. Cartwright (Eds.), *Effective educational leadership* (pp. 173-185). Paul Chapman.

Roach, V., Smith, L. W., & Boutin, J. (2011). School leadership policy trends and developments: Policy expediency or policy excellence? *Educational Administration Quarterly, 47*(1), 71–113. doi:10.1177/0011000010378611

Robinson, A. M. (2018). Let's talk about stress: History of stress research. *Review of General Psychology, 22*(3), 334–342. doi:10.1037/gpr0000137

Robles, F. M. R. (2006). *Learner characteristic, interaction and support service variables as predictors of satisfaction in Web-based distance education.* The University of New Mexico.

Roehling, P. V., Roehling, M. V., & Moen, P. (2001). The relationship between work-life policies and practices and employee loyalty: A life course perspective. *Journal of Family and Economic Issues, 22*(2), 141–170. doi:10.1023/A:1016630229628

Rogers, C. R. (2002). The interpersonal relationship in the facilitation of learning. In R. Harrison, R. Reeve, A. Hanson, & J. Clarke (Eds), Supporting lifelong learning (pp. 25-29). Perspectives on learning, 1. Routledge-Farmer and Open University.

Rogers, C. R. (1951). *Client-centered therapy.* Houghton-Mifflin.

Rogers, C. R. (1961). *On become a person.* Houghton-Mifflin.

Rogers, C. R. (1969). *Freedom to learn.* Merrill.

Rogers, E. (1962). *Diffusion of innovation.* Free Press.

Rogers, E. (2003). *Diffusion of innovations.* The Free Press.

Rollermation [screen name]. (2011). *Metagogy I.* http://elearningportal,drupalgardens.com.comment/6

Rong, P., & Liu, S. (2021). Top management team knowledge hiding and enterprise innovation performance: A moderated mediation model. *Frontiers in Psychology, 12*, 783147. Advance online publication. doi:10.3389/fpsyg.2021.783147 PMID:34987451

Rosenberg, W. (1984). *Bolshevik visions: The first phase of the cultural revolution in Soviet Russia.* Ardis Publishers.

Rosenstock-Huessy, E. (1970). *I am an impure thinker.* Argo Books.

Rospigliosi, P. A. (2022). Metaverse or simulacra? Roblox, Minecraft, Meta, and the turn to virtual reality for education, socialization, and work. *Interactive Learning Environments, 30*(1), 1–3. doi:10.1080/10494820.2022.2022899

Rossett, A. (Ed.). (2002). *ASTD e-learning handbook.* McGraw-Hill.

Roux, M. (2020). Leadership 4.0. In J. Reams (Ed.), *Maturing leadership: How adult development impacts leadership* (pp. 7–35). Emerald Group Publishing Limited. doi:10.1108/978-1-78973-401-020201003

Ruan, N. (2021). Accumulating academic freedom for intellectual leadership: Women professors' experiences in Hong Kong. *Educational Philosophy and Theory, 53*(11), 1097–1107. doi:10.1080/00131857.2020.1773797

Rudolph, J., Tan, S., & Tan, S. (2023). ChatGPT: Bullshit spewer or the end of traditional assessments in higher education? *Journal of Applied Learning & Teaching*, *6*(1).

Rumberger, R. W. (1984). The job market for college graduates, 1960-1990. *The Journal of Higher Education*, *55*(4), 433–454. doi:10.2307/1981442rr

Rychen, D. S., & Salganik, L. H. (2003). *Key Competencies for a Successful Life and a Well-Functioning Society*. Hogrefe Publishing.

Saadatmand, M., Uhlin, L., Hedberg, M., Åbjörnsson, L., & Kvarnström, M. (2017). Examining learners' interaction in an open online course through the community of inquiry framework. *The European Journal of Open and Distance Learning*. doi:10.1515/eurodl-2017-0004

Sabga, N. I. (2017). *Leaders who learn: The intersection of behavioral science, adult learning and leadership* (Publication No. 10636182) [Doctoral dissertation, University of South Florida]. ProQuest Dissertations Publishing.

Sadler, W. A. (1969). *Existence and love: A new approach in existential phenomenology*. Charles Scribner's Sons.

Sáenz, M. (1927). Integrating Mexico through education. *Bulletin of the Pan-American Union*, *61*, 762–777.

Sáenz, M. (1932). The School and culture. *Progressive Education*, *9*(2), 99–111.

Salmon, G. (2019). May the fourth be with you: Creating education 4.0. *Journal of Learning Development*, *6*(2), 95–115. doi:10.56059/jl4d.v6i2.352

Sa, M. J., & Serpa, S. (2020). The COVID-19 pandemic as an opportunity to foster the sustainable development of teaching in higher education. *Sustainability*, *12*(20), 8525. Advance online publication. doi:10.3390u12208525

Samba, C., Williams, D. W., & Fuller, R. M. (2022). The forms and use of intuition in top management teams. *The Leadership Quarterly*, *33*(3), 101349. doi:10.1016/j.leaqua.2019.101349

Sanchez, G. I. (1936). Mexico: A revolution by education. Viking.

Sánchez, G. I. (1944). *The development of higher education in Mexico*. King's Crown Press. doi:10.7312anc93304

Sandholtz, J., Ringstaff, C., & Dwyer, D. (1997). *Teaching with technology*. Teachers College Press.

Sapienza, A. M. (2004). *Managing scientists: Leadership strategies in scientific research* (2nd ed.). Wiley. doi:10.1002/047166359X

Satterwhite, R., McIntyre Miller, W., & Sheridan, K. (2015). Leadership for sustainability and peace: Responding to the wicked challenges of the future. In M. Sowcik, A. C. Andenoro, M. McNutt, & S. E. Murphy (Eds.), *Leadership 2050: Critical challenges, key contexts, and emerging trends* (pp. 59–74). Emerald Group Publishing Limited.

Saunders, L. (2017). Connecting information literacy and social justice: Why and how. *Communications in Information Literacy*, *11*(1), 15. doi:10.15760/comminfolit.2017.11.1.47

Savicevic, D. M. (1991). Modern conceptions of andragogy: A European framework. *Studies in the Education of Adults*, *23*(3), 179–201. doi:10.1080/02660830.1991.11730556

Saxon, E., Baker, B., Hargrove, W., Hofman, F., & Zganjar, C. (2004). Mapping environments at risk under different global climate change scenarios. *Ecology Letters*, *8*(1), 53–60. doi:10.1111/j.1461-0248.2004.00694.x

Scandura, T. A., & Lankau, M. J. (1996). Developing diverse leaders: A leader-member exchange approach. *The Leadership Quarterly*, *7*(2), 243–263. doi:10.1016/S1048-9843(96)90043-0

Scanlon, E., McAndrew, P., & O'Shea, T. (2015). Designing for educational technology to enhance the experience of learners in distance education: How open educational resources, learning design and MOOCs are influencing learning. *Journal of Interactive Media in Education*, *6*(1), 1–9. doi:10.5334/jime.al

Schaffhauser, D. (2016). Report: Education Tech Spending on the Rise. *The Journal.* https://thejournal.com/articles/2016/01/19/report-education-tech-spending-on-the-rise.aspx?m=1

Schein, E. H. (2010). *Organizational culture and leadership* (Vol. 2). John Wiley & Sons.

Schejter, A. M., & Tirosh, N. (2015). "Seek the meek, seek the just": Social media and social justice. *Telecommunications Policy*, *39*(9), 796–803. doi:10.1016/j.telpol.2015.08.002

Schejter, A., & Tirosh, N. (2014). New Media Policy: The Redistribution of Voice 1. In Y. Liu & R. Picard (Eds.), *Policy and marketing strategies for digital media* (pp. 73–86). Routledge. doi:10.4324/9781315794303-6

Scherer, R., Rohatgi, A., & Hatlevik, O. E. (2017). Students' profiles of ICT use: Identification, determinants, and relations to achievement in a computer and information literacy test. *Computers in Human Behavior*, *70*, 486–499. doi:10.1016/j.chb.2017.01.034

Schippers, M. C., West, M. A., & Dawson, J. F. (2012). Team reflexivity and innovation: The moderating role of team context. *Journal of Management*, *41*(3), 769–788. doi:10.1177/0149206312441210

Schley, T., & van Woerkom, M. (2014). Reflection and reflective behavior in work teams. *Discourses on professional learning: On the boundary between learning and working*, 113-139.

Schmalfeldt, T. (2022). Design thinking in computer science edcuation. In I. Schmidberger, S. Wippermann, S. Stricker, & U. Müller (Eds.), *Design thinking im bildungsmanagement* (pp. 147–159). Springer. doi:10.1007/978-3-658-36951-4_9

Schmidt, S. W., English, L. M., & Carr-Chellman, A. (2022). Conversations with leaders: Sharing perspectives on the impact of and response to COVID-19 and other crises. In *The Covid-19 Pandemic and Other Ongoing Crises: Reflections, Innovations, and Adaptations.* Wiley. doi:10.1002/ace.20455

Schmidt-Crawford, D. A., Thompson, A. D., & Lindstrom, D. L. (2021). Condolences and congratulations: COVID-19 pressures on higher education faculty. *Journal of Digital Learning in Teacher Education*, *37*(2), 84–85.

Schön, D. (1983). *The reflective practitioner: How professionals think in action.* Basic Books.

Schramm, W. (1954). *The process and effects of mass communication.* University of Illinois Press.

Schröder, S. H., Baldegger, U., & Klösel, K. (2021). Empowering leadership als erfolgsfaktor in der innovation. *Leadership, Education, Personality. An Interdisciplinary Journal*, *3*, 9–20.

Schwarzmuller, T., Brosi, P., Duman, D., & Weloe, I. M. (2018). How does the digital transformation affect organizations? Key themes of change in work design and leadership. *Management Review*, *29*(20), 114–138.

Schyns, B., & Sczesny, S. (2010). Leadership attributes valence in self-concept and occupational self-efficacy. *Career Development International*, *15*(1), 78–92. doi:10.1108/13620431011020907

Scottish Community Education Council. (1990). *CeVe Scotland: Preservice training for community education work.* Scottish Community Education Council.

Scott, S., & Webber, C. F. (2008). Evidence-based leadership development: The 4L framework. *Journal of Educational Administration*, *46*(6), 762–776. doi:10.1108/09578230810908343

Searby, P. (1993). Edward Thompson as a teacher: Yorkshire and Warwick. In J. Rule & R. Malcolmson (Eds.), *Protest and aurvival: Essays for E. P. Thompson*. Merlin.

Seemiller, C. (2021). Preparing leaders of tomorrow: An analysis of leadership competencies within accredited academic program learning outcomes. *The Journal of Leadership Studies*, *15*(1), 6–27. doi:10.1002/jls.21737

Seifert, C. F., & Yukl, G. (2010). Effects of repeated multi-source feedback on the influence behavior and effectiveness of managers: A field experiment. *The Leadership Quarterly*, *21*(5), 856–866. doi:10.1016/j.leaqua.2010.07.012

Sendjaya, S., Eva, N., Butar, I., Robin, M., & Castles, S. (2017). SLBS-6: Validation of a short form of the servant leadership behavior scale. *Journal of Business Ethics*, *156*(4), 941–956. doi:10.100710551-017-3594-3

Senge, P. M., Cambron-McCabe, N., Lucas, T., Smith, B., Dutton, J. & Kleiner, A. (2000). *Schools that learn: A fifth discipline field book for educators, parents, and everyone who cares about education.* Doubleday/Currency.

Senge, P. M. (2006). *The fifth discipline: The art and practice of the learning organization*. Doubleday Currency.

Senge, P., Hamilton, H., & Kania, J. (2015). The dawn of system leadership. *Stanford Social Innovation Review*, *13*(1), 27–33.

Serenko, A., & Bontis, N. (2016). Understanding counterproductive knowledge behavior: Antecedents and consequences of intra-organizational knowledge hiding. *Journal of Knowledge Management*, *20*(6), 1199–1224. doi:10.1108/JKM-05-2016-0203

Sharif, A. (2015). Quality assurance in e-learning programs. *Teaching and learning in digital worlds: Strategies and issues in higher education*, 63-70.

Sharp, L. A. (2018). Collaborative digital literacy practices among adult learners: Levels of confidence and perceptions of importance. *International Journal of Instruction*, *11*(1), 153–166. doi:10.12973/iji.2018.11111a

Shields, C. M. (2009). Transformative Leadership: A Call for Difficult Dialogue and Courageous Action in Racialised Contexts. *International Studies in Educational Administration, 37*(3).

Shields, C. M. (2004). Dialogic Leadership for Social Justice: Overcoming Pathologies of Silence. *Educational Administration Quarterly*, *40*(1), 109–132. doi:10.1177/0013161X03258963

Shields, C. M. (2010). Transformative leadership: Working for equity in diverse contexts. *Educational Administration Quarterly*, *46*(4), 558–589. doi:10.1177/0013161X10375609

Shields, C. M., & Hesbol, K. A. (2020). Transformative leadership approaches to inclusion, equity, and social justice. *Journal of School Leadership*, *30*(1), 3–22. doi:10.1177/1052684619873343

Shriram, K., & Schwartz, R. (2017, March). All are welcome: Using VR ethnography to explore harassment behavior in immersive social virtual reality. *IEEE Virtual Reality*, 225-226.

Shum, S., De Liddo, A., & Klein, M. (2014). *DCLA meet CIDA: Collective intelligence deliberation analytics*. Paper presented at 2nd International Workshop on Discourse-Centric Learning Analytics, LAK14: 4th International Conference on Learning Analytics & Knowledge, Indianapolis, IN.

Shuttler, R., & Burdick, J. (2006). Creating a unified system of assessment. In M. Hricko & S. L. Howell (Eds.), *Online assessment and measurement: Foundations and challenges* (pp. 165–173). Information Science Reference. doi:10.4018/978-1-59140-720-1.ch008

Siddique, A., Aslam, H. D., Khan, M., & Fatima, U. (2011). Impact of academic leadership on faculty's motivation, and organizational effectiveness in higher education system. *International Journal of Academic Research*, *3*(3).

Siddiquei, A., Asmi, F., Asadullah, M. A., & Mir, F. (2021). Environmental-specific servant leadership as a strategic tool to accomplish environmental performance: A case of China. *International Journal of Manpower*, *42*(7), 1161–1182. doi:10.1108/IJM-07-2020-0350

Siemens, G., GaSevil, D., & Dawson, S. (2015). Preparing for the digital university: A review of the history and current state of distance, blended, and online learning. Link Research Lab.

Sierra, J. (2017). Discourse at the inauguration of the National University. In C. Sanchez & R. E. Sanchez Jr., (Eds.), *Mexican philosophy of the 20th Century: Essential readings* (pp. 17–26). Oxford University Press.

Sifuna, D. N. (2012). Leadership in Kenyan public universities and the challenges of autonomy and academic freedom: An overview of trends since independence. *Journal of Higher Education in Africa/Revue de l'enseignement supérieur en Afrique, 10*(1), 121-137.

Simmel, G. (1904a). The sociology of conflict I. *American Journal of Sociology*, *9*(4), 490–525. doi:10.1086/211234

Simmel, G. (1904b). The Sociology of Conflict II. *American Journal of Sociology*, *9*(5), 672–689. doi:10.1086/211248

Simmons, N., & Taylor, K. (2019). Leadership for the scholarship of teaching and learning: Understanding bridges and gaps in practice. *The Canadian Journal for the Scholarship of Teaching and Learning*, *10*(1), 1. doi:10.5206/cjsotl-rcacea.2019.1.7995

Simonds, T. A., & Brock, B. L. (2014). Relationship between age, experience, and student preference for types of learning activities in online courses. *Journal of Educators Online*, *11*(1), 1. doi:10.9743/JEO.2014.1.3

Simonson, M., Schlosser, C., & Orellana, A. (2011). A distance education research: A review of the literature. *Journal of Computing in Higher Education*, *23*(2-3), 124–142. doi:10.100712528-011-9045-8

Sinclair, E. (2017). A case study on the importance of peer support for e-learners. *Proceedings of the 9th International Conference on Computer Supported Education*, 280–284. 10.5220/0006263602800284

Siyav, A., & Jo, G. S. (2021). Neuro-symbolic speech understanding in aircraft maintenance metaverse. *IEEE Access : Practical Innovations, Open Solutions*, *9*, 154484–154499. doi:10.1109/ACCESS.2021.3128616

Skinner, B. F. (1968). *The technology of teaching*. Appleton-Century-Crofts.

Sloane, P. F. E., & Dilger, B. (2005). The competence clash – Dilemmata bei der übertragung des 'Konzepts der nationalen Bildungsstandards' auf die berufliche Bildung. *bwpat, 8*.

Smidt, E., Bunk, J., McGrory, B., Li, R., & Gatenby, T. (2014). Student attitudes about distance education: Focusing on context and effective practices. *The IAFOR Journal of Education*, *2*(1), 40–64. doi:10.22492/ije.2.1.02

Smith, J. (2021). *Q&A: future pandemics are inevitable, but we can reduce the risk*. https://ec.europa.eu/research-and-innovation/en/horiz on-magazine/qa-future-pandemics-are-inevitable-we-can-reduce-risk

Smith, R. O. (2012). Fostering transformative learning online. In E. W. Taylor & P. Cranton (Eds.), The handbook of transformative learning: Theory, research, and practice (pp. 408-422). Jossey-Bass.

Smith, S.U., Hayes, S., & Shea, P. (n.d.). A critical review of the use of Wenger's Community of Practice (CoP) theoretical framework in online and blended learning research. *Online Learning*, 209-237.

Smith, Ch., Hoderi, M., & McDermott, W. (2019). A preliminary study of students' perception and learning from different delivery methods. *Academy of Educational Leadership Journal*, *23*(2). https://www.abacademies.org/articles/a-preliminary-studyof-students-perception-andlearning-from-different-delivery-methods-8235.html

Smith, D., Kelly, D., & Allard, C. (2017). Dialogic spaces: A critical policy development perspective of educational leadership qualifications. *International Journal of Leadership in Education*, *20*(4), 393–415. doi:10.1080/13603124.2015.1116608

Smith, H. L. (1942). *Education in Mexico*. Indiana U School of Education Press.

Soricut, R., & Brill, E. (2006). Automatic question answering using the web: Beyond the factoid. *Information Retrieval*, *9*(2), 191–206. doi:10.100710791-006-7149-y

Southwick, S. M., Douglas-Palumberi, H., & Pietrzak, R. H. (2014). Resilience. In M. J. Friedman P. A. Resick, & T. M. Keane (Eds.), Handbook of PTSD: Science and practice (2nd ed., pp. 590-606). Guilford Press.

Sowcik, M., Andenoro, A. C., McNutt, M., & Murphy, S. E. (2015). *Critical challenges, key contexts, and emerging trends*. Emerald Group Publishing Limited.

Spencer, L. M., McClelland, D. C., & Spencer, S. M. (1992). *Competency assessment methods history and state of the art* (1st ed.). Hay McBer Research Press.

Sperka, M., & Stolár, A. (2005). Graphic Design in the Age of Interactive Media. In *3rd International Symposium of Interactive Media Design*. Yeditepe University.

Spigelberg, H. (1982). *The phenomenological movement: A historical introduction* (3rd ed.). Martinus Nijhoff.

Spillane, J. (2012). *Distributed leadership*. Jossey-Bass.

Spillane, J. P. (2004). Educational leadership. *Educational Evaluation and Policy Analysis*, *26*(2), 169–172. doi:10.3102/01623737026002169

St. Clair, R., & Kapplinger, B. (2021). Alley or autobahn? Assessing 50 years of the Andragogical Project. *Adult Education Quarterly*, *71*(3), 272–289. doi:10.1177/07417136211027879

Standley, N. (2023, January 6). *Nearly a third of university courses still have hybrid teaching*. https://www.bbc.com/news/education-64130367

Stanton, F., & Grant, J. (1999). Approaches to experiential learning, course delivery and validation in medicine. A background document. *Medical Education*, *33*(4), 282–297. doi:10.1046/j.1365-2923.1999.00272.x PMID:10336759

Stead, V., & Elliott, C. (2013). Women's leadership learning: A reflexive review of representations and leadership teaching. *Management Learning*, *44*(4), 373–394. doi:10.1177/1350507612449504

Stein, D., & Glazer, H. R. (2003). Mentoring the adult learner in academic midlife at a distance education university. *American Journal of Distance Education*, *17*(1), 7–23. doi:10.1207/S15389286AJDE1701_2

Stephenson, N. (1992). *Snow crash*. Bantam Books.

Stern, S. E. (1999). Addiction to technologies: A social psychological perspective of Internet addiction. *Cyberpsychology & Behavior*, *2*(5), 419–424. doi:10.1089/cpb.1999.2.419 PMID:19178214

Stewart, C., Bachman, C., & Johnson, R. (2010). Predictors of faculty acceptance of online education. *Journal of Online Learning and Teaching*, *6*(3), 597–616.

Stoller, J. K. (2009). Developing physician-leaders: A call to action. *Journal of General Internal Medicine*, *24*(7), 876–878. doi:10.100711606-009-1007-8 PMID:19455370

Strauss, A., & Corbin, J. (1998). *Basics of qualitative research: Techniques and procedures for developing Grounded Theory*. SAGE Publications, Inc.

Strohschen, G. (2020). By means of critical theory: Informed emancipatory education—An essay on realities and possibilities. In V. Wang (Ed.), *Handbook of research on ethical challenges in higher education leadership and administration* (pp. 296–308). Information Science Reference. doi:10.4018/978-1-7998-4141-8.ch016

Strom, K., & Porfilio, B. (2019). Critical hybrid pedagogies: A self-study inquiry into faculty practices in a blended educational leadership EdD program. *E-Learning and Digital Media, 16*(1), 1–14. doi:10.1177/2042753018811871

Su, Z., Zhang, N., Liu, D., Luan, T. H., & Shen, X. (2022). A survey on metaverse: fundamentals, security, and privacy. arXiv:2203.02662. doi:10.48550/arXiv.2203.02662

Suderman, J. L., & Foster, P. A. (2015). Envisioning leadership in 2050: Four future scenarios. In M. Sowcik, A. C. Andenoro, M. McNutt, & S. E. Murphy (Eds.), *Leadership 2050: Critical challenges, key contexts, and emerging trends* (pp. 23–38). Emerald Group Publishing Limited.

Sung, S. Y., & Choi, J. N. (2012). Effects of team knowledge management on the creativity and financial performance of organizational teams. *Organizational Behavior and Human Decision Processes, 118*(1), 4–13. doi:10.1016/j.obhdp.2012.01.001

Sun, P.-C., Tsai, R. J., Finger, G., Chen, Y.-Y., & Yeh, D. (2008). What drives successful e-Learning? An empirical investigation of the critical factors influencing learner satisfaction. *Computers & Education, 50*(4), 1183–1202. doi:10.1016/j.compedu.2006.11.007

Surrette, T., & Johnson, C. (2013). Assessing the ability of an online environment. *School Science and Mathematics, 115*(6), 260–270. doi:10.1111sm.12132

Swan, K. (2002). Building learning communities in online courses: The importance of interaction. *Education, Communication & Information, 2*(1), 23-49.

Swan, K., & Shih, L. F. (2005). On the nature and development of social presence in online course discussions. *Journal of Asynchronous Learning Networks, 9*(3), 115–136.

Sweeney, D., & Mausbach, A. (2019). Creating Powerful Principal and Coach Partnerships. *Educational Leadership, 76*(6), 30–35.

Szeto, E. (2015). Community of Inquiry as an instructional approach: What effects of teaching, social and cognitive presences are there in blended synchronous learning and teaching? *Computers & Education, 81*, 191–201. doi:10.1016/j.compedu.2014.10.015

Taguma, M., Feron, E., & Lim, M. H. (2018). *Future of education and skills 2030: Conceptual framework.* Organization of Economic Co-operation and Development.

Tahir, L. M., Musah, M. B., Al-Hudawi, S. H. V., & Daud, K. (2020). Becoming a Teacher Leader: Exploring Malaysian In-Service Teachers' Perceptions, Readiness and Challenges. *Education & Science/Egitim ve Bilim, 45*(202).

Tajfel, H. & Turner, J. C. (2004). *The social identity theory of intergroup behavior.* Psychology Press. . doi:10.4324/9780203505984-16

Talmage, C. A., Hansen, R. J., Knopf, R. C., Thaxton, S. P., McTague, R., & Moore, D. B. (2019). *Unleashing the value of lifelong learning institutes: Research and practice.* Academic Press.

Tang, M., & Schmidt, J. T. (2021). Fostering creativity in interdisciplinary and intercultural teams: The importance of team building and team management. *Creative Success in Teams*, 55–79. doi:10.1016/B978-0-12-819993-0.00004-7

Tannenbaum, F. (1933). *Peace by revolution: Mexico after 1910.* Columbia University Press. doi:10.7312/tann92264

Tannenbaum, S. I., Mathieu, J. E., Salas, E., & Cohen, D. (2012). Teams are changing: Are research and practice evolving fast enough? *Industrial and Organizational Psychology: Perspectives on Science and Practice*, *5*(1), 2–24. doi:10.1111/j.1754-9434.2011.01396.x

Tarcan, E., Varol, E. S., & Toker, B. (2010). A study on the acceptance of information technologies from the perspective of academicians in Turkey. *Ege Academic Review*, *10*(3), 791–812. doi:10.21121/eab.2010319615

Taylor, J. E. (2022). Motivational immediacy. [electronic resource]: fostering engagement in adult learners. Stylus Publishing, LLC.

Taylor, E. W., & Cranton, P. (2012). *The handbook of transformative learning: theory, research and practice*. Jossey-Bass.

Taylor, J. S., & Machado-Taylor, M. de L. (2010). Leading strategic change in higher education: The need for a paradigm shift toward visionary leadership. In T. Claes & D. S. Preston (Eds.), *Frontiers in Higher Education* (pp. 167–194). Brill. doi:10.1163/9789042030978_011

Taylor, J., & Machado, M. D. L. (2006). Higher education leadership and management: From conflict to interdependence through strategic planning. *Tertiary Education and Management*, *12*(2), 137–160. doi:10.1080/13583883.2006.9967164

Taylor, S. E., Lerner, J. S., Sage, R. M., Lehman, B. J., & Seeman, T. E. (2004). Early environment, emotions, responses to stress, and health. *Journal of Personality*, *72*(6), 1365–1394. doi:10.1111/j.1467-6494.2004.00300.x PMID:15509286

Teine, M. (2018). A prototypical participatory design-process: Bringing digital learning and user experience together. In I. Bouchrika & P. Vu (Eds.), *Learner experience and usability in online education* (pp. 36–60). IGI Global. doi:10.4018/978-1-5225-4206-3.ch002

Teine, M. (2022). *Zur partizipativen gestaltung von (digitalen) lernangeboten in der betrieblichen (weiter-)bildung. Eine empirische modellentwicklung*. Universität Paderborn.

Thakhathi, A., Le Roux, C., & Davis, A. (2019). Sustainability leaders' influencing strategies for institutionalizing organizational change towards corporate sustainability: A strategy-as-practice perspective. *Journal of Change Management*, *19*(4), 246–265. doi:10.1080/14697017.2019.1578985

Theoharis, G. (2007). Social justice educational leaders and resistance: Toward a theory of social justice leadership. *Educational Administration Quarterly*, *43*(2), 221–258. doi:10.1177/0013161X06293717

Thompson, E. P. (1968). *The making of the English working class*. Penguin. https://www.academia.edu/30385746/E_P_Thompson_Making_of_the_English_Working_Class_Vintage_1966_1_

Thompson, E. P. (1970). *Warwick university ltd: Industry, management and the universities*. Penguin. https://files.libcom.org/files/2022-09/warwick_uni_ltd_ep_thompson.pd

Thompson, E. P. (1955). *William Morris: Romantic to revolutionary*. Lawrence & Wishart.

Thompson, E. P. (1975). *The poverty of theory and other essays*. Merlin.

Thompson, E. P. (1975). *Whigs and hunters: The origins of the Black act*. Allen Lane.

Thompson, E. P. (1978). *The sykaos papers*. Bloomsbury.

Thompson, P., Vogler, J. S., & Xiu, Y. (2017). Strategic tooling: Technology for constructing a community of inquiry. *Journal of Educators Online*, *14*(2), n2. doi:10.9743/jeo.2017.14.2.10

Thoring, K., Mueller, R. M., Desmet, P., & Badke-Schaub, P. (2022). Toward a unified model of design knowledge. *Design Issues*, *38*(2), 17–32. doi:10.1162/desi_a_00679

Tingle, E., Corrales, A., & Peters, M. L. (2019). Leadership development programs: Investing in school principals. *Educational Studies*, *45*(1), 1–16. doi:10.1080/03055698.2017.1382332

Tipton, W., & Wideman, S. (2021). Toward an invitational andragogy: Articulating a teaching philosophy for the andragogic classroom. *Journal of Communication Pedagogy*, *5*, 156–163. doi:10.31446/JCP.2021.2.16

Tjosvold, D. (1991). *Team organization: An enduring competitive advantage*. John Wiley & Sons.

Toffler, A. (1980). *The third wave*. Bantam Books.

Tolstoy, L. (1904). *The complete works of Count Tolstoy: Pedagogical articles*. Dana Estes & Company.

Tong, J., & Mitra, A. (2009). Chinese cultural influences on knowledge management practice. *Journal of Knowledge Management*, *13*(2), 49–62. doi:10.1108/13673270910942691

Torrance, D., & Humes, W. (2015). The shifting discourses of educational leadership: International trends and Scotland's response. *Educational Management Administration & Leadership*, *43*(5), 792–810. doi:10.1177/1741143214535748

Torre, T., & Sarti, D. (2020). The "Way" toward e-leadership: Some evidence from the field. *Frontier Psychology*. https://www.ncbi.nlm.nih.gov/pmc/articles/PMC7685990/

Torres-Coronas, T., & Arias-Oliva, M. (2006). *Encyclopedia of human resources information systems: Challenges in e-HRM*. http://www.etse.urv.es/~ttorres/

Torrisi-Steele, G. (2014). Factors predisposing academics towards the use of blended strategies: A model. *International Journal of Adult Vocational Education and Technology*, *5*(3), 1–20. doi:10.4018/ijavet.2014070101

Tough, A. (1987). Potential futures: Implications for adult educators. *Spons Agency*, 13.

Tough, A. (1967). *Learning without a teacher*. Institute for Studies in Education.

Tough, A. (1967). *Learning without a teacher: Educational Research Series 3*. Ontario Institute for Studies in Education.

Tough, A. (1971). *The adult's learning projects: A fresh approach to theory and practice in adult learning*. Ontario Institute for Studies in Education.

Tough, A. (1979). *The adult's learning project*. Institute for Studies in Education.

Tracy, S. J. (2020). *Qualitative research methods: collecting evidence, crafting analysis, communicating impact* (2nd ed.). Wiley Blackwell.

Trow, M. (2010). *Twentieth-century higher education: Elite to mass to universal*. Johns Hopkins. doi:10.56021/9780801894411

Tsai, W. H. S., & Men, L. R. (2017). Social CEOs: The effects of CEOs' communication styles and parasocial interaction on social networking sites. *New Media & Society*, *19*(11), 1848–1867. doi:10.1177/1461444816643922

Tsang, M. C., & Levin, H. (1985). The economics of overeducation. *Economics of Education Review*, *4*(2), 93–104. doi:10.1016/0272-7757(85)90051-2

Tsang, W. K. (2003). Journaling from Internship to Practice Teaching. *Reflective Practice*, *4*(2), 221–240. doi:10.1080/14623940308269

Turner, R., & Topalian, A. (2002). *Core responsibilities of design leaders in commercially demanding environments*. Design Leadership Forum, organized by Alto Design.

Turner, J. K. (1969). *Barbarous Mexico* (2nd ed.). University of Texas Press.

Tusaie, K., & Dyer, J. (2004). Resilience: A historical review of the construct. *Holistic Nursing Practice, 18*(1), 3–8. doi:10.1097/00004650-200401000-00002 PMID:14765686

Tyler, R. (1949). *Basic principles of curriculum and instruction.* University of Chicago Press.

U. S. Department of Labor. (1991). *Secretary's Commission on Achieving Necessary Skills.* Government Printing Office.

U.S. Department of Education, National Center for Education Statistics. (2019). *Adult literacy in the United States.* https://nces.ed.gov/datapoints/2019179.asp

Uğur, B., Akkoyunlu, B., & Kurbanoğlu, S. (2011). Students' opinions on blended learning and its implementation in terms of their learning styles. *Education and Information Technologies, 16*(1), 5–23. doi:10.100710639-009-9109-9

Ulla, M., & Perales, W. (2022). Hybrid teaching: Conceptualization through practice for the post COVID19 pandemic education. *Frontiers in Education, 7.*

UNESCO. (2022). *Making lifelong learning a reality: A handbook.* UNESCO.

United Nations. (2015). *Agenda for sustainable development.* United Nations.

University College London (UCL). (2014). *Local action on health inequalities: Building children and young people's resilience in schools.* Public Health England.

Ur Rehman, Z., Shafique, I., Khawaja, K. F., Saeed, M., & Kalyar, M. N. (2023). Linking responsible leadership with financial and environmental performance: Determining mediation and moderation. *International Journal of Productivity and Performance Management, 72*(1), 24–46. doi:10.1108/IJPPM-12-2020-0626

US Department of Education. (2009). *Guidance on Standards, Assessments, and Accountability.* https://www2.ed.gov/policy/elsec/guid/ standardsassessment/guidance_pg5.html

Valera, P., Acuna, N., Alzate-Duque, L., Liang, L. E., Cupertino, P., & Merulla, J. (2021). The development and prototype feedback of digital cancer 101 videos to enhance cancer education for marginalized communities with limited health literacy. *Cancer Control, 28,* 1–10. doi:10.1177/10732748211006055 PMID:33784858

Valmont, W. (2003). *Technology for literacy teaching and learning.* Houghton Mifflin.

van der Gaag, M., Hoffman, T., Remijsen, M., Hijman, R., de Haan, L., van Meijel, B., van Harten, P., Valmaggia, L., de Hert, M., Cuijpers, A., & Wiersma, D. (2006). The fivefactor model of the positive and negative syndrome scale II: A ten-fold cross-validation of a revised model. *Schizophrenia Research, 85*(1), 280–287. doi:10.1016/j.schres.2006.03.021 PMID:16730429

van Dierendonck, D. (2011). Servant leadership: A review and synthesis. *Journal of Management, 37*(4), 1228–1261. doi:10.1177/0149206310380462

Van Outvorst, F., Vistker, C., & de Waal, B. (2017). Digital leadership: The consequences or organizing and working in a digital society. *Proceedings of the 5th International Conference on Management, Leadership and Governance,* 463-471.

VanGronigen, B. A., Cunningham, K. M., & Young, M. D. (2018). How exemplary educational leadership preparation programs hone the interpersonal-intrapersonal (i2) skills of future leaders. *Journal of Transformative Leadership & Policy Studies, 7*(2), 1–11. doi:10.36851/jtlps.v7i2.503

Vasconcelos, J. (1926). *Aspects of Mexican civilization.* The University of Chicago Press.

Vasconcelos, J. (1979). *The cosmic race/La raza cósmica* (Vol. 1925). Johns Hopkins University Press.

Vasconcelos, J. (1981). *Antología de textos sobre educación* [Anthology of texts on education] (A. Molina, Ed.). Fondo de Cultura Económica.

Vaughan, M. K. (1982). *The state, education, and social class in Mexico, 1880-1928.* Northern Illinois University Press.

Vaughan, M. K. (1997). *Cultural politics in revolution: Teachers, peasants, and schools in Mexico, 1930-1940* (2nd ed.). University of Arizona Press. doi:10.2307/j.ctv1d6q40z

Veletsianos, G. (2011). Designing opportunities for transformation with emerging technologies. *Educational Technology, 51*(2), 41–46.

Veletsianos, G. (Ed.). (2016). *The emergence and innovation in digital learning.* AU Press.

Venkatesh, V., & Bala, H. (2008). Technology acceptance model 3 and a research agenda on interventions. *Decision Sciences, 39*(2), 273–315. doi:10.1111/j.1540-5915.2008.00192.x

Vergne, J. P. (2021). *The future of trust will be dystopian or decentralized: Escaping the metaverse.* Revista de Occidente on the Future of Trust.

Verkroost, M., Meijerink, L., Linsten, H., & Veen, W. (2008). Finding a balance in dimensions of blended learning. *International Journal on E-Learning Academic Research Library, 7*(3), 499–522.

Volini, E., Schwartz, J., Eaton, K., Mallon, D., Van Durme, Y., Hauptmann, M., Scott, R., & Poynoton, S. (2021). *The social enterprise in a world disrupted. Leading the shift from survival to thrive. 2021 Deloitte Global Capital Trends.* Deloitte.

Vukovich, C. (2022). *As if Covid wasn't enough: A continuous assault on student's rights and education.* ACLU. https://www.acluohio.org/en/news/if-covid-wasnt-enough-continuous-assault-students-rights-and-education

Vygotsky, L. (1930). Socialisticheskaja peredelka cheloveka. *VARNITSO, The Journal of the All-Union Association of Workers in Science and Technics for the Furthering of the Socialist Edification in the USSR.*

Vygotsky, L. (1999). Consciousness as a problem in the psychology of behavior [1925]. In Undiscovered Vygotsky: Etudes on the pre-history of cultural-historical psychology (European studies in the history of science and ideas, Vol. 8). Peter Lang Publishing.

Vygotsky, L. (1986). *Thought and Language.* Massachusetts Institute of Technology Press. (Original work published 1934)

Vygotsky, L. S. (1978). *Mind in Society.* Harvard University Press.

Wai, J. (2014). Experts are born, then made: Combining prospective and retrospective longitudinal data shows that cognitive ability matters. *Intelligence, 45*(July–August), 74–80. doi:10.1016/j.intell.2013.08.009

Waits, T., & Lewis, L. (2003). *Distance education at degree-granting postsecondary institutions: 2000–2001.* U.S. Department of Education, National Center for Education Statistics.

Walker, B. (2019). *Finding resilience: Change and uncertainty in nature and society.* CSIRO PUBLISHING.

Walker, B. (2020). Resilience: What it is and is not. *Ecology and Society, 25*(2), 11. doi:10.5751/ES-11647-250211

Wals, A., Mochizuki, Y., & Leicht, A. (2017). Critical case-studies of non-formal and community learning for sustainable development. *International Review of Education, 63*(6), 783–792. Advance online publication. doi:10.100711159-017-9691-9

Wamuyu, P. (2017). Bridging the digital divide among low income urban communities. Leveraging use of community technology centers. *Telematics and Informatics, 34*(8), 1709–1720. doi:10.1016/j.tele.2017.08.004

Wang, F., Fong, J., & Kwan, R. R. (2010). Handbook of research on hybrid learning models: Advanced tools, technologies, and applications. Information Science Reference.

Wang, S.-C. (2009). University instructor perceptions of the benefits of technology use in e-learning. In K. Jusoff, S. S. Mahmoud, & R. Sivakumar (Ed.), *Proceedings of the Second International Conference on Computer and Electrical Engineering* (pp. 580-585). IEEE Computer Society Washington. 10.1109/ICCEE.2009.275

Wang, V., & Bain, B. (2016). Andragogical leadership and management in the workplace. In V. Wang B. Bain, J. Hope, & C. Hansman (Eds.), Educational leadership and organizational management (pp. 249-273). Information Age Publishing.

Wang, V. (2005). Perceptions of teaching preferences of online instructors. *Journal on Excellence in College Teaching*, *16*(3), 33–54.

Wang, V. (2006). *Essential elements for andragogical styles and methods: How to create andragogical modes in adult education*. PEARSON Education.

Wang, V. (2007). How contextually adapted philosophies and the situational role of adult educators affect learners' transformation and emancipation. In K. P. King & V. Wang (Eds.), *Comparative adult education around the globe: International portraits and readings of the history, practice philosophy, and theories of adult learning* (pp. 139–150). Zhejiang University Press.

Wang, V. (2008). *Facilitating adult learning: A comprehensive guide for successful instruction* (Rev. ed.). Pearson Education.

Wang, V. (2023). *Introduction (from call for chapters). In Handbook of research on andragogical leadership and technology in a changing society*. Information Science Reference.

Wang, V., & Bain, B. (2017). Andragogical leadership and management in the workplace. In V. Wang (Ed.), *Theory and practice of adult and higher education* (pp. 249–273). Information Age Publishing.

Wang, V., & Bott, P. A. (2003-2004). Modes of teaching of Chinese adult educators. *The New York Journal of Adult Learning*, *2*(2), 32–51.

Wang, V., & Farmer, L. (2008). Adult learning methods in China and Bloom's taxonomy. *International Journal for the Scholarship of Teaching and Learning*, *2*(2), 1–15. doi:10.20429/ijsotl.2008.020213

Wang, V., & Hansman, C. A. (2017). Pedagogy & andragogy in higher education. In V. Wang (Ed.), *Theory and practice of adult and higher education* (pp. 87–111). Information Age Publishing.

Wang, V., Hitch, L., & Torrisi-Steele, G. (2022, January). Active learning online: Necessity, faculty role, and a concept model for course design. *International Journal of Online Pedagogy and Course Design*, *12*(1), 1–13. doi:10.4018/IJOPCD.2022010105

Wang, V., & King, K. (2010). Transformative learning and ancient Asian educational perspectives. In V. Wang (Ed.), *Assessing and evaluating adult learning in career and technical education* (pp. 13–26). IGI Global. doi:10.4018/978-1-61520-745-9.ch002

Wang, V., & Torrisi-Steele, G. (2015). Online teaching, change, and critical theory. *New Horizons in Adult Education and Human Resource Development*, *27*(3), 18–26. doi:10.1002/nha3.20108

Wang, V., & Torrisi-Steele, G. (2019). Transformation of higher education in China: A teaching methods perspective. In M. Khosrow-Pour (Ed.), *Handbook of research on challenges and opportunities in launching a technology-driven international university* (pp. 131–160). Information Science Reference. doi:10.4018/978-1-5225-6255-9.ch008

Wang, V., & Torrisi-Steele, G. (2021). The gap between theory and practice: A scenario from higher education. In K. N. LaVenia & J. J. May (Eds.), *Case studies in leadership and adult development applying theoretical perspectives to real world challenges* (pp. 167–176). Routledge.

Wang, V., & Torrisi-Steele, G. (2022). Leading adult learning in organizations. In V. Wang (Ed.), *Handbook of Research on Educational Leadership and Research Methodology* (pp. 1–15). IGI Global. doi:10.4018/978-1-6684-4144-2.ch001

Wang, V., Torrisi-Steele, G., & Reinsfeld, E. (2021). Transformative learning, epistemology and technology in adult learning. *Journal of Adult and Continuing Education*, *27*(2), 324–340. doi:10.1177/1477971420918602

Wan, S. W. Y., Law, E. H. F., & Chan, K. K. (2018). Teachers' perception of distributed leadership in Hong Kong primary schools. *School Leadership & Management*, *38*(1), 102–141. doi:10.1080/13632434.2017.1371689

Wanstreet, C. E. (2006). Interaction in online environments: A review of the literature. *Quarterly Review of Distance Education*, *7*(4), 399–411.

Ward, K. (2003). Faculty Service Roles and the Scholarship of Engagement. ASHE-ERIC Higher Education Report. Jossey-Bass Higher and Adult Education Series. Jossey-Bass.

Warren, T. (2023). Microsoft is looking at OpenAI's GPT for Word, Outlook, and PowerPoint. *The Verge*. https;//www.theverge.com/2023/1/9/23546144/micorsoft-openai-word-powrpoint-outlook-gpt-integration-rumor

Wasserberg, M. (2000). Creating the vision and making it happen. In H. Tomlinson, H. Gunter, & P. Smith (Eds.), *Living headship* (pp. 154–160). Paul Chapman Publishing.

Watzlawick, P., Beavin, J. H., & Jackson, D. D. (1967). *Pragmatics of Human Communication*. Norton.

Webel, S., Bockholt, U., Engelke, T., Gavish, N., Olbrich, M., & Preusche, C. (2013). An augmented reality training platform for assembly and maintenance skills. *Robotics and Autonomous Systems*, *61*(4), 398–403. doi:10.1016/j.robot.2012.09.013

Weimer, M. (2012). Learner-centered teaching and transformative learning. In E. W. Taylor & P. Cranton (Eds.), *The handbook of transformative learning: Theory, research, and practice* (pp. 439–454). Jossey-Bass.

Weis, T. M., Benmayor, R., O'Leary, C., & Eynon, B. (2002). Digital technologies and pedagogies. *Social Justice*, *29*(4), 153-167.

Weitkamp, E. W. (2006). *Teaching style and technology: The role of educational values in the adoption of information technology innovations in the classroom.* http://gateway.proquest.com/openurl%3furl_ver=Z39.88-2004%26res_dat=xri:pqdiss%26rft_val_fmt=info:ofi/fmt:kev:mtx:dissertation%26rft_dat=xri:pqdiss:3183581

Welton, M. R. (Ed.). (1995). Introduction. In M. R. Welton (Ed.), In defense of the lifeworld (pp. 1-10). SUNY Press.

Wenger, E. (1998). *Communities of practice: Learning, meaning, and identity*. Cambridge University Press. doi:10.1017/CBO9780511803932

Wenner, J. A., & Campbell, T. (2017). The theoretical and empirical basis of teacher leadership: A review of the literature. *Review of Educational Research*, *87*(1), 134–171. doi:10.3102/0034654316653478

Whitehead, A. N. (1931). On foresight. In W. B. Graham (Ed.), *Business adrift*. McGraw-Hill.

White, R. P., Hodgson, P., & Crainer, S. (1996). *The future of leadership: Riding the corporate rapids into the 21st century*. Pitman Publishing.

White, R. P., & Shullman, S. L. (2010). Acceptance of uncertainty as an indicator of effective leadership. *Consulting Psychology Journal, 62*(2), 94–104. doi:10.1037/a0019991

Whittaker, J., Looney, S., Reed, A., & Votta. F. (2021). Recommender systems and the amplification of extremist content. *Internet Policy Review, 10*(2), 1–29.

Wicks, D. A., Craft, B. B., Mason, G. N., Gritter, K., & Bolding, K. (2015). An investigation into the community of inquiry of blended classrooms by a Faculty Learning Community. *The Internet and Higher Education, 25*, 53–62. doi:10.1016/j.iheduc.2014.12.001

Widisatria, D., & Nawangsari, L. C. (2021). The influence of green transformational leadership and motivation to sustainable corporate performance with organizational citizenship behavior for the environment as a mediating: Case study at PT Karya Mandiri Sukses Sentosa. *European Journal of Business and Management Research, 6*(3), 118–123. doi:10.24018/ejbmr.2021.6.3.876

Wilson, J. H. (2019). *Andragogy and the learning-tech culture revolution.* https://www.igi-global.com/chapter/andrafogy-and-the-learning-tech-culture-revoution/207508

Wilson, E., Lee, L., Klas, R., & Nesbit, K. (2020). Technology and rehabilitation training for community health workers: Strengthening health systems in Malawi. *Health & Social Care in the Community, 28*(3), 833–841. doi:10.1111/hsc.12914 PMID:31808218

Wilson, J. H. (2014). *Investing in English skills: The limited English proficient workforce in US metropolitan areas.* The Brookings Institution.

Wisconsin Department of Public Instruction. (2008). *Wisconsin's components of communication education.* Wisconsin Department of Public Instruction.

Wiseman, T. (1996). A concept analysis of empathy. *Journal of Advanced Nursing, 23*(6), 1162–1167. doi:10.1046/j.1365-2648.1996.12213.x PMID:8796464

Wittrock, M. (1991). Generative teaching of comprehension. *The Elementary School Journal, 92*(2), 167–182. doi:10.1086/461686

Witzel, A. (2000). The problem-centered interview. *Forum Qualitative Social Research, 1*(1), 22.

Wolff, K. H. (1984). Surrender-and-catch and phenomenology. *Human Studies, 7*(2), 191–210. doi:10.1007/BF00144932

Wolfson, N. E., Cavanagh, T. M., & Kraiger, K. (2014). Older adults and technology-based instruction: Optimizing learning outcomes and transfer. *Academy of Management Learning & Education, 13*(1), 26–44. doi:10.5465/amle.2012.0056

Wolin, S. S. (2017). *Democracy incorporated: Managed democracy and the specter of inverted totalitarianism - new edition.* Princeton University.

Woods, P. A., & Roberts, A. (2019). Collaborative school leadership in a global society: A critical perspective. *Educational Management Administration & Leadership, 47*(5), 663–677. doi:10.1177/1741143218759088

Wooldridge, A. (2015). The Icarus syndrome meets the wearable revolution. *Korn/Ferry Briefings Talent Leaders, 6*, 27-33.

World Economic Forum. (2020). *The Future of Jobs Report 2020.* Author.

World Health Organization. (2022). *Enactment guide for the Yeonsu declaration for learning cities.* World Health Organization.

Woulfin, S. L. (2018). Mediating instructional reform: An examination of the relationship between district policy and instructional coaching. *AERA Open*, *4*(3). doi:10.1177/2332858418792278

Wright, P. (2016). Helping older adults conquer digital tablets. *Ger ontechnology, 14*(2), 78-88. doi:10.4017/gt.2016.14.3.005.00

Wrigley, C., & Straker, K. (2015). Design Thinking pedagogy: The Educational Design Ladder. *Innovations in Education and Teaching International.*

Wu, X., Siu, K. W. M., & Bühring, J. (2022b). Empathy between designers in the design synthesis stage. In C. Stephanidis, M. Antona, S. Ntoa, & G. Salvendy (Eds.), *HCI International 2022 – Late Breaking Posters. HCII 2022. Communications in Computer and Information Science* (Vol. 1654). Springer. doi:10.1007/978-3-031-19679-9_27

Wu, X., Siu, K. W. M., Bühring, J., & Villani, C. (2022a). The relationship between creative self-efficacy, achievement motivation, and job burnout among designers in China's e-market. *Social Sciences*, *11*(11), 509. doi:10.3390ocsci11110509

Xu, D., & Jaggars, S. S. (2013). The impact of online learning on students' course outcomes: Evidence from a large community and technical college system. *Economics of Education Review*, *37*, 46–57. doi:10.1016/j.econedurev.2013.08.001

Yang, J. (2021). Comparing Chinese and US Educational Leadership Styles: A Review of the Literature. *Performance Improvement Quarterly*, *34*(3), 247–260. doi:10.1002/piq.21371

Yang, Y., Yeh, H., & Wong, W. (2010). The influence of social interaction on meaning construction in a virtual community. *British Journal of Educational Technology*, *41*(2), 287–306. doi:10.1111/j.1467-8535.2009.00934.x

Yeh, J. (2017). The effects of successful ICT-based smart city services: From citizens' perspectives. *Government Information Quarterly*, *34*(3), 556–565. doi:10.1016/j.giq.2017.05.001

Yeo, G. (1982). "Eldergogy": A special approach to education for elders. *Lifelong Learning*, *5*(5), 4–7.

Yermish, I. (2000). A case for case studies via video-conferencing. In L. Lau (Ed.), *Distance learning technologies: Issues, trends and opportunities* (pp. 208–217). Idea Group Publishing. doi:10.4018/978-1-878289-80-3.ch015

Ying, M., Faraz, N. A., Ahmed, F., & Raza, A. (2020). How does servant leadership foster employees' voluntary green behavior? A sequential mediation model. *International Journal of Environmental Research and Public Health*, *17*(5), 1792. doi:10.3390/ijerph17051792 PMID:32164222

Yocco, V., Danter, E., Heimlich, J., Dunckel, B., & Myers, C. (2011). Exploring use of new media in environmental education contexts: Introducing visitors' technology use in zoos model. *Environmental Education Research*, *17*(6), 801–814. doi:10.1080/13504622.2011.620700

Young, G. (2002). 'Hybrid' Teaching Seeks to End the Drive Between Traditional and Online Instruction. *The Chronicle of Higher Education*, A33.

Yuan, L., Powell, S., & Cetis, J. (2013). *MOOCs and open education: Implications for higher education.* Cetis White Paper Creative Commons Attribution 3.0 UK. http://publications.cetis.ac.uk/2013/667

Yukl, G. (2009). *Leadership in organizations.* Pearson Education.

Yusuf, B. N. (2020). Are we prepared enough? A case study of challenges in online learning in a private higher learning institution during the Covid-19 outbreaks. *Advances in Social Sciences Research Journal*, *7*(5), 205–212. doi:10.14738/assrj.75.8211

Zaccaro, S. J. (2007). Trait-based perspectives of leadership. *The American Psychologist*, *62*(1), 6–16. doi:10.1037/0003-066X.62.1.6 PMID:17209675

Zacher, H., Rosing, K., & Frese, M. (2011). Age and leadership: The moderating role of legacy beliefs. *The Leadership Quarterly*, *22*(1), 43–50. doi:10.1016/j.leaqua.2010.12.006

Zaim, H., Muhammed, S., & Tarim, M. (2018). Relationship between knowledge management processes and performance: Critical role of knowledge utilization in organizations. *Knowledge Management Research &Amp. Practice*, *17*(1), 24–38. doi:10.1080/14778238.2018.1538669

Zaleznik, A. (1992). *The Harvard business review on leadership: ''Managers and leaders: Are they? different?* Harvard Business School Press.

Zallio, M., & Clarkson, P. J. (2021). Inclusion, diversity, equity, and accessibility in the built environment: A study of architectural design practice. *Building and Environment*, *206*, 206. doi:10.1016/j.buildenv.2021.108352

Zamenopoulos, T., & Alexiou, K. (2007). Toward an anticipatory view of design. *Design Studies*, *28*(4), 411–436. doi:10.1016/j.destud.2007.04.001

Zanjani, N., Edwards, S., Nykvist, S., & Geva, S. (2017). The important elements of LMS design that affect user engagement with e-learning tools within LMSs in the higher education sector. *Australasian Journal of Educational Technology*, *33*(1), 19–31.

Zawacki-Richter, O., Marin, V. I., Bond, M., & Gouverneur, F. (2019). A systematic review of research on artificial intelligence applications in higher education-where are the educators? *International Journal of Educational Technology in Higher Education*, *16*(1), 1–27. doi:10.118641239-019-0171-0

Zdjelar, R., & Zajdela Hrustek, N. (2021). Digital divide and e-inclusion as challenges of the information society – research review. *Journal of Information and Organizational Sciences*, *45*(2), 601–638. doi:10.31341/jios.45.2.14

Zhai, X. (2022). *ChatGPT user experience: Implications for education*. https://papers.ssrn.com/sol3/papers.cfm?abstract_id=4312418

Zhang, K., & Aslan, A. (2021). AI technologies for education: Recent research & future directions. *Computers and Education: Artificial Intelligence*, *2*, 100025. doi:10.1016/j.caeai.2021.100025

Zhang, R. (2020). Exploring blended learning experiences through the community of inquiry framework. *Language Learning & Technology*, *24*(1), 38–53.

Zhang, X., & Bartol, K. M. (2010). Linking empowering leadership and employee creativity: The influence of psychological empowerment, intrinsic motivation, and creative process engagement. *Academy of Management Journal*, *53*(1), 107–128. doi:10.5465/amj.2010.48037118

Zuckerberg, M. (2021) The metaverse and how we will build it together. *Connect 2021*. https://www.youtube.com/watch?v=Uvufun6xer8

Zusman, A. (2005). Challenges facing higher education in the twenty-first century. *American higher education in the twenty-first century: Social, political, and economic challenges, 2*, 115-160.

Zwart, R., Wubbels, T., Bergen, T., & Bolhuis, S. (2009). Which characteristics of a reciprocal peer coaching context affect teacher learning as perceived by teachers and their students? *Journal of Teacher Education*, *60*(3), 243–257. doi:10.1177/0022487109336968

About the Contributors

Viktor Wang is an associate professor at the Department of Educational Leadership & Technology, California State University, San Bernardino, USA, where he teaches CTE courses and doctoral courses in educational leadership and oversees accreditation. Dr. Wang has brought leadership to the study of education. He has solidified scholars' understanding of how to research the complexities of the learning process. Alongside the production of 240+ refereed publications, Dr. Wang has provided many opportunities for his peers and students to develop their scholarly capabilities and stimulated the research agendas of numerous colleagues. He received the 2016 national Presidential Award for Exceptional and Innovative Leadership from American Association For Adults and Continuing Education, in addition to multiple institutional awards at home and abroad.

* * *

Marius Boboc is a Professor of Education at Cleveland State University (CSU) in Cleveland, OH. Marius earned his M.A. in Teacher Leadership from Roosevelt University in Chicago, and his Ed.D. in Curriculum and Instruction from the University of Northern Iowa in Cedar Falls. His work experience at CSU includes being the Director of the Office of Student Learning Assessment, serving as chair of the Department of Curriculum and Foundations in the College of Education and Human Services, and, most recently, providing oversight to the Office of Academic Planning. In that capacity, he has managed strategic planning, student learning assessment, academic program review and institutional research and analysis, e-Learning, Instructional Technology and Distance Learning, space management, General Education as well as technology fee allocation, among other responsibilities. Marius' research interests relate to online instruction (ranging from design to implementation and evaluation), institutional effectiveness, assessment and accreditation in higher education, postmodern education, leadership in higher education, teacher preparation reform, adult learning, and change management in higher education. His work has been published as books by Palgrave Macmillan, Sage Publications, IGI Global, Information Age Publishing, and Cengage Learning, as book chapters by Information Age Publishing and IGI Global, in journal articles in Brock Education Journal, TechTrends, International Journal of Learning and Development, International Journal of Education, Journal of Engaged Learning, Academic Exchange Quarterly, Electronic Journal of eLearning, Science Scope, and the Journal of Technology and Teacher Education, as well as book reviews by Teachers College Record. Research currently under contract by Sense Publishers focuses on the use of assessment data to inform curriculum development in higher education, as well as postmodernism in education, both representing book expected to be in print by the end of 2023. His numerous peer-reviewed international and national conference presentations support the same scholarship agenda.

Jörn Bühring is an Assistant Research Professor at the Hong Kong Polytechnic University, School of Design. He earned a BA-equivalent degree in the field of Tourism Management (HH), an M.B.A. (VU) majoring in Entrepreneurship and Innovation, and a Ph.D. in futures Consumer Engagement Innovation at Swinburne University of Technology (SWIN) in Australia. In his current position, he is developing a collaborative approach (Design Economies, Ignite Innovation Futures) that brings academia, designers, and business stakeholders together to develop high-impact design knowledge, concepts, and solutions. His research explores Design Foresight, Vision, and Fiction techniques as emerging processes within Strategic Design and Innovation Management. An honorary Adjunct Associate Professor within the College of Business at RMIT University, a Visiting Professor at the Aalborg University in Copenhagen, and a Distinguished Scholar at the Abu Dhabi University, he collaborates with leading Universities in research and design-led innovation. He gained international work experiences in corporate leadership throughout various industry, company, and project initiatives, including innovative start-ups, new business formations, and new market entries. He has worked across a wide range of global brand icons (e.g. Bang & Olufsen A/S, Toyota Motor Corporation, Four Seasons Hotels & Resorts), organizational cultures, and international markets.

Howard Doughty has lectured at a number of colleges and universities in Canada and the USA - notably McMaster University, the Ontario Insitute for Studies in Education, St. Mary's University, and the Universities of Hawai'i and Toronto. He has also taught at in the MA program in Diplomacy and Military Studies and the Department of Sociology at Hawai'i Pacific University and is currently completing his 53rd year as a Professor at Seneca College in Toronto. A former editor of Bridges: Explorations in Science, Technology and Society (1986-1991), The College Quarterly (1992-2017), and The Innovation Journal (1998-2020), he is currently a regular contributor to Critical Links, a monthly publication of The Centre for Inquiry Canada. He has published widely on the theory and practice of democracy and critical pedagogy. His most recent books in include Soft Science: Notes and Readings in Cultural Anthropology, Culture and Difference, and Discourse and Community (the latter two edited with Marino Tuzi). Howard Doughty is also a community and trade union activist with over 40 years as an elected Local and occasional provincial Executive Member of the Ontario Public Service Employees Union, a 180,000-member union representing blue-collar, white-collar, and professional public-sector workers. In 2017, he received a "lifetime achievement" award from his union and, perhaps counter-intuitively, was presented with a "Canada 150 Citizenship Award by the Parliament of Canada on the occasion of the nation's Sesquicentennial.

Linda Ellington conducted her doctoral research in Higher Educational Leadership at the University of Oxford and the University of Warwick in the United Kingdom. She is a faculty member at Southern New Hampshire University. She worked for IBM for 23 years in various leadership positions. She sits on the Board for Keiser University. She joined IJAVET as an Associate Editor and is a peer reviewer for several journals and textbooks. She is a doctorate committee member for several universities, including the University of Johannesburg, South Africa. In addition to teaching and researching, she has written 8 children's books in the genre of bullying, being good stewards of our planet, and being a family that consists of several different critters!

Karen Escalante is an Assistant Professor in the College of Education at California State University San Bernardino where she teaches pedagogical courses grounded in social justice and humanizing pedagogy. Karen is the president-elect for the California Council on Teacher Education (CCTE), serves as part of the research team and advisory board for UCLA's Center for the Transformation of School's California Educator Diversity Project, and serves as a member of AACTE's Program Advisory Committee on Educator Diversity (PACED). Her research focuses on the through line between Culturally Responsive Teaching and teaching performance assessments, teacher candidate / in-service teacher belonging, and operating within the third space (boundary spanning) to support P-12 and teacher preparation programs.

Lesley Farmer, Professor at California State University (CSU) Long Beach, coordinates the Teacher Librarianship program, and was named as the university's Outstanding Professor. She also manages the CSU ICT Literacy Project. She earned her M.S. in Library Science at the University of North Carolina Chapel Hill, and received her doctorate in Adult Education from Temple University. A frequent presenter and writer for the profession, she is a Fulbright scholar and has garnered several professional awards and honors. Dr. Farmer's research interests include digital citizenship, information and media literacy, and data analytics. Her most current books are Fake News in Context (2021, Routledge) and Reference and Information Resources and Services for Children and Young Adults (2022, Rowman & Littlefield).

Wendy Green specializes in the area of learning and development with additional expertise in organizational culture and diversity. She works as an Associate Professor at Cleveland State University in the Levin College of Public Affairs and Education. She teaches in the medical education for health professionals and adult learning and development master's programs as well as the Adult, Continuing, and Higher Education doctoral program. Her research has focused on the role of social identity affinity groups in higher education and organizations, particularly how they intersected with the organization's culture and facilitated learning, engaged in advocacy, and created social networks. In addition, she works as a consultant in the areas of organizational culture, diversity, and international medical education. Dr. Green earned a PhD and MS in Education from the University of Pennsylvania and a MS in Human Resource Development from Florida International University.

Catherine Hansman, Ed.D., is Professor Emeritus of Adult Learning and Development at Cleveland State University and has over 30 years in teaching and administrative roles in higher education institutions. She served as the program coordinator of the graduate Adult Learning and Development (ALD) program, and she was also the director of the Master of Education in Health Professions Education (MEHPE), a partner program with the Cleveland Clinic. Dr. Hansman has written and edited numerous books, book chapters, and articles in refereed journals concerning her research interests in adult and higher education, mentoring, health professions education, context-based learning, communities of practice, qualitative research, history of adult education, and issues of diversity. Her most recent edited books, The Covid-19 Pandemic and Other Ongoing Crises: Reflections, Innovations, and Adaptations (2022) and Mentoring, Learning, and Leadership in Adult and Continuing Education (2020) focus on supporting and mentoring adult learners in higher education contexts. She was a co-editor of the Adult Education Quarterly from 2014 to 2019. She was awarded the Cyril O. Houle Emerging Scholar in Adult and Continuing Education Scholarship and was conferred a Distinguished Faculty Award for Research by Cleveland State University. In 2016, Dr, Hansman was inducted into the International Adult and Continuing Education Hall of Fame. She is a Past President of the American Association for Adult and Continuing Education (AAACE), and she also chaired the Commission of Professors of Adult Education (CPAE) and served on its executive committee for six years.

Yasemin Kırkgöz holds a B.A. in English language teacher education from Cukurova University in Turkey, and postgraduate degrees in applied linguistics from the University of Aston in the UK. She has published widely in foreign language policy and practice, English medium education in higher education, curriculum renewal, and teacher education in a number of international journals including Teaching and Teacher Education, Educational Policy, Teaching in Higher Education, Current Issues in Language Planning and System, and wrote book chapters in some leading publishers including IGI, Routledge, Multilingual Matters, and Springer. She has taught English for Academic/Specific Purposes and teacher education courses at Cukurova University in Turkey since 1985. Currently, at the undergraduate level, the courses she delivers are Curriculum and World Englishes, and at the postgraduate level Foreign Language Policy, Management and Leadership in ELT, ESP, and Problem-Based Learning in ELT. She received the Third Annual David E. Eskey Memorial Award for Curricular Innovation for her publication Innovation as a Curriculum Renewal Process in a Turkish University in 2006, and the Leadership and Management Special Interest Group (SIG) award from IATEFL in 2013. Her co-edited books 'Key Issues in English for Specific Purposes in Higher Education', and 'English as the Medium of Instruction in Turkish Higher Education: Policy, Practice and Progress' were published by Springer.

Anil Lalwani is the Post Doc Fellow for Assessment at the University of Vermont. Previously, Anil completed doctoral work in Adult, Continuing and Higher Education at Cleveland State University. Anil has served on the leadership team of International Adult Education (CIAE) -- a Commission of the American Association of Adult and Continuing Education (AAACE). Anil's current research is aimed at promoting access and equity among postsecondary learners, by addressing barriers to student engagement and help-seeking on campus, especially international and other minoritized groups of postsecondary learners.

Steve Lambert is an associate professor of educational leadership in the School of Education, at the University of Chester, UK. He teaches modules related to leadership theory and the traits and behaviours that leaders in education need. He is an experienced doctoral supervisor and an active researcher. His research focuses on the development of future leaders. He has a MA in Educational Leadership, an MSc in Psychology and an EdD in educational leadership, policy and values.

Julius Lukwago holds a BSc. in Engineering from Makerere University in Uganda, an MBA in strategic planning from Edinburgh Business School/Herriot-Watt University, Scotland, a Chartered Postgraduate Diploma in Marketing, from the Chartered Institute of Marketing, an MSc in Public Health, from the University of London and a Ph.D. in Leadership Studies from University of KwaZulu-Natal, South Africa. He has consulted widely in strategy, leadership, and organizational development for corporations in the banking, media, telecom, legal, FMCG, government, development, manufacturing, and oil and gas sectors. He has taught policy, planning, finance, and economics at Uganda Christian University School of Medicine since 2014. In 2018, on the occasion of the Mandela 100 Celebration, he was honored with the Distinguished Leadership Award by the South African ambassador to Uganda, as nominated by coaches at the Leadership Academy South Africa, for his contribution to leadership development in Africa.

Ana Martins holds a Ph.D. in Leadership, Organizational Behaviour, and Information Science from the University of Fernanda Pessoa, a Competent Manager diploma from the Open University, and a

Bachelor of Arts degree in Communication Sciences, History of Art, and English Literature from the University of South Africa. She is a Professor at the University of KwaZulu-Natal, Graduate School of Business & Leadership, South Africa. Her scholarship spans over 30 years and encompasses a range of countries, namely, the UK, South Africa, Hong Kong, China, Germany, Portugal, including the Middle East. Her research includes over 100 indexed international peer-reviewed academic journals and book chapters with Emerald, Springer, SAGE Publications, Taylor & Francis, and Palgrave Macmillan Publishing. Her key research topics embrace Emerging Concerns Related to Humanizing Leadership and Organizations. She has supervised to completion, numerous Doctoral and Masters students. She also serves as a member of the International Accreditation Agencies for Higher Education.

Melissa Meetze-Hall, in addition to prior teaching experience at the elementary and high school levels, and administrative experience at the elementary and county office levels, also guides Master's and Doctoral students in their leadership development as they work toward advanced degrees and their California Administrative Services Credential. A Leading-Edge Certified digital educator and administrator, with a Masters in e-learning and technology and a Doctorate in Higher Education, Melissa's research studies focus on the power of video for reflection and the collaborative conditions which support educator growth. She is the author of numerous publications and co-author of one book: The CalTPA Handbook: Unpacking Equitable Practices in the California Teaching Performance Assessment.

Li Ming, , Ph.D. in Foreign Linguistics and Applied Linguistics, is presently a graduate in Faculty of Foreign Studies, Beijing Language and Culture University. She has already published around 20 academic papers in distinguished journals and participated in four scientific research projects based on rich teaching and research experience. Her research interest includes English pragmatics, language acquisition, and computer-assisted language learning.

Judith Parker is an adjunct Assistant Professor in the Organization and Leadership Department at Teachers College/Columbia University in New York teaching graduate courses on the theory and practice of adult learning that are totally online, totally classroom, and blended delivery formats. She has earned a doctorate, Ed.D. degree, and an M.S. degree in Adult Learning and Leadership from Teachers College/Columbia University, an M.S. degree in physics from Purdue University in Indiana, and a B.S. degree in physics and mathematics from Notre Dame College in Ohio. Judith is a co-author of the book "Pedagogical and Andragogical Teaching and Learning with Information Communication Technologies" and is on the editorial review board for the "International Journal of Adult Vocational Education and Technology". She has authored numerous chapters and articles on cross-cultural workplace training and learning and the history, philosophy, practice, and theories of adult learning and technology integration for effective teaching and learning. She has presented numerous papers at conferences globally including most recently the International Seminar on Democracy, Education, and Development at the National University of Educational Planning and Administration in New Delhi, India, and The International Transformative Learning Conference in Athens, Greece. Judith has over 20 years of experience in education at all levels and in leadership positions within global business organizations. She has worked extensively with technical managers and technical employees in Asia and Europe in leadership development and technical training skill development. She is also an instructor for the American Museum of Natural History in New York teaching online courses "Space, Time and Motion" and "The Solar System". She has received the American Association of Physics Teachers Innovative Teaching Award, and the 3M Community Volunteer Award, and has been elected a Fellow of the American Association for the Advancement of Science.

Mattius Rischard was a graduate teaching associate for the English and Political Science departments, as well as a Graduate Assistant Director conducting research for the UA Writing Program. He currently works at the intersections of African American and Urban Anglophone literatures as they relate to the politics of cultural production and representation. As a PhD Fellow, he is a contributor to journals on writing andragogy and pedagogy, as well as the redefining volume on intersectional studies of politics in Anglophone literature, The Routledge Companionto Literature and Class (2021), and currently completing a monograph on the literary history and stakes of American street literature. Mattius has published widely in journals and edited volumes in the fields of Comparative Urban Literature, Postcolonial Studies, Film Criticism, Performance Studies, and Digital Humanities.

Kin Wai Michael Siu is Eric C. Yim Professor in Inclusive Design and Chair Professor of Public Design. His research interests are in public design, inclusive design, user reception, and design education. He has published over 400 journal papers and solely owned 50 US and international patents.

Jonathan E. Taylor is an Associate Professor at Auburn University (USA). He holds a Ph.D. in Educational Psychology and Research from the University of Tennessee. Prior to his academic career, he spent over a decade as a training coordinator and instructor in workplace contexts. His areas of interest and scholarship are learning resistance and engagement, applied philosophy, conceptual change, and workplace training and development.

Orthodox Tefera is an Adjunct Lecturer at the Graduate School of Business and Leadership at the University of Kwa-Zulu Natal Durban.

Matthias Teine received his Ph. D. in 2022 from the Paderborn University for his research on the participatory design of digital learning offers in vocational education. Currently, he is working as an expert for leadership development and cultural change at a Germany-based logistics provider with more than 23.000 international associates. Since 2016, Matthias works in expert positions for renowned international companies ranging from 2.500 to more than 400.000 employees and wrote his doctoral thesis extra-professionally. Hence, it is important to him that his research helps improving practice and that his practical work is in-line with the latest research findings. His focus topics are adult education, cultural change in organization, digital learning, leadership development, and participatory design.

Geraldine Torrisi-Steele is an associate professor in the School of Information and Communication Technology at Griffith University, Australia. She is a Senior Fellow of the Higher Education Academy (UK). Her interest in learning and teaching, learning theory, technology for learning, human-computer interaction, and leadership distills from her 'pre-academia' experiences in secondary school mathematics and physics teaching, and the design, development, and project management of digital interactive media for learning in higher education. Cutting across her interests is a belief in design with 'the human at the center.

Caterina Villani (Ph.D.) is a distinguished postdoctoral fellow at the School of Design, The Hong Kong Polytechnic University, and a qualified architect. Her work lies at the intersection of urban policy research, spatial planning, and sharing practices, aiming to combine multi-disciplinary approaches to envision more inclusive urban forms and processes. Caterina's research investigates the production of public space, spatial justice, and migrant urbanism, primarily focusing on Asian global cities. Caterina's doctoral research was supported by the Hong Kong Research Grants Council and received international awards. She holds a Bachelor's and Master of Architecture from Iuav University of Venice. Recent studies were published in Landscape Research, the Journal of Urban Design, Transport Policy, and Cities.

Sandra Williamson-Leadley is a Senior Lecturer and the Programme Coordinator for the GradDip/PGDip/MTchLn (Primary) in the Faculty of Education. She was a Primary school teacher for 17 years before becoming a staff member at the University of Otago College of Education and then at the University of Canterbury. She teaches Professional Learning and Inquiry papers and specializes in Technology Education and e-Learning/Digital Technologies. She is a Professional Practice Lecturer and works with undergraduate and graduate/postgraduate students. She is also a supervisor for MEd and EdD/Ph.D. students. Dr. Williamson-Leadley's doctoral research focused on New Zealand primary school teachers' ICT professional development and classroom practices. Her other research activities have included exploring the use of tablet technology for assessment in mathematics; investigating mainstream primary teachers' use of digital technology to support their indigenous students' learning and the lessons that can be learned from teachers' use of digital technology in indigenous schools. This has also led to the development of interest in indigenous research methodologies. Her research interests include the preparation of pre-service teachers, technology education, teachers' classroom practices using digital technology, and using digital technology to support indigenous students.

Wu Xiuxiu is a doctoral candidate from the School of Design, The Hong Kong Polytechnic University. Her research focuses on design knowledge, knowledge management, and collective creativity in the context of designers or design teams, with the goal of contributing to the field of design practice in the organization using her more than ten years of work experience across industries and psychological background.

Zhonggen Yu, , Professor (distinguished) and Ph.D. Supervisor in Department of English Studies, Faculty of Foreign Studies, Beijing Language and Culture University, Ph.D. in English language, a dual Master-degree holder in applied linguistics and law, and a post-doctoral researcher in psycho-linguistics, has already published over 110 academic papers in distinguished journals based on rich teaching and research experiences. He is Editor in Chief of International Journal of Technology-Enhanced Education and Academic Editor of Education Research International. His research interest includes educational technologies, language attrition, and language acquisition.

Yuan Zhang is a graduate student at Auburn University (USA), currently pursuing her Ph.D. in Adult Education. She holds a master's degree in educational leadership and a master's degree in foreign language teaching and learning. Her previous experience includes world language teaching in both school and community settings. Her current areas of interest are learning theories and educational implications.

Index

N

O

P

Q

R

S

T

V

W

www.ingramcontent.com/pod-product-compliance
Lightning Source LLC
LaVergne TN
LVHW062106300926
849161LV00010B/2382